Contributions to Management Science

More information about this series at http://www.springer.com/series/1505

Jeffrey Yi-Lin Forrest • Yong Liu

Value in Business

A Holistic, Systems-based Approach to Creating and Achieving Value

Springer

Jeffrey Yi-Lin Forrest
Department of Accounting, Economics and Finance
Slippery Rock University of Pennsylvania
Slippery Rock, PA, USA

Yong Liu
School of Business
Jiangnan University
Wuxi, Jiangsu, China

ISSN 1431-1941 ISSN 2197-716X (electronic)
Contributions to Management Science
ISBN 978-3-030-82897-4 ISBN 978-3-030-82898-1 (eBook)
https://doi.org/10.1007/978-3-030-82898-1

This Springer imprint is published by the registered company Springer Nature Switzerland AG
The registered company address is: Gewerbestrasse 11, 6330 Cham, Switzerland

Synopsis

The objective of this book is of two folds. First, it attempts to provide marketers, managers, and entrepreneurs with a scientific tool of analysis to assist them with decision-making, while knowing that their decisions are mostly reliable. Second, it offers scholars with a brand-new approach to exploring opportunities of value creation and capture. Instead of data- and anecdote-based analysis, this book accomplishes this practically significant and theoretically important objective by establishing results based on systems science and a logical reasoning that is parallel to that commonly used in mathematics and natural science. By employing such an approach, all limitations of econometric methods can be avoided.

This book is composed of six parts, addressing various key issues related to value creation and capture. The first part introduces the basics of systems science necessary for the rest of this book, develops a general theory on how a market of free competition evolves and how the resource-based theory of the firm can be established through axiomatization. The second part, entitled *Demand- and Supply-Side Strategies*, looks at the supply-chain ecosystem of a firm, how its upstream and downstream affect a firm's performance, and when synergetic innovations, involving, for example, simultaneous utilities and multi-sided markets, appear. The third part explores how to generalize the well-known Porter's value-chain framework by first investigating the systemic structure of the mind, how consumers classify product information, and how value can be created out of innovation and resources and captured along with memberships in strategic networks and blocks.

The fourth part, entitled *Customer Values*, examines when effective consumer value propositions (CVPs) emerge, how CVPs can lead to values and competitive advantages, and why market-sensing capability is vitally important to firm performance. The fifth part, entitled *Some Roles of Manufacturing and Artificial Intelligence*, addresses why manufacturing is important in the present transformation of industries and how artificial intelligence affects technological innovations. The sixth part, entitled *Government Policies and Supports*, looks at such questions as how the government can help stimulate economic growth and when government policies can lead to improvements in firm performance. As the conclusion of this book, the afterword, entitled *Need for A Multi-Approach Methodology and Economic Induction*, fills several important methodological holes that emerge throughout the book and outlines the next steps for future research based on what has been established here.

The existent literature on value creation and capture is mostly dominated by data and/or anecdote analysis. Although such works establish potential facts and provide managerial suggestions, consequent decisions have been time and again shown to be problematic. That surely is one of the reasons why one magnificent success in a particular value-related effort cannot be readily copied to another effort although various conjectures or theories have been developed to explain how the initial success appeared. That is the very reason why this book opens up a brand-new territory valuable for scholarly research and practical decision-making, while providing reliable managerial recommendations that are expected to produce real-life benefits.

This book aims particularly at graduate students, researchers, and practitioners in areas related to the creation and capture of value. By referencing to what this book presents, the reader will be able to employ systems methods and holistic thinking to resolve various demanding issues in his/her life and career. By masterfully employing the systemic intuition—the yoyo mode, he/she will be able to make decisions fairly swiftly without unnecessarily spending other resources of limited availability.

Preface

Among all research areas of business and economic studies, value creation and capture stand for one of the central topics intensively considered by both scholars and practitioners. Because of this reason, there appears the need to organize relevant concepts and conclusions in a systematic fashion. However, instead of proposing another seemingly plausible theory, like many other theories in social science, based on conjectures suggested empirically, this book develops the planned theory in a scientific way in the language of systems science by using a logical reasoning that is parallel to the one commonly employed in mathematics and natural science.

The motivation for us to take this particular approach in our effort to develop the planned theory is the sharp difference between studies of mathematics/natural science and those on value creation and capture. The former establishes scientifically sound results, which can be widely employed to design and produce useful commercial products, while the latter derives theories, from isolated events and processes that cannot be generally applied successfully to different situations or different times. By carefully comparing the difference, it can be seen clearly that mathematics and natural science develop new conclusions, such as theorems, based on a few postulates or laws through using rigorous logical reasoning, while studies on value creation and capture tend to be data and anecdotes driven through using econometrical models. Even for theoretical studies of value creation and capture, they generally derive conclusions based on earlier empirical discoveries, which, of course, do not constitute a solid ground on which to build reliable theories. Therefore, methodologically speaking, studies in mathematics and natural science develop conclusions and principles that can be widely employed so that as long as one knows about the functionality of a product, he/she is able to reproduce it without any need of knowing exactly how the earlier ones are made. However, the same is not true with studies of value creation and capture, where recognized successes, no matter how well they are studied, cannot be readily reproduced in another business setting at a different location. To potentially help resolve such a problem with studies of value creation and capture, this book attempts to identify a few axioms, instead of empirical discoveries, on which the planned theory is reliably constructed.

Another important realization about the difference between studies of mathematics and natural science and those on value creation and capture is that the logical reasoning of the former references heavily on some kinds of intuitive playgrounds, while the latter simply does not have one such playground. For example, Euclidean plane geometry relies heavily on the two-dimensional plane; calculus and statistics make frequent use of the Cartesian coordinate system; concepts of set theory are illustrated conveniently by using Venn diagrams; to overcome this deficit for studies of value creation and capture, this book adopts the systemic yoyo model as its intuitive playground on which many conclusions can be figuratively seen first and logically shown second. Because of the particular approach of analysis and intuitive playground employed, the theory developed in this book is able to avoid influences of individually different backgrounds and knowledge structures of all involved scholars. And because of this reason, one can expect that conclusions established in this book will be independent of scholars' individually different experiences and background knowledge and will be generally employable in real-life applications.

When the world is seen as piles of isolated objects, and when natural processes are cut into unrelated subprocesses, as done conventionally, the concepts of numbers and numerical variables are introduced. On top of such backgrounds, calculus-based approaches and statistics-based methods are developed and conventionally adopted to investigate the natural world, human organizations, and mutual interactions between the world and men and between organizations. However, what is clearly missing in all of these efforts of scholarly endeavors is the organizational nature of the world and the organic and holistic feature of human organizations and their interactions. That is the very reason why this book adopts systems science as its basic way of thinking and reasoning so that the conventionally ignored features of organization and wholeness can be brought back to the center stage and purposefully emphasized in order for us to develop more realistic conclusions than previous studies.

This holistic approach is more appropriate than the conventional ones because studies of value creation and capture focus on how various economic agents, be they large or small, interact with each other, while the agents generally possess their respectively different, yet rich, internal structures. For instance, each business enterprise has its explicit history, organizational culture, and operational processes, all of which jointly constitute the unique system of resources the enterprise can employ to innovatively understand market signals and then take appropriate actions. And each individual enjoys the kinds of consumptions fundamentally determined by his/her deeply rooted values and beliefs about the world, which are formulated gradually over time through the person's interactions with the environment. In other words, all economic agents generally possess their individually different internal structures (or organizations); and it is due to the natural existence of these internal organizational structures that this book readily adopts systems science to investigate issues of value creation and capture, believing that this approach is more adequate than any other tools developed on numbers or numerical variables, such as calculus and statistics. More specifically, calculus-based approaches generally extrapolate the present situation (or the initial value) into the situation for which a prediction is needed, while statistics-based methods attempt to extend the past (or data/anecdotes). Other than these epistemological limitations, the former requires such conditions as continuity and differentiability, while the latter imposes strict requirements on the quality of samples. However, these conditions and requirements are generally not satisfied, not met in practical applications, making derived conclusions not very reliable. That explains why there is a need for our planned theory of value creation and capture to go beyond the epistemological boundaries of the conventional calculus-based approaches and statistics-based methods.

Indeed, both the concept of numbers and that of systems stem from the same natural world, to which humans and human organizations belong. They harmonically characterize this world in two different angles. The concept of numbers emerges when a business enterprise is seen as a collection of people, properties, investments, etc. The concept of systems comes into being when the organization of the enterprise is examined as a functional whole underlying the enterprise, where the isolated components are joined together through various associations to form a visible being, known as a firm. With the existence of the firm, it is the associations among the component parts that make the firm visible and identifiable. With this understanding in place, it can be seen naturally that most issues considered in the studies of value creation and capture are fundamentally problems about organizations (or systems) and their interactions. Here, the organizations can take the form of individuals, business enterprises, markets, industries, economies, and others.

Other than characterizing the natural world in two different angles, the concepts of numbers and systems represent two different aspects of that same world. In particular, numbers and relevant approaches focus more on small-scale and local phenomena, while systems and methods developed on the basis of systems emphasize on large-scale organizational features of the world. Another major difference between these two concepts is that numbers exist after existence, and systems come into being at the same time when an existence is still in the process of emerging. This difference vividly illustrates why number-based methodologies, such as those developed on either calculus or statistics, tend to be limited with their abilities when prediction, be it long term or short term, is concerned with, and why system-based methodologies are appropriate tools for the investigation of value creation and capture since the internal structures of the involved economic agents in such studies cannot be ignored.

To potentially make our planned theory of value creation and capture theoretically relevant and practically useful, we try to make this theory satisfy the following four conditions in order for it to have a glorious and long-lasting life, as argued by Y. Lin (2009) in the monograph *Systemic Yoyo: Some Impacts of the Second Dimension* (CRC Press, New York):

1. It can be read readily without much difficulty.
2. It concurs with such an intuition that people can easily imagine.
3. It possesses a beauty that can be easily felt when people learn or apply it.
4. It provides meaningful theoretical results and practical insights.

Specifically, to make our theory satisfy Condition 1, we try to phrase each conclusion and related argument in nontechnical terms as much as possible. In this regard, although Chapter 3 is the most technical part of the entire book, accompanied with all the necessary symbolic proofs, each of the following chapters rephrases results from Chapter 3 in nontechnical terms and gathers needed symbolic proofs in chapter-specific appendixes. So, for readers who are not mathematically inclined, they can conveniently skip over all technical arguments and symbolic proofs without interrupting their enjoyment of the entire book. In order to meet the need of Condition 2, systemic intuitions are provided as frequently as possible for readers to intuitively understand concepts and conclusions. The beauty of our theory (Condition 3) is constructed by various figures of relevant yoyo model representations. And the usefulness of this theory (Condition 4) is demonstrated with the large array of topics investigated in this book and by managerial recommendations provided at the end of each chapter.

It is our hope that you, the reader, will have an enjoyable time reading this book and consequently find it useful in your real-life endeavors. We love to hear from you and learn about your comments and suggestions. We can be reached at jeffrey.forrest@sru.edu (Dr. Jeffrey Forrest) and clly1985528@163.com (Dr. Yong Liu).

Slippery Rock, PA, USA — Jeffrey Yi-Lin Forrest
Wuxi, Jiangsu, China — Yong Liu

Acknowledgments

This book contains many research results previously published in various sources. We are grateful to the copyright owners for permitting us to use the material. They include

Emerald Publishing
Gordon and Breach Science Publishers (Yverdon, Switzerland, and New York)
Hemisphere (New York)
IGI Global (Hershey, Pennsylvania)
Inderscience (Genèva, Switzerland)
International Association for Cybernetics (Namur, Belgium)
International Federation for Systems Research (Vienna, Austria)
International Institute for General Systems Studies, Inc. (Slippery Rock, Pennsylvania)
Kluwer Academic and Plenum Publishers (Dordrecht, Netherlands, and New York)
MCB University Press (Bingley, UK)
Meteorological Press (Beijing, China)
Northeastern Association of Business, Economics and Technology
Northeast Business & Economics Association
Pennsylvania Association of Economics
Pergamon Journals Ltd. (Oxford)
Science Press (Beijing, China)
Sciendo (Warsaw, Poland)
Scientific Research—An Academic Publisher
Springer Nature
Taylor and Francis Ltd.
World Scientific Press
Wroclaw Technical University Press (Wroclaw, Poland)

Contents

About the Authors

Jeffrey Yi-Lin Forrest also known as Yi Lin, holds all his educational degrees in pure mathematics and had one-year post-doctoral experience in statistics at Carnegie Mellon University. He had been a guest professor of economics, finance, mathematics, and systems science at several major universities in China, including Nanjing University of Aeronautics and Astronautics. And currently, he is a professor of mathematics and research coach for the School of Business at Slippery Rock University, Pennsylvania, and the president of the International Institute for General Systems Studies, Inc., Pennsylvania. He serves either currently or in the past on the editorial boards of 13 professional journals, including *Kybernetes: the International Journal of Systems, Cybernetics and Management Science*, *Journal of Systems Science and Complexity*, *International Journal of General Systems*, *The Journal of Grey Systems*, etc. Currently, Dr. Forrest serves as the editor-in-chief of three book series, *Systems Evaluation, Prediction, and Decision-Making (CRC Press, New York)*, *Communications in Cybernetics, Systems Science and Engineering (CRC Press, Balkema)*, and *Communications in Cybernetics, Systems Science and Engineering—Proceedings (CRC Press, Balkema)*.

Some of Dr. Forrest's research was funded by United Nations, State of Pennsylvania, National Science Foundation of China, and German National

Research Center for Information Architecture and Software Technology. As of the end of 2019, he has published well over 500 research works, including over 50 monographs and special topic volumes. Some of these monographs and volumes were published by such prestigious publishers as Springer, Taylor and Francis, World Scientific, Kluwer Academic, and Academic Press. Over the years, Dr. Forrest's scientific achievements have been recognized by various professional organizations and academic publishers. In 2001, he was inducted into the Honorary Fellowship of the World Organization of Systems and Cybernetics. His research interests are wide ranging, covering areas like economics, finance, management, marketing, data analysis, predictions, mathematics, systems research and applications, philosophy of science, etc.

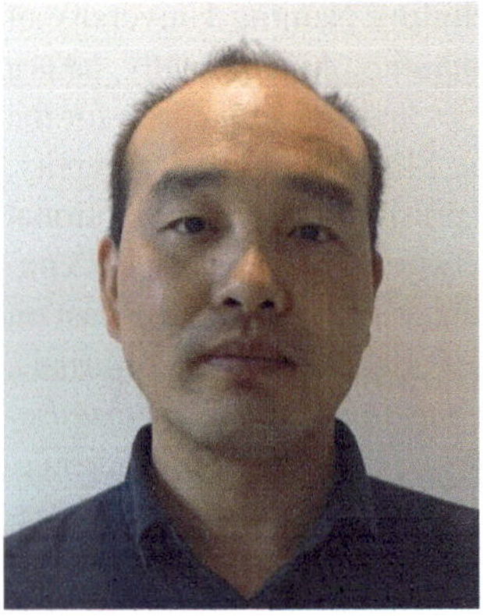

Yong Liu earned a B.S. degree in civil engineering and his M.S. and Ph.D. degrees in System Engineering and Management Science and Engineering from Nanjing University of Aeronautics and Astronautics. As a visiting professor, he spent one year at Slippery Rock University of Pennsylvania, USA. Currently, he is a full professor at School of Business, Jiangnan University. As of the end of 2019, Dr. Liu had published over 60 research papers and one monograph. His research activities are mostly in areas of the science and technology of management and are recognized, respectively, by Jiangsu Provincial Department of Education, Wuxi Municipal People's Government, China Business Federation, and Jiangsu Province Social Science Application Research Excellent Project. Over the years, Dr. Liu's works have been financially sponsored by various funding agencies, such as National Social Science Foundation, National Natural Science Foundation, Jiangsu Social Science Foundation, and Jiangsu Natural Science Foundation.

Chapter 1
Some Challenges Encountered in Value Creation and Capture

Abstract As the title suggests, this chapter introduces the reader to the exciting journey this book is about to embark on by outlining how this volume contributes to the existent knowledge on value creation and capture. It describes important gaps that exist in the knowledge of value and deficiencies the commonly and widely employed methods of analysis and reasoning experience. Collectively, these gaps and deficiencies point to the need to adopt systems science and methodology as the next best tool for analyzing situations of value and for developing a cohesive theory of value. After all these objectives are accomplished, this chapter turns its attention to briefly introduce some of the fundamentals of systems thinking and relevant methodologies needed for the rest of this volume.

Keywords Blind men and an elephant · Gaps in literature · Implementing best practices · Methodological deficits · Systems

This chapter is organized as follows. Section 1.1 details the gaps in the literature this book attempts to fill. Section 1.2 analyzes the deficiencies that exist with the methods and approaches of reasoning widely used in the literature. Section 1.3 demonstrates why systems thinking and methodology are needed in the study of value creation and capture. And then this chapter concludes in Sect. 1.4 that outlines the contents of the book.

1.1 Gaps in Literature this Book Attempts to Fill

The subsections detail the gaps in literature this book attempts to fill by developing a general theory of value creation and capture based on a set of rigorously developed game-theoretic results.

J. Y.-L. Forrest, Y. Liu, *Value in Business*, Contributions to Management Science,
https://doi.org/10.1007/978-3-030-82898-1_1

1.1.1 The Theoretical Foundation

The first gap that exists in the foundation of the theory this book attempts to develop is the lack of a practically useful theory on how a market of free competition evolves with price as the primary factor, followed by secondary factors, such as business models, technologies, and government policies. In particular, by a primary factor, it means such a factor that human can control and make adjustment. And by a secondary factor, it stands for one that somehow directly or indirectly appears naturally as a consequence of the functional effect of primary factors. In the marketplace, price drives the levels of consumption and production, which in turn encourages technological development so that productivity can be desirably improved. The competitive need to steadily improve productivity forces firms to adopt different business models in order to improve their managerial efficiency. To guarantee the smooth operation of market exchanges, the government plays its role by providing laws, regulations, and relevant reinforcement.

Clearly separating primary and secondary factors is very important in the development of our planned theory for the following reason: although studies in business and economics often involve many interacting factors, scientifically speaking, neither natural science nor mathematics knows how to deal with the general mutual interaction of three factors, which is the well-known three-body problem (Lin, 2009), letting alone scenarios involving more than three variables. This end explains why developmental economics has provided inconsistent conclusions regarding how a nation could kick-start a self-sustaining momentum of economic growth (e.g., Lipton, 1977; Rostow, 1960; Studwell, 2013) until Wen (2016) and Forrest et al. (2018) arrive at the scene. In particular, the conventional studies in developmental economics attempt to explore the mutual interactions of many factors, way more than three, simultaneously so that the roles of primary and secondary factors are entangled together and confused with one another, leading to varied and inconsistent conclusions that are dependent on which specific angle researchers take. On the other hand, these recent works (Wen, 2016; Forrest et al., 2018) anecdotally discover and theoretically confirm, respectively, that for a nation to develop a self-sustained momentum of economic growth, it needs to keep its market exchanges open to free competition, which is rooted in market prices.

Regarding this gap, this book presents a systematic theory on how market price drives competition and invites innovation based on game theory. And on top of this rigorous development of market dynamics, the rest of the theory of value creation and capture is constructed.

The second gap in the foundation of the theory this book attempts to develop is concerned with resources. Since the time when Penrose (1959) recognizes how important organizational resources are in terms of a firm's success, the concept of resources has been employed by many scholars to empirically elucidate and envisage what underlies the competitive advantage and performance of a firm (Crook et al., 2008; Kozlenkova et al., 2014). However, empirical analyses and hypothesis tests suffer from severe constraints. So, to convert empirically confirmed claims into reliable facts, scientifically speaking, they need to be shown by using rigorous logical reasoning with a set of solid starting axioms.

To fill this gap in the literature, this book introduces three basic axioms about firms and their resources, followed by the development of a series of generally true formal propositions regarding associations between resources and the emergence of sustainable competitive advantages, resources and capabilities, and resources and performance of firms. This approach is similar to Rathod et al. (2019) where an axiomatic approach and process variable are utilized to build their model of agile system in supply-chain design.

The third gap in the foundation of the theory this book attempts to develop is related to the effort of management and marketing scholars devoted to developing the capability of explaining and predicting the bases of a firm's competitive advantage and performance (Crook et al., 2008; Kozlenkova et al., 2014). In the past decades, although these scholars have confirmed a good number of discoveries and theories, mostly based on summarizing anecdotes or analyzing data, they are still generally limited to providing managerial suggestions instead of recommendations. This fact explains why business decisions, especially those of mid- to long-terms, tend to be not very reliable (Forrest et al., 2020; Lin & OuYang, 2010).

To fill this gap of the literature developed on the resource-based view of the firm, this book generates a cohesive theory of resources on top of the afore-described axioms by referencing some of the thought-provoking empirical discoveries of the past. Because of the novelty of the methodology used in this book—systems reasoning (Lin, 1999) and the systemic yoyo model (Lin, 2009), this book is able to develop general results on such important issues, among others, as when a firm is likely or certainly to develop sustainable competitive advantages, how the capability-rigidity paradox can be avoided, what could potentially help improve firm performance, how relational resource impacts the innovativeness of the firm, when interactions of resources could lead to undesirable outcomes, when and how a capability would evolve or cease to exist, and when a firm would fail to consistently exploit its resources.

1.1.2 Understanding of Supply-Chain Ecosystems

It is well recognized (Adner et al., 2013) that when a firm comprehends a market invitation innovatively, other than internal reasons other players in the ecosystem of the firm's supply chain greatly affect how well the firm can capture the opportunity to develop a significant competitive advantage over its competitors. One good example to illustrate this end is the current development of flying cars (Lemoussu et al., 2018). It is surely one possible solution to the difficult daily commuting problem confronting those individuals who live in a suburban area and commute to work inside a major US metropolitan. Although the manufacturing of such cars is readily possible based on modern physics and engineering, the true practicality of this idea depends on the availability of such key complements as road conditions and air traffic controls so that flying cars can move around freely in an orderly manner. So, a gap in the relevant literature appears as follows: After a firm comprehends a market invitation innovatively, how will the success of the firm be consequently dependent on the innovation and technological capabilities of other players in its ecosystem? To fill this both theoretically and practically important gap, this book first distinguishes suppliers and complementors so that suppliers provide their outputs for the focal firm of concern to integrate into its market offer and complementors facilitate necessary conditions for customers to fully utilize the offer.

Another gap in the literature is well illustrated by the following questions. When a firm innovatively deciphers a market invitation, how will the consequent design of the firm's new product(s) post challenges to other players within its ecosystem? And how can the firm systemically manage its vertical interdependence within its ecosystem? In its temptation to address these questions, this book correspondingly establishes the following main conclusions among others: (1) the performance advantage of a focal firm over its competitors is positively correlated to the level of challenges the firm's suppliers face, and negatively to that of the firm's complementors; (2) to successfully ride waves of transient competitive advantages, a firm has to introduce such innovative products that suppliers can possibly provide necessary components and complementors can readily facilitate needed complements; and (3) when contracting with upstream suppliers, a firm has to deal with technological and behavioral uncertainty, where the former affects the firm's creation of value and the latter impacts the firm's capture of value.

McGrath (2013) and others demonstrate that firms need to develop organizational cultures and capabilities necessary for them to effectively ride fast-changing waves of transient competitive advantages. To do so successfully, a firm can look inwardly or outwardly to see what values it can create for consumers (e.g., Barney, 1991, Eisenhardt & Martin, 2000), and to find out where consumer synergies are located and what can be potentially offered (Drucker, 1954). In this regard, a firm can identify potentials of different competitive advantages through economies of scope by diversification at either the producer side (Porter, 1985; Santalo & Becerra, 2008) or the demand side (Ye et al., 2012) or a combination of both. Corresponding to this approach of creating competitive advantages, there are two gaps in the relevant literature. One is about how a firm can create producer-side synergies if the strategy of economies of scope is employed and the other is concerned with the development of demand-side synergies through using simultaneous consumer utilities and multisided markets; after all, it is consumers who determine a firm's success (Penrose, 1959). To potentially fill these gaps and address these questions, this book rigorously develops a cohesive theory on how to potentially develop synergistically innovative ideas at either the producer side or the demand side.

By continuing the thinking logic of effectively riding fast-changing waves of transient competitive advantages (e.g., McGrath, 2013), this book addresses the following natural questions, derives results on how to produce consumer synergies through developing simultaneous consumer utilities, two-sided markets, and other related ideas. (1) What fundamental decisions can a retailer make in terms of its offers to consumers? (2) How can a retailer create simultaneous consumer utilities by mutual forbearance, interorganizational networks, platforms that directly connect sellers and buyers, and how information flows affect the emergence of creative destruction.

1.1.4 Studies of Consumer Value Propositions

By methodically evaluating market demands, Sonoco Products was able to successfully formulate its adequate customer value propositions (CVPs) on which it achieved consequent phenomenal performance in the marketplace (Anderson et al., 2007). Along with this story of success, it has been repeatedly confirmed (Payne & Frow, 2005; Webster, 2002) that companies first organize themselves effectively on particular CVPs and then create values for customers on top of their effective CVPs.

The idea of CVPs has been noted since over 100 years ago. For example, in the area of marketing, Starch (1914) studied the concept of propositions. However, the necessity to communicate created values to customers, as a practically important way to potentially capture value for a company, is not accentuated until the 2010s (Marketing Science Institute, 2010). Even with such much delayed recognition of the need to proactively communicate created values to customers, as recently as of this writing, the concept of CVPs is still poorly defined, unconcernedly talked about, and discussed inconsequentially in both theory and practice (Lanning, 2003; Payne et al., 2017). It is in 2017 that Payne et al. finally introduce a workable definition for the concept of CVPs.

Considering the theoretical and practical importance of CVPs, as just described, a clearly visible gap existing in theory and practice that urgently needs to be filled is to develop a widely useful theory of CVPs and successes of relevant applications. Evidently, to make it widely useful, the theory has to be established on some foundation of rigor with conclusions derived in such a way that they are not constrained by particular samples of data, by the limitations of econometrical methods, and by the required conditions of calculus-based tools. To this end, by utilizing the intuition of systems research and the rigor of game theory this book presents such an imagined theory that

- Includes sufficient conditions under which a firm can formulate an effective CVP and acquire additional profit beyond the case without the CVP;
- Can practically reveal how CVPs play their parts in general aspects of business operation and how CVPs assist the creation and capture of values for individual companies; and
- Demonstrates how CVPs affect the development of competitive advantages for a company, how adopted CVPs could increase values for shareholders, and how a CVP's impacts can be materialized.

Because of the characteristics of how relevant conclusions are established within this theory based on a rigorous foundation, it is anticipated that these conclusions can practically produce visible economic benefits for managers and entrepreneurs.

In terms of market sensing, Day (1994), Forrest et al. (2017), and McGrath (2013) either confirm or demonstrate that it is an important capability for a company to first survive and then succeed in increasingly fast-evolving markets. However, Ardyan (2016) finds that such capability does not have any positive effect on companies' profitability. Lindblom et al. (2008) reveal some effects although not significant. As for its impacts on the quality of market entry and the creation of knowledge, Sugiyarti and Ardyan (2017) find that this capability has only some positive impacts, while Alshanty and Emeagwali (2019) show that it has significant impacts. That is, the existent literature points to inconsistent findings. To help sort through these inconsistencies, this book theoretically looks at what the inconsistent results mean in theory and in practice.

The importance of such effort cannot be overlooked because no matter whether it is for a market leader or for a follower, knowing the future direction of consumer demands is essential for a firm to plan itself strategically. Specifically, the current world of business presents such a case where once sustainable competitive advantages have become transient (McGrath, 2013). And, beyond sorting through the inconsistencies existing in the literature, this book also enables us, as case studies, to look at profit opportunities in stagnant industries that experience little or no market growth and show how market knowledge and relevant innovative understanding of the knowledge can assist a company to construct effective CVPs.

1.1.5 Comprehension of Manufacturing and Artificial Intelligence

With the maturing technology of Internet and fast emergence of artificial intelligence, a number of nations from around the globe have been engaging in a new round of industrial transformation. These forward looking nations include China (State Council of the PRC, 2015), France (Marc et al., 2018), Germany (Industry 4.0), the Great Britain (Hall & Pesenti, 2017), Japan (Government of Japan, 2015), and the USA (OWH, 2016a). They planned and implemented their efforts by focusing on the manufacturing sector based on a domineering lesson of the past successes of industrial revolutions (Wen, 2016). Such nearly simultaneous efforts of the leading powerful nations naturally lead to theoretical curiosity about the role manufacturing plays in a nation's effort to generate and advance its self-sustained momentum of economic growth. The curiosity is considered natural because recent literature suggests that the advancement of the service sector would be the basis underlying the coming rounds of economic booms (e.g., Szirmai & Verspagen, 2015). In particular, recent data from both developed and developing countries indicate that services account for more than 50% to over 80% of the economy.

That makes it increasingly difficult for developing economies to industrialize through purposefully growing their manufacturing because that sector steadily provides fewer employment opportunities.

Although this curiosity has been explored by many scholars from different angles, all of them produce either inconclusive or inconsistent results (Wen, 2016) due to the employment of data- or anecdotes-based approaches. In order to face this challenge satisfactorily, Rostow (1960) points to the need for new methods of analysis and different logics of thinking.

On top of this 60-year-old call, this book theoretically confirms the rationale behind various nations' attempts to maintain their leading positions in the world by focusing on the manufacturing sector. At the same time, it introduces systems thinking and methodology, and the logical reasoning that has won victories one after another for mathematics and natural science (Kline, 1972) into the study of industrial revolutions. Because of the novelty of the methodology employed, this book is able to develop sharp theoretically important and practically meaningful conclusions and insights regarding the role manufacturing plays in economic development.

The current political and geographical developments around the world have altered patterns of economic growth in unprecedented ways. To adjust appropriately to the fast-changing environment, many nations adopted different strategies by zooming in on artificial intelligence (AI) as their next direction of economic development. For example, the government of Japan recommended (Government of Japan, 2015) the establishment of a national R&D promotion mechanism to center on Internet of things, big data, AI, and other technologies, and the realization of a super-intelligent society through extending AI to all aspects of life. In 2016, the Obama White House put out seven chief stratagems for the advancement of AI in the United States of America in two documents, entitled respectively "Preparing for the Future of Artificial Intelligence" and "The National Artificial Intelligence Research and Development Strategic Plan" (OWH, 2016a, 2016b). To keep pace with Japan and the United States, the term of AI appeared in the 2017 Report of the 19th National Congress of the Communist Party of China (Xi, 2017); and the European Commission (EC, 2018) submitted in May of 2018 its document, "Artificial Intelligence: A European Perspective." It pronounced the EU's place in the international AI competition and presented a plan for relevant actions.

Although the importance of AI has been widely recognized and AI technologies have been increasingly employed, scholarly research on how AI impacts technological innovation is still scant and in the stage of infancy. The focus of published works is mainly on how to define relevant concepts. Still open and waiting to be addressed are many questions regarding how AI and technological innovation are related to each other. This is exactly the place where this book makes its contribution in the following two aspects: (1) At the height of theory it explores the inherent mechanism underneath how AI affects technological innovation and (2) based on the panel data of the provincial level from China, it empirically confirms the impact of AI on technological innovation.

1.1.6 Understanding the Roles Government Policies and Supports Play

Other than making daily living convenient, the development of communication technology changes how enterprises conduct their business. Such changes at the same time challenge some ambitious nations, either developed or developing, to think about how they can employ policy tools to help transform their states of industries in order for them to maintain and/or to acquire their leadership in the world economy and international politics. These nations crave to help their industries to rise from automated manufacturing, the current state, to intelligent manufacturing, an expected future state, such as Industry 4.0 (Hermann et al., 2016). Along with this trend of development from around the world, this book employs the theory it develops in the following chapters to explain (1) why government policies can actually play their roles in economic systems, (2) why policy tools and governmental supports are fundamentally necessary to practically stimulate economic growth, and (3) how a government uses policies to affect the performance of the manufacturing sector.

These explanations help on a scientific basis fill a gap that exists in the existent literature regarding why policies and supports from the government are needed in stimulating economic development. In particular, regarding the second why mentioned above, the existent literature provides both yes and no answers simultaneously. For example, in support of a yes answer, Andreoni and Chang (2019) developed a framework to coordinate interactive industrial policy measures based on Howell et al.'s (2019) argument (for the case of the US and the industry of microelectronics) that for a nation to maintain its future competitiveness internationally, it needs to introduce complementary government, industry, and company policies. In support of a no answer, Jomo (2019) examines the eight Asian economies—Japan, South Korea, Taiwan, Singapore, Hong Kong, Malaysia, Thailand, Indonesia—that with government interventions achieved rapid economic growth during 1965–1990, a statistically unlikely event. Jomo finds it ambiguous as to whether or not adopted industrial policies actually led to the miracle gains. Similarly, for the concerted effort of the US government to coordinate the Advanced Manufacturing Initiative for America's Future, Hemphill (2014) finds that such effort would not blend well with the Obama administration.

Beyond filling the previously described gap in the literature, evidently, explanations provided in this book are also important for all nations in the world. In particular, due to historical momentum, developed nations like to maintain their leading and/or dominating positions in the world. Some representatives of these nations include the Great Britain (Hall & Pesenti, 2017), the USA (OWH, 2016a), Germany (Industry 4.0), Japan (Government of Japan, 2015), and France (Marc et al., 2018). With their failed attempts of modernization in the past 100 years (Wen, 2016), all other nations, either developing or undeveloped, face an increasing possibility of plummeting further behind the developed ones. To potentially avoid such likelihood, some developing nations, such as China (State Council of the PRC,

2015), also actively assist their domestic economies to make necessary leaps through policy means and governmental supports.

1.2 Methodological Deficits of the Literature

This section outlines all the major methodological challenges facing the study of value creation and capture, and points to what this book needs to overcome methodologically in its attempt to develop a general theory of value.

1.2.1 Construction and Development of Theories

Compared to studies of mathematics and natural science, there are a few very noticeable methodological deficits in the literature of value creation and capture. The first is that, contrary to the case of mathematics and natural science, new conclusions in the knowledge of value creation and capture are not always consistent with the ones developed earlier. As a matter of fact, in a lot of cases, they are either inconsistent with or even contradictory to each other. For example, in natural science, first established is a set of elementary laws, along with which a few very fundamental concepts are introduced as the base for studying all widely varied scenarios often encountered in the natural world, although these scenarios can be very different from one case to another. As the laws and fundamental concepts are employed to investigate more diversified problems, new concepts are introduced and relevant new conclusions are derived using rigorous logical reasoning based on the initial laws and fundamental concepts. Within such expansion of knowledge, new concepts are introduced in such a way that the earlier laws and concepts are mostly kept unchanged except that they may be enriched in occasional cases with a wider scope of understanding or further generalized to cover additional scenarios. To this end, the study of conservation laws in physics is a good representative of such a forever expansion scenario. Although in those few epoch-making cases throughout the history, one or more basic laws are shown to be incorrect, modifications are first done and the entire edifice of knowledge is then consequently reestablished as a more or less consistent hierarchy of theories. The case regarding studies in mathematics is exactly the same as what is just described. To see how successful such an approach of knowledge development is, one only needs to look at the rapid development of natural science and technology over the past several 100 years. The answer is: incredibly magnificent!

In comparison, the knowledge of value creation and capture does not have any elementary laws and fundamental concepts on which the rest of the knowledge is logically developed. Instead, empirically developed claims are mostly employed as starting foundation for a new theory to appear and to take shape. And when such a theory is employed to explain a situation about value in more or less practical terms, scholars tend to speak like, "The (name) theory maintains . . .. . .; so if this theory could be applied to the current case of concern, then . . .. would . . .. . ." What is most noticeable in these logical statements is the usage of such words as "believe," "might," "could," "should," "would," and other similar terms or phrases of uncertainty. That is very different from statements derived in mathematics and natural science, where affirmative terms are generally used to state conclusions. This end leads to the second methodological deficit in the literature of value creation and capture: Empirically confirmed phenomena tend to be utilized as theories and starting foundations of new theories.

In this literature, although there are studies that make use of conventional analytical approaches, such as calculus-based tools (e.g., Melitz, 2003; Melitz & Redding, 2014), game-theoretic methods (e.g., Forrest et al., 2019), and utility-function technologies (e.g., Ye et al., 2012), most studies employ the format of hypothesis testing—a statistics-based approach, where hypotheses are logically developed on either prima facie bases or empirically revealed phenomena previously (e.g., Srivastava et al., 1999). Then, sets of particular data are employed to test whether the hypotheses can be confirmed with statistical significance. Once statistically confirmed as significant, the relevant hypotheses will then be treated as somehow established propositions that can be readily applied to produce new theories as bases of next round of hypothesis testing and to provide suggestions for decision-making managers and entrepreneurs for practical purposes.

The entire process that consists of first hypotheses development, second data-based statistical testing, and third the establishment of general propositions can be very well mapped onto the situation of the proverb of "the blind men and an elephant" (Goldstein, 2010, p. 492). This proverb describes how a group of blind men without any prior knowledge try to conceptualize what an elephant is like by physically touching the elephant. In this proverb, it assumes that each of the blind men can only feel a different but a designated part of the elephant's body. For example, one man can only feel the side, a second man the tusk, the third man the leg, the fourth man the ear, the fifth man the nose, the sixth man the back, and the seventh man the tail. With their newly acquired first-hand experience about the elephant, they then hypothesize how the elephant looks like. Evidently, these men produce their imaginations about the elephant very differently from one another due to their limited experiences, even under the assumption that other than visions their sensing abilities are perfect, Fig. 1.1. This parallelism between hypothesis testing and the blind men and an elephant illustrates very well why data-based and anecdote-based approaches tend to lead to inconsistent or even contradictory theories. For example, these blind men will produce the following very inconsistent images of an elephant:

- To the man on the back of the animal, the elephant seems to be like a small-scale mountain ridge;
- To the man who feels the tail of the animal, the elephant seems to be like a hairy rope that naturally hanging in the air;

Fig. 1.1 The proverb of blind men and an elephant

- To the man who has the honor to feel a leg of the animal, the elephant seems to be a large, soft pole;
- Etc.

In short, the elephant seems to take seven completely different shapes simultaneously. Are the men all incorrect? The answer to this question is: No, none of them is incorrect except that as a group, they did not acquire the *holistic* view of the elephant.

This proverb vividly illustrates why in natural science and mathematics neither anecdotes nor data analyses are recognized as reliable ways to produce generally true conclusions and scientifically sound theories beyond finding potential facts.

1.2.2 Language- and Calculus-Based Analysis and Reasoning

The previous subsection discussed two methodological deficits in the literature of value creation and capture. One is about the absence of elementary laws (or postulates) and fundamental concepts that underly all studies on value creation and capture. The other is the issue of utilizing empirically confirmed phenomena as theories and starting foundations of new theories. The third methodological deficit experienced by studies of value creation and capture is the lack of flexible and holistic employment of all available methods and approaches.

By glancing through the relevant literature, it is quite obvious that scholars in the area of value creation and capture tend to respectively favor their individually different methodologies of analysis. Most commonly employed methodologies include, but not limited to, those that are ordinary language-based (e.g., Andreoni & Chang, 2019; Srivastava et al., 1999), calculus-based (e.g., Melitz & Redding, 2014), anecdote-based (e.g., McGrath, 2013; Rostow, 1960; Wen, 2016), statistics-based, microeconomics-based (e.g., Ye et al., 2012), and game theory-based (e.g., Forrest et al., 2019). However, every single methodology suffers from its unique set of limitations. Speaking differently, when in doubt, the general validity of one conclusion, derived by using a particular method, needs to be reconfirmed by using another different approach. In the following, let us look at these listed methodologies in some details.

In terms of analysis and reasoning that are ordinary language-based, the advantage is that they tend to be most accessible by scholars and practitioners. However, although written statements represent a form of articulation of logic, no matter how logical and precise they may be, their inferences are often inconclusive due to the linearity embedded in the underlying reasoning. The linearly sequential nature makes written statements in ordinary language unable to control for the simultaneous effect of several arguments in combination. Such statements in general cannot pinpoint one optimal outcome—"equilibrium"—out of many. One good example to this end is the literature on industrial revolutions (Rostow, 1960; Wen, 2016). Out of the same historical event and process underlying the Industrial Revolution of England, scholars take their individually different angles and perspectives to draw inconsistent and even contradictory conclusions on what actually caused the appearance of that particular Industrial Revolution. What is very fascinating is that such analysis that is based on ordinary language even confuses primary and secondary determinants as all being the fundamental causes (Wen, 2016), although the latter naturally appear when the former is in place.

Another good example of how ordinary language-based analysis and reasoning can lead to inconclusions is those rounds of so-called debates, which appear one after another, on whether the industrial policy actually works in real life. In particular, industrial policy has been one of the most debated subject since the earliest days of economic studies, although the exact term was not widely employed until after WWII (Reinert, 1995). For several decades in the second half of the last century, both developing and developed nations treated industrial policy as the key for advancing their existing economies. Then the rise of neoliberalism in the 1970s discredited industrial policy. And recently, accompanying the governmental pushes for industrial transformation of several leading nations from around the world, industrial policy resurfaces in the mainstream economic debates. For details, see Andreoni and Chang (2019) and Part VI in this book. Compared to studies in mathematics and natural science, the word "debate," as used in these heated back-and-forth arguments, very well describes the situation: Claims made are not derived on solid scientific grounds by using rigorous logic. Instead, they are ordinary language-based, see, for example, Andreoni and Chang (2019).

Regarding calculus-based analysis and reasoning, they indeed won victories for natural science one after another over the years. However, when they are applied to studies of value creation and capture, they encounter difficulties before any conclusion is derived in the stage of modeling, which matches calculus-based models with real-life-like scenarios. In particular, methods and results established on the basis of calculus require some fundamental conditions to hold true. Such conditions include, but not limited to, the following, see Afterword of this book for relevant details:

- All variables of concern need to take values on some sorts of continuums at least to a certain degree;
- Relationships between variables need to be expressible in functional form;
- All functional expressions of associations between variables need to be continuous and differentiable on most parts of their domains.

In applications, the first condition is often violated. For instance, it is obvious that the size of a population or the number of employees of a firm can only take natural-number values so that it is a discrete variable. However, to employ methods and results of calculus, scholars often treat such variables as continuous in order to take advantage of the tools available in calculus. Not only so, the second and third conditions listed above are also simply assumed to be true without any supporting evidence. For example, in studying the existence of different industry sizes, the number $n = n(p)$ of units of certain product sold at price p per unit—a discrete variable taking only natural-number values—is assumed to be differentiable (Lin, 2009, p. 219); in the study of interindustry wage differentials, market demand for a certain product, and the staffing need of a company (Lin, 2009, p. 227–228)—both discrete variables—are also assumed to be differentiable.

Another good example is Melitz's (2003) work on the impact of trade on intra-industry reallocations and aggregate industry productivity. Before a dynamic industry model is developed, several key assumptions are given, including, among others (1) the collection of all goods (indexed by ω) considered by a representative consumer is a continuum and (2) there is a continuum of firms, each of which produces a different variety ω of product. Evidently, the total number of goods considered by any consumer has to be a natural number, while the same is true for the number of firms that produce the products a consumer is interested in consuming. The reason for Melitz to introduce these assumptions is to fit the situation of concern into the framework of the calculus-based methodology.

Due to these listed and other unlisted issues with the modeling of business scenarios using the calculus-based methodology, derived have to be suggestions instead of recommendations for decision-making managers and entrepreneurs to consider in real life.

1.2.3 Analysis and Reasoning Based on Methods of Microeconomics

As for microeconomics-based analysis and reasoning, such as those using utility functions, they have provided means for scholars to develop very important conclusions. On the other hand, when specific scenarios are concerned with, some of the most powerful tools of microeconomics become limited in their use values. This fact suggests that different tools and methods need to be introduced to complement what is available in microeconomics. To illustrate this end, let us examine the work by Ye et al. (2012). The reason why this particular work is chosen to illustrate our point is that it is an excellent piece of research on how mundane assets can be employed to simultaneously create value for consumers and capture value for retailers. However, when it is time to establish a theoretical conclusion, the most appropriate tool of microeconomics fails to satisfy what initially seems to be an easy demand. The rest of this subsection will detail what it is meant here.

The following conclusion is conjectured by Ye et al. (2012) through constructing the example that will be briefed below:

> A demand-side synergy, where the willingness for a consumer to pay for a product or service is increased, can be produced by either the simultaneous consumer utility effect or the two-sided market effect.

In particular, a state university in the northern USA has a student population of over 29,000, most of whom are undergraduates, unmarried living on tight budgets. Because most of the rental properties targeting the students do not offer laundry facilities, self-servicing coin-operated laundromats near the campus compete for the students' business. At the same time, indoor tanning is also a successful college-town business that attracts mostly college women. Seeing the potential, two entrepreneurs provide their independent, single-location businesses by jointly providing coin-operated self-service laundry facility and indoor tanning service.

From this example, the previous conjecture is observed as follows: In the constructed scenario, these business people create demand-side synergies by combining the two specific services, although they are mundane, due to the following reasons: (1) a simultaneous consumer utility effect is produced, where clothes washing and skin tanning can be accomplished concurrent, saving the otherwise unproductive and costly waiting time; and (2) a two-sided market effect is developed, where an opportunity is created for students to establish relationships because the single location of combined services helps males and females to meet, while the greater participation of women in the tanning service attracts a larger number of men to the laundry service.

To convert the previous conjecture into a generally true conclusion, Ye and colleagues outline the following base microeconomics model with a representative consumer who maximizes her utility

$$\max_{q_1, q_2, \ldots, q_n} U(q_1, q_2, \ldots, q_n; \delta) \ \ s.t. \sum_{i=1}^{n} p_i q_i \le Y \tag{1.1}$$

where q_i is the consumed amount of good i at unit price p_i, Y the consumer's total financial ability, and δ all other consumption-related parameters.

If each consumed amount q_i can be obtained from the previous maximization problem as

$$q_i = D_i(p_1, p_2, \ldots, p_n; \delta). \tag{1.2}$$

then retailer i maximizes its profit:

$$\max_{p_1, p_2, \ldots, p_n} \pi_i = (p_i - c_i) D_i(p_1, p_2, \ldots, p_n; \delta). \tag{1.3}$$

where c_i stands for the unit cost of retailer i.

From this model, the Nash equilibrium price $p^*_{(\delta)} = \left(p^*_{1(\delta)}, p^*_{2(\delta)}, \ldots, p^*_{n(\delta)}\right)$ emerges. From

$$\frac{\partial p^*_{(\delta)}}{\partial \delta} = \left(\frac{\partial p^*_{1(\delta)}}{\partial \delta}, \frac{\partial p^*_{2(\delta)}}{\partial \delta}, \ldots, \frac{\partial p^*_{n(\delta)}}{\partial \delta}\right) = 0, \tag{1.4}$$

One is able to reveal the effect of δ on prices. In particular, if δ means shopping time spent on consuming goods 1, 2, ..., n, then what is produced is how shopping time affects prices.

Although this model is theoretically beautiful, it, unfortunately, cannot be mathematically solved. To get around this difficulty, Ye and his colleagues have to specify all involved functions into simple cases, such as linear cost function for each supplier with constant margin, quadratic utility function for the consumer, and others. So, the derived conclusions can once again be merely suggestions instead of generally true conclusions.

To summarize, many methods widely used in microeconomics can be beautifully employed to develop insightful conclusions for general cases. However, when they are applied to confirm specific situations, as in the case just described above, the specifically constructed models are most likely not solvable mathematically. And every time when such a situation arises, the initially expected generally true conclusions deteriorate into mere theoretical conjectures and managerial suggestions.

1.2.4 Empirical Analysis and Reasoning

The term empirical analysis and reasoning can be naturally divided into two categories: one is anecdote-based and the other data-based. By anecdote, it means an event or a process, while data can be numerical or categorical. In this subsection, let us look at these two situations separately.

For anecdote-based analysis and reasoning, a specific set of events or processes is normally employed to draw first inferences and then managerial suggestions. There are many anecdote-based publications in the area of business and economics. For instance, Rostow (1960) provides an example of how conjectures are derived in terms of what causes the appearance of one industrial revolution in one location based on a few success stories since the first one that appeared in England several 100 years ago. Riding on the back of such studies, Wen (2016) recognized the inconsistencies existing in previously derived lists of causes. And then he details the process of how modern China went through its three rounds of industrial revolutions one after another quickly without suffering from major drawbacks. After comparing China's story with those of England, the United States, and Japan, he then produces his list of causes, which as expected is different from any of the previous lists derived by different scholars over time. Similarly, on a different but related topic, McGrath (2013) finds a set of characteristics the most successful companies from around the world share. She then suggests that these characteristics represent the fundamental reasons for their success and that other companies, including individual persons, which (respectively who) like to achieve magnificent performance in the business world or professional careers, should more or less implement strategies to acquire such characteristics.

All such anecdote-based studies naturally lead to the possible suggestion or compulsion of real-life implementation of highlighted and identified best practices (e.g., Gliva-McConvey et al., 2020; White, 2017). However, as pointed out by Forrest et al. (2020, Chaps. 8 and 9), before implementing such so-called best practices, the necessary conditions under which these practices are fundamental to the success of a general business enterprise or an individual professional need to be first identified. That is, it is natural for people to see that in most cases there must be constraints within which certain practices will be considered the best. This end is figuratively illustrated in Fig. 1.2, where the apple likes to be as slender as a banana. However, she cannot simply implement what the banana does by fitting herself into a set of banana's clothes because her body structure is different from that of a banana.

Next, let us examine statistics-based analysis and reasoning. In Sect. 1.2.1, the process of hypothesis testing is figuratively illustrated by using the proverb of blind men and an elephant. In this illustration, the elephant represents the population of concern, and by touching the elephant, the blind men experience anecdotes, on which they develop hypotheses. After that, these men go back to where they are allowed to touch the elephant to collect additional evidences or data and then confirm whether or not their hypotheses are sufficiently supported. Due to their limitations, they make respectively completely different inferences about how the

Fig. 1.2 Implementing best practices might not work out successfully

elephant looks like. Evidently, based on what is available to each of them, these men can be seen as obtaining correct local answers, even though they produce seven different imagines. However, holistically, none of them is right. This end implies that to obtain a more complete answer, these men need to somehow combine their individual findings into a systemic imagine reflecting everyone's discoveries.

At this junction, the imposed constraint—each man can only touch the elephant at a designated location—seems to be questionable, because in any scholarly study the researcher surely wants to explore the elephant as completely as possible before making his/her inferences. However, in real life, what is desired and what is available are two totally different things. That is, what is desired may not be available at all and what is available may not be what is desired. For instance, when an anecdote is presented or when data are collected, one always faces such problems as sampling error, missing representation, etc. That is, the observed anecdote might not be indicative of anything; the collected data might not be representative of the population of concern. For example, established conclusions on market entry and entry timing are derived on available data, while there is no sufficient amount of data on failed attempts (Zachary et al., 2015). Although empirically confirmed conclusions on firms' innovativeness are universally stated, the very concept of innovation used in the confirmations means very dissimilar things partially due to data availability (Becheikh et al., 2006). Such and similar issues about data widely exist in the literature. Therefore, in Fig. 1.1, the "blind men" represent researchers of an empirical study, who can only "feel" particular parts of the population of concern. Therefore, in terms of value creation and capture, both theorists and practitioners have to question how much they can trust the validity of the "general" conclusions derived empirically.

Other than issues with the availability of data and anecdotes, as discussed above, the analysis of data and anecdotes can also readily produce "realities" that only exist within the employed sets of data (Lin & OuYang, 2010). The situation is similar to

Fig. 1.3 An impressive cloud pattern in the sky

that of finding cloud patterns in the sky. For example, if one looks into the sky and sees the pattern in Fig. 1.3, he/she can conclude that two heavenly beings are in love because there is not only an anecdote (the image) but also data (measurements of the pattern).

Beyond what is discussed above, two additional issues with empirical analysis and reasoning are the following: (i) from the same set of evidences different managerial inferences can be drawn depending on individually different background and knowledge structures, and (ii) without any exception, tools of econometrics, although widely employed in the literature, are only available under their respectively strict requirements.

1.3 Need for Systems Thinking and Methodology

Based on the previous two sections, this section illustrates the reason why systems thinking and methodology are appropriate for us to use in the development of our planned theory of value creation and capture. In particular, our theory will be constructed in a similar fashion as a scientific theory by using the systemic yoyo model as our intuition and playground, and a set of axioms and rigorously established game-theoretic results as the starting points. Here, such a logical reasoning as the one commonly used in mathematics and natural science will be used to derive generally true conclusions by relating each newly introduced concept to the starting points or previously established conclusions.

1.3.1 Numbers and Numerical Variables

The concept of numbers appeared initially and naturally in daily lives, as people pursued activities for survival, such as hunting, where recording and counting were necessary. And then in ancient Greece, Pythagoras treated numbers as the origin of everything, while believing that through numbers and their properties, the mankind can comprehend all things under heaven. Such belief in the west led to the development of the traditional science as the religious foundation (Kenny, 2012). At roughly the same time, in ancient China of the east, Zhan Yin, a scholar of the warring states time, realized that the use of numbers is limited because there are things numbers just cannot describe (Qu, 1985). Such recognition about the limitations of numbers led to how later generations of Chinese people treated numbers and why they did not appreciate numbers as much as Westerners.

Beyond their limited power to describe things, such as organizations, events, and processes (Wu & Lin, 2002), numbers cannot appear before existence, reflecting their original purpose—measurement and recording. This reality greatly affects how well and useful numbers can be employed to predict the future and study the evolution and interaction of organizations, two key issues in the research of value creation and capture. In particular, since the time of Pythagoras, numbers have been identified as points on the real-number line, which is an imaginary linear axis, leading to the appearance of the infinity ∞. That explains why analysis and reasoning based on numbers or numerical variables must be of bounded validity because the natural world is curved so that it never reaches this imaginary ∞ (Lin, 2009, p. 21–22). That indicates that such a science and methodology that can be utilized to describe organizations need to be introduced to the study of value creation and capture. While systems science meets this requirement, it needs to be adopted as the tool of analysis and language of reasoning for our planned theory. In particular, by the system, it means an organization or a structure where components are related somehow to each other to form an organic whole.

To know how consumer preference evolves and what is going to be favored next, one has to consider the attributes and systemic structures of the forever changing market condition and how they interact with one another. However, in the language of natural science, interactions of different products, market forces, and organizations of parts belong to non-inertial systems. That separates the traditional science of the first push, which heavily uses quantities and inertial systems, from systems science of the second stir, which focuses on structures, organizations, and non-inertial systems (Lin & OuYang, 2010).

For things with or without internal structures, their evolution and interaction demonstrate different characteristics. For example, when two objects without internal structure collide, the law of action and reaction applies. However, when the colliding objects possess their individually different internal structures, the consequence of the collision can be very different, where the third law of motion does not apply. This end explains why nations in real life need to have their own security systems to protect and maintain their peace and stability. Hence, the logic of thinking and methodologies of analysis needed to handle these distinct evolutions and interactions will have to be different.

As what physics has shown, methods of analysis, such as calculus, statistics, and consequent theories, developed on the formality and generality of quantities, can be reasonably useful when applied to investigate the evolution and interaction of things that have no internal structure. However, in terms of value creation and capture, where all economic entities involved do possess internal structures, quantities and methodologies developed on quantities have experienced great difficulties, for details, see Sect. 1.2 above. To coordinate these two methodological approaches—one quantity-based and the other system-based—one can associate them by using the concept of rotational stirring energies or the systemic yoyo model. Doing so naturally leads to the overall system of a two-dimensional science with the first dimension focusing on the study of things without internal structure and the second dimension on matters involving internal organization (Lin, 2009). For relevant details, see Chap. 2.

By carefully examining the world systemically or through the lens of systems, one can readily see that the fundamental form of movement in the universe is rotation, no matter whether what is considered is physical, intellectual, or economic (Lin, 2009). To this end, it is Vilhelm Bjerknes (1898) who first recognized that the fundamental form of atmospheric or oceanic motion is rotation, then it is Shoucheng OuYang (Wu & Lin, 2002) who generalized this result to solids. Based on this fundamental, rotational form of movement of fluids and solids, the concepts of stirring energy and second stir (Lin & OuYang, 2010) and the yoyo model (Lin, 2009) are introduced. For a quick introduction to the yoyo model, see Sect. 1.3.3.

In short, methods of systems science are developed to deal with evolutions and interactions of organizations; the systemic yoyo model helps managers and entrepreneurs intuitively see how economic entities evolve and interact with each other. Hence, it can be expected that jointly these tools will assist decision-makers to effectively process market information and make appropriate decisions (Forrest, 2018; Forrest et al., 2020)

1.3.2 Reflexive Relationship and Systems Science

To possibly resolve the problem of how to innovatively comprehend market information (or cues or signals), let us first examine the concept of information. It was Shannon (1949) who first officially defined information by using probability. Since then, this concept has been jointly investigated with those of uncertainty and greatly helped the debate between determinacy and indeterminacy within the world of learning. While the school of determinacy imposes strong conditions, such as stability or continuity and differentiability, to smooth out specifics that appear with each particular application, the school of indeterminacy removes small-probability events by focusing on stable time series. That is, scholars arrive at the same destination although they take different approaches: they acquire the desirable generality by eliminating either specifics or complexities of the events and processes under consideration. For instance, to produce conclusions of desired generality, one widely used method is to take averages, known as expected values. However, in a lot of real-life cases, such expected values are guaranteed to be not reachable, contradicting the meaning of expected values. The following is a good example to demonstrate this end.

> The experiment of concern consists of flipping a fair coin once. Let X be 1, if the head shows up, or 0, if the tail is on the top, assuming that the coin does not land on its edge. Then the expected outcome of the experiment is $E(X) = 50\ \% \cdot 1 + 50\ \% \cdot 0 = .5$, an impossible value for X to take or the impossible situation that the coin is expected to land on its edge.

Speaking differently, to innovatively understand market information involved specifics cannot be ignored by decision-making managers and entrepreneurs although quantitative methods cannot handle them. On the other hand, various unique understandings of the market cue entice managers and entrepreneurs to take their individually different actions, which further alters the meaning of the original market cue. Such reflexive nature between a piece of market information and innovative understandings of it naturally motivates scholars and practitioners to employ systems science and methodology to deal with events and processes (or systems) and explore such issues as how to innovatively understand market information with relevant specifics and how to practically apply such understanding to make more reliable predictions and decisions.

From the angle of systems, it can be readily seen that an innovative understanding of market information is closely related to the interaction of various economic events and agents (Lin & OuYang, 2010). So, by purposely examining the product market from the angle of interacting economic events and agents, one will be able to discover value potentials in consumers' forever changing preferences. In other words, specifics of market information represent the dynamic mechanism of evolving underlying structures; their importance in terms of value creation and capture goes beyond the formality of quantities; and innovatively understanding them relies directly on the attributes and properties of the systemic structure of the interacting economic entities and agents. The current effort devoted to big data uses the name of quantities to uncover hidden systemic structures. That signals the fact that human exploration of knowledge has gone beyond the limitations of numbers and numerical variables and into the era of systems thinking of events and processes.

In principle, specifics of market information reveal changes taking place in the marketplace, while knowledge and technologies of big data provide a new approach and capability for dealing with changes and for expecting what is coming up next.

Such an approach represents another methodology different from those that are calculus-based and statistics-based. It marks a new development in the epistemological system of knowing the world through analyzing the underlying systems and their interactions.

Because of the duality of action and reaction and that of spin directions, when a particular piece of market information appears, changes in the original market structure have already occurred. So, considering the fact that the present world of business is in the era of transient competitive advantages (McGrath, 2013), there appears an urgent need for decision-makers to generate innovative comprehensions of the market by employing systems science. In this regard, systemic thinking and intuition—the yoyo model—can expediently help managers and entrepreneurs walk out of the realm of quantities and enter into that of interacting organizations and non-inertial systems (or systems science).

All methodologies listed in Sect. 1.2 have not produced managers and entrepreneurs who can consistently pinpoint to potentials of value creation and capture with appropriate reliability. Specifically, presently available knowledge, developed on either calculus-based or statistics-based approaches, has not dependably helped managers and entrepreneurs to generate innovative understandings of market cues, from which potentials of value creation and capture appear. No matter how this situation is looked at, it has to be a serious challenge that needs to be met urgently due to its theoretical and practical importance and the fact that the present business world is in the era of transient competitive advantages (McGrath, 2013). To answer this challenge, our planned theory comes into being: it attempts to fill the gaps outlined in Sect. 1.1 and to avoid the deficits presented in Sect. 1.2 by employing systems methodology and by considering the specific attributes of the scenario at hands. This book demonstrates clearly that systems are more adequate than quantities when managers and entrepreneurs need to discover potentials and opportunities of value. Additionally, this book shows how the advantages of the methodologies listed in Sect. 1.2 can be utilized to address value-related issues and challenges based on the guidance of an underlying systemic analysis.

1.3.3 Systems Thinking and Methodology

After the previous two subsections demonstrated the need for scholars and practitioners in the area of value creation and capture to adopt systems thinking and methodology, this subsection provides a quick glance of relevant basics of systems science.

Systems science is developed to enable scholars to look at the world holistically by focusing on the investigation of organizations and structures, each of which is known as a system (Lin, 1999; Klir, 1985). Based on the discussions in the previous subsections and the fact that entities involved in studies of value creation and capture generally possess their respectively rich, yet different, internal structures, this book adopts systems science as its way of reasoning and intuitively sees how various value potentials evolve either individually or collectively.

As is defined earlier, a system represents an organization or a structure whose component parts are associated with each other to form an organic whole, where the whole is a greater sum of parts (Lin, 1999). Hence, the system is everywhere in the world, especially in investigations of value creation and capture. For example, each consumer is both a complex biological system and an input and output economic system, depending on in which angle the person is seen. At the same time, other than being a member of many larger systems, such as a family, neighborhoods, and communities, he/she also interacts with various man-made systems, such as a car and an ATM machine, and with different social systems, such as the company of employment and sports clubs.

Historically, scholars in different disciplines over the recorded history have either consciously or unconsciously employed the concept of systems by using their respective languages (von Bertalanffy, 1968). And in our modern time, an increasing number of scholars in the area of business, such as Forrest et al. (2020), Klir (1985), Lin (2009), and Porter (1985), also demonstrate the power of holistic thinking and relevant methodology in terms of producing reliable and usable conclusions regarding business entities and how they interact with each other.

Similar to the situation of how numbers item from the physical world, systems can also be abstracted out of any and every physically or socially existing object, event, and process. However, beyond their common origin, numbers and systems represent two different aspects of the world. In particular, when a firm is seen as a collection of isolated component parts, then the firm can be described with numbers—*m* staff members, *n* office rooms, etc. And when the firm is investigated as a potential place from where values emerge, then it becomes a system with its particular culture and philosophical values, specifically associated talents, operational routines, among others. It is the invisible whole that makes the firm exist both physically and intellectually. Of course, the firm considered in this discussion can be readily replaced by an individual (seen as an economic agent whose behavior is dictated by his/her value system), a market, an industry, an economy, etc. That implies that behind each business entity, there is an abstract, theoretical system within which the relevant whole, component parts, and their interconnectedness are emphasized. Conversely, it is the whole and interconnected parts that the totality is known as a firm, or a market, or an industry, or an economy. That is, other than specific circumstances where numbers play their roles, the business world is made up of dominantly systems (or structures or organizations).

In comparison, numbers, if used in relationships, describe small-scale and nonstructural phenomena (Wu & Lin, 2002), while the concept of system is large-scale and organizational. That is another reason why the systems methodology is more appropriate than all theories developed on numbers and numerical variables for the investigation of value creation and capture when the internal structures of involved economic entities cannot be ignored.

By systems science, it means the collection of all studies of systems and organizations. It investigates the systemhood of objects, structures (or organizations), and evolutionary processes. That is what makes systems science different from the traditional science that studies the thinghood of objects. Jointly, systems science and traditional science constitute a two-dimensional spectrum of knowledge with the latter being the first dimension and the former the second dimension (Klir, 2001). With systems science—an additional dimension of knowledge—in place, human ability to learn has gained additional strength so that more problems that have been challenging the very survival of the mankind since the beginning of time can be more adequately addressed than ever before. The materialization of such great expectation of systems science relies on the particular speaking language and intuition—the (systemic) yoyo model (Lin, 2007)— Fig. 1.4. That is similar to how the Cartesian coordinate system, consisting of intersecting real-number lines, plays its role in mathematics and traditional science (Kline, 1972). Specifically, any system of concern, be it physical or intellectual, tangible or intangible, a living being or an organization, a culture or a civilization, can be intuitively treated as a spinning field structure. When such structure is seen in the three-dimensional space we live in, the structure in Fig. 1.4 appears. It remains in a continual spin motion, which stands for the internal working of the system; otherwise, it will no longer exist as an identifiable system. For more in-depth discussions, see Chap. 2.

As expected, this yoyo model has successfully played the role of intuition and playground for scholars who investigate the world and explore new knowledge holistically, just as what the Cartesian coordinate system did for the traditional science (Lin, 2009; Lin & Forrest, 2011; Forrest, 2013, 2014; Forrest & Tao, 2014; Ying & Forrest, 2015). In particular, this yoyo model of general systems has been successfully applied in the investigation of Newtonian physics of motion, the concept of energy, economics, finance, history, foundations of mathematics, small-probability disastrous weather forecasting, civilization, business organizations, the mind, among others. Along this same line of thinking, this book uses this systemic intuition to establish generally true conclusions on value creation and capture.

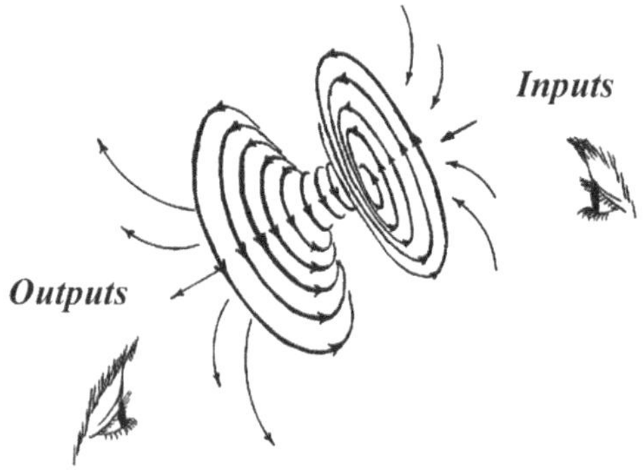

Fig. 1.4 The systemic yoyo model is shown in three-dimensional space

1.4 Organization of Contents in this Book

This book advances a unified general theory of value creation and capture that applies systemic thinking in two different ways. One is to employ such a novel way of thinking to resolve unsettled problems and fill gaps existing in the literature. The other is to help mobilize various available methodologies to investigate issues of concern in order to take advantage of the strengths of these different approaches. Beyond this Chap. 1 introduces the reader to the rest of this volume with all the related background information, the entire book consists of six parts and 19 additional chapters and an afterword.

The first part lays the foundation for the rest of the book by offering the basics of systems science, how market competition plays out, and how the resource-based view of the firm can be made rigorous through axiomatization. In particular, this part consists of four chapters. Chapter 2 looks at the origin of the concept of systems, how this concept is made operational in the development of theories, and the systemic yoyo model. Chapter 3 presents a general theory on how a market of free competition appears, grows, and matures by using game theory. Chapter 4 axiomatizes the well-known resource-based view of the firm by establishing a set of basic axioms, on which the chapter looks at when sustainable competitive advantages can appear, a firm's profitability, and other issues. Chapter 5 continues what is developed in the previous chapter and looks at several resource-related key issues, such as how resources evolve, capability rigidity, knowledge competence, technological opportunism, and firms' innovativeness.

Part II studies strategies that are applicable to either the demand side or the supply side of market exchanges. This part consists of four chapters. In particular, Chap. 6 looks at the composition and the systemic structure of a firm's supply-chain ecosystem and how a firm's success and challenges affect the rest of the ecosystem. Chapter 7 considers how the upstream suppliers and downstream complementors impact a firm's performance and how players of the ecosystem are vertically interdependent on each other. Chapter 8 establishes sufficient conditions for synergistic innovations to emerge in both the demand side and the supply side. Chapter 9 examines consumer synergies in terms of simultaneous utilities and multisided markets and introduces a mechanical method to produce synergistic innovations.

Part III pays a revisit to Porter's value-chain framework and generalizes it so that the upgraded version of the framework can be equally employed to evaluate companies that produce physical products, or offer services, or provide informational goods. In particular, this part contains four chapters. Chapter 10 presents the systemic, hierarchical structure of the mind. By developing the systemic foundation for the orientation of a representative individual, this chapter is able to demonstrate the existence of a systemic field within a person's recognition through the four human endowments—self-awareness, imagination, conscience, and free will. Chapter 11 continues this work on the mind and looks at the questions of how consumers prefer certain taxonomies and how they classify product information. Chapter 12 studies values created out of innovations and available resources.

Chapter 13 presents the generalized value-chain framework by looking at how values can be captured through memberships in strategic networks and blocks.

Part IV presents the concept of customer value propositions (CVPs) and their fundamental properties that are not individual-case specific. In particular, this part consists of three chapters. Chapter 14 examines when effective CVPs emerge by first modeling the concept systemically and then looking at a real-life case analysis. Chapter 15 focuses on the relationship between values and CVP-based competitive advantages while looking at the association between CVPs and shareholder values, and signs, risks, and values of implemented CVPs. Chapter 16 is concerned with the concept and properties of market-sensing capabilities, followed by two case studies.

Part V investigates the roles manufacturing and artificial intelligence play in industrial transformation and technological innovations. In particular, this part consists of two chapters. Chapter 17 shows the existence of a feedback mechanism between product markets and the manufacturing sector and demonstrates how a self-sustaining economic growth can be acquired. Chapter 18 focuses on how artificial intelligence affects technological innovations by specifying the individual paths of influence, followed by an empirical confirmation.

Part VI studies the contribution of government policies and supports to value creation and capture. In particular, this part consists of two chapters. Chapter 19 addresses the question of how the government can help stimulate economic growth and examines how policy tools potentially work and why government policies and supports are necessary for economic development. Chapter 20 studies conditions under which government policies can help improve firm performance. It looks at how government policies influence market competition and confirms established generally true conclusions by using an empirical case analysis.

At the conclusion of this book, an afterword is given in order to fill some logical holes that appear throughout the book. First, it shows how human sensing organs can produce incorrect conclusions, making results out of consequent logical reasoning also wrong. However, such difficult situations can be potentially detected and corrected by using analytical means. Second, it demonstrates how calculus-based analytical analysis can produce contradictory conclusions due to hidden misunderstanding of the concept of limits. Jointly, what is shown here is that in terms of the employment of various methodologies, a systemic approach needs also be employed so that all available tools of analysis and reasoning can be utilized systemically according to the characteristics of the problems of concern.

What might be shown to be the most thought-provoking question in the literature of business and economic studies this afterword raises is the potential of introducing a principle of economic induction. With such an induction, potential facts revealed by individual empirical studies will be possibly converted into generally true conclusions. And at the end of this afterword, outlined are the next steps scholars and practitioners can take based on what is presented in this book.

Because of the particular way the conclusions in this book are established, we expect that scholars, interested in the study of value creation and capture, and managers and entrepreneurs, who are involved in actual creation and capture of value, will find this book timely and beneficial in their works. And to make this book friendly to readers, we tried our best to develop each chapter as self-contained as possible so that readers do not need to flip through the pages to look up relevant concepts or results.

Acknowledgment Many colleagues have contributed to the success of this volume. Specifically, both Jeffrey Yi-Lin Forrest (Slippery Rock University, USA) and Yong Liu (Jiangnan University, China) threaded and edited all relevant contents together into this organized book. Jeffrey Yi-Lin Forrest (Slippery Rock University, USA) and Yong Liu (Jiangnan University, China) wrote Chap. 1. Yong Liu (Jiangnan University, China) and Jeffrey Yi-Lin Forrest (Slippery Rock University, USA) wrote Chap. 2. Chapter 3 is very technical, representing the joint effort of the following colleagues: Jeffrey Yi-Lin Forrest (Slippery Rock University, USA), Zhen Li (Texas Woman's University), Roger A. Solano (Slippery Rock University, USA), David Doorn (West Chester University), Joachim Wagner (Leuphana University Lueneburg, Germany), Melanie Anderson (Slippery Rock University, USA), John Lipinski (Indiana University of Pennsylvania), Yong Liu (Jiangnan University, China), and Xiaoguang Tian (Purdue University Fort Wayne).

Jeffrey Yi-Lin Forrest (Slippery Rock University, USA), Lawrence Shao (Slippery Rock University, USA), Ralph McKinney (Marshall University), Dale Shao (Marshall University) wrote Chap. 4.

Jeffrey Yi-Lin Forrest (Slippery Rock University, USA), Yong Liu (Jiangnan University, China), Eric Swift (Slippery Rock University, USA), Tooraj Karimi (University of Tehran, Iran) wrote Chap. 5. Jeffrey Yi-Lin Forrest (Slippery Rock University, USA), Yong Liu (Jiangnan University, China), Theresa A. Wajda (Slippery Rock University, USA), Erkan köse (Nuh Naci Yazgan University, Turkey), Oğuzhan A. Arik (Nuh Naci Yazgan University, Turkey) wrote Chap. 6. Jeffrey Yi-Lin Forrest (Slippery Rock University, USA), Yong Liu (Jiangnan University, China), Theresa A. Wajda (Slippery Rock University, USA), Erkan köse (Nuh Naci Yazgan University, Turkey), Oğuzhan A. Arik (Nuh Naci Yazgan University, Turkey) wrote Chap. 7. Jeffrey Yi-Lin Forrest (Slippery Rock University, USA), Yaoguo Dang (Nanjing University of Aeronautics and Astronautics, China), Larry McCarthy (Slippery Rock University, USA), Sifeng Liu (Nanjing University of Aeronautics and Astronautics, China), Yong Liu (Jiangnan University, China) wrote Chaps. 8 and 9. Theresa A. Wajda and Jeffrey Yi-Lin Forrest (Slippery Rock University, USA), Michael Y. Hu (Kent State University), Zhen Li (Texas Woman's University), Yong Liu (Jiangnan University, China) wrote Chaps. 10 and 11.

Jeffrey Yi-Lin Forrest (Slippery Rock University, USA), Sailau Baizakov (Institute of Mathematics and Mathematical Modeling, Kazakhstan), Yohannes G. Haile (Penn State Behrend), Abel Gyan (Slippery Rock University, USA), and Yong Liu (Jiangnan University, China) wrote Chap. 12. Jeffrey Yi-Lin Forrest (Slippery Rock University, USA), Yong Liu (Jiangnan University, China) John Golden (Slippery Rock University, USA), Yaya Sissoko (Indiana University of Pennsylvania), Qiaoxing Li (Guizhou University, China) wrote Chap. 13.

Jeffrey Yi-Lin Forrest (Slippery Rock University, USA), Abdykappar Ashimov (Kazakh National Research Technical University, Kazakhstan), Louie Ren (University of Houston—Victoria), Shahriar Gias (Slippery Rock University, USA), Yuriy Borovskiy (Kazakh National Research Technical University, Kazakhstan) wrote Chap. 14. Jeffrey Yi-Lin Forrest (Slippery Rock University, USA), Dmitry A. Novikov (Russian Academy of Sciences, Moscow), Stephen Larson (Slippery Rock University, USA), Fen Wang (Central Washington University), Jing Yang (State University of New York at Oneonta) wrote Chap. 15. Jeffrey Yi-Lin Forrest (Slippery Rock University, USA), Zaiwu Gong (Nanjing University of Information Science and Technology, China), Edward Scott (Slippery Rock University, USA), Gerardo Eloy Soto Ruiz (Autonomous University of the West, Mexico) wrote Chap. 16.

Jeffrey Yi-Lin Forrest (Slippery Rock University, USA), Jun Liu (Nanjing University of Information Science and Technology, China), Gérard Martorell (Lock Haven University), Liang Xu (Slippery Rock University, USA), Yong Liu (Jiangnan University, China), Davood Darvishi (Payame Noor University, Iran) wrote Chap. 17. Jun Liu and Huihong Chang (Nanjing University of Information Science and Technology, China), Jeffrey Yi-Lin Forrest (Slippery Rock University,

USA), Baohua Yang (Jiangsu Normal University, China) wrote Chap. 18. Jeffrey Yi-Lin Forrest (Slippery Rock University, USA), Abdou K. Jallow (Slippery Rock University, USA), Chao Wen (Ohio Northern University), Juehui Shi (Angelo State University), Huan Guo (Jianghan University, China) wrote Chap. 19. Jeffrey Yi-Lin Forrest (Slippery Rock University, USA), Jun Liu (Nanjing University of Information Science & Technology, China), Huan Guo (Jianghan University, China), Yong Liu (Jiangnan University, China), Theresa M. Phipps (Slippery Rock University, USA), Yuanjun Yang and Yaru Cao (Nanjing University of Information Science & Technology, China) wrote Chap. 20. Jeffrey Yi-Lin Forrest (Slippery Rock University, USA) wrote Afterword.

References

Adner, R., Oxley, J. E., & Silverman, B. S. (Eds.). (2013). *Collaboration and competition in business ecosystems*. Emerald.

Alshanty, A. M., & Emeagwali, O. L. (2019). Market-sensing capability, knowledge creation and innovation. *Journal of Innovation & Knowledge, 4*(3), 171–178.

Amit, R., & Zott, C. (2001). Value creation in e-business. *Strategic Management Journal, 22*(6/7), 493–520.

Anderson, J. C., Kumar, N., & Narus, J. A. (2007). *Value merchants: Demonstrating and documenting superior value in business markets*. Harvard Business Press.

Andreoni, A., & Chang, H. J. (2019). The political economy of industrial policy: Structural interdependencies, policy alignment and conflict management. *Structural Change and Economic Dynamics, 48*(March), 136–150.

Ardyan, E. (2016). Market sensing capability and SMEs performance. *DLSU Business & Economics Review, 25*(2), 79–97.

Barney, J. (1991). Firm resources and sustained competitive advantage. *Journal of Management, 17* (1), 99–120.

Becheikh, N., Landry, R., & Amara, N. (2006). Lessons from innovation empirical studies in the manufacturing sector: A systematic review of the literature from 1993 to 2003. *Technovation, 26* (5), 644–664.

Bjerknes, V. (1898). Uber einen hydrodynamischen Fundamentalsatz und seine Anwendung besonders auf die Mechanik der Atmosphare und des Weltmeeres. *Kungliga Svenska Vetenskapsakademiens Handlingar, 31*, 1–35.

Chiou, J. S., Hsiao, C. C., & Chiu, T. Y. (2018). The credibility and attribution of online reviews: Differences between high and low product knowledge consumers. *Online Information Review, 42*(5), 630–646.

Chowdhury, T. G., Murshed, F., & Khare, A. (2018). Flexible flexibility! Food categorization flexibility and utilitarian preference. *Journal of Consumer Marketing, 35*(1), 1–10.

Crook, T., Ketchen, D., Combs, J., & Todd, S. (2008). Strategic resources and performance: A meta-analysis. *Strategic Management Journal, 29*(11), 1141–1154.

Day, G. S. (1994). The capabilities of market-driven organizations. *Journal of Marketing, 58*(4), 37–52.

Drucker, P. F. (1954). *The practice of management*. Harper & Row.

EC. (2018). http://publications.jrc.ec.europa.eu/repository/bitstream/JRC113826/ai-flagship-report-online.pdf, Accessed on April 16, 2019.

Eisenhardt, K., & Martin, J. (2000). Dynamic capabilities: What are they? *Strategic Management Journal, 21*(10/11), 1105–1121.

Forrest, J. Y. L. (2013). *A systemic perspective on cognition and mathematics*. CRC Press, an imprint of Taylor and Francis.

Forrest, J. Y. L. (2014). *A systems perspective on financial systems*. CRC Press, an imprint of Taylor and Francis.

Forrest, J. Y.-L. (2018). *General systems theory: Foundation, intuition and applications in business decision making*. Springer.

Forrest, J. Y.-L., Buttermore, J., & Theresa Wajda, T. (2017). At Nash equilibrium when new market competitions appear? *Kybernetes: The International Journal of Cybernetics, Systems and Management Sciences, 46*(2), 256–271.

Forrest, J. Y.-L., Nicholls, N., Schimmel, K., & Liu, S. F. (2020). *Managerial decision making: A holistic approach*. Springer.

Forrest, J. Y.-L., & Tallapally, P. (2018). Customers are less patient, sustainable advantage becomes transient and the key for firms to succeed in fast changing markets. *Journal of Business, Economics and Technology, 21*(1), 1–14.

Forrest, J. Y. L., & Tao, L. X. (2014). *Investment and employment opportunities in China*. CRC Press, an imprint of Taylor and Francis.

Forrest, J. Y.-L., Trebing, M. E., Chatterjee, A., & Wagner, J. (2019). Domestic firms, exporters and importers: A systemic perspective. *Theoretical Economics Letters, 9*, 649–674.

Forrest, J. Y.-L., Zhao, H. C., & Shao, L. (2018). Engineering rapid industrial revolutions for impoverished agrarian nations. *Theoretical Economics Letters, 8*, 2594–2640.

Forrester Research Report. (2000). Global eCommerce approaches Hypergrowth. April 2000.

Gliva-McConvey, G., Nicholas, C. F., & Clark, L. (Eds.). (2020). *Comprehensive healthcare simulation: Implementing best practices in standardized patient methodology*. Springer.

Goldstein, E. B. (2010). *Encyclopedia of perception*. SAGE Publications.

Government of Japan. (2015). https://www8.cao.go.jp/cstp/kihonkeikaku/5basicplan_en.pdf, Accessed on April 16, 2019.

Hall, W., & Pesenti, J. (2017). Growing the artificial intelligence industry in the UK. https://assets.publishing.service.gov.uk/government/uploads/system/uploads/attachment_data/file/652097/Growing_the_artificial_intelligence_industry_in_the_UK.pdf, Accessed on October 15, 2017.

Hemphill, T. A. (2014). The US advanced manufacturing initiative: Will it be implemented as an innovation – Or industrial – Policy? *Innovation: Organization & Management, 16*(1), 67–70.

Hermann, M., Pentek, T., & Otto, B. (2016). Design principles for industry 4.0 scenarios. *The 49th Hawaii international conference on system sciences (HICSS)*, Koloa, HI (pp. 3928–3937).

Hitt, M., & Ireland, R. D. (2017). The intersection of entrepreneurship and strategic management research. In D. L. Sexton, H. Landstrom, & C. Jones (Eds.), *The Blackwell handbook of entrepreneurship* (pp. 45–63). Blackwell.

Howell, T. R., Noellert, W. A., MacLaughlin, J. H., & Wolff, A. W. (2019). *The microelectronics race: The impact of government policy on international competition*. Routledge.

Jomo, K. S. (2019). *Southeast Asia's misunderstood miracle: Industrial policy and economic development in Thailand, Malaysia and Indonesia*. Routledge.

Kenny, A. (2012). *A new history of Western philosophy*. Oxford University Press.

Kline, M. (1972). *Mathematical thought from ancient to modern times*. Oxford University Press.

Klir, G. (1985). *Architecture of systems problem solving*. Plenum Press.

Klir, G. (2001). *Facets of systems science*. Springer.

Kozlenkova, I. V., Samaha, S. A., & Palmatier, R. W. (2014). Resource-based theory in marketing. *Journal of the Academy of Marketing Science, 42*, 1–21.

Lanning, M. (2003). An introduction to the market-focused philosophy, framework and methodology called delivering profitable value. Retrieved February 24, 2012 from http://www.exubrio.com/whitepapers/DPVIntro-eXubrio.pdf.

Lemoussu, S., Chaudemar, J. C., & Vingerhoeds, R. A. (2018). Systems engineering and project management process modeling in the aeronautics context: Case study of SMEs. *International Journal of Mechanical and Mechatronics Engineering, 12*(2), 88–96.

Lin, Y. (1999). *General systems theory: A mathematical approach*. Plenum and Kluwer Academic Publishers.

Lin, Y. (2007). Systemic yoyo model and applications in Newton's, Kepler's laws, etc. *Kybernetes: The International Journal of Cybernetics, Systems and Management Science, 36*(3–4), 484–516.

Lin, Y. (2009). *Systemic yoyos: Some impacts of the second dimension*. Auerbach Publication, an imprint of Taylor and Francis.

Lin, Y., & Forrest, B. (2011). *Systemic structure behind human organizations: From civilizations to individuals*. Springer.

Lin, Y., & OuYang, S. C. (2010). *Irregularities and prediction of major disasters*. CRC Press, an imprint of Taylor and Francis.

Lindblom, A., Olkkonen, R., Mitronen, L., & Kajalo, S. (2008). Market-sensing capability and business performance of retail entrepreneurs. *Contemporary Management Research, 4*(3), 219–236.

Lipton, M. (1977). *Why poor people stay poor: Urban Bias in developing countries*. Temple Smith.

Mandler, G. (1982). The integration and elaboration of memory structures. In F. Klix, J. Hoffman, & E. van der Meer (Eds.), *Cognitive research in psychology* (pp. 33–40). North-Holland.

Marc, S., Yann, B., Charly, B, et al. (2018). Mission Villani: for a meaningful artificial intelligence towards a French and European strategy. https://www.aiforhumanity.fr/pdfs/MissionVillani_Report_ENG-VF.pdf, Accessed on March 8, 2018.

Marketing Science Institute. (2010). *MSI research priorities 2010–12*. MSI.

McGrath, R. G. (2013). *The end of competitive advantage: How to keep your strategy moving as fast as your business*. Harvard Business Review Press.

Melitz, M. J. (2003). The impact of trade on intra-industry reallocations and aggregate industry productivity. *Econometrica, 71*(6), 1695–1725.

Melitz, M. J., & Redding, S. J. (2014). Heterogeneous firms and trade. *Handbook of International Economics, 4*, 1–54.

Meyers-Levy, J., & Tybout, A. M. (1989). Schema congruity as a basis for product evaluation. *Journal of Consumer Research, 16*(1), 39–54.

Moss, G. (2009). *Gender, design and marketing: How gender drives our perception of design and marketing*. Routledge.

Nedungadi, P. (1990). Recall and consumer consideration sets: Influencing choice without altering brand evaluations. *Journal of Consumer Research, 17*(3), 263–276.

OWH. (2016a). *The National Artificial Intelligence Research and Development Strategic Plan*. https://www.nitrd.gov/PUBS/national_ai_rd_strategic_plan.pdf, Accessed on April 16, 2016.

OWH. (2016b). *Preparing for the Future of Artificial Intelligence*. https://obamawhitehouse.archives.gov/sites/default/files/whitehouse_files/microsites/ostp/NSTC/preparing_for_the_future_of_ai.pdf, Accessed on April 20, 2019.

Payne, A., & Frow, P. (2005). A strategic framework for customer relationship management. *Journal of Marketing, 69*(4), 167–176.

Payne, A., Frow, P., & Eggert, A. (2017). The customer value proposition: Evolution, development, and application in marketing. *Journal of the Academy of Marketing Science, 45*, 467–489.

Penrose, E. (1959). *The theory of the growth of the firm*. Blackwell.

Porter, M. E. (1985). *Competitive advantage: Creating and sustaining superior performance*. Free Press.

Priem, R. L., Wenzel, M., & Koch, J. (2018). Demand-side strategy and business models: Putting value creation for consumers center stage. *Long-Range Planning, 51*(1), 22–31.

Qu, Y. (1985). *The songs of the south: An anthology of ancient Chinese poems by Qu Yuan and other poets* (translated by David Haykes). Penguin Classics.

Rathod, G. H., Puranik, V. S., & Hiremath, V. (2019). Developing agility system in supply chains. *International Journal of Business and Systems Research, 13*(2), 247–258.

Reinert, E. (1995). Competitiveness and its predecessors: A 500-year cross-national perspective. *Structural Change and Economic Dynamics, 6*(1), 23–42.

Rostow, W. W. (1960). *The stages of economic growth: A non-communist manifesto*. Cambridge University Press.

Sahni, N. S. (2016). Advertising spillovers: Evidence from online field experiments and implications for returns on advertising. *Journal of Marketing Research, 53*(4), 459–478.

Santalo, J., & Becerra, M. (2008). Competition from specialized firms and the diversification-performance linkage. *Journal of Finance, 63*(2), 851–883.

Schrift, R. Y., Parker, J. R., Zauberman, G., & Srna, S. (2018). Multistage decision processes: The impact of attribute order on how consumers mentally represent their choice. *Journal of Consumer Research, 44*(6), 1307–1324.

Shannon, C. E. (1949). *The mathematical theory of communication*. Illinois University Press.

Srivastava, R. K., Shervani, T. A., & Fahey, L. (1999). Marketing, business processes, and shareholder value: An organizationally embedded view of marketing activities and the discipline of marketing. *Journal of Marketing, 63*(Special issue), 168–179.

Stabell, C. B., & Fjeldstad, O. D. (1998). Configuring value for competitive advantage: On chains, shops, and networks. *Strategic Management Journal, 19*(5), 413–437.

Starch, D. (1914). *Advertising: Its principles, practice, and technique*. Scott, Foresman.

State Council of the PRC. (2015). Notice of State Council of the PRC on printing and distributing "Made in China 2025". http://www.gov.cn/zhengce/content/2015-05/19/content_9784.htm, Accessed on May 8, 2015.

Studwell, J. (2013). *How Asia works: Success and failure in the World's Most dynamic region*. Grove Press.

Sugiyarti, G., & Ardyan, E. (2017). Market sensing capability and product innovation advantages in emerging markets: The case of market entry quality and marketing performance of Batick industry in Indonesia. *DLSU Business & Economics Review, 27*(1), 175–189.

Sujan, M. (1985). Consumer knowledge effects on evaluation strategies mediating consumer judgments. *Journal of Consumer Research, 12*(June), 31–46.

Szirmai, A., & Verspagen, B. (2015). Manufacturing and economic growth in developing countries, 1950-2005. *Structural Change and Economic Dynamics, 34*, 46–59.

Vendrell-Herrero, F., Bustinza, O. F., Parry, G., & Georgantzis, N. (2017). Servitization, digitization and supply chain interdependency. *Industrial Marketing Management, 60*(January), 69–81.

von Bertalanffy, L. (1968). *General systems theory*. George Braziller.

Webster, F. E. (2002). *Market-driven management: How to define, develop and deliver customer value* (2nd ed.). Wiley.

Wen, Y. (2016). *The making of an economic superpower: Unlocking China's secret of rapid industrialization*. World Scientific.

White, M. (2017). Keep calm and simulate on: Faculty experiences and insights into implementing best practices in simulation. *Teaching and Learning in Nursing, 12*(1), 43–49.

Wu, Y., & Lin, Y. (2002). *Beyond nonstructural quantitative analysis: Blown-ups, spinning currents and modern science*. World Scientific.

Xi, J .P. (2017). http://www.gov.cn/zhuanti/2017-10/27/content_5234876.htm, Accessed on April 16, 2019.

Ye, G. L., Priem, R. L., & Alshwer, A. A. (2012). Achieving demand-side synergy from strategic diversification: How combining mundane assets can leverage consumer utilities. *Organization Science, 23*(1), 207–224.

Ying, Y. R., & Forrest, J. Y. L. (2015). *Capital account liberation: Methods and applications*. CRC Press, an imprint of Taylor and Francis.

Zachary, M. A., Gianiodis, P. T., Payne, G. T., & Markman, G. D. (2015). Entry timing: Enduring lessons and future directions. *Journal of Management, 41*(5), 1388–1415.

Part I
Preparation

Chapter 2
Introduction to Systems Research and Systemic Reasoning

Abstract Because the systems research and systemic reasoning employed herein are not commonly used by the scholars who work in areas this book touches, this chapter introduces the relevant basics in order to make this volume self-contained. On the other hand, for those readers who are not mathematically oriented, those of the following sections that are heavy with mathematical symbols, notations, and terminologies, can be skipped without causing any trouble with the reading of the rest of this book.

Keywords Cognitive system · Feedback · Ordered pair · Organization · Relation · Systemic yoyo

This chapter is organized as follows. Section 2.1 provides a brief development history of the concept of systems. Section 2.2 focuses on a particular definition of systems for the purpose of developing theoretically significant and practically useful conclusions. Section 2.3 looks at the component elements of the systemic yoyo model and examples on how this model can be utilized to produce useful intuitions in the development of generally true conclusions. Section 2.4 concludes this preparation chapter with a few remarks.

2.1 Systems: The Concept

By system, it means an organic whole that consists of a collection of objects and a set of relations of the objects. Without the relations, the collection of objects will not be considered as a system, or simply a trivial system introduced to make the consequent theory plausible, just as the reason why zero is needed in mathematics (Lin, 1999). Thus, systems are everywhere especially in investigations of economic issues and business decision-making. For instance, each individual person is a biological system that is composed of numerous smaller systems, such as internal organs, body parts, etc. Moreover, individual people are also members of various social and economic systems (e.g., families, neighborhoods, communities, occupational

J. Y.-L. Forrest, Y. Liu, *Value in Business*, Contributions to Management Science,
https://doi.org/10.1007/978-3-030-82898-1_2

groups) that compose their social identities. Beyond social identity, individuals regularly interact with different systems (e.g., vehicles, smart devices, retail stores, and organizations). Because of the existence of such individuals that are part of various systems simultaneously, systems are constantly interacting with one another. Hence, when conducting scholarly studies, other than using numbers and variables to investigate problems and issues of business and economics, which is what is mostly done in the literature of traditional areas of research, there is an urgent need to employ the concept of systems and relevant methods to study economic phenomena in order to obtain new understandings and conclusions that are closer to the real situations than those obtained through employing numbers and variables.

Historically, the concept of systems has been either directly or indirectly introduced by scholars from different disciplines; and studies on various systems can be traced back as far as the recorded history goes. For example, the idea of systems has been used by many great thinkers to study different problems using their respective languages throughout history. For example, Aristotle's statement that "the whole is greater than the sum of its parts" is an early definition of a basic systems problem. Nicholas of Cusa, a profound thinker of the fifteenth century, linking medieval mysticism with the first beginnings of modern science, introduced the notion of the *coincidentia oppositorum*. Leibniz's hierarchy of monads looks quite like that of modern systems. Gustav Fechev, known as the author of the psychophysical law, elaborated, in the way of the naive philosophers of the nineteenth century, supraindividual organizations of higher order than the usual objects of observation thus romantically anticipating the ecosystems of modern parlance. For more details of such a historical account, see von Bertalanffy (1972). For our purpose here, we list only a few of the great thinkers of the past.

Although the concept of systems enjoys such a long history, it is not until the 1920s that von Bertalanffy (1924) officially introduced the concept in biology by pointing out that "Because the fundamental character of living things is their organization, the customary investigation of individual parts and processes cannot provide a complete explanation of the phenomenon of life." After nearly 100 years of theoretical studies and practical applications, this concept has been widely accepted by scientists and scholars in all disciplines (Blauberg et al., 1977), including business-related studies. For example, Rostow (1960) in economics wrote:

> The classical theory of production is formulated under essentially static assumptions ... to merge classical production theory with Keynesian income analysis ... introduced the dynamic variables: population, technology, entrepreneurship, etc. But ... do so in forms so rigid and general that their models cannot grip the essential phenomena of growth ... We require a dynamic theory ... which isolates not only the distribution of income between consumption, savings, and investment (and the balance of production between consumers and capital goods) but which focuses directly and in some detail on the composition of investment and on developments within particular sectors of the economy.

Others (cf., Forrest, 2018; Klir, 1985; Lin, 2009; Porter, 1985) also demonstrate how powerful systemic thinking and relevant methodology can be in terms of producing conclusions, regarding organizations, that are realistically reliable and practically usable. In particular, a gap that exists in the widely employed methodologies in business and economics studies is that a method that can be effectively employed to investigate interactions of business entities of different magnitudes, economies of various scales, and markets of whatever kinds, has been missing until the time when one brings systems research and systemic reasoning into her works. Since the 1920s, such a systemic and holistic view of nature, organizations, and social events has permeated the entire spectrum of knowledge exploration, see Lin (2009) and references found there for details on how specific areas of learning have benefitted from systems research.

Although the concepts of numbers and systems are abstracted out of the same physical world, they are fundamentally different in two perspectives. In particular, when an organization is perceived as a collection of unrelated people, properties, and assets, numbers come into play (e.g., n employees, m copy machines, $\$X$ invested). Such treatment of an organization divides the whole of that organization into pieces without considering the relationships among and between the component units and factors of that organization. That is where and how difficulties of studying interactions of organizations emerge. In other words, to understand how an organization, as an organic whole composed of component parts, operates and how it interacts with other organizations, its internal operational connections of the parts must be considered. In other words, when an organization is seen holistically, as it should be in order to understand how the business world actually functions, the concept of systems needs to be employed, where such elements as employees, capital, and properties, form an organic whole through various relationships, without which the organization simply does not exist. That is, studies in business-related disciplines are essentially about relationships of systems, be they firms, markets, industries, or economies. In other words, major differences between the concepts of numbers and systems are that:

1. The former is a small-scale local concept, while the latter is a large-scale organizational concept (Lin, 1988, 1999) and
2. Numbers appear only post existence, while systems emerge at the same time when physical or intellectual existence comes into being (Lin, 2009).

Item (2) above is the reason why systems methodology is a more appropriate tool than theories developed on numbers and variables for investigating economic entities when their internal structures are concerned with and why market participants still cannot successfully make advanced predictions for imminent economic disasters (for more in-depth explanation see Lin & OuYang, 2010).

To make the systems science, the organized systematic investigation of general systems and various specific kinds of systems, practically useful, scholars within the last century have established many valuable approaches of systems analysis, developed in different disciplines. Of course, as with any scientific theory, each of these approaches has to start from how a system is defined, leading to many different descriptions of what a system is. To establish a theoretical foundation for all these approaches of systems analysis, Mesarovic, in the early 1960s, introduced the mathematical definition of (general) systems, based on Cantor's set theory, as follows (Mesarovic & Takahara, 1975): A (general) system S is a relation on nonempty sets V_i:

$$S \subseteq \prod \{V_i : i \in I\}, \tag{2.1}$$

where I is an index set.

This structure of general systems reflects the unification of "isolated" objects, a relation between the objects, and the structure of layers. Specifically, elements in the nonempty sets V_i, $i \in I$, are the objects of the system, S represents a relation between objects, and elements of the sets V_i, $i \in I$, can be systems, too.

To make this definition of systems more symbolically operational for the purpose of developing a practically useful systems theory, Lin (1987) defines the concept of general systems as follows: S is a (general) system provided that S is equal to an ordered pair (M, R) of sets. That is, symbolically, we have $S = (M, R)$, where R is a set of some relations defined on the set M. Each element in M is called an object of the system S, and M and R are called the object set and the relation set of S, respectively.

Upon consideration of the interrelationship between the systems of concern and some of their environments, Bunge (1979) considered a model of systems as follows: For nonempty set T, the ordered triple $W = (C, E, S)$ is a system over T if and only if C and E are mutually disjoint subsets of T and S is a nonempty set of relations on $C \cup E$; C and E are called the composition and an environment of the system W, respectively.

Klir (1985) introduced a philosophical concept of general systems, which reads: "a system is what is distinguished as a system." Theoretically, Klir's definition contains the most general meaning of the concept of systems originally posed by von Bertalanffy.

Starting in 1976, Xuemou Wu and his followers have studied many different theories, covered by the overarching title of pansystems analysis under the name "pansystems." This pansystems analysis is a new research of multilevels across all disciplines. The theory deals with general systems, relations, symmetry, transformation, generalized calculus, and *shengke* (means survival and vanquishing), called the emphases of pansystems. Based on the research of these emphases, the theory blends philosophical reasoning, mathematical logic, and mechanical structures into one solid body of knowledge (Wu, 1990).

The discussion here gives rise naturally to the following question: Why was the concept of systems not studied more in-depth before now? One answer is that the development of modern technology, e.g., communication technology, designs of satellites, climate control of giant buildings, etc., reveals one fact: History is presently in a special moment, where the accumulation of knowledge has reached such a level that each discovery of a relation between different areas of knowledge will materially produce a useful or consumable product. That is the reason behind the claim (Forrest, 2018) that the next scientific revolution will be system-based.

Closely related to the concept of systems are the following questions:

1. What is the meaning of a system with contradictory relations, e.g., $\{x > y, x \leq y\}$ as part of the relation set?
2. How can we know whether there exist contradictory or inconsistent relations in a given realistic system?

From Klir's definition of systems, which is one of the definitions with the widest meaning among all definitions of systems, it follows that generally these two questions cannot be answered. Concerning this end, a theorem of Gödel shows that it is impossible to show whether the systems representation of mathematics, developed on the ZFC axioms is consistent. For a detailed discussion of the systems representation of mathematics, see (Lin, 1989). This fact implies that there are systems, for example, the system of set theory on ZFC, so that we do not know if propositions with contradictory meanings exist or not. This means that maybe not every system is consistent or that not every system contains no contradictory relations. This end is quite easy to understand in terms of business organizations because conflicts or inconsistencies of various kinds always coexist in business decision-making and consequent implementations.

Magnificent successes of the traditional science, for the most part, are achieved on the basis of the works of Descartes and Galileo who contributed the needed methods of reasoning and administration. In particular, Descartes suggested to divide the problem under consideration into as many small parts as possible and study each isolated part (Kline, 1972), and Galileo recommended to simplify the complicated phenomenon of concern into basic parts and processes (Kuhn, 1962). In the history of science and technology, these methods of reasoning and administration have been very successfully applied, leading to great victories one by one (von Bertalanffy, 1972). Beyond so, these methods are still currently widely employed in research activities of natural and social sciences.

Within the world currently considered with increasing complexity, scholars and managerial decision-makers are forced to study problems and issues with many cause-effect chains (i.e., systems), where internal organizations and structures cannot be ignored. So, relevantly adequate logics of thinking and methodologies need to be introduced in order to successfully deal with such problems whose focus is on systems, organizations, and structures. In other words, in the study of such problems, Descartes' and Galileo's methods have to be modified and improved, because these methods emphasize separating the whole of concern into parts and isolated processes instead of being a whole. But according to systemic thinking and real-life practice of managerial decision-making (e.g., Porter, 1979, 1985), one needs to recognize that other than a pile of innumerable isolated "parts," each business decision is more about how these parts work together organically as a whole. The basic characteristic of business decisions is about organization(s), interactions between internal component units and processes, and exchanges with and among external organizations. Thus, the chief task of managerial decision-making is about knowing systemically a company, a market, an economy, and the world filled with a myriad of interacting wholes.

2.2 Systems: An Operational Definition

For our purpose of developing theoretical significant and practically useful results, a system is identified with an ordered pair $S = (M, R)$ of an object set M and a relation set R (Lin, 1999). In particular, M consists of all the isolated objects of system S and R a set of relations that connect the objects in M. For example, each business firm is composed of a set of employees, properties, equipment, assets, etc. This set is collectively known as that of objects of the organizational system of the firm. More importantly, these objects are associated with each other through a set of particular relations. It is in these particular relations that the whole is acknowledged as a functional firm.

For theoretical completeness, a system $S = (M, R)$ is discrete if it satisfies:

$$R = \varnothing \text{ or } R = \{\varnothing\} \text{ and } M \neq \varnothing, \tag{2.2}$$

where $\varnothing$ stands for the empty set. In other words, a system S is discrete when it consists of isolated objects only.

A system $S = (M, R)$ is said to be trivial, if $M = \varnothing$. The concepts of discrete and trivial systems are needed for making the consequent theory of general systems theoretically complete and practically usable. The situation is exactly the same as for why and how the concept of zero is necessary for the theory of numbers.

For a given system $S = (M, R)$, another system $S_1 = (M_1, R_1)$ is said to be a subsystem of S, if M_1 is a subset of M and R_1 is a subset of R. And the system $S_1 = (M_1, R_1)$ is said to be a partial system of S, if M_1 is a subset of M and for any relation $r_1 \in R_1$ there is a relation $r \in R$ such that

$$r_1 = r \mid M_1. \tag{2.3}$$

That is, the relation r_1 of the system S_1 is the restriction of a relation r in the system S. In other words, for each relation r_1 in the subsystem S_1, there is a relation r in R of the original system S such that r_1 consists of all relationship descriptions in r of objects in M_1 except those descriptions involving objects outside the subset M_1. For convenience, we write the subsystem S_1 as $S \mid M_1$ and the relation set R_1 as $R \mid M_1$, consisting of relations in R but restricted on M_1.

Let $S_i = (M_i, R_i)$, $i = 1, 2$, be systems and h: $M_1 \to M_2$ a mapping from the object set M_1 to the other object set M_2. For each relation $r \in R_1$ and any object sequence $(x_1, x_2, x_3, \ldots) \in r$, let

$$h(x_1, x_2, x_3, \ldots) = (h(x_1), h(x_2), h(x_3), \ldots) \tag{2.4}$$

and

$$h(r) = \{h(x_1, x_2, x_3, \ldots) : (x_1, x_2, x_3, \ldots) \in r\}. \tag{2.5}$$

Without confusion, write h: $S_1 \rightarrow S_2$. A mapping h: $S_1 \rightarrow S_2$ is said to be partial, if for some object $x \in M_1$, $h(x)$ is not defined. The identity mapping $id : S_1 \rightarrow S_1$ is defined as

$$id(x) = x, \text{for } x \in M_1. \tag{2.6}$$

For two systems $S_i = (M_i, R_i)$, $i = 1, 2$, a mapping h: $S_1 \rightarrow S_2$ is said to be one-to-one if $h(M_1) = M_2$ and for any $x, y \in M_1$, $x \neq y$ implies $h(x) \neq h(y)$. The systems S_1 and S_2 are said to be similar, if there is a one-to-one mapping h: $S_1 \rightarrow S_2$ such that $h(M_1) = M_2$ and $h(R_1) = R_2$. In this case, the mapping h is known as a similarity mapping from system S_1 to system S_2.

Theorem 2.1 If h: $S_1 \rightarrow S_2$ is similarity mapping from system S_1 to system S_2, then the inverse function h^{-1}: $S_2 \rightarrow S_1$ is also a similarity mapping but from system S_2 to system S_1.

Proof Because h: $S_1 \rightarrow S_2$ is similarity mapping, it follows that h is one-to-one and $h(M_1) = M_2$. So, for any $y \in M_2$, there is a unique $x \in M_1$ such that $h(x) = y$. Therefore, the inverse mapping h^{-1}: $S_2 \rightarrow S_1$ is a well-defined one-to-one mapping by $h^{-1}(y) = x$. From $h(R_1) = R_2$, it follows that for any relation $s \in R$, there is a relation $r \in R_1$ such that

$$\begin{aligned} h^{-1}(s) &= \left\{h^{-1}(y) : y = (y_1, y_2, y_3, \ldots) \in s\right\} \\ &= \left\{\left(h^{-1}(y_1), h^{-1}(y_2), h^{-1}(y_3), \ldots\right) : y_1, y_2, y_3, \ldots \in M_2\right\} \\ &= \left\{(x_1, x_2, x_3, \ldots) : x_1, x_2, x_3, \ldots \in M_1\right\} = r \in R_1 \end{aligned} \tag{2.7}$$

Therefore, the inverse mapping h^{-1} is a similarity mapping from S_2 to S_1.

Let $\{S_i = (M_i, R_i) : i \in I\}$ be a set of systems, where I is an index set. If for any two systems S_i and S_j, $i \neq j \in I$, their objects sets are disjoint, then the free sum of this collection of systems, denoted $\bigoplus\{S_i : i \in I\}$ or $S_1 \bigoplus S_2 \bigoplus \ldots \bigoplus S_n$, if $I = \{1, 2, \ldots, n\}$ is finite, is simply defined as the system whose object set is equal to the totality of the objects in the given systems, and whose relation set that of all the relations in the individual systems. That is,

$$\bigoplus\{S_i : i \in I\} = S_1 \bigoplus S_2 \bigoplus \ldots \bigoplus S_n = (\cup_{i \in I} M_i, \cup_{i \in I} R_i). \tag{2.8}$$

Assume that $S - \{S_i = (M_i, R_i) : i \in I\}$ is a set of systems, where I is an index set. Define another set of systems $S' = \left\{S'_i = \left(M'_i, R'_i\right) : i \in I\right\}$ as follows:

$$M'_i = M_i \times \{i\}\ R'_i, \tag{2.9}$$

for any relation $r \in R_i$

$$r' = \{((x_1, i), (x_2, i), (x_3, i), \ldots) : (x_1, x_2, x_3, \ldots) \in r\}, \tag{2.10}$$

and $R_i' = \{r' : e \in R\}$. In this general case, the free sum $\bigoplus\{S_i : i \in I\}$ of the systems in S is defined as $\bigoplus\{S_i' : i \in I\}$.

Theorem 2.2 For each set of systems, its free sum exists uniquely up to a similarity. In other words, for any set of systems, although there are many different free sums, they are all similar to each other.

Proof This result follows directly from the fact that for a set of systems $S = \{S_i : i \in I\}$, any two free sums of these systems only differ by different index sets used to label the individual systems in the given set. If two free sums ${}_1S$ and ${}_2S$ of the set S involve respectively two index sets I and J, then there must be a one-to-one mapping $h: I \to J$ satisfying $h(I) = J$. This mapping h can help induce a similarity mapping from ${}_1S$ to ${}_2S$ as follows: For any object (x, i) in ${}_1S$,

$$h(x, i) = (x, h(i)), \text{for any } i \in I. \tag{2.11}$$

A system $S = (M, R)$ is connected if the system satisfies the following condition: for any m_1 and $m_2 \in M$, there are a sequence of objects $n_1, n_2, \ldots, n_k \in M$, and a sequence of relations $r_1, r_2, \ldots, r_{k+1} \in R$, for some natural number k, satisfying that r_1 relates m_1 and n_1, r_2 relates n_1 and n_2, …, r_k relates n_{k-1} and n_k, and r_{k+1} relates n_k and m_k. In other words, a system is connected if any two objects in the system are related through a finite number of relations of the system. Otherwise, the system S is known as a disconnected system, meaning that the system is really equal to the free sum of two or more subsystems, defined on pairwise disjoint object sets.

Theorem 2.3 A system $S = (M, R)$ is connected, if and only if for any two objects $x, y \in M$, there exist a natural number $n > 0$ and n relations $r_i \in R$ such that $x \in Supp(r_1)$ and $y \in Supp(r_n)$ and $Supp(r_i) \cap Supp(r_{i+1}) \neq \varnothing$, for each $i = 1, 2, \ldots, n - 1$, where for any $r \in R$,

$$\text{Supp}(r) = \{m \in M: m \text{ appears in the relation } r\}. \tag{2.12}$$

Proof ($\Rightarrow$) By contradiction, assume that the system S is connected and there exist two objects x and $y \in M$ such that there do not exist relations $r_i \in R$, $i = 1, 2, \ldots, n$, for any natural number $n \geq 1$, such that $x \in \text{Supp}(r_1)$ and $y \in \text{Supp}(r_n)$ and $\text{Supp}(r_i) \cap \text{Supp}(r_{i+1}) \neq \varnothing$, for each $i = 1, 2, \ldots, n - 1$. Hence, there must be relations r_1, $s_1 \in R$ such that $x \in \text{Supp}(r_1)$, $y \in \text{Supp}(s_1)$, and $\text{Supp}(r_1) \cap \text{Supp}(s_1) = \varnothing$. Let $U_0 = \text{Supp}(r_1)$ and $V_0 = \text{Supp}(s_1)$ and for any natural number n define

$$U_n = \cup\{\text{Supp}(r): r \in R \,\&\, \text{Supp}(r) \cap U_{n-1} \neq \varnothing\}$$

and

$$V_n = \bigcup\{\text{Supp}(s) \colon s \in R \& \text{Supp}(s) \cap V_{n-1} \neq \varnothing\}.$$

Then two subsystems $S_i = (M_i, R_i)$, $i = 1, 2$, of S emerge such that $M_1 = \bigcup_{n=0}^{\infty} U_n$, $M_2 = M - M_1$, $R_1 = \{r \in R : \text{Supp}(r) \cap M_1 \neq \varnothing\}$, and $R_2 = \{r \in R : \text{Supp}(r) \subseteq M_2\}$. It can be readily shown that $S = (M, R) = (M_1 \cup M_2, R_1 \cup R_2) = S_1 \bigoplus S_2$. This end contradicts the fact that S is a connected system.

($\Leftarrow$) Again by contradiction, assume that S is disconnected. Then S is equal to the free sum $S_1 \bigoplus S_2$ of two nontrivial subsystems $S_1 = (M_1, R_1)$ and $S_2 = (M_2, R_2)$. Pick an object $m_i \in M_i$, $i = 1, 2$. Then there are no relations $r_j \in R$, $j = 1, 2, \ldots, n$, for any fixed $n \in \mathbb{N}$, such that $m_1 \in \text{Supp}(r_1)$ and $m_2 \in \text{Supp}(r_n)$ and $\text{Supp}(r_j) \cap \text{Supp}(r_{j+1}) \neq \varnothing$, for each $j = 1, 2, \ldots, n - 1$, a contradiction.

A system $S_0 = (M_0, R_0)$ is multileveled, if there is at least one object $S_1 \in M_0$ such that $S_1 = (M_1, R_1)$ is a system, and there is at least one object $S_2 \in M_1$ such that $S_2 = (M_2, R_2)$ is a system, ..., where S_1 is known as a first-level object system of S_0, S_2 a second-level object system of S_0, ...

Theorem 2.4 (ZFC). Each chain $S_1 = (M_1, R_1), S_2 = (M_2, R_2), S_3 = (M_3, R_3), \ldots$, of object systems must be finite, where $S_{i+1} \in M_i$, for $i = 1, 2, 3, \ldots$

Proof By contradiction, assume that there is an infinite chain of object systems $S_i = (M_i, R_i)$ such that $S_{i+1} \in M_i$, i is a natural number. Define

$$X = \{M_i \colon i = 1, 2, 3, \ldots\} \cup \{S_i \colon i = 1, 2, 3, \ldots\} \cup \{\{M_i\} \colon i = 1, 2, 3, \ldots\}.$$

From the Axiom of Regularity, it follows that there exists $Y \in X$ such that $Y \cap X = \varnothing$. There now exist three possibilities: (1) $Y = M_i$ for some i; (2) $Y = S_i$ for some i; and (3) $Y = \{M_i\}$ for some i.

If possibility (i) holds true, then we have $S_{i+1} \in Y \cap X$. Therefore, $Y \cap X \neq \varnothing$, a contradiction. If (ii) holds true, then we have $Y = S_i = (M_i, R_i) = \{\{M_i\}, \{M_i, R_i\}\}$ and $\{M_i\} \in Y \cap X \neq \varnothing$, a contradiction. If (iii) holds true, then we have $M_i \in Y \cap X \neq \varnothing$, a contradiction. These contradictions show that no chain of object systems is infinite.

A system $S = (M, R)$ is said to be centralized if each object in S is a system and there exists a system $C = (M_C, R_C)$ such that $M_C \neq \varnothing$ and for any distinct objects $x, y \in M$, say $x = (M_x, R_x)$ and $y = (M_y, R_y)$,

$$M_C = M_x \cap M_y \text{ and } R_C \subseteq R_x|M_C \bigcap R_y|M_C, \text{d} \tag{2.13}$$

where $R_x|M_C = \{r|M_C : r \in R_x\}$ and $R_y|M_C = \{r|M_C : r \in R_y\}$. System C is called a center of the centralized system S. Speaking in nonmathematical terms, a system S is centralized, if and only if there is another non-discrete system C that is a partial system of every object system in S. It can be shown that under given cardinality conditions, each multilevel system has a partial system that is centralized.

Theorem 2.5 (ZFC). Let κ be an infinite cardinality and $\theta > \kappa$ a regular cardinality so that for any $\alpha < \theta$, $|\alpha^{<\kappa}| < \theta$. If $S = (M, R)$ is a system with each object m also a

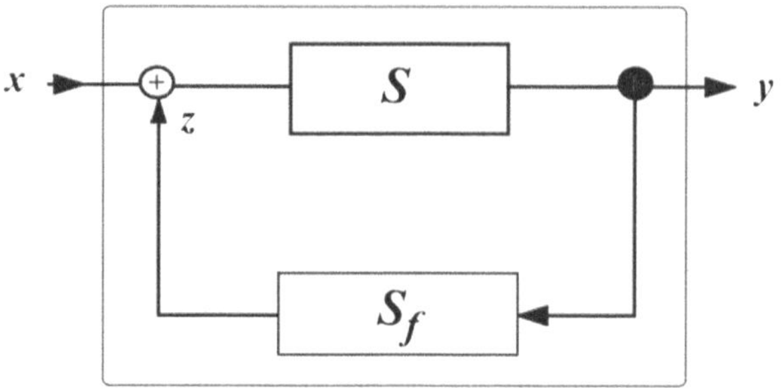

Fig. 2.1 The structure of a feedback system. (**a**) The yoyo model of the general system. (**b**) The meridian field of the yoyo model. (**c**) Typical trajectory of how matters return

system $m = (M_m, R_m)$ so that $|M| \geq \theta$, $|M_m| < \kappa$, for each object m in S, and there exists an object G that belongs to at least θ object systems in M, then there exists a partial system $B = (M_B, R_B)$ of S such that $|M_B| \geq \theta$, $G \in M_1$ for each $S_1 = (M_1, R_1) \in M_B$, and B forms a centralized system.

Proof Without loss of generality, assume that $|M| = \theta$ and that there is a common element in all the object systems in M, which implies $|\cup\{M_z: x = (M_z, R_z) \in M\}| \leq \theta$. And for each object $z = (M_z, R_z) \in M$, assume $\cup\{M_z: z = (M_z, R_z) \in M\} \subseteq \theta$. So, for each $z = (M_z, R_z) \in M$, the order type $< \kappa$ of M_z is a subset of θ. Because θ is regular and $\theta > \kappa$, there exists a $\rho < \kappa$ so that $M_1 = \{z \in M$: M_z has order type $\rho\}$ has cardinality θ. Let us fix such a ρ and deal only with the partial system $S_1 = (M_1, R_1)$ of S, where R_1 is the restriction of the relation set R on M_1.

For each $\alpha < \theta$, $|\alpha^{<\kappa}| < \theta$ implies that less than θ objects of the partial system S_1 have object sets as subsets of α and $\bigcup\{M_z: z = (M_z, R_z) \in M_1\}$ is cofinal in θ. If $z \in M_1$ and $\xi < \rho$, let $M_z(\xi)$ be the ξth element of M_z. Because θ is regular, there is some ξ such that $\{M_z(\xi): z \in M_1\}$ is cofinal in θ. Let ξ_0 be the least such ξ. Then the condition that there exists a common element in each system in M_1 implies that $\xi_0 > 0$. Let $\alpha_0 = \cup\{M_z(\eta) + 1: z \in M_1$ and $\eta < \xi_0\}$. Then $\alpha_0 < \theta$ and $M_z(\eta) < \alpha_0$ for all $z \in M_1$ and all $\eta < \xi_0$.

By transfinite induction on $\mu < \theta$, pick $z_\mu \in M_1$ so that $M_{z_\mu}(\xi_0) > \alpha_0$ and $M_{z_\mu}(\xi_0) > \max\{\alpha_0, \cup\{M_{z_\nu}(\eta): \eta < \rho \text{ and } \nu < \mu\}\}$. Let $M_2 = \{z_\mu: \mu < \theta\}$. Then $|M_2| = \theta$ and $M_{z_1} \bigcap M_{z_2} \subseteq \alpha_0$ whenever $z_1 = (M_{z_1}, R_{z_1})$, $z_2 = (M_{z_2}, R_{z_2}) \in M_2$ are distinct objects. Because for each $\alpha < \theta$, $|\alpha^{<\kappa}| < \theta$, there exists an $r \subset \alpha_0$ and a $B \subset M_2$ with $|B| = \theta$ and for each $z \in B$, $M_z \bigcap \alpha_0 = r$, $S_2 = (B, R_B)$ forms a centralized system, where R_B is the restriction of the relation set R on B.

Let X and Y be two linear spaces over the same field $\mathcal{A}$ and $S: X \to Y$ and $S_f: Y \to X$ linear functions. Then the feedback system of S with a feedback component S_f is defined as the input-output system $S' \subseteq X \times Y$ such that

$$(x, y) \in S' \leftrightarrow (\exists z \in X)\big((x + z, y) \in S \text{ and } (y, z) \in S_f\big). \tag{2.14}$$

A flow-chart depiction of the concept of feedback systems is given in Fig. 2.1, where the feedback component system S_f plays its role in the feedback loop. To help understand this concept in the figurative form, one can imagine that the input x is a particular government policy and y the output produced by implementing the policy. In real life, the output y is really a combined effect of the policy x and market reaction $S_f(S(x))$ to the policy, where $S(x)$ stands for the expected output of the market by the policymakers. This is similar to Soros' (2003) idea of reflexivity.

Theorem 2.6 Suppose that $S \subset X \times Y$ is a linear system and S_f: $Y \rightarrow X$ a linear functional system. Then a subset $S' \subset X \times Y$ is the feedback system of S with feedback component S_f, if and only if $S' = \{(x - S_f(y), y) : (x, y) \in S\}$.

Proof ($\Rightarrow$). Suppose that S' is the feedback system of S with feedback component S_f. Then, by the definition of feedback systems, for each $(x, y) \in S'$, $(x + S_f(y), y) \in S$. This implies that $(x, y) \in \{(z - S_f(y), y)$: $(z, y) \in S\}$. Again, Eq. (2.14) implies that for each element $(x, y) \in S$, $(x - S_f(y), y) \in S'$. That is, $\{(z - S_f(y), y)$: $(z, y) \in S\} \subset S'$.
($\Leftarrow$). It is clear because the assumed $S' = \{(x - S_f(y), y)$: $(x, y) \in S\}$ implies Eq. (2.14).

Before we move on to introduce the systemic yoyo model, the playground and intuition that aid decision-makers to see how systems evolve and interact with each other, let us emphasize the theoretical and practical beauty of applying the concept of systems in business-related studies. In particular, if the business world and scenarios of decision-making are studied by employing approaches of system-problem solving (Klir, 1985), then the following observation can be readily confirmed: seemingly unrelated business decisions are actually related to each other in one way or another; different areas of the business landscape—large or small, global or regional—evolve and develop in concert. When the operation of a market segment changes, as the consequence of a newly made decision, the developments of many other seemingly unrelated market segments also change accordingly. That once again confirms the fact that evolutions in the business world need to be seen as a whole, and the whole evolution of each and every business phenomenon needs to be considered if one wants to understand how business entities or processes evolve as wholes and how they associate with each other.

To conclude this section, let us note that this ordered pair representation of the general system $S = (M, R)$ is of a very wide range of methodological coverage. For example, relations in the relation set R can be systems of algebraic equations, inequalities, differential equations, statistics-based expressions, etc. For related details, see Forrest (2018). At the same time, relations in R do not have to be numerical or quantitative. For example, it will not be appropriate to express the relationship that two people x and y are siblings quantitatively (Liu & Lin, 2010), unless one only likes to capture this relationship partially.

Additionally, let us emphasize that beyond adding additional explanatory power to the existing knowledge on value creation and capture, applying systems methodology enables the discovery of new conclusions and provides general and practically applicable recommendations. Unmistakably, such outcomes stand for an academic endeavor that is worthy of further pursuing, within which the true power of systems methodology lies (Forrest et al., 2020).

2.3 Systemic Yoyo: The Intuition of General Systems

The development of mathematics and natural science reveals the fact that it is important to have a playground available for researchers to intuitively see facts and foresee potential facts before they can successfully construct necessary proofs to support the truthfulness of some of the potential facts or counterexamples to disprove the correctness of others. As far back as the recorded history of mathematics goes, general concepts of mathematics are always studied alongside with relevant intuitions. To this end, one good example is geometry. And, in terms of numbers and variables, each and every number is intuitively seen as a point on the real-number line or a point in a high-dimensional Euclidean space. Each and every quantitative variable is treated as a moving point on the real-number line or in the Euclidean space. However, in terms of business and economics, studies tend to involve many factors (or variables). That implicitly means that such studies tend to deal with great masses of moving points in a Euclidean space, the playground of these scholarly explorations. For example, in studies on what determines the innovativeness of a firm, over 60 internal and external factors are identified (Becheikh et al., 2006). So, to investigate the concept of innovation along the traditional path, one has to deal with over 60 moving points, where other than each of them moving individually with their respectively different directions and speeds, their movements also mutually affect each other. That is simply an impossible scientific project. As a matter of fact, as of this writing, natural science and mathematics still do not know how to deal with the mutual interaction of three factors, the well-known three-body problem (Lin, 2009), letting alone over 60 variables.

Other than demonstrating the need to introduce approaches beyond number-based methodologies, the discussion above also shows the necessity for investigators of business and economics problems to employ unconventional methods and approaches, such as systems science and methodology, in their works in order to avoid scientific impossibilities. Corresponding to systems science and methodology, of course, one needs to have a playground within which she is able to intuitively observe facts and foresee potential truths. To this end, the following question naturally arises: Can we still use the conventional Euclidean spaces as the desired systemic intuition?

The answer to this question is NO. Other than the issues, as just discussed above, of having to deal with masses of moving points, an impossible scientific problem, many recent and age-old challenges, such as nonlinearity, chaos, etc., facing the traditional science are really consequences of how Euclidean spaces are composed of. In particular, because each Euclidean space is composed of crossing real-number lines, which represent linear measures, nonlinear phenomena, of course, emerge naturally. In other words, conventional academic studies, including all calculus- and statistics-based ones in business and economics, attempt to investigate “curvatures” using tools developed linearly in fictitious spaces (for more in-depth discussion, see Lin & OuYang, 2010).

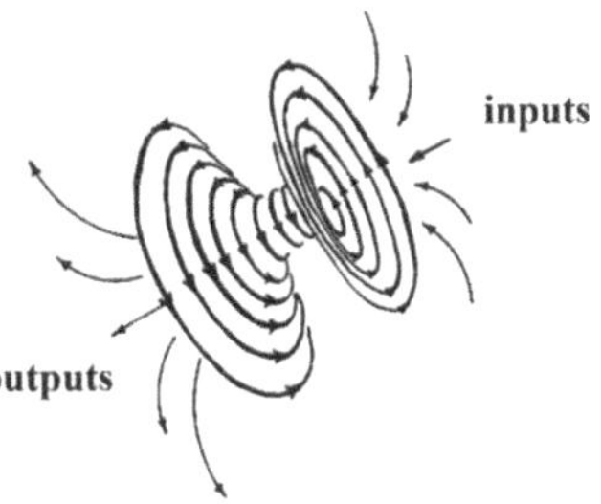

(a) The yoyo model of the general system

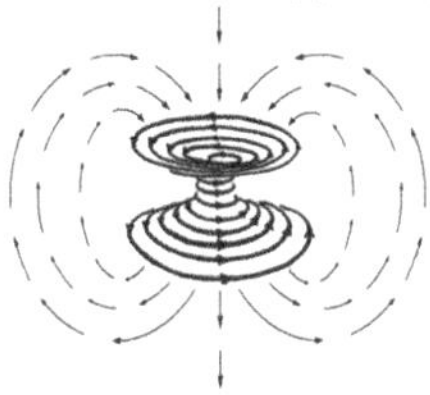

(b) The meridian field of the yoyo model

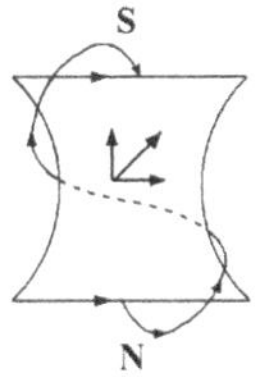

(c) Typical trajectory of how matters return

Fig. 2.2 The Systemic yoyo model

The intuition underneath the previous question is that because systems science, as the second dimension of science, studies systemhood, while the traditional science, as the first dimension of science, studies thinghood, see Klir (1985) for details, the difficulty one faces when dealing with open, complex, giant systems, such as many business and economic situations, is really a difficulty one experiences in the first dimension of science. Hence, the desired playground and intuition in the second dimension should be more manageable than the intuitive situation in the first dimension where one has to deal with masses of moving points in order for one to carry out large-scale tasks of entrepreneurial decision-making.

Based on various approaches of systems thinking and systems analysis developed in the past 90 some years, the needed playground and systemic intuition are the systemic yoyo model (Lin, 2007) in Fig. 2.2. It is figuratively shown in the three-dimensional Euclidean space in which we live. Specifically, on the basis of the blown-up theory (Wu & Lin, 2002), a general theory of development and evolution, and the discussion on whether or not the world can be seen from the viewpoint of systems (Lin, 1988; Lin et al., 1990), the concepts of inputs, outputs, and converging and diverging eddy motions are joined together in the model shown in Fig. 2.2 for each system imaginable. In other words, each system is a multidimensional entity that spins about its axis. If such a spinning entity is fathomed in our three-dimensional space, such a structure, as artistically shown in Fig. 2.2(a), appears. The input side pulls in "*things*" (e.g., materials, information, investment, and human talents). After funneling through the body, things are spit out in the form of outputs. Some of the outputs do not return to the other side and some will, Fig. 2.2(b). Due to its general shape, such a structure is referred to as a *yoyo*.

This systemic model indicates that each entity in the universe, be it physical or intellectual, and be it a tangible or intangible object, a living being, an organization, a market, or an economy, can be intuitively seen as a kind of realization of a certain multidimensional spinning yoyo with an eddy and meridian field around. It stays in a spinning motion as depicted in Fig. 2.2a. If it stops its spinning, it will no longer exist as an identifiable system. What Fig. 2.2c shows is that due to the interaction between the eddy field, which spins perpendicularly to the axis of spin, of the model, and the meridian field, which rotates parallel to the axis of spin, things that are either new to the yoyo body or returning to the input side travel along a spiral trajectory.

Comparing to the construction of Euclidean spaces (or Cartesian coordinate systems), one can tell that this yoyo model is more widely existing in the universe than the fictitious crossings of real-number lines. In particular, the phenomena of spin exist in all levels of physical existence, in business interactions, and in political struggles. And to comprehend geometrically why this systemic intuition might work in terms of dealing with business and economic scenarios, let us imagine a two-dimensional city surrounded by a circular steady wall (or an enclosed bounded area in the two-dimensional Cartesian coordinate plane). If the wall has no gap in the plane, then it will be difficult or impossible for any army to break into the city from within the two-dimensional space. Now, by going beyond the accustomed thinking logic, assume that one smart engineer designs an air strike by making use of a third dimension, then the forces of his or her side can easily parachute into the city along the third dimension. That is, difficulties one faces when making decisions in business situations is partially and mainly due to the reason that he/she keeps himself or herself within the traditional way of thinking or the first dimension of science without truly taking advantage of systems science, the newly found second dimension of science.

To illustrate this geometrical explanation, let us now look at a few examples that demonstrate how the yoyo model plays its intuitive role for scholars in their efforts to establish new facts, which in turn assist decision-makers to face various challenges.

Example 2.1 For each individual person, the hierarchical structure of his/her mind and the evolutionary hierarchy of his/her environment jointly form a cognitive system with the yoyo structure shown in Fig. 2.3. In particular, the person's imagination exhibits a network structure that expands indefinitely to the right on the first layer, labeled by Ps, of life's physiological needs and the second layer, labeled by Vs, of specific philosophical principles of value. His/her mind consists of interacting blocks of various knowledge and experiences about things and events, be they physical, intellectual, conceptual, etc.

Simultaneously, his/her seemingly infinite environment is both visually and conceptually seen as a hierarchical network with the family setting as the bottom layer, as labeled by Fs, where by family it means the place in which the person grew up. With the person growing older, he/she receives and internalizes more information from the environment, initially limited to the family setting F within which the person's life starts, then expanded into a larger and wider environment E with time and growth. As the person's vision multiplies further outward (to the left in Fig. 2.3),

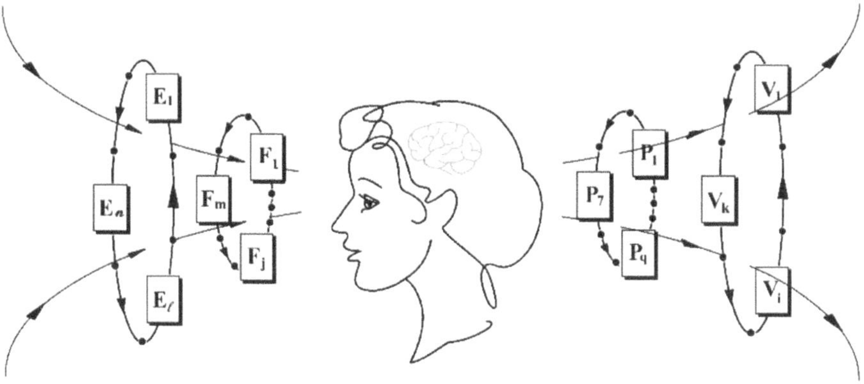

Fig. 2.3 The yoyo structure of a person's cognitive system

the network structure of his/her mind becomes further enriched with depth (towards the right in Fig. 2.3). Such expansions in the reaches of vision and the complexity of the structure of mind reinforce each other so that further a person's vision reaches, the richer his/her mind becomes. That in turn helps the person see even further than before. Such a mutual enrichment cycle continues until such a time when the person is no longer learning due to whatever reason(s).

In the graphical expression of Fig. 2.3, the circular movements of each level, such as E, F, P, and V, are employed to model the fact that factors within each of these levels are actually more or less associated with each other. The long and arrowed curves from left to right model how different levels of the environment and layers of the mind interact with each other.

Based on this systemic yoyo intuition, the following result can be logically produced:

Proposition 2.1 No category of the highest level of abstraction can ever exist, no matter which particular taxonomy of things, events, and thoughts is considered.

In this conclusion, used is Rosch's (1978) taxonomy, which is a system that permits one category to be related to another by means of class inclusion, denoted $\subseteq$. Specifically, for two classes X and Y, the expression $X \subseteq Y$ means that X is a subclass of Y. That is, the more members a category accommodates (i.e., is more inclusive), the higher abstraction level the category is of. For related in-depth discussion, please go to Chap. 10.

Example 2.2 The concept of systemic holes is defined as follows: an object in a connected system is a hole, if removing the object from the system makes the originally connected system partition into at least two separate connected systems. By using the systemic model in Fig. 2.4, one can readily comprehend the significance of a systemic hole within the supply-chain ecosystem of a focal firm. Assume that the systemic structure of the entire supply chain is described as a system $S = (M, R)$, where the top most upstream players are the providers of raw materials and the

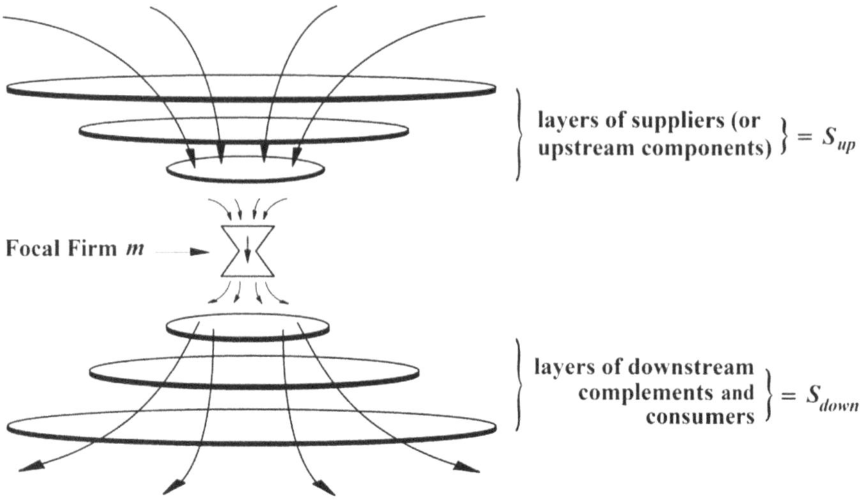

Fig. 2.4 How a systemic hole appears within a supply-chain ecosystem

bottom most downstream players are the consumers or the end users. That is, the object set M of this model consists of the set of all players, such as suppliers, producers, retailers, and consumers, in the specified supply chain, and the relation set R describes how these players are associated with each other based on how their corresponding inputs and outputs are connected into each other. For example, we say that two players m_1 and m_2 are associated, if the outputs of m_1 are the inputs of m_2 either directly or indirectly.

In this systemic model, a particular player $m \in M$ is a hole in this system S or supply chain, if when this player m is removed from this system, this supply chain S falls apart into two disjoint subsystems, neither of which constitutes a complete supply chain, as follows:

$$S_{\text{up}} = \left(M_{\text{up}}, R_{\text{up}}\right) \text{ and } S_{\text{down}} = \left(M_{\text{down}}, R_{\text{down}}\right), \tag{2.15}$$

where the object set M_{up} of the upstream system S_{up} contains all the components that are upstream to m and the object set M_{down} of the downstream system S_{down} all the downstream complements to m.

The reason why these two sub-supply chains S_{up} and S_{down} are not complete is that the outputs of the former cannot eventually reach the demands of the latter, while the latter cannot produce its product(s) without the input from the former. In business terms, in the customer-supplier network of an industry, if two trading partners can only complete their transactions through a specific focal industry, then this particular industry is a systemic hole in the entire supply-demand network. Figuratively, Fig. 2.4 demonstrates the concepts of systemic hole m, supply-chain system S, and the disjoint subsystems S_{up} and S_{down}. The arrowed curves represent the direction of the input-output flows of the overall supply chain S.

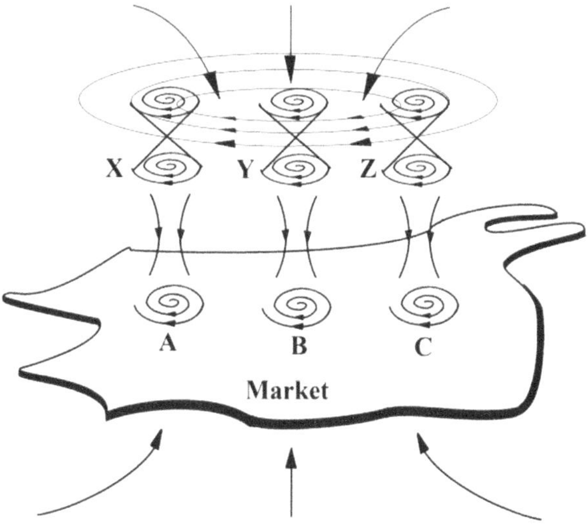

Fig. 2.5 Meeting three market demands simultaneously

Based on this systemic yoyo intuition, the following result can be seen readily.

Proposition 2.2 Each systemic hole naturally carries power and profitability within the supply-demand network which it is a part of due to its location advantage.

For a related in-depth discussion, please go to Chap. 13.

Example 2.3 If two or more market demands can be met simultaneously, while saving valuable resources for consumers, then the effort of meeting different market demands at the same time is said to have created a simultaneous consumer utility effect. By employing the yoyo model, each simultaneous consumer utility effect can be figuratively described, for example, Fig. 2.5.

In this systemic depiction, three market demands A, B, and C can be satisfied by either three different business entities X, Y, and Z separately or by one business firm whose yoyo field encompasses all those of the three entities X, Y, and Z. The idea is kind like, but not exactly the same, as that of one-stop shopping. The reason why consumers are willing to pay extra is demonstrated by the fact that the conflict zones, existing in areas between the yoyo fields X, Y, and Z, if they are separate from one another, no longer exist, when these small yoyo fields are combined into a much larger field that encompasses those of X, Y, and Z.

By carefully analyzing this systemic intuition, combined with logical reasoning, the conclusion below follows.

Proposition 2.3 When collocating diverse products and/or services leads to a simultaneous consumer utility effect, consumers will be willing to pay additionally for a product or service beyond the market norm.

For a related in-depth discussion, please go to Chap. 8.

2.4 A Few Remarks

As a preparation for the rest of this book, after introducing the reader to the concept of systems, this chapter demonstrates how generally true theorems can be established by using the language of naïve set theory. And after having described the composition of the systemic yoyo model for general systems, we then look at three examples of actually employing the model to produce tangle results.

What needs to be noted at this junction is that this chapter attempts to introduce a different methodology than all the ones, which are either calculus- or statistics-based, widely used in the literature this book contributes to. Because this new methodology is not constrained by data or anecdotes, developed results are universally true unless given if-conditions are violated.

As is well-known, although a figure is worth thousands of words, it can also easily lead to misconceptions and incorrect conclusions. To this end, in Afterword of this book, we use examples from mathematics to demonstrate that when intuitions and analytical analysis work jointly, they can help each other to overcome difficulties.

In terms of references, Sect. 2.1 is mainly based on Forrest (2018) and Forrest et al. (2020); Sect. 2.2 is extracted from Forrest (2018) and Lin (1999); and the beginning part of Sect. 2.3 is composed on the basis of Lin (2009).

References

Becheikh, N., Landry, R., & Amara, N. (2006). Lessons from innovation empirical studies in the manufacturing sector: A systematic review of the literature from 1993 to 2003. *Technovation, 26* (5), 644–664.

Blauberg, I. V., Sadovsky, V. N., & Yudin, E. G. (1977). *Systems theory, philosophical and methodological problems*. Progress Publishers.

Bunge, B. (1979). *Treatise on Basic Philosophy*. Vol. 4 : A World of Systems. Reidel, Dordrecht, Holland.

Forrest, J. Y.-L. (2018). *General systems theory: Foundation, intuition and applications in business decision making*. Springer.

Forrest, J. Y.-L., Nicholls, J., Schimmel, K., & Liu, S. F. (2020). *Managerial decision making: A holistic approach*. Springer.

Kline, M. (1972). *Mathematical thought from ancient to modern times*. Oxford University Press.

Klir, G. (1985). *Architecture of systems problem solving*. Plenum Press.

Kuhn, T. (1962). *The structure of scientific revolutions*. University of Chicago Press.

Lin, Y. (1987). A model of general systems. *Mathematical Modelling: An International Journal, 9* (2), 95–104.

Lin, Y. (1988). Can the world be studied in the viewpoint of systems?? *Mathematical and Computer Modeling, 11*, 738–742.

Lin, Y. (1989). A multi-relation approach of general systems and tests of applications. *Synthese - An International Journal for Epistemology, Methodology and Philosophy of Science, 79*, 473–488.

Lin, Y. (1999). *General systems theory: A mathematical approach*. Kluwer Academic and Plenum.

Lin, Y. (2007). Systemic yoyo model and applications in Newton's, Kepler's laws, etc. *Kybernetes: The International Journal of Cybernetics, Systems and Management Science, 36*(3–4), 484–516.
Lin, Y. (2009). *Systemic yoyos: Some impacts of the second dimension*. Auerbach Publication, an imprint of Taylor and Francis.
Lin, Y., Ma, Y., & Port, R. (1990). Several epistemological problems related to the concept of systems. *Mathematical and Computer Modeling, 14*, 52–57.
Lin, Y., & OuYang, S. C. (2010). *Irregularities and prediction of major disasters*. CRC Press, an imprint of Taylor and Francis.
Liu, S. F., & Lin, Y. (2010). *Grey systems: Theory and applications*. Springer.
Mesarovic, M. D., & Takahara, Y. (1975). *General systems theory: Mathematical foundations*. Academic Press.
Porter, M. (1979). How competitive forces shape strategy. *Harvard Business Review, 57*(2), 137–145.
Porter, M. E. (1985). *Competitive advantage: Creating and sustaining superior performance*. Free Press.
Rosch, E. (1978). Principles of categorization. In E. Rosch & B. B. Lioyd (Eds.), *Cognition and categorization* (pp. 27–48). Lawrence Erlbaum.
Rostow, W. W. (1960). *The stages of economic growth: A non-communist manifesto*. Cambridge University Press.
Soros, G. (2003). *The alchemy of finance*. Wiley.
von Bertalanffy, L. (1924). Einfuhrung in Spengler's werk. Literaturblatt Kolnische Zeitung, May.
von Bertalanffy, L. (1972). The history and status of general systems theory. In G. Klir (Ed.), *Trends in GeneralSystems theory* (pp. 21–41). Wiley.
Wu, X. M. (1990). *The Pansystems view of the world*. People's University of China Press.
Wu, Y., & Lin, Y. (2002). *Beyond nonstructural quantitative analysis: Blown-ups, spinning currents and modern science*. World Scientific.

Chapter 3
Evolution of a Market of Free Competition: A Symbolic Approach

Abstract The mainstream developmental economics has conventionally provided inconsistent lists of factors considered crucial for kick-starting a lasting economic growth (e.g., Lipton, 1977, *Why Poor People Stay Poor: Urban Bias in Developing Countries*. London: Temple Smith; Rostow, 1960, *The Stages of Economic Growth: A Non-communist Manifesto*. Cambridge, UK: Cambridge University Press; Studwell, 2013, *How Asia Works: Success and Failure in the World's Most Dynamic Region*. New York: Grove Press). For example, the most common among these identified factors include high rates of investment in modern technology, democracy, universal suffrage, property right, and others. However, by anecdotally studying the respective successes of economic development in England, the United States of America, Japan, and modern China during different times of history, Wen (2016, *The Making of an Economic Superpower: Unlocking China's Secret of Rapid Industrialization*. Singapore: World Scientific) discovers that these factors are actually not relevant much. Instead, what really matters is the sequence and process of economic buildup; and this anecdotal discovery is then theoretically confirmed by Forrest et al. (2018, *Theoretical Economics Letters*, *8*,2594–2640) through employing systemic yoyo model and game theory. Other than these recent theoretical developments which invalidate earlier claims of developmental economics, the practical front also empirically validates this end. For example, many countries' failures in their attempts to modernize and to industrialize themselves in the twentieth century are vivid evidence confirming that the conventional wisdom of developmental economics is incorrect. In particular, some of the trying nations, based on what was concluded in the mainstream developmental economics, attracted foreign direct investments and established modern industries in major cities and commercial centers based on most current technologies (Acemoglu & Robinson, 2012, *Why nations fail—the origins of power, prosperity, and poverty*. New York: Crown Business; Breslin, 2009, *The Beijing Forum*; Chang, 2003, *Kicking Away the Ladder: Development Strategies in Historical Perspective*. London: Anthem Press). But, instead of achieving desired sustaining economic growth, they failed to emulate the successes of the developed nations (Rostow, 1960, *The stages of economic growth: A non-communist manifesto*. Cambridge, UK: Cambridge University Press). Examples of such failures include many Latin American and East and Southeast Asian nations. Although these nations enjoyed annual economic growth of

J. Y.-L. Forrest, Y. Liu, *Value in Business*, Contributions to Management Science,
https://doi.org/10.1007/978-3-030-82898-1_3

around or above 5% for decades in the middle of the twentieth century, their phenomenal growths quickly dissipated within short periods of time (Studwell, 2013, *How Asia works: Success and failure in the world's most dynamic region*. New York: Grove Press).

Keywords Emergence of a market · Fultz's dishpan · Incumbent firms · Market characteristics · Mutual forbearance · Primary factor

Speaking differently, what is anecdotally discovered by Wen (2016) and theoretically confirmed by Forrest et al. (2018) is that the healthy development of various markets open to free competition is fundamental for a nation to develop its self-sustained momentum of economic growth. Riding on the back of this shown, generally true result, this chapter develops a theory on how a market of free competition evolves in the perspective of price as the only determinant primary factor, followed by other secondary factors, such as business models, technologies, and government policies. Such perspective is crucially important for us to establish a practically useful theory, as mentioned in one earlier chapter. In particular, studies in business and economics tend to involve many interacting factors (or variables). However, scientifically speaking, neither natural science nor mathematics knows how to deal with the general mutual interaction of three factors, the well-known three-body problem (Lin, 2009), letting alone scenarios involving more than three variables. Here, the buzz words are "primary" and "secondary." Specifically, by a primary factor, it means such a factor human can control and make adjustment; by a secondary factor, it means one that somehow directly or indirectly appears naturally as a consequence of the functional effect of the primary factor.

The rest of this chapter is organized as follows: Sect. 3.1 shows how a market initially emerges. Section 3.2 studies properties of a market that is in a state of mutual forbearance. Section 3.3 looks at how incumbent and entering firms interact with each other. Section 3.4 examines applications of the established results in the previous sections. And Sect. 3.5 concludes this chapter.

3.1 Initial Emergence of a Market

Let us start by looking at the start of time when people still live in primeval conditions. Based on natural conditions and resources readily available within the environment they live in, people make their living by harvesting whatever granted to them by nature and by adopting similar tools for production, conquering hardships occurring in their surroundings, and other endeavors (Marchall, 2016). The supply of natural resources is abundant compared to the low density of the human population; and people live well, although their tools available for survival are limited and inefficient. Limitations of technology divide people on the same land into small tribes. Because these tribes, although separated from each other due to obstacles of nature, are really located in the same natural environment, they naturally develop similar approaches to handle routine works, to overcome hardships, and established methods to manage members in their individual tribes. As time goes on, improved tools for more efficient production, communication, and transportation, and better managerial practices of governance are designed and employed in individual tribes.

The natural human desire for better living helps people design and adopt various improved and new tools and technologies. With improving efficiency of communication and transportation, the information and knowledge of new consumable products, including the underlying production how-to and relevant managerial routines, begin to pass around the land through words of mouth. That leads to the formation of circulating masses of people with special skills and knowledge. Other than skilled people and knowledge, the circulating masses also carry commercial goods. In other words, over the lightly populated land a circulation appears. So, Bjerknes' Circulation Theorem, for technical details, see the Appendix of this chapter, guarantees the appearance of abstract eddy motions over the land, each of which consists of migration of people, spread of knowledge and information, and transportation of goods. This explains how initial customer demands start to appear although only locally in today's standard.

To satisfy these initial and local demands, family-based workshops are formed to produce what is needed by neighbors, members of the same tribes, and occasionally affiliates of other tribes of neighboring areas. At this stage in time, the nonexistent market can be systemically depicted as a non-spinning yoyo field in Fig. 3.1. In particular, the overall circular pool stands for the fledgling market. Because the regional commerce does not really exist the overall circular dish in Fig. 3.1 does not spin. In other words, the totality of regional businesses cannot be seen as a living, functional system; and no real market is visible or existent.

Proposition 3.1 In the aforementioned, independent family-based economy, each financially viable workshop sells its offer on the average for a price between the production cost and customer reservation value.

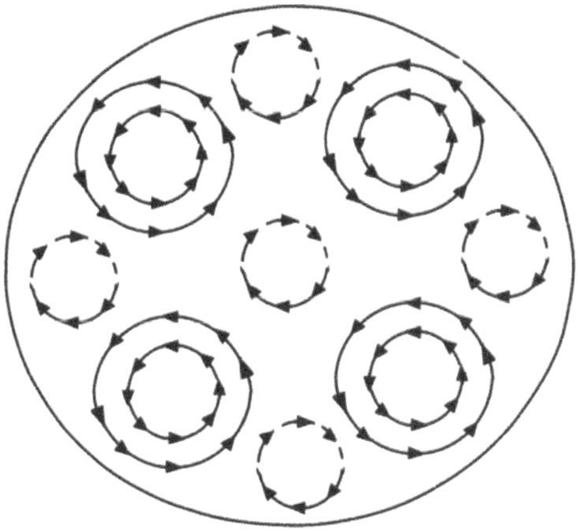

Fig. 3.1 The family-based market

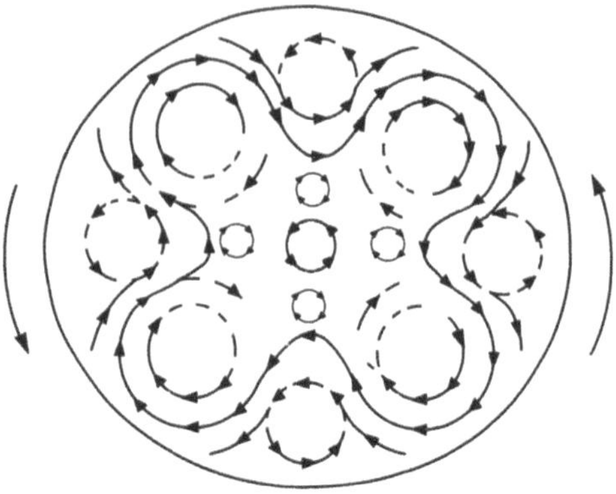

Fig. 3.2 Growing appearance of a market

The key assumptions of this proposition include "financial viability" and "on the average." It is because if a workshop does not care about whether or not it can last financially, then it can simply produce its offer as the end goal. In such a case, the workshop will sooner or later close its operation when it runs out of cash flow. As for the assumption of "on the average," it is because a workshop can give away some units of its offer in order to acquire other desired nonfinancial gains. In such a case, the production cost of these units will need to be cover by those sold units.

With continuous advancement of technologies and tools, the region becomes more connected through long-distance transportation of goods, communication of information, and sharing of knowledge. Hence, initially independently operated family-based workshops start to associate with one another; the products of one workshop become supplies of another workshop. That is, supply chains start to emerge and to help bind workshops into more and more mutually dependent business entities. The systemic structure underneath the association of family-based workshops is shown in Fig. 3.2. Specifically, the counterclockwise spin of the overall dish models this second stage scenario that throughout the region goods, information, and knowledge are now moving and shared. At the same time, the local eddy pools, representing family-based workshops, also rotate around with the overall dish. That means that the family-based workshops, initially serving only locally, are also serving or ready to serve the entire market with much increased demands from all over the market. Along with this increased market connectivity, the initially family-based workshops gradually develop into factories of larger scales, while employing relatively more advanced technologies. On the flip side, the greatly increased supplies spoil customers, making them become less patient than before and their preferences and tastes constantly evolving.

At the beginning of this development stage of the regional economy, all workshop-turned factories can be seen as new entrants of the fledgling market. And customers are trying out horizontally differentiated offers from different suppliers without any preference to and discrimination against one supplier over another. In other words, no factory has a relatively defined base of loyal customers. In this case, all the factories compete for business by using price per unit value of their offers sold. In particular, although factory offers of concern are considered horizontally differentiated or mutually substitutable, in real life they still entail individually different features and functionalities, such as differences in appearance, in craftmanship, and in quality of service. For example, the conventional walkie-talkies and modern iPhones can be seen as substitutable products if used in the limited situations for which the former was designed. Because the latter possesses much more values and functionalities beyond the said limited use scope than the former, the price of an iPhone is much higher than that of a walkie-talkie. Now, if one compares the price of a unit of the value (or unit value) of a walkie-talkie to that of an iPhone, she will most likely find that the unit-value price of an iPhone is cheaper than that of a walkie-talkie. So, in real life, how can the factories compete with each other by using price? The following result provides an answer.

Proposition 3.2 In the aforementioned market, the average price P_i of factory i can be normalized into an equivalent price α_i, satisfying $0 \leq \alpha_i \leq 1$, while the corresponding cost and maximum expected selling price are respectively normalized to 0 and 1.

Proof For factory i, its price P_i, on the average (Proposition 1), satisfies $C_i \leq P_i \leq M_i$, where C_i stands for the cost of production and M_i the maximum expected selling price (or the reservation price). Then the actual average selling price can be written as follows:

$$P_i = C_i + \alpha_i(M_i - C_i), i = 1, 2, \ldots, m \tag{3.1}$$

for some parameter α_i from the interval [0,1] and assume that there are m factories of concern. In other words, no offer, on the average, is sold at a price below its cost. Now, instead of competing against each other by using the absolute terms of the selling prices P_i, the factories compete by adjusting the parameters α_i, $i = 1, 2, \ldots, m$, to see who is more competitive in terms of unit value of their offers.

Without loss of generality, in the rest of this chapter, each factory's price P_i will be normalized as outlined in Proposition 3.2. And the normalized price α_i will still be written as P_i.

Assume that during this stage of development, the fledgling market is served by m firms, named Firm i (= 1, 2, ...) with their horizontal differentiated products. To attract customers, the firms, of course, employ the novelty and freshness of their products and then price. As analyzed before, instead of being concerned with the variety of features and functionalities of individual products, let us focus on their unit-value prices, as normalized in Proposition 3.2. These unit-value prices of the products' absolute prices help normalize away differences in the products' features and functionalities. In other words, these firms utilize adjustable (normalized) prices to attract customers who make purchase decisions based on whose price is more competitive in terms of the unit value of all offered features and functionalities within each available product. In the rest of this chapter, such customers will be referred to as switchers; and the totality of these customers is assumed to have the magnitude β. In this stage of market development, β is equal to the sum of all customers and potential customers.

To make our analytical reasoning possible, assume that to serve the fledgling market well, the management of these firms are aware of each other's pricing strategies; and they respectively and individually respond by playing Nash equilibria through pure self-analyses.

Because these m firms are correctly seen as entrants who are still in the process of entering into a brand-new and unoccupied market, the firms' successes in this market can be naturally measured by whether or not they actually make profits. Speaking realistically, that means that each of these m firms uniformly randomizes its selling price P over the interval $[C, M]$ as long as the firm can generate a profit on the average, where C stands for the production cost, M the reservation price customers are willing to pay for the product. Proposition 3.2 implies that P, C, and M can be respectively normalized so that $0 \leq P \leq 1$, $C = 0$ and $M = 1$.

Proposition 3.3 In the aforementioned fledgling market, the profit Π of each competing firm is equal to

$$\Pi = \beta P(1 - P)^{m-1}. \tag{3.2}$$

Proof If $m = 1$, that is, only one firm is serving the market, then the profit Π the firm makes is equal to the product of the firm's market occupancy and price. Symbolically, we have

$$\Pi = \beta P = \beta(1 - P)^0 P = \beta P(1 - P)^{m-1}.$$

If ($m =$) 2 firms serve the market, then each of these firms' profit Π is equal to the product of its market occupancy and price. That is, we have

$$\Pi = \beta(1 - P)^1 \times P = \beta P(1 - P)^{m-1},$$

where the reason why these firms set their selling price P the same is because if not, customers will make their purchases from the firm of lower price, say Firm 1. So, the demand for Firm 1's product will be the same as the entire market demand, while that for Firm 2's product completely diminishes. Through such pricing dynamics, the prices charged by these firms have to eventually be the same and equal to each other if they want to stay in business within this fledgling market.

Continuing this reasoning leads to that if three firms ($m = 3$) simultaneously serve the market, then each firm's profit Π is equal to the product of its market occupancy and price:

$$\Pi = \beta(1 - P)^2 \times P = \beta P(1 - P)^{m-1}.$$

So, by using mathematical induction, the general Eq. (3.2) follows.

3.2 Mutual Forbearance of Incumbent Firms

The next development stage of the fledgling market is generally caused by the initial disruption to the prevailing industrial order, along with advances of related technologies. The simultaneous development in markets and organizational networks is accompanied consequently by increasing severity of rivalry among competing firms (Gustafsson et al., 2016) in order to stabilize their market shares or bases of loyal customers. This development stage is characterized by a stabilizing market pattern, as depicted in Fig. 3.2, although additional firms still attempt to enter the emerging market. In other words, in this stage of development, some firms thrive, some barely survive and most others fail (Markman & Waldron, 2014). Except for those failed firms, some of the m firms, as discussed above, attracted the following of their respective bases of loyal customers. Assume that n firms have survived the previous stage of development and still stay competitive within the current stage. Let N_i stand for the magnitude of the established Firm i's base of loyal customers and β the magnitude of the market segment of switchers, known as consumer surplus, who make purchase decisions based on which factory's unit-value price is lower.

As for boundary conditions, each of the n factories is generally limited by its unique set of constraints. Such uniqueness in boundary conditions for the individual factories is realistically caused by differences in resources available to these factories. Each factory can utilize its resources, which are defined as tangible or intangible assets (Harmancioglu et al., 2009), to design and implement its business strategies (Barney & Arikan, 2001). For example, each factory has its unique set of missions, organizational culture, history, administrative structure, capital-flow pattern, and managerial style. Symbolically, assume that for the n factories, the factors involved in the production and marketing of their offers are $f_1, f_2, f_3, \ldots$ Then, for factory i $(= 1, 2, 3, \ldots, n)$, the magnitude N_i of its base of loyal customers reflects how available resources are mobilized in ways that are accepted both internally and externally. It must be naturally constrained by its lower boundary condition $L_i(f_1, f_2, f_3, \ldots)$ and upper boundary condition $U_i(f_1, f_2, f_3, \ldots)$ as follows:

$$L_i(f_1, f_2, f_3, \ldots) \leq N_i \leq U_i(f_1, f_2, f_3, \ldots). \tag{3.3}$$

Therefore, for factory i, $i = 1, 2, \ldots, n$, there is a unique number β_i in the interval [0,1] satisfying.

$$N_i = L_i(f_1, f_2, f_3, \ldots) + \beta_i[U_i(f_1, f_2, f_3, \ldots) - L_i(f_1, f_2, f_3, \ldots)]. \tag{3.4}$$

Similar to the previous analysis of Proposition 3.2, it can be concluded that the magnitude N_i of the base of factory i's loyal customers, for fixed $f_1, f_2, f_3, \ldots$, can be normalized to the parameter β_i $(\in[0, 1])$. Hence, the result below holds true.

Proposition 3.4 For the n aforementioned factories, their lower and upper boundary conditions can be normalized to 0 and 1, respectively. And, the magnitudes of their bases of loyal customers can be normalized to parameters that take values between 0 and 1 inclusively.

In the rest of this chapter, the magnitude of factory i's loyal customers will be denoted as β_i, satisfying $0 \leq \beta_i \leq 1$, $i = 1, 2, \ldots, n$, and the magnitude of the market segment of switchers as β such that $\beta = 1 - \sum_{i=1}^{n} \beta_i$ and $0 \leq \beta \leq 1$.

Proposition 3.5 In the aforementioned developing and stabilizing market, if $F_i(P)$ is the price distribution of factory i, $i = 1, 2, \ldots, n$, then factory j's profit Π_j, $j = 1, 2, \ldots, n$, and $j \neq i$, is equal to

$$\Pi_j = \beta_j P + \beta P \prod_{\substack{i=1 \\ i \neq j}}^{n} [1 - F_i(P)]. \tag{3.5}$$

Proof If $n = 1$, there is only one factory that serves the said market. The profit Π of the factory is equal to the sum of its profit $\beta_j P$ from the loyal customers and that from the consumer surplus βP. Symbolically, we have

$$\Pi = \beta_j P + \beta P.$$

If $n = 2$, the profit Π_i of factory i is determined by the sum of the profit $\beta_j P$ from its loyal customers and that from its portion of the consumer surplus $\beta P(1 - F_j(P))$, for $i, j = 1, 2$, $i \neq j$. Symbolically, we have

$$\beta_j P + \beta P \prod_{\substack{j=1 \\ j \neq i}}^{2} [1 - F_j(P)].$$

So, Eq. (3.5) follows by using mathematical induction on the number n of factories that serve the said market.

In this stage of market development, a mutual forbearance among the current market-serving factories appears. In particular, these factories mitigate rivalry by dividing the market in proportion to their respective strengths (Bernheim & Whinston, 1990) so that a so-called mutual forbearance ensues. With such mutual forbearance, the factories submit dominance to the stronger competitors in their less efficient market segments (Greve, 2000; Li & Greenwood, 2004). As these factories compete in a large number of segments of the market, they learn quickly their competitors' strengths and weaknesses. They become aware of their codependence within the market ecosystem. That helps them gradually de-escalate rivalry (Yu & Cannella Jr., 2012), leading to decreasing entry and exit rates in the market (Fuentelsaz & Gómez, 2006; Stephan et al., 2003) and decreasing interfirm hostility

(Baum & Korn, 1999; Haveman & Nonnemaker, 2000; Jayachandran et al., 1999). Therefore, a state of mutual forbearance among the market-serving factories rules the entire market so that the incumbents are able to more or less work together to fight off new entry attempts while competing among themselves by using pricing strategies. As a consequence of the formed mutual forbearance among incumbent factories, the government is one way or another forced to step in to regulate the market operation so that fair and free competition can be assured, if that is desired by the people of the land. For the rest of this chapter, assume that the market of our concern operates under the condition of free competition.

Proposition 3.6 In Nash equilibrium of the aforementioned market of free competition, no pure strategy of pricing exists for the n incumbent factories.

Proof Assume that $(p_1, p_2, \ldots, p_m)$ is a pure strategy portfolio of prices. If there is a unique $i \in \{1, 2, \ldots, n\}$ such that

$$p_i < p_j, \tag{3.6}$$

for any $j \in \{1, 2, \ldots, m\}$ so that $j \neq i$, then factory i has successfully won over all the switchers in the marketplace and can therefore slightly raise its price p_i to bring in additional profits as long as the new price is still the lowest, Eq. (3.6), while all other regional economies hold their respective strategies unchanged. So, the chosen strategy portfolio $(p_1, p_2, \ldots, p_n)$ is not a Nash equilibrium.

Other than the case just considered above, if there are k factories that charge the lowest price, then these factories have already jointly absorbed all switchers. Hence, when one of these factories lowers its price slightly, it will be able to create additional profits by attracting all switchers. In other words, the pure strategy portfolio $(p_1, p_2, \ldots, p_n)$ is not a Nash equilibrium.

If the discussion in the previous paragraph applies to the case of $k = n$, the argument leads to the conclusion that in Nash equilibrium, the said market does not have any symmetric pure strategy portfolio.

Proposition 3.7 In Nash equilibrium, the magnitude β_k of the loyal-customer base within the aforementioned market and the price distribution $F_k(P)$ of factory k satisfy

$$F_k(P) = 1 - \frac{1}{\beta_k}\left(\frac{(1-P)\prod_{i=1}^{n}\beta_i}{\beta P}\right)^{\frac{1}{n-1}}, k = 1, 2, \ldots, n, \tag{3.7}$$

such that

$$F_k(1) = 1 \text{ and } F_k\left(\frac{\beta_k}{\beta_k + \beta}\right) = 0. \tag{3.8}$$

Proof With the normalizations described in Propositions 3.2 and 3.4 in place, the n factories only need to compete against each other by using prices. Proposition 3.6

indicates that there is no pure strategy portfolio for the n-player game. Then, factory k's profit from its loyal customers are $\beta_k P$ and those from its share of the switchers are $\prod_{i=1,i\neq k}^{n}[1 - F_i(P)]\beta P$. Hence, the objective function for factory k is to maximize the following expected profit by adjusting $F_k(P)$:

$$E(\Pi_k) = \int_{-\infty}^{+\infty} \left\{\beta_k P + \prod\nolimits_{i=1,i\neq k}^{n}[1 - F_i(P)]\beta P\right\} dF_k(P), \tag{3.9}$$

where Π_k is factory k's profit, $E(\Pi_k)$ its expected profit. The Nash equilibrium indifference condition for factory k is

$$\beta_k P + \prod\nolimits_{i=1,i\neq k}^{n}[1 - F_i(P)]\beta P = \beta_k, k = 1, 2, \ldots, n. \tag{3.10}$$

Equation (3.10) implies the following system of equations:

$$\{[1 - F_1(P)]\ldots[1 - F_{k-1}(P)][1 - F_{k+1}(P)]\ldots[1 - F_n(P)]\beta P = \beta_k(1 - P), \tag{3.11}$$

for any $k = 1, 2, \ldots, n$. Let $k = i$ and $j = 1, 2, \ldots, n$, respectively, $i \neq j$. Then dividing the ith equation by the jth equation from Eq. (3.11) produces

$$\frac{1 - F_j(P)}{1 - F_i(P)} = \frac{\beta_i}{\beta_j}, \text{or } 1 - F_j(P) = \frac{\beta_i}{\beta_j}[1 - F_i(P)]. \tag{3.12}$$

By letting $i = 1$, Eq. (3.12) leads to

$$1 - F_j(P) = \frac{\beta_1}{\beta_j}[1 - F_1(P)], j = 2, 3, \ldots, n. \tag{3.13}$$

Substituting equations in Eq. (3.13) into Eq. (3.10) leads to

$$\beta_1[1 - F_1(P)] = \left(\frac{(1 - P)\prod_{i=1}^{n}\beta_i}{\beta P}\right)^{\frac{1}{n-1}}. \tag{3.14}$$

In this analysis of Eqs. (3.13) and (3.14), letting $i = k$, $j = 1, 2, \ldots, m$, and $j \neq k$ produces

$$\beta_k[1 - F_k(P)] = \left(\frac{(1 - P)\prod_{i=1}^{n}\beta_i}{\beta P}\right)^{\frac{1}{n-1}}, k = 1, 2, \ldots, n, \tag{3.15}$$

from which Eq. (3.7) follows. As for Eq. (3.8), the first part can be verified directly. For the second part, because factory k tries to attract as many switchers as possible in order to potentially increase its profits from the guaranteed level β_k from its loyal consumers by charging them the reservation value 1, the following holds true

$$\beta_k P + \beta P \geq \beta_k.$$

So, $P \geq \beta_k/(\beta_k + \beta)$; and the second part in Eq. (3.8) follows.

Corollary 3.1 In Nash equilibrium, a symmetric mixed strategy exists for n factories in the aforementioned market, if and only if the magnitudes of the loyal-customer bases of the factories are equal to each other.

In fact, Eq. (3.13) implies that $F_i(P) = F_j(P)$, if and only if $\beta_i = \beta_j$, for all $i, j = 1, 2, \ldots, n$.

3.3 Interaction Between Newly Entering and Incumbent Firms

As the aforementioned market continues to grow with its scale and depth, all factories of concern correspondingly evolve into firms, as triggered by intensifying competitions. Other than producing physical products, these firms also start to make other kinds of consumable market offers, such as services and informational goods. With such expanding scope of economic activities in the marketplace, additional profit opportunities appear and additional firms want to enter the market.

To technically develop useful results, regarding how new entrants interact with incumbent firms, the following proposition demonstrates that the individually different magnitudes of the incumbent firms' loyal-customer bases can also be normalized to a same value.

Proposition 3.8 Assume that incumbent firms' selling prices, production costs, loyal customers' reservation values within the aforementioned market, lower and upper boundary conditions, and magnitudes of loyal-customer bases are all, respectively, normalized to P_i, 0, 1, 0, 1, and β_i, $i = 1, 2, \ldots, n$. Then, in Nash equilibrium, the loyal-customer bases of the incumbent firms have the same magnitude. That is

$$\beta_k = \alpha, \text{for } k = 1, 2 \ldots, n. \tag{3.16}$$

Proof By contradiction, assume that there are i and j $(=1, 2, \ldots, n)$ such that $\beta_i \neq \beta_j$. Let

$$\beta_k = \max\{\beta_1, \beta_2, \ldots \beta_n\} \tag{3.17}$$

for some k satisfying $1 \leq k \leq n$. And there is at least one k_0 satisfying $1 \leq k_0 \leq m$ and.

$$\beta_k > \beta_{k_0}. \tag{3.18}$$

From Eq. (3.9), the expected profit of firm k in Nash equilibrium is

$$E(\Pi_k) = \int_{-\infty}^{+\infty} \left\{\beta_k P + \prod_{j=1,j\neq k}^{n} [1 - F_j(P)]\beta P\right\} dF_k(P) \tag{3.19}$$

$$= \int_{\frac{\beta_k}{\beta_k+\beta}}^{1} \beta_k dF_k(P) = \beta_k F_k(P)\Big|_{\frac{\beta_k}{\beta_k+\beta}}^{1} \tag{3.20}$$

$$= \beta_k \left\{ \frac{1}{\beta_k} \left[\frac{\left(1 - \frac{\beta_k}{\beta_k+\beta}\right)\prod_{j=1}^{n}\beta_j}{\frac{\beta\beta_k}{\beta_k+\beta}} \right]^{\frac{1}{n-1}} \right\} \tag{3.21}$$

$$= \left(\prod_{j=1,j\neq k}^{n}\beta_j\right)^{\frac{1}{n-1}}. \tag{3.22}$$

The transition from Eq. (3.19) to Eq. (3.20) is based on Eqs. (3.10) and (3.7). The assumptions about the loyal consumers implies

$$E(\Pi_k) = \left(\prod_{j=1,j\neq k}^{n}\beta_j\right)^{\frac{1}{n-1}} \geq \beta_k = \left(\underbrace{\beta_k\beta_k\ldots\beta_k}_{n-1 \text{ times}}\right)^{\frac{1}{n-1}}. \tag{3.23}$$

This end is impossible because Eq. (3.17) implies

$$\left(\prod_{j=1,j\neq k}^{n}\beta_j\right)^{\frac{1}{n-1}} < \left(\underbrace{\beta_k\beta_k\ldots\beta_k}_{n-1 \text{ times}}\right)^{\frac{1}{n-1}} = \beta_k, \tag{3.24}$$

for every $j = 1, , 2, \ldots, n, j \neq k, j \neq k_0, \beta_j \leq \beta_k$ and $\beta_{k_0} < \beta_k$. So, the magnitudes of the loyal-customer bases of the incumbent firms must be equal to each other.

Corollary 3.2 In Nash equilibrium of the aforementioned market, a symmetric mixed strategy exists for the incumbent firms so that the price distribution $F_k(P)$ of firm k is

$$F_k(P) = F(P) = 1 - \left[\frac{(1-P)\alpha}{\beta P}\right]^{\frac{1}{n-1}}, k = 1, 2, \ldots, n, \tag{3.25}$$

where α is the constant magnitude of any loyal-customer base such that

$$F(P) = 1 \text{ and } F(P)\left(\frac{\alpha}{\alpha+\beta}\right) = 0. \tag{3.26}$$

Proof This result follows readily from Corollary 3.1 and Proposition 3.7.

In the effort of striving for increasingly better living conditions, people realize the great importance of industrialization and how technologies play crucial roles in the relevant processes. As suggested by studies of development economics, throughout the twentieth century many countries from around the world devoted energies, resources, and efforts to modernize and industrialize by investing heavily and by introducing various policies (e.g., Gerschenkron, 1962; Lipton, 1977; Studwell, 2013). Although these countries were successful in attracting foreign direct investments, in importing most current technologies, and in establishing industries in major cities and commercial centers (Acemoglu & Robinson, 2012; Breslin, 2009; Chang, 2003), they however did not succeed in achieving what was desired for—industrialization and modernization (Forrest et al., 2018; Wen, 2016). Examples of such attempts and consequent failures include nations in Latin America, East, and Southeast Asia. Although these nations achieved around 5% annual economic growth year after year, they eventually quickly lost their wealth accumulated over the prior decades (Studwell, 2013). On top of this backdrop of reality, a theoretically significant and practically important question arises: What really happened here in these efforts? According to Forrest et al. (2018), the root reason underneath these failures is that the literature these trying nations based their efforts on did not really capture the fundamental cause behind the kick-start of a self-sustaining momentum of economic growth. In other words, the accepted wisdom does not fully grasp the systemic drive that underlies the occurrence of industrialization and modernization. To develop the missing link in the literature, let us see the following theorems that describe how different economic entities compete against each other.

Theorem 3.1 In Nash equilibrium, at least one firm will profitably enter this market with its version of substitute offer, as a competitor of the n incumbent firms so that the total expected profit of all entering firms will be directly proportional to the normalized magnitude of market surplus.

Proof First of all, because consumer preferences and tastes change constantly, in the said market there are always switchers, who might once be loyal but now are dissatisfied customers. That is, it is reasonable to assume that there is always a consumer surplus in the marketplace.

Next, assume that only one new firm enters into the said market in such a way that it uniformly randomizes its price P over the interval [0,1] (or the interval from the cost of production to the expected maximum selling price, Proposition 3.2) for as long as it makes profits on the average while establishing itself within the market. In other words, the entering firm provides its offering at any price as long as no losses can be averaged out. As discussed above, this firm's marginal cost and maximum expected selling price are normalized to 0 and 1, respectively.

To face the emerging market disruption, the n incumbent firms sets its price by considering the price of the entrant and those of other incumbent firms. So, based on Eq. (3.10) and Proposition 3.7, the equilibrium indifference condition of incumbent firm k

$$\alpha P + \beta P(1-P)\prod\nolimits_{j=1, j\neq k}^{n}\left[1 - F_j(P)\right] = \alpha. \tag{3.27}$$

can be simplified into

$$\alpha P + \beta P(1-P)[1 - F(P)]^{n-1} = \alpha. \tag{3.28}$$

Solving this expression for $F(P)$ produces:

$$F(P) = 1 - \left(\frac{\alpha}{\beta P}\right)^{\frac{1}{n-1}} \tag{3.29}$$

which is defined for all price P satisfying $1 \geq P \geq \alpha/\beta$, obtained from the fact that $F(P) \geq 0$.

The expected profit of the entering firm is equal to

$$E_e(\Pi) = \int_{-\infty}^{\alpha/\beta} \beta P dP + \int_{\alpha/\beta}^{+\infty} \beta P[1 - F(P)]^n dP \tag{3.30}$$

Because price P is limited to the interval [0,1] and $F(P)$ has a mass point of size $(\alpha/\beta)^{\frac{1}{n-1}}$ at the reservation value $P = 1$, Eq. (3.27) can be rewritten as:

$$E_e(\Pi) = \beta\left\{\int_{0}^{\alpha/\beta} P dP + \int_{\alpha/\beta}^{1} P[1 - F(P)]^n dP + \left(\frac{\alpha}{\beta}\right)^{\frac{n}{n-1}}\right\}, \tag{3.31}$$

Because the expression within the braces is a positive real number, Eq. (3.31) implies that the entering firm expects to make a profit and the amount of profit is positively proportional to the magnitude of consumer surplus.

Theorem 3.2 If the market described below is in Nash equilibrium, then the following hold true:

1. If the market's consumer surplus is larger than the loyal-customer base of one incumbent firm, then the market calls for new innovation and additional competition.
2. If some firms answer the call by entering the market with their versions of substitute offer, then the expected profit of all these firms can be potentially larger than that of at least one incumbent firm.

The market mentioned in Theorem 3.2 satisfies the following conditions:

- It is served by a fixed number of firms with their horizontally differentiated offers;
- Its operation is only affected by market forces, such as demand and supply, and consumers' forever evolving preferences and tastes;
- Each incumbent firm has a base of loyal customers if its price is not more than its reservation price;
- There are switchers, collectively known as consumer surplus, who make purchases based on which firm's unit-value price is lower;
- All firms' pricing strategies are known to the incumbent companies who respond by playing Nash equilibria through untainted self-analyses.

For such a market, as described above, it generally means that the technology employed and managerial processes followed have been standardized. Hence, for a firm to profitably enter the market, it must have introduced a more efficient technology and/or managerial routine that can significantly reduce its business expenditure in providing its market offer of increased sophistication and functionality. For empirical confirmation of this statement, see Christensen et al. (1998).

Proof Assume that the market is occupied by n incumbent firms, $n = 1, 2, 3, \ldots$, respectively named Firm i, $i = 1, 2, \ldots, n$. Then, the propositions developed above can be applied to normalize individually different boundary conditions of these firms so that the production cost is 0, loyal customers' reservation price 1, price P satisfies $0 \leq P \leq 1$, loyal-customer bases of the incumbent firms are of the same percentage scale α with $\beta = 1 - n\alpha$ being the scale of the consumer surplus. Assume that there is only one firm that enters into this market. For the case of more than one firm that enters the market, a similar argument can be given. All relevant details are omitted here. Assume that the entering firm randomizes its average price between its business cost 0 and consumer reservation price 1.

To protect their established territories and hold on to the established mutual forbearance in the marketplace, the incumbent firms set their prices by considering all competitors. So, from Eqs. (3.27) and (3.28), the incumbent firms adopt their symmetric equilibrium pricing given in Eq. (3.29).

The assumption $\beta \geq \alpha$ implies that Eq. (3.29) defines a mixed strategy for each incumbent for price P, satisfying $\alpha/\beta \leq P \leq 1$. Because $F(P)$ experiences a jump of $[\alpha/\beta]^{\frac{1}{n-1}}$ at $P = 1$, the expected profit of the entrant is:

$$E_e(\Pi) = \int_0^{\alpha/\beta} \alpha P dP + \int_{\alpha/\beta}^{+\infty} \beta P[1 - F(P)]^n dP \tag{3.32}$$

$$= \begin{cases} \dfrac{-n\alpha^2}{2\beta(n-2)} + \dfrac{n-1}{(n-2)}\left(\dfrac{\alpha^n}{\beta}\right)^{\frac{1}{n-1}} + \beta\left(\dfrac{\alpha}{\beta}\right)^{\frac{n}{n-1}}, & \text{if } n \geq 3 \\ \dfrac{\alpha^2}{2\beta} - \dfrac{\alpha^2}{\beta}\ln\dfrac{\alpha}{\beta} + \beta\left(\dfrac{\alpha}{\beta}\right)^{\frac{n}{n-1}}, & \text{if } n = 2 \end{cases} \tag{3.33}$$

The first term on the right-hand side of Eq. (3.32) is equal to the entrant's expected profit when it charges a price lower than the incumbent firms and captures all switchers. The second term stands for the entrant's expected profit when it directly competes with the incumbent firms. The expected profit of an incumbent firm is

$$E_m(\Pi) = \int_{u/v}^{1} \left\{ \alpha P + \beta P(1-P) \prod_{j \neq i}^{n} [1 - F(P)] \right\} dF(P) + \alpha \left(\frac{\alpha}{\beta} \right)^{\frac{1}{n-1}} = \alpha. \quad (3.34)$$

Because $\frac{\partial}{\partial \alpha}[E_e(\Pi) - E_m(\Pi)] > 0$, and when $\beta = 1/(n+1) = \alpha$, $E_e(\Pi) - E_m(\Pi) > 0$, there is $\alpha^* \in (0, 1/(n+1))$ so that when $\alpha = (1-\beta)/n \geq \alpha^*$, $E_e(\Pi) > E_m(\Pi)$. That is, the entrant can actually expect to make more profits in the said market than at least one incumbent firm.

3.4 Market Characteristics that Signal New Opportunities

When the market is monopolistic, the number of incumbent firms in Theorems 3.1 and 3.2 is equal to 1; that is, $n = 1$. In this case, switchers do not actually exist due to the monopolistic nature of the market; and the severity of customer dissatisfaction and the scale of dissatisfied customers represent the signal of market invitation for new innovation and additional competition. In such a scenario, entrepreneurs with different knowledge structures and background experiences will produce different versions of offers to potentially satisfy the customer demand. So, a surplus of producers appears to answer the market call, which, of course, depends on what a profit potential is presented in the marketplace. Hence, another way to sense a market invitation for new innovation and additional competition is the appearance of producer surplus.

The market economy is characterized by widely existing competition in the marketplace. To continuously locate profitable market niches, entrepreneurial behaviors are an important key for success in running business enterprises, although they intensify market competition. To this end, Theorems 3.1 and 3.2 above help all entrepreneurs address their question of what characterizes a market that is signaling forthcoming new opportunities and competition.

As discussed earlier in this chapter, formed mutual forbearance among firms that are incumbent in the marketplace may very well spoil the existent environment of fair and free competition. To prevent such possible systemic market deterioration, the government will need to step in to regulate the market operation so that a desired open market can be potentially assured. Along this same line of thought, the existing market, as assumed in Theorems 3.1 and 3.2, implicitly indicate that the government might occasionally need to provide supports for the formation of a fledgling market,

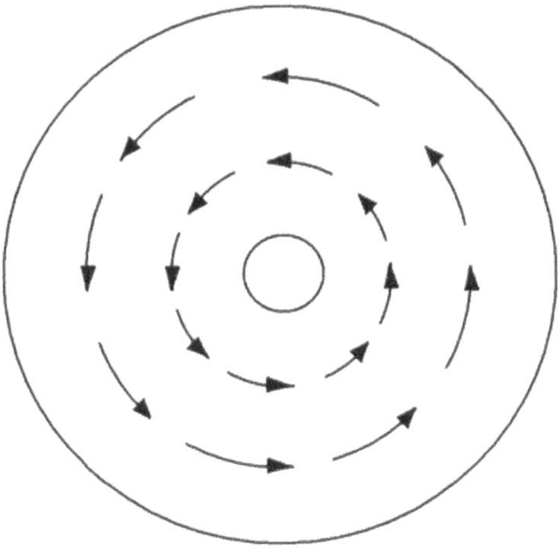

Fig. 3.3 The systemic view of our marketplace

for the smooth and orderly operation of existing markets, and for the protection of markets against aggressions of foreign entities.

In the rest of this section, let us see how the systemic yoyo model helps us intuitively foresee the main results established above before their rigorous proofs are developed. First, let us look at Proposition 3.8, which says that when all boundary conditions are normalized to the scenario of symmetry, then the loyal-customer bases of the incumbent firms must be of the same magnitude.

In particular, let us fathom the said market as an abstract yoyo field so that one can look at this imaginary multidimensional yoyo body at a distance from above either the input side or the output side. For the sake of communication convenience, assume that everything takes place in our three-dimensional space. Hence, one is looking at a pool of spinning fluid with the word "fluid" representing an abstraction of all kinds of economic factors, such as goods, information, knowledge, money, and credit, that appear and exist throughout business activities. Graphically one is looking at the market of concern as the pool of spinning fluid shown in Fig. 3.3, where the core in the center stands for the government that exerts overall control of the market through introducing various policies and employing law reinforcing systems. The circular periphery models the fact that all individually firm-specific, different boundary conditions have been normalized to symmetry. So, no matter where a firm is located in the market (the dish), its surroundings are the same as those of another firm.

When the development of this market accelerates, economic entities (particles of the fluid) within the dish start to speed up their circular motion as indicated by the arrowed curves. To see what happens next, let us introduce the well-known dishpan experiment, which was initially conducted successfully by Raymond Hide (1953) of Cambridge University, England, and then by Dave Fultz and his colleagues of University of Chicago (1959). This experiment shows that when the movement of the fluid within the rotational dish is under enough pressure created by either the sufficient speed of rotation or sufficient difference in the temperature between the

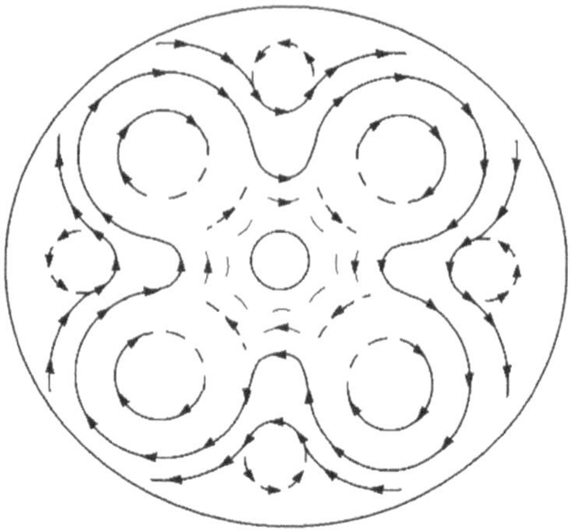

Fig. 3.4 Asymmetric flow observed in Fultz's dishpan experiment

center and the periphery of the dish, the pattern of uniform movement, as shown in Fig. 3.3, will develop into the chaos, as shown in Fig. 3.4. The number of local eddy leaves is determined either by the rotational speed or by the temperature difference or both and increases with the speed and the temperature difference. The physical reason why those local eddy leaves appear is that under the effect of the same force, particles of the fluid travel the same curved distance within a unit time. So, particles located closer to the periphery seem to move against the overall spin direction of the dish when compared to the particles near the center, causing the appearance of local pools all over the entire dish.

Now, the perfect symmetry that exists in the setup of the dishpan experiment, which also holds true for our afore-described marketplace, achieved by normalizations of boundary conditions, does not really help prevent the occurrence of the local eddy leaves in the spinning fluid. This fact, as tested in laboratories, implies that no matter how many local pools that spin in the same direction appear within the dish, they have to be the same size. If these local pools are systemic models of loyal-customer bases of incumbent firms, then these bases must be of the same scale. That is the conclusion conveyed by Proposition 3.8.

Similarly, the conclusion of Theorem 3.2 can be intuitively seen by using the yoyo model. In particular, let us model the entire market of our concern as the entire spinning dish in Fig. 3.4, and each of the n incumbent firms as a largest local eddy pool, assuming all firm-specific boundary conditions have been normalized. Then, the symmetry that exists in the marketplace, achieved by the normalization, is just like that that exists in the dishpan experiment, suggesting that it is impossible that

1. A large blank space, which is at least as big as the area occupied by one of the largest local eddy leaves, can appear within the circular chain of the largest local eddy leaves.
2. The bordering areas between adjacent largest local eddy leaves and between the periphery of the dish and circular chain of these local eddy leaves are too big.

The reason for impossibility 1 is due to the distribution of the forces that move the fluid particles. In particular, the forces acting on fluid particles are uniform in terms of strength; and they are uneven along any radius of the dish. So, the largest local eddy leaves have to be evenly distributed within the spinning dish along the periphery and around the center of the dish. And, the reason underneath impossibility 2 is that the appearance of the local eddy leaves is caused by fluid particles traveling equal linear distances, which translate into short curved distances for particles closer to the periphery than those closer to the center.

By comparing against the literature on market entry timing, defined as the timing of entry as the order of entry into a new or existing space, for a good review, see (Zachary et al., 2015), we can vividly see the theoretical and practical significance of Theorems 3.1 and 3.2. In particular, Lieberman and Montgomery (1988) consider the first mover advantage (FMA) by showing how first or early movers gain extra time over later entrants. So, they have time advantage to capture customers and build capabilities that help advance and perpetuate their early gains. However, Lieberman and Montgomery (2013) and others suggest that later entrants also have time advantage so that they are able to improve their risk estimations, clear uncertainties, improve capability gaps, and facilitate learning from the costly mistakes of earlier entrants. These works jointly confirm the appearance of the following dilemma: An identical logic is able to explain two opposite scenarios equal well (Lévesque et al., 2013). Scientifically speaking, such a dilemma is representative of empirical conclusions drawn on available data, while some essential data are missing. For example, in the studies just mentioned here, not available are the data of failed first movers and those outside North America. As discussed earlier in this chapter, when data and anecdotes from the same environment are used to draw conclusions, biased beliefs are established. The situation is similar to that underneath why people residing in environments of similar landscape hold similar philosophical beliefs and value systems (Lin & Forrest, 2011).

Since the FMA was proposed in 1988 by Lieberman and Montgomery, various scholars have confirmed that entry timing does not make a difference, while contingencies and antecedents could dictate exact timings (Fosfuri et al., 2013; Szymanski et al., 1995; VanderWerf & Mahon, 1997). Now, Theorems 3.1 and 3.2 imply that as long as the market shows consumer surplus sufficient to cover costs, new competition will naturally appear even without considering firms that have additional recourses and capabilities to mobilize. That is, no matter what contingencies and antecedents there are in the marketplace, market itself begets new innovations and competitions, while promising good levels of profits.

What is practically significant about Theorem 3.2 is the indication that established firms in the business world do not need to rush into an emerging market as first or early movers. Other than they need to invest resources to maintain their established status quo in the mutual forbearance of their current markets, they can, if necessary, expect to become one of the top performers in a newly formed market after it has become stable and sufficiently deep.

The discussion in the previous paragraphs confirms the empirical findings of Golder and Tellis (1993). By analyzing 500 products in 50 different product

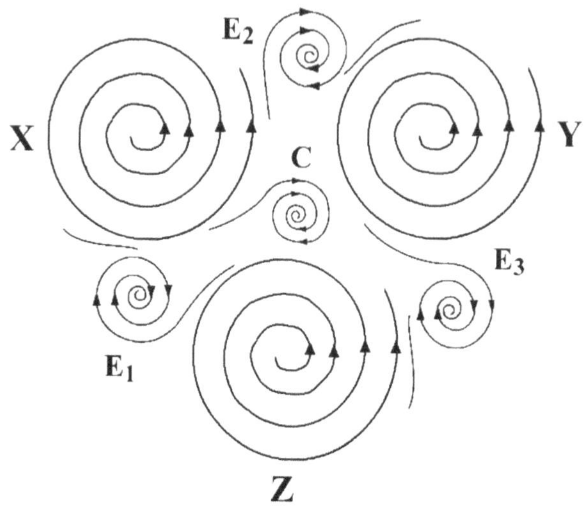

Fig. 3.5 The interaction of three markets

categories these scholars find that pioneers' average market share is 10%, while nearly 50% of them fail outright. In contrary, early followers enjoyed around 28% of the market while on about 8% of them fail. In other words, early followers tend to be more successful than pioneers for the reason that long-term survival and performance are closely dependent on the ability to acquire and mobilize resources and the capability for large-scale production, as supported by Dobrev and Gotsopoulos (2010). Additionally, Theorem 3.2 theoretically approves the conjecture that large, multi-business enterprises exploit slack resources and complementary capabilities to scale up and bypass earlier market entrants (Markides & Geroski, 2005). That is, having a size and/or diversity advantage can be a sign of fitting for latter market entries (Zachary et al., 2015).

By continuing the systemic yoyo intuition developed above, we can see how markets evolve through the natural life cycle of birth-growth-mature-death as all other forms of life. In particular, Fig. 3.5 depicts the interaction of three markets, served respectively by industries X, Y, and Z. The divergencies of these markets model the fact that offers are provided by the underlying industries to satisfy three individually different market demands. Then within the adjacent zones, sub-eddy leaves, labeled as E_1, E_2, E_3, and C, are naturally created by the opposing field directions of the neighboring large yoyo fields. These sub-eddy leaves model such market offers, provided by more specialized suppliers, that jointly possess the features and functionalities of the offers available in the neighboring large fields. For example, sub-eddy leaf E_1 models the totality of all such offers that possess the features and functionalities of those offers respectively available in fields of X and Z. Specifically, sub-eddy leaf C includes all offers that have all the features and functionalities of the offers respectively available in markets of X, Y, and Z. If the technology of mass production underlying any of these sub-eddy leaves becomes available, then the industries X, Y, and Z will gradually die out and their respective

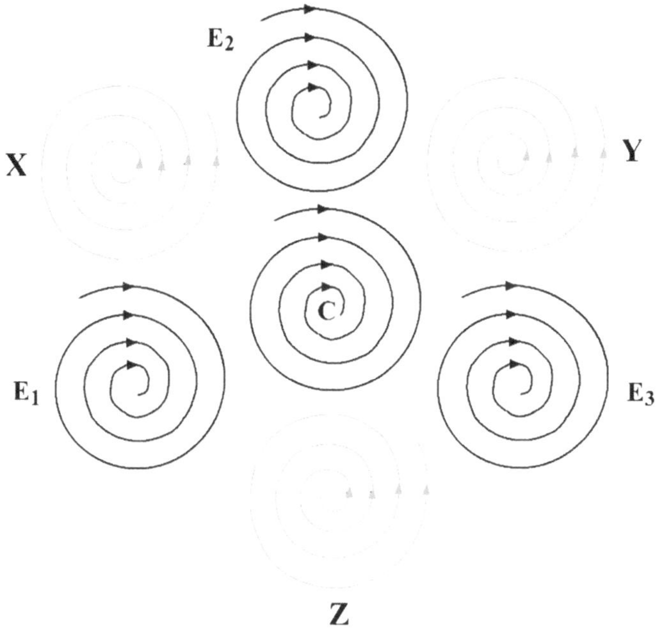

Fig. 3.6 The growth and emergence of new markets

markets will be replaced by one or several new markets growing out of the sub-eddy leaves. And the consumers of the existing markets will be absorbed by one or several emerging new markets, Fig. 3.6, where the greatly expanded yoyo fields E_1, E_2, E_3, and C will eventually combine into one large market due to their congruent, converging flow patterns.

3.5 Final Words

Aiming at avoiding the situation of having to deal with multi-body problems, which are so far unsolvable in natural science and mathematics, this chapter establishes a rigorous theory on how an open market of free competition evolves with price being the only determinant primary factor. Because the analytical reasoning employed in this chapter is based on game theory and the systemic intuition, all established results herein are generally true without suffering from the constraints of statistics- and anecdote-based approaches.

This chapter contributes to the existent literature by establishing the fact that for an economy to grow sustainably, beyond the requirement of being open for free competition, the government needs to introduce policies to nurture fairness of competition and encourage uninterruptive innovations to satisfy forever evolving consumer preferences. This result indirectly clears up the inconsistencies existing in the suggested lists of crucial factors that help kick-start a lasting economic growth (e.g., Lipton, 1977; Rostow, 1960; Studwell, 2013). It also explains why such factors as high rates of investment in modern technology, democracy, universal suffrage, property right, and others, seem to be very important, although they are only secondary factors.

In terms of the literature, the content of this chapter is mainly based on Forrest, Wagner, et al. (n.d.) (to appear) and Forrest, Li, et al. (n.d.) (to appear).

Appendix Bjerknes' Circulation Theorem

At the end of 1800s, Bjerknes (1898) discovered eddy effects of changing density in the media within movements of the atmosphere and ocean. Afterward, this well-known eddy effect was named after him. By circulation, it stands for a closed contour in a fluid. Analytically, each circulation Γ is defined as the line integral about the contour of the component of the velocity vector locally tangent to the contour, Fig. 3.7.

In symbols, Let $\vec{V}$ represent how fast a fluid moves, S a closed curve within the fluid, and $\delta\vec{r}$ the difference vector of two nearby points on the curve S (Fig. 3.7), then a circulation Γ is defined as follows:

$$\Gamma = \oint_S \vec{V}\delta\vec{r} \tag{3.35}$$

By employing a series of ingenious analytical manipulations, the rate of change in this circulation with respect to time t can be written as the following accelerating circulation:

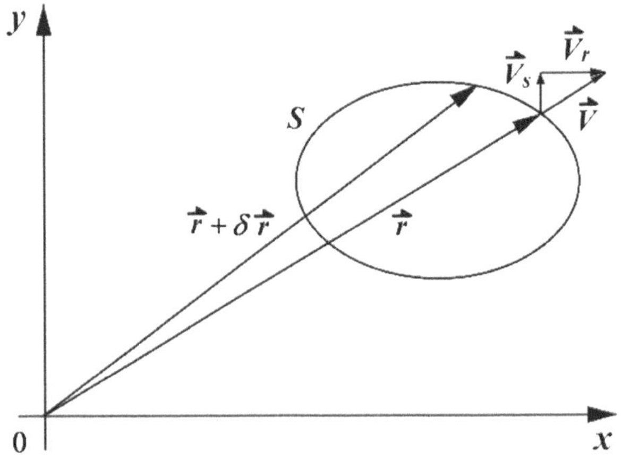

Fig. 3.7 How a circulation S looks like

$$\frac{d\Gamma}{dt} = \oint_S \frac{d\vec{V}}{dt} \delta \vec{r} \tag{3.36}$$

By applying the Euler equation of fluid mechanics in this Eq. (3.36), one can obtain the following well-known Bjerknes' Circulation Theorem:

$$\frac{d\vec{V}}{dt} = \int_\sigma \nabla\left(\frac{1}{\rho}\right) \times (-\nabla p) \cdot \delta\sigma - 2\Omega \frac{d\sigma}{dt}, \tag{3.37}$$

where σ is the projection area on the equator plane of the area enclosed by the closed curve S, p the atmospheric pressure, ρ the density of the atmosphere, and Ω the earth's rotational angular speed.

According to Newton's second law of motion, the left-hand side of Eq. (3.37) is equivalent to the force acting on the fluid. The first term on the right-hand side is known as a solenoid term in meteorology. It is caused by the interaction between density and pressure (or between the p- and ρ-planes due to the unevenness existing in density ρ so that a twisting force is created). Hence, the fluid spins with its spinning direction determined by the equal p- and ρ-plane distributions (Fig. 3.8). The second term in Eq. (3.37) comes from the rotation of the carrying system—the earth.

To conclude this appendix, let us justify the scientific validity for us to employ the Bjerknes' Circulation Theorem in business settings, as we did earlier in this chapter and will do in the following chapters in this book because in theory, this theorem holds true only for fluids.

First, when each human organization is modeled as a spinning systemic yoyo, the organization is seen as a spinning pool of "fluid," consisting of flows of such things as energy, information, knowledge, human talents, and materials. These things circulate within the inside of, are absorbed into, and are given off from the organization. Analytically, the reason why the flows of these things can be treated as "fluids" is because they are generally and widely expressed as continuous and

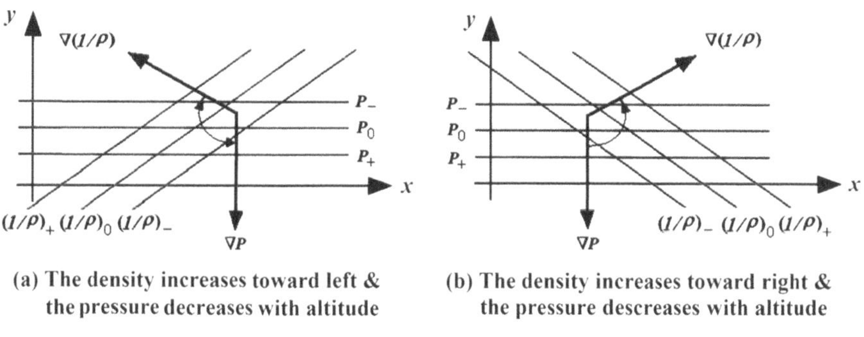

Fig. 3.8 Solenoid circulations. (**a**) The density increases towards left and the pressure decreases with altitude. (**b**) The density increases towards right and the pressure decreases with altitude

differentiable functions. Second, the systemic yoyo model of systems naturally implies that the universe is a huge ocean of eddy motions, changing and evolving constantly. That is, one can study the totality of the physically and intelligently existing world as a huge pool of interacting fluids. Third, as described in the previous sections, people help circulate information, knowledge, goods, etc., all of which are modeled by utilizing continuous or differentiable functions in social sciences in general and economics in particular. So, when such factors of a market are modeled by using functions of such characteristics, they are generally seen in physics and mathematics as flows of fluids and the expressions are widely known as flow functions. In particular, the circulations of these general factors within an economy make operational markets emerge, where people are simply local "impurities" existing within the "fluids"; and these "impurities" carry concentrated amount of "energy," information, knowledge, etc., around the economy.

References

Acemoglu, D., & Robinson, J. (2012). *Why nations fail – The origins of power, prosperity, and poverty*. Crown Business.

Barney, J., & Arikan, A. (2001). The resource-based view: Origins and implications. In M. Hitt, R. Freeman, & J. Harrison (Eds.), *Handbook of strategic management* (pp. 124–185). Blackwell.

Baum, J. A. C., & Korn, H. J. (1999). Dynamics of dyadic competitive interaction. *Strategic Management Journal, 20*(3), 251–278.

Bernheim, B. D., & Whinston, M. D. (1990). Multimarket contact and collusive behavior. *RAND Journal of Economics, 21*, 1–26.

Bjerknes, V. (1898). Uber einen hydrodynamischen Fundamentalsatz und seine Anwendung besonders auf die Mechanik der Atmosphare und des Weltmeeres. *Kungl. Svenska vetenskapsakademiens handlingar, 31*, 1–35.

Breslin, S. (2009). State led development in historical perspective: From Friedrich list to a Chinese mode of governance? Presented at Beijing university, *the Beijing forum*, Beijing, November 2009.

Chang, H. J. (2003). *Kicking away the ladder: Development strategies in historical perspective*. Anthem Press.

Christensen, C. M., Suárez, F. F., & Utterback, J. M. (1998). Strategies for Survival in Fast-Changing Industries. *Management Science, 44*, S207–S220.

Dobrev, S. D., & Gotsopoulos, A. (2010). Legitimacy vacuum, structural imprinting, and the first mover disadvantage. *Academy of Management Journal, 53*, 1153–1174.

Forrest, J. Y-L., Li, Z., Solano, R. A., & Doorn, D. (n.d.). (to appear). Technological and disruptive innovations: Their roles in a nation's industrialization and international leadership development. Submitted for publication.

Forrest, J. Y-L., Wagner, J., Anderson, M., Lipinski, J., Liu, Y., & Tian, X.G. (n.d.). (to appear). A profit-based view of economic granularity in the perspective of market competition. Submitted for publication.

Forrest, J. Y.-L., Zhao, H. C., & Shao, L. (2018). Engineering rapid industrial revolutions for impoverished agrarian nations. *Theoretical Economics Letters, 8*, 2594–2640.

Fosfuri, A., Lanzolla, G., & Suarez, F. F. (2013). Entry-timing strategies: The road ahead. *Long Range Planning, 46*, 300–311.

Fuentelsaz, L., & Gómez, J. (2006). Multipoint competition, strategic similarity and entry into geographic markets. *Strategic Management Journal, 27*(5), 477–499.

Fultz, D., Long, R.R., Owens, G.V., Bohan, W., Kaylor, R., & Weil, J. (1959). *Studies of thermal convection in a rotating cylinder with some implications for large-scale atmospheric motion.* : Meteorol. Monographs (American Meteorological Society): 21(4).

Gerschenkron, A. (1962). *Economic backwardness in historical perspective.* Belknap Press of Harvard University Press.

Golder, P. N., & Tellis, G. J. (1993). Pioneer advantage: Marketing logic or marketing legend? *Journal of Marketing Research, 30*, 158–170.

Greve, H. R. (2000). Market niche entry decisions: Competition, learning, and strategy in Tokyo banking, 1894–1936. *Academy of Management Journal, 43*(5), 816–836.

Gustafsson, R., Jaaskelainen, M., Maula, M., & Uotila, J. (2016). Emergence of industries: A review and future directions. *International Journal of Management Reviews, 18*(1), 28–50. https://doi.org/10.1111/ijmr.12057

Harmancioglu, N., Droge, C., & Calantone, R. (2009). Strategic fit to resources versus NPD execution proficiencies: What are their roles in determining success? *Journal of the Academy of Marketing Science, 37*(3), 266–282.

Haveman, H. A., & Nonnemaker, L. (2000). Competition in multiple geographic markets: The impact on growth and market entry. *Administrative Science Quarterly, 45*(2), 232–267.

Hide, R. (1953). Some experiments on thermal convection in a rotating liquid. *Quarterly Journal of the Royal Meteorological Society, 79*, 161.

Jayachandran, S., Gimeno, J., & Varadarajan, P. R. (1999). The theory of multimarket competition: A synthesis and implications for marketing strategy. *Journal of Marketing, 63*(3), 49–66.

Lévesque, M., Minniti, M., & Shepherd, D. (2013). How late should Johnny-come-lately come? *Long Range Planning, 46*, 369–386.

Li, S. X., & Greenwood, R. (2004). The effect of within industry diversification on firm performance: Synergy creation, multi-market contact and market structuration. *Strategic Management Journal, 25*, 1131–1153.

Lieberman, M. B., & Montgomery, D. B. (1988). First-mover advantages. *Strategic Management Journal, 9*, 41–58.

Lieberman, M. B., & Montgomery, D. B. (2013). Conundra and progress: Research on entry order and performance. *Long Range Planning, 44*, 312–324.

Lin, Y. (2009). *Systemic yoyos: Some impacts of the second dimension.* Auerbach Publication, an imprint of Taylor and Francis.

Lin, Y., & Forrest, B. (2011). *Systemic structure behind human organizations: From civilizations to individuals.* Springer.

Lipton, M. (1977). *Why poor people stay poor: Urban Bias in developing countries.* Temple Smith.

Marchall, T. (2016). *Prisoners of geography: Ten maps that explain everything about the world.* Scribner.

Markides, C. C., & Geroski, P. A. (2005). *Fast second: How smart companies bypass radical innovation to enter and dominate New markets.* Jossey Bass/Wiley.

Markman, G. D., & Waldron, T. L. (2014). Small entrants and large incumbents: A framework of micro entry. *Academy of Management Perspectives, 28*(2), 179–197.

Rostow, W. W. (1960). *The stages of economic growth: A non-communist manifesto.* Cambridge University Press.

Stephan, J., Murmann, J. P., Boeker, W., & Goodstein, J. (2003). Bringing managers into theories of multimarket competition: CEOs and the determinants of market entry. *Organization Science, 14*(4), 403–421.

Studwell, J. (2013). *How Asia works: success and failure in the world's most dynamic region.* Grove Press.

Szymanski, D. M., Troy, L. C., & Bharadwaj, S. G. (1995). Order of entry and business performance: An empirical synthesis and reexamination. *Journal of Marketing, 59*(4), 17–33.

VanderWerf, P. A., & Mahon, J. F. (1997). Meta-analysis of the impact of research methods on findings of first-mover advantage. *Management Science, 43*, 1510–1519.

Wen, Y. (2016). *The making of an economic superpower: Unlocking China's secret of rapid industrialization*. World Scientific.

Yu, T., & Cannella, A. A., Jr. (2012). A comprehensive review of multimarket competition research. *Journal of Management, 39*, 76–109.

Zachary, M. A., Gianiodis, P. T., Payne, G. T., & Markman, G. D. (2015). Entry timing: Enduring lessons and future directions. *Journal of Management, 41*(5), 1388–1415.

Chapter 4
Axiomatization of the Resource View: The Firm and Markets

Abstract Resource-based theory or resource-based view of the firm has been in recent decades widely applied by scholars in areas of marketing and management in their studies of various topics. Because of the implicit power of this view, relevant studies are able to establish useful explanations and predictions about competitive advantages and outcomes of firms' performance. Instead of continuously relying on such a prima facie view for reasoning, this chapter, mainly based on (Forrest et al., 2020, *International Journal of Business and Systems Research*), attempts to axiomatize it by using systems science in general and the systemic yoyo model in particular so that organizations and their interactions can be more adequately modeled and investigated. Specifically, this chapter shows, among others, when a firm advances its competitive advantage, why capabilities are also resources, how by including time capabilities become dynamic capabilities naturally, when a firm can improve its performance, and why the risk of expanding into new markets is worth taking. Because the approach used in this chapter is generally valid without suffering from the constraints of statistics- and anecdote-based methods, results and conclusions established in this chapter can be generally employed in practical applications unless given conditions are violated.

Keywords Capability · Competitive advantage · Coordinately monopolized market · Effect of environmental or industry factors · Economy of scale · Market entry · Systems · Technology

4.1 Introduction

It is Penrose (1959) who initially recognizes the importance of organizational resources to a firm's success. And then in the past several decades, various researches of management and marketing have witnessed a drastically increasing use of the concept of resources for the purpose of elucidating and envisaging what underlies the competitive advantage and performance of a firm (Crook et al., 2008; Kozlenkova et al., 2014). However, most of the relevant literature is abound with analyses based on either anecdotes or data (or data mining). Developing conclusions

J. Y.-L. Forrest, Y. Liu, *Value in Business*, Contributions to Management Science,
https://doi.org/10.1007/978-3-030-82898-1_4

by using such an approach is known to suffer from severe constraints in every step of the way, from data collection to variable coding to eventual analysis. In other words, such works with limited validity indeed help uncover potential facts, known as theories in areas of management and marketing. But, these potential facts, scientifically speaking, can only become established facts after they are successfully shown based on rigorous logic of reasoning with a set of solid starting axioms.

In order to strengthen the afore-described hole in the literature, related to the concept of resources, in management and marketing, this chapter advances a consistent and rigorous theory of resources so that what has been established more or less empirically in the past decades can be placed on a solid foundation. In particular, this chapter adopts the systems thinking (Lin, 1999) in general as the logic of reasoning and the systemic yoyo model (Lin, 2009) in particular as the intuition so that the evolution and interaction of organizations can be figuratively fathomed. This established cohesive theory is of the following organization: it first starts with three basic axioms about firms and their resources, which are vindicated in various ways. Second, it derives a series of generally true formal propositions regarding the associations between resources and the emergence of sustainable competitive advantages, resources and capabilities, and resources and performance of firms. This chapter is similar to what Rathod et al. (2019) do when these scholars employ an axiomatic approach and process variable to build their model of agile system in supply-chain design.

The rest of the chapter is organized as follows. Section 4.2 reviews the literature and shows how this chapter makes its contribution. Section 4.3 introduces three basic axioms, along with necessary justifications, that will underlie the cohesive theory of resources. Section 4.4 looks at some scenarios when sustainable competitive advantages can potentially appear. Section 4.5 examines the association between capabilities and resources and how capabilities can naturally become dynamic as soon as time is concerned with. Section 4.6 considers conditions that help boost a firm's performance, and risks and opportunities associated with a firm's effort to expand into new markets. This chapter concludes in Sect. 4.7.

4.2 Literature Review

Penrose (1959) is the first person who recognized how important organizational resources are in terms of the success of firms. However, it is not until the 1980s that the resource-based view of the firm begins to take shape (Porter, 1979; Wernerfelt, 1984). Since then, various scholars, see, for example, Barney et al. (2011), Crook et al. (2008), Kozlenkova et al. (2014), Slotegraaf et al. (2003), and Vorhies and Morgan (2005), demonstrate the importance of the concept of resources and the consequent resource-based theory. The power of this concept of resources lies in its ability to offer a framework (Fang et al., 2011) for integrating multiple, dissimilar resources to explain its synergistic, differential effects on firms' performance and the contingencies associated with each effect.

Dutta et al. (1999) and Moorman and Slotegraaf (1999) maintain that a firm is able to accrue the greatest benefits when its externally focused, market-based resources are supplemented with internal resources. This end agrees with Day's (1994) belief that to make use of outside-in capabilities, a firm has to match these capabilities with inside-out ones. By treating customers as a resource that participates in events of a firm's ecosystem, Marakhimov et al. (2018) produce an experimental framework to combine system dynamics modeling and fuzzy sets. The consequent simulation suggests that customers positively influence the financial performance, innovation capability, and the business ecosystem's health of the ecosystem leader. Barney and Hesterly (2012) claim that if a firm's resources are valuable, rare, imperfectly imitable, then the firm is capable of exploiting the potential of its available resources, and generating a sustained competitive advantage. Ombaka et al. (2015) observe that resources, either tangible or intangible, have a statistically significant influence on nonfinancial performance of insurance companies. However, resources' individual influences on various indicators of a firm performance are mixed. By applying the insights of resource-advantage theory, Cacciolatti and Lee (2016) analyze moderators of the relationship between capabilities and firm performance, such as market orientation, marketing strategy, and organizational power. They then explain how various types of marketing capabilities directly contribute to the performance of a firm.

Auh and Menguc (2009) look at leadership and market orientation and find that (1) a CEO's marketing background and politics on marketing decision-making influence the development and deployment of resources; (2) a CEO's marketing background helps increase his transformational leadership capability but not his market orientation; and (3) politics underneath marketing decisions reduce the transformational leadership and market orientation of a CEO. Evanschitzky (2007) empirically finds that resources amount to about two times the variation in firm performance than industrial structure and that a network's market orientation increases the effectiveness of resources regarding the development of sustained competitive advantages. Mamun et al. (2018) confirm that both entrepreneurial and market orientations create significant positive effects on consumer engagement and the performance of small- and medium-sized manufacturing firms.

De Oliveira (2017) investigates the ideal competency profile that leaders must possess in terms of IT resources by employing a descriptive analysis of the perception of administrators and hospital area managers. This author finds that the ideal profile would contain attributes related to the knowledge and abilities related to strategic and operational planning, leadership ability, and ethics attitude. Hooley et al. (2005) conclude that human resources and customer connections are positively impacted by managerial capabilities and market-based resources, and that reputational assets, such as brand, firm reputation, and credibility, are increased with managerial capabilities. These factors, underlying the growth in firm performance, serve as cues and signals that may connect to external observers, such as customers and competitors, on various aspects, e.g., integrity, strength, success, of a firm. Hence, they can influence decisions such as conducting business or competing with a firm (Helm & Mark, 2007). By strategically mobilizing human resources, a competitive advantage, which sustains customer relationships and increases firm value, can be created, if the management focuses on creating positive synergies among employees (Huselid et al., 1997; Richard & Johnson, 2001).

By emphasizing market orientation, entrepreneurship, innovativeness, and organizational learning, a firm can improve its long-term performance (Hult & Ketchen, 2001). Such performance is reliant on a firm's organizational responsiveness if the firm facilitates positive joint effects of market orientation, market information processing, and continuous improvement of performance (Hult et al., 2005). These joint effects depend also on the strategic mobilization of human resources to direct the long-term productivity of tangible assets with respect to a firm's market orientation and industry as its capitalization of tangible assets and its need for such can significantly vary (Richard & Johnson, 2001). For example, the literature on the hospitality industry looks mostly at intangible resources and capabilities and their impacts on the performance of hotels, where knowledge plays the driving role while tangible resources are rarely considered (Božič & Cvelbar, 2016).

The literature of market entry shows that pioneering firms tend to enjoy a preliminary advantage of profit for a period of about 12–14 years; and then this advantage develops into a disadvantage when compared to later entrants (Boulding & Christen, 2003). That is, the order of market entry in itself cannot be counted on as a sustainable advantage (Cui & Lui, 2005). As for the connection between firm performance and market entry timing, it is found that pioneers' marketing capabilities help improve firm performance; early entrants' managerial/technological capabilities help boost firm performance; and late entrants' technological/marketing capabilities provide support in lifting firm performance (Ruiz-Ortega & García-Villaverde, 2008). Timsit et al. (2015) demonstrate, by using multiagent simulation, that among all entrants of a new market, market-pull firms perform about 4% higher than resource-push firms, while approximately 25%, far less than that of resource-push firms, of market-pull firms survive.

Corporate performance can improve through lifting the effectiveness of resources, dynamic capabilities, and sustainable competitive advantage (Supriyadi, 2017). For firms that reside in a homogenous industry with shared resources that are neither rare nor heterogeneous across firms, Kabue and Kilika (2016) investigate the problem of how these firms develop their sustainable competitive advantages. They propose a model to link available resources with competencies and sustainable competitive advantage with the role of the environment simultaneously considered.

Comparing the discussion above, this chapter contributes to the existent literature in various ways. (1) It establishes generally true results, labeled as propositions, that can be practically applied without suffering from the constraints of uncertainty that appear with empirical studies. (2) It shows how systems science and systemic logic can be effectively applied to the study of important issues on firm performance and market expansion from the angle of resources. And (3) this chapter develops a theory of resources with validity way beyond what can be rigorously established based on anecdotes and data analysis. In other words, all results established in the following sections are universally true within the axiomatic system of our cohesive theory constructed on three very intuitive postulates. More specifically, by employing the logic of systems thinking and logical reasoning, this chapter shows.

- Under what conditions a sustainable competitive advantage will likely appear,
- How a firm's capabilities can be seen as resources of the firm,
- How capabilities and dynamic capabilities are really the same except the latter involves time while the former does not,
- When a firm can expect its performance to go higher,
- Whether internal resources or external ones have greater effects on a firm's performance,
- What risks can a firm expect from expanding into new market(s) by relying on existing resources and leveraging the expected economies of scale, and
- What benefits can a firm expect from expanding into new markets in terms of generating competitive advantages.

4.3 The Basic Axioms: The Starting Theoretical Points

By resource, it means an asset, either tangible or intangible (Harmancioglu et al., 2009), of a firm that can be deployed through strategy implementations for the firm to gain competitive advantages over its rivals (Barney & Arikan, 2001). Speaking differently, a resource is such a thing, which can be physical, financial, intellectual, or organizational, owned by a firm that the firm can utilize to accomplish its business goals. Hence, in terms of firms, resources, and ownership of resources, the following three axioms appear naturally.

4.3.1 Axiom 4.1 (Resource Heterogeneity)

No matter which industry a firm is a part of, the firm owns its unique system of resources.

This axiom implies that any two firms possess their respectively different systems of resources, even if they operate within the same industry or are composed of the same set of resources. The concept of systems naturally appears here, because how resources are mobilized differs from one firm to another.

The truthfulness of this axiom is quite systemically intuitive. In fact, each existing business firm is naturally an input-output system with its unique supplies and offers. So, we can imagine the firm as a spinning yoyo field (Lin, 2009), as shown in Fig. 2.2. This system consists of people, equipment, capital assets, investments, etc., as its hard component parts, organized organically into an operational whole in a specific way through particular relationships among thc component parts. It is because of these organizational and cultural relationships that the firm comes into being and is recognized as a business entity. This systemic modeling and intuition explain why any two firms can easily separate from each other by their individually different systems of resources even when they are composed of the same set of component parts. For example, most definitely, the firms likely have different sets of employees; and even if the employees are the same, the firms will have different sets of missions, indicating how the employees are mobilized differently. In particular, different employees naturally lead to different levels of capabilities—a form of resources; and different ways of mobilizing employees of course lead to different outcomes. That is, aiming at materializing different sets of missions, the employees will behave differently.

This axiom is formulated based on Peteraf and Barney's (2003) work. It emphasizes on the systemness of the resources that are under the control of a firm because in reality, firms are able to provide their market offers only by jointly and successfully mobilizing their resources. In particular, two firms' systems of their respectively available resources can be different in two ways: available resources are different or although available resources are the same, the ways they work together are different. Hence, this axiom both explicitly and implicitly implies that firms are different in their abilities of carrying out activities and accomplishing goals due to their unique systemic setups of available resources. Therefore, beyond the fact that respective sets of resources could be different, which is mostly the case in real life, what materially separates one firm from another is how their available resources are specifically joined to implement different strategies and to accomplish different business goals. In other words, firms naturally enjoy their individually different strengths and suffer from their respective weaknesses.

4.3.2 Axiom 4.2 (Resource Immobility)

Due to practical difficulties of trading resources between firms and formulating similar associations among available resources within individual firms, differences in the systemness of firms' available resources endure over time.

In fact, from the systemic logical reasoning of Axiom 4.1, one can readily see that for any two firms, their structural (both organizational and cultural) aggregates, including their sets of available resources, are different. These differences persist over time as long as the firms are different.

This axiom of resource immobility represents another theoretical progress of the work by Barney and Hesterly (2012). It explicates why firms are different in many aspects, even along the dimension of time. For example, firms generally hold determinedly dissimilar systems of resources, where the dissimilarity can be about the sets of available resources or in how differently firms associate their resources in their respectively varied ways to achieve their individually specific business goals.

4.3.3 Axiom 3 (Different Levels of Efficiency)

Different firms enjoy their individually different levels of performance, no matter whether the firms belong to the same industry or not.

The truthfulness of this axiom naturally follows from the facts that firms possess different sets of resources that are related to each other in firm-specific ways, which are generally determined by firm-specific missions and ambitions (Forrest et al., 2019), and that firms have different efficiencies of utilizing their resources (Forrest & Orvis, 2016).

To see this axiom intuitively, let us model the performance of each firm of concern as a spinning motion with the overall being of the firm as an abstract yoyo body, Fig. 2.2. Then, for any chosen set of firms, the characteristics of the spinning motion, such as spinning direction, strength, and density, of the yoyo fields of these firms are naturally different from each other. This end illustrates why firms have intrinsically different levels of efficiency in terms of how to mobilize their available resources.

This axiom of different levels of efficiency is theoretically formulated based on the work of Peteraf and Barney (2003). It reflects the real-life fact that firms deploy their individually available resources in strategically different ways. The strategic differences lead naturally to different levels of efficiency.

By the statement that a firm has a competitive advantage, it means that by deploying a specific strategy and/or utilizing a particular routine of operation, production, and/or management the firm is able to create more value for consumers and capture more economic value from the marketplace than other firms that do not implement the strategy and/or routine (Peteraf & Barney, 2003). With its competitive advantage, a firm generally can innovatively and effectively organize its resources for defined business ends. If other firms cannot replicate the rewards of this advantage in the marketplace, then this particular competitive advantage is said to be sustained (Barney & Clark, 2007). And if the advantage of a strategy can be maintained into the future, then the advantage is said to be sustainable (McGrath, 2013).

4.4 Conditions that Lead to Sustainable Competitive Advantages

By a competitive advantage of a firm, it means a special way, such as a strategy or way of operation, for the firm to mobilize its resources so that the firm generates more economic value than other firms, considered marginal or just break even in its product market (Peteraf & Barney, 2003). Listed below are three primary ways for a competitive advantage to occur (Hunt, 2007):

1. An advantaged efficiency so that more value is created and captured more efficiently than competitors;
2. An advantaged effectiveness so that more value is created and captured more effectively than competitors; and
3. An advantaged efficiency-effectiveness so that more value is more efficiently and effectively produced than competitors.

Assume that a firm is able to create more economic value than marginal firms in its industry by applying a competitive advantage, while other firms cannot duplicate the outcome of this advantage. Then this competitive advantage is said to be sustained (Barney & Clark, 2007). If the firm's advantageous situation can continue into the future, then the advantage is said to be sustainable (McGrath, 2013).

Proposition 4.1 If a particular aspect of a firm's system of resources is not shared by other firms due to various constraints, such as high cost and intellectual or physical difficulty to replicate, then the firm holds the potential to develop a sustainable competitive advantage.

A specific version of this proposition is given by Barney and Hesterly (2012), and the truthfulness of this result is systemically straightforward. In particular, let us model the firm of concern as a spinning yoyo, Fig. 2.2. Then what is assumed in the proposition implies that for some reason the underlying force, potentially making the yoyo field of the firm spin strongly, is not shared by the yoyo fields of other firms. Hence, the yoyo field of our firm of concern consequently has a chance to spin stronger than those of other firms. Of course, what actually happens to the firm of our concern depends on internally how the system of the firm, such as its ability to implement necessary changes, to acquire needed knowledge, etc., is composed.

The truthfulness of this proposition comes from an assumed scarcity of some aspect of the firm's resource system. That directly implies that if a firm could make use of its scarce resources, such as human talents and a particular path of history, then few other firms have the potential to compete with it successfully. The actual development of any sustainable competitive advantage, of course, is a function of whether or not the firm can find a means to take advantage of its resources or particular aspect of its resource system.

This proposition describes the likelihood for a firm with a particular system of resources to develop a sustainable competitive advantage. In real life, the firm still has to innovatively adopt an implementable method to mobilize its particular resources and to develop a sustainable competitive advantage. For example, although the Eastman Kodak had very specific resources and advantages available to itself when compared to Fuji Films, Kodak's organizational culture prohibited itself from developing a sustainable competitive advantage against Fuji Films through innovatively mobilizing its available resources (for a more in-depth discussion, see McGrath, 2013).

A resource is seen as valuable to a firm (Barney, 1996) if it enables the firm to introduce and to implement a particular strategy that helps increase that firm's performance beyond what would be the case without the resource. Such achievement can be accomplished by lowering the net costs, by increasing the net revenue, or a combination of these strategies. Speaking differently, the aggregate resources are seen as valuable to a firm, if they empower the firm to either take advantage of an opportunity or neutralize a threat. A resource is seen as rare, if only a few competing firms possess it. A resource is seen as exploitable by a firm, if that firm holds an organizational culture, processes, polices, and procedures that are appropriate for exploiting the competitive potential of the resource.

Proposition 4.2 If a firm possesses simultaneously valuable and rare resource (s) that are exploitable by the firm, then the firm will generate a sustainable competitive advantage out of the resource(s).

A specific version of this result was initially formulated by Barney and Hesterly (2012). This conclusion is really a corollary of Proposition 4.1. To see the details, let us continue the systemic modeling given in the analysis of Proposition 4.1. The assumption that a resource is exploitable by a firm means that the composition of the firm system permits the firm to take advantage of the resource.

To see why this result holds true based on Proposition 4.1, let us first note that the valuableness of a resource to a firm is not enough for the firm to generate a competitive advantage. In fact, in this case, other firms can possibly have the same resource too; or when a firm recognizes a resource to be valuable, the firm is not culturally and organizationally suitable to make use of the resource. For example, as the head of Kodak's Emulsion Research Division, Tom Whiteley in 1979 attended talks presented by using overhead projector slides generated on Xerox Alto. Although Tom recognized the value of Xerox Alto, an early version of a personal computer, he did not see any need or urgency to take advantage of this new technology. Speaking differently, at that very moment in time, Kodak was not in any position ready to monetize this new recognition and relevant technician of the firm by introducing and mobilizing a time-changing resource. This inability to embrace change propelled Kodak to continuously exploit and strengthen its long-standing advantages in film. At the same time, Fuji Photo foresaw and launched into a future without film. For details with this rivalry between these two companies, see (McGrath, 2013).

If a resource is valuable to a firm but not rare in the marketplace, then exploiting and manipulating such a resource will normally lead to intensified competition; all firms with this resource will most possibly have the capability to exploit and manipulate it, too. For example, in the economic sector of electronics, market competition has greatly reduced the prices of electronic products to such low levels that can barely support the manufacturers financially. If a resource is both valuable and rare, then it is also imperfectly imitable, meaning that it is impossible or difficult for competing firms to either imitate the resource or substitute it by another resource due to various constraints, such as high cost, historical conditions, intellectual limitations, physical inabilities, and organizational complexity. In fact, if a resource is valuable and perfectly imitable, then it cannot be rare, because as soon as its value is recognized, competing firms will be encouraged to imitate the resource, making the original rarity, if any, disappear quickly.

Lastly, for a firm to generate a (either sustainable or not) competitive advantage out of a valuable and rare resource, the firm must be able to exploit the resource, because such an ability is not always available to a firm. Such ability is deeply rooted in appropriate organizational culture and value system of the firm (Forrest et al., 2019). So, when a firm is able to exploit a resource, the firm will be capable of profitably exploiting the resource. Speaking systemically, the yoyo structure and its spinning orientation of the firm represent two essential factors for the firm to first identify the potential value of a resource, second design a genuinely unique way to mobilize the resource by considering prevalent market opportunities, and then materialize the benefits embodied in the resource (Barney & Clark, 2007; McGrath, 2013).

4.5 Capabilities and Dynamic Capabilities

By capability, it stands generally for an information-based process, either tangle or intangible, which empowers a firm to accomplish by deploying its available resources its business goals, such as improved productivity, increasing efficiency, and better service. In real life, each firm has its specific capabilities, deeply embedded in its organizational structure (Makadok, 2001). With this concept and recognition in place, the result below follows naturally. For a slightly different version of this result, see Makadok (2001):

Proposition 4.3 A firm's capabilities are also resources of the firm that can be utilized to accomplish the firm's business goals.

When the concept of time is combined with those of resources and capabilities, that is, when each resource and capability are seen as time-dependent, as they should be in real life, a general evolution theory of firms emerges. For example, the value of a piece of capital asset is generally a function of time. The rarity of any resource dissipates over time. The initial imperfect imitability of a resource becomes perfectly imitable with the elapse of time.

If a capability is seen as a function of time, then the concept of dynamic capability (Teece et al., 1997) as a special case of that of capability naturally appears. In particular, for a capability to remain as a capability over time, the underlying process of the capability has to advance accordingly with time so that the constantly improving process continuously assists a firm to create new things, extend what has been achieved, upgrade to newly desired features and functionalities, protect what the firm has occupied, and maintain the firm relevant in the marketplace. In other words, the unique system of that firm's resources evolves promptly in the constantly changing business environment.

Treating a capability as a function of time is really a simple mind exercise. However, the practical implication of such theoretical treatment can be rather intricate and complex. For example, to preserve a prevailing process as a worthwhile capability, a firm has to introduce well-established routines for product development and production, channels of transfers, instruments of resource allocation, constantly improving alliances and acquisitions, and processes of knowledge discovery and creation (Forrest, 2018).

Because in the present business world competitive advantages that were once sustainable have become short-lived (McGrath, 2013), treating each capability as a function of time is both theoretically and practically important. As a matter of fact, in the present dynamically changing environment, it has become infrequent occurrence for any firm to develop sustainable competitive advantages based on relatively static resources. And if such a sustainable competitive advantage is established, it will be transient and become fruitless very soon. Rapid changes within the present business environment render many non-time dependent resources obsolete because firms have to quickly and constantly reconfigure, acquire, and dispose of their static resources in order to stay in pace with market shifts (Eisenhardt & Martin, 2000). This analysis in fact leads to the following result:

Proposition 4.4 If a capability is a function of time, meaning that the capability reconfigures over time so that it stays as a capability no matter when the time is, then the capability has to be effectively adaptive so that the firm that possesses the capability can anticipate market trends and events before these trends and events are fully apparent.

4.6 Firms' Profitability, Market Share, and Return on Investment

This section studies firm performance in terms of either profitability, or market share, or return on investment.

4.6.1 When Firms' Performance Can Be Enhanced

The following result is a restatement of Proposition 4.2. A slightly more specific version of this conclusion is given by Kozlenkova et al. (2014) based on a relevant literature review.

Proposition 4.5 If a firm adopts and implements policies and procedures that help exploit the potentials of its valuable and rare resources, then the firm will be able to improve its performance.

Empirically, Morgan et al. (2009), for example, have indirectly shown this result. These scholars confirmed the positive effect of brand management resources on a firm's performance; the effect increases from negative to positive when the firm's operation centers around managing critical customer relationship and around exploiting the potential of brand resources. Additionally, such sustainably increasing positive effect on firm performance may be contingent on integrating policies and procedures into the organizational culture with "knowledge-friendly environment" driving the organizational structures and strategies (Zheng et al., 2010, p.769).

Proposition 4.6 If a firm exploits its internal resources within the controlling influence of external factors, such as those environmental or industrial, then the firm's performance is affected more by internal factors than external ones.

In fact, if the business world is intuitively seen as a large systemic yoyo field (Chap. 2), then it can be seen as an ocean of many smaller spinning yoyo fields. These smaller fields fight against each other for first survival, second obtaining as much input as possible, and third lodging as much space as possible. Hence, within the same set of environmental boundaries, which particular yoyo fields expand and which ones shrink in their competition with one another is materially determined by how their corresponding internal strengths and factors work systemically together, while in harmony with the superseding effects of environmental factors.

This result is also empirically supported. For example, Evanschitzky (2007) confirms that internal resources elucidate nearly twice the variation in firm performance than industrial structure does.

4.6.2 Additional Advantages and New Markets

This subsection looks at several issues related to doing business with new partners and expansion into new product markets. To this end, it employs the concept of resources from two different angles:

- How existing resources can be used to develop additional advantages; and
- How new opportunities can be employed to generate additional resources and advantages in both existing and new markets.

Proposition 4.7 If existing resources are deployed to expand into new market(s) for the purpose of leveraging expected economies of scale, then the effort is risky.

The systemic reasoning for the previous proposition naturally leads to this conclusion. Specifically, the assumed expansion into new market(s) generally means that the firm of our concern is performing well within the market it has already occupied (Wagner, 2016). Hence, the assumed expansion implies that the firm's organizational culture, internal strength, and various factors, have jointly and beneficially exploited the firm's existing resources and work well holistically under the overriding influence of the external market environment. When such a working mechanism is extended into another market, it is most likely the case that the new environmental boundary conditions might not allow the mechanism to continue the way it operates before.

Speaking systemically in terms of the systemic yoyo model, the spin direction, orientation, and strength of the firm's yoyo field could very possibly be inconsistent with those of the new market's yoyo field. For example, when the spin directions of the yoyo fields of the firm and the market are opposite of each other, the yoyo field of that firm's systemic body, when interacting with the new market, can be easily destroyed by this new market. That will inevitably create losses to that firm. In such a case, the expected economies of scale just will not appear for the firm.

In the actual business world, there are many real-life cases of failed expansions into new markets. For example, Walmart failed its expansion into the German market (Landler & Barbaro, 2006). Such failure seems to be unthinkable, considering how successful Walmart has been in the US market through well exploiting its internal resources, such as logistics and low-cost processes. But, Walmart just could not have successfully entered the German market by exploiting its internal resources, because German consumers prefer specialized grocery stores while Walmart pushed for the experience of one-stop shopping. That is, the boundary conditions of the US market and the German market are opposite of each other.

By entry timing, it means the timing of when an entry into either a new or existing space takes place relative to competitors, technology development, product life cycle, or other contextual referents (Zachary et al., 2015). The word "space" can be a market, an industry, or a geographic region. In terms of this concept of entry timing, a well-known so-called first mover advantage regarding entry into a new market is proposed by Lieberman and Montgomery (1988). These scholars show that first (or early) movers in a new market can capture customers and develop capabilities; and that the extra time the first (or early) movers acquired over later comers helps the former establish themselves in the marketplace and prolong their early gains. However, what is theoretically interesting is that it soon found (cf. Lieberman & Montgomery, 2013; Zachary et al., 2015) that later entrants can benefit from their delayed entries because they can improve their risk estimations, clear uncertainties, fill gaps in capabilities, and gain knowledge from earlier entrants' costly mistakes. So, collectively, these studies seem to suggest that the order of entry into a new market does not matter much, since both early and later entrants enjoy their respectively different benefits. This end in fact is supported by findings of Boulding and Christen (2003) and Cui and Lui (2005)—the order of entry does not offer long-lasting competitive advantages. As is obverse from the angle of logical reasoning, different contingencies and antecedents can well determine the entry timing into a market (Fosfuri et al., 2013; Szymanski et al., 1995; VanderWerf & Mahon, 1997), one natural question still arises: are there any differences between early and later entrants in a new market? To this end, the following result, which is a restatement of Theorem 3.2, implies that absent of any interference of contingency and antecedent, a product market generally calls for new innovations and competitors with a promise of potentially making at least as much profit as some of the incumbent firms. For more detailed discussions, see Forrest et al. (2017).

Theorem 4.1 If the managements of incumbent firms of a product market respond to market competition by playing Nash equilibria with pricing strategies, then by showing a consumer surplus as large as one of the incumbent firms' loyal-customer bases the market.

- Signals its invitation for new innovation and additional competition, and
- Promises the potential of making at least as much aggregate profit as one of the incumbent firms.

Because of the inevitable costly mistakes and other difficult issues early entrants have to bear and experience, this theorem actually explains why established firms tend to enter a new market after the market has been well-formed with shown depth.

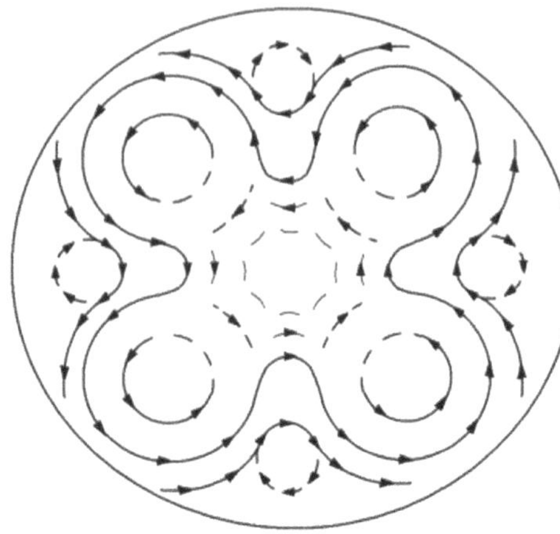

Fig. 4.1 A disturbed fluid in a spinning dishpan

Other than the game-theoretic argument of this theorem in Chap. 3, the truthfulness of Theorem 4.1 can be seen readily from the systemic yoyo model, if one imagines the entire market of concern as the entire spinning dish in Fig. 4.1, and each of the incumbent firms a local eddy pool. Then, after necessary normalizations of all the boundary conditions, for details see Chap. 3, the achieved symmetry, as is implicitly assumed in the dishpan experiment, in the marketplace suggests that it is impossible that:

1. A blank or undisturbed space, which is at least as large as one of the local eddy leaves, could appear within spinning dish. In other words, all local eddy leaves are physically distributed evenly within the dish along the periphery and around the center of the dish. Moreover,
2. The area between local eddy leaves and the dish's periphery is too big, because the local eddy leaves are caused to appear by uneven distributions of forces that act on the fluid at different distances from the dish's center.

Now, Theorem 4.1 also provides an explanation for why the incumbent firms have to compete over the consumer surplus, consisting of switchers whose purchases are determined by which firm's price is the lowest, although each incumbent enjoys a steady share of loyal consumers. In particular, although the incumbent firms can be risk neutral and want to continuously reap in their respective profits by securely defending their established market turfs, they still have to compete over the switchers in order to reduce the size of the switcher segment. Otherwise, as Theorem 4.1 indicates, when the size of the market segment of switchers grows greater than the size of the loyal consumers' base of one incumbent firm, new competition(s) will definitely enter the market. Not only so, the entering firm can potentially make as much profit as some of the incumbent firms.

Expanding into a new market brings two possible scenarios for the entering firm: the new market is either advanced or less advanced. By an advanced market, the former scenario, it represents such a market that products imported into it have to have the best possible quality and functionality compared to other options available to consumers. Hence, for the purpose of expanding into such a new market, a firm has to be able to cover the substantial sunk costs. Beyond the costs of transportation, distribution, marketing, and personnel with the necessary skill to manage new networks, the firm also has to cover the additional costs of improving current products for consumption of the advanced new market. At the same time, within the advanced new market, the expanding firm's products are just new alternatives additional to the existing market offers. They do not possess any advantage of being new to the market. By a less advanced market, it stands for such a market that is not as well developed economically and/or is not as well serviced with quality products as is the currently occupied market. In such a new market the functionality and quality of the products can be reversely innovated. That is, they need to be adjusted downwards according to the depth or purchasing power of the new market. So, the expanding firm can in this case reap in savings that are sufficient to cover the sunk costs of expanding into this new market.

Speaking generally, sunk costs tend to be an entry barrier for any firm, especially for less resourceful one, to enter new markets. That indicates that if a firm plans to expand its products into a new market, either advanced or less advanced, the firm has to have a better economic performance with a higher level of productivity than non-expanding similar (e.g., size and industry) firms before it actually starts its expansion (Bernard & Jensen, 1995; Wagner, 2012). The word "productivity" here means the efficiency for a firm to turn available resources (e.g., labor, physical capital, energy, materials, and managerial know-how) into market offers (i.e., physical products, informational goods, and/or services).

Proposition 4.8 By entering new market(s), a firm acquires opportunities to derive new resources and new competitive advantages in both existing and new markets.

The afore-described scenarios naturally lead to this proposition. In particular, for the first case that the new market a firm intends to enter is an advanced one, the firm needs to painstakingly inspect the successful products (and joint services) within the new market to guarantee that its products are at least as good as other competing products. As a result, the created knowledge will help that firm to improve its production efficiency, its managerial effectiveness, and its design and introduction of better versions of its products.

For the second case that the new market a firm intends to enter is a less advanced one, the firm needs to carefully plan out on how to cut its production cost by reversely adjusting the functionalities and features of its products while upholding the overall quality acceptable by the new market. Therefore, the acquired know-how on adjustment can naturally be applied to make new versions of products that can be offered to consumers within that firm's currently existing market of lower socioeconomic status.

Speaking systemically, venturing into a new market profitably means that the yoyo field of the venturing firm needs to well withstand the "tortures" of struggle against a bunch of new yoyo fields of the systemic structures of new business rules, new competitors, and new customers with completely different preferences and tastes. If successful, the venture will surely make, in both theory and in practice, the systemic field of that venturing firm spin more powerfully and oriented in a more optimized direction than before. Such increased field strength and orientational solidarity can surely be felt by customers and competitors in the existing and the new market.

4.7 A Few Final Words

This chapter aims at filling a void between the need of developing conclusions generally applicable in practice and specific suggestions advanced on anecdotes and data. This objective is accomplished successfully by employing the thinking logic of systems science and the rigor of game theory, leading to a series of generally true propositions. Specifically, the significance of this chapter is that through the holistic view of systems and the rigor of symbolic computation, derived conclusions are of theoretical and practical certainty and reliability. Other than showing a different way of thinking and reasoning from those used in the literature, this chapter illustrates methodologically that the holistic thinking of systems science and the theory of symbolic games can indeed be fruitfully employed to studying resources and their relationships with competitive advantages, with firm performance, and with market territory expansion.

Although the resource-based theory or resource-based view of the firm has been shown to be powerful in studies of various topics of management and marketing, most of the important results in the literature are affected adversely in one way or another by uncertainties involved in data collection and analysis. Hence, when such results of limited validity are used as a basis to make practical business decisions, the outcomes are subject to unexpected surprises. Opposite to the effort of continuing the landscape of relevant literature, where theoretical and practical uncertainties widely exist, this chapter establishes reliable results by using logical reasoning on the basis of three axioms.

At the same time, all conclusions in this chapter are developed on the formal definition of resources, each of which stands for an asset a firm can employ in its design and implementation of its adopted strategies (Barney & Arikan, 2001). However, in practice, which systemic part(s), either hard or soft, of a firm that can be treated as assets is not universally clear. What constitutes an asset and how much an asset is worth generally vary significantly from one individual to another and from one firm to another, depending on their individually different preferences and organizational goals and objectives. And, this challenge is made all more difficult by the fact that in the design and implementation of a firm's strategy, there are generally several assets that collectively contribute to the eventual success with some assets playing more important roles than others. So, there is a need to fully consider the degrees of importance of resources.

References

Auh, S., & Menguc, B. (2009). Broadening the scope of the resource-based view in marketing: The contingency role of institutional factors. *Industrial Marketing Management, 38*(7), 757–768.

Barney, J., & Arikan, A. (2001). The resource-based view: Origins and implications. In M. Hitt, R. Freeman, & J. Harrison (Eds.), *Handbook of strategic management* (pp. 124–185). Blackwell.

Barney, J., & Clark, D. (2007). *Resource-based theory: Creating and sustaining competitive advantage*. Oxford University Press.

Barney, J., & Hesterly, W. (2012). *Strategic management and competitive advantage: Concepts and cases* (4th ed.). Pearson.

Barney, J., Ketchen, D., & Wright, M. (2011). The future of resource-based theory: Revitalization or decline? *Journal of Management, 37*(5), 1299–1315.

Barney, J. (1996). The resource-based theory of the firm. *Organization Sicience, 7*(5), 469–592.

Bernard, A. B., & Jensen, J. B. (1995). Exporters, jobs, and wages in U.S. manufacturing: 1976–1987. *Brookings Papers on Economic Activity, Microeconomics, 1995*(1), 67–119.

Boulding, W., & Christen, M. (2003). Sustainable pioneering advantage? Profit implications of market entry order. *Marketing Science, 22*(3), 371–392.

Božič, V., & Cvelbar, L. K. (2016). Resources and capabilities driving performance in the hotel industry. *Tourism and Hospitality Management, 22*(2), 225–246.

Cacciolatti, L., & Lee, S. H. (2016). Revisiting the relationship between marketing capabilities and firm performance: The moderating role of market orientation, marketing strategy and organisational power. *Journal of Business Research, 69*(12), 5597–5610.

Crook, T., Ketchen, D., Combs, J., & Todd, S. (2008). Strategic resources and performance: A meta-analysis. *Strategic Management Journal, 29*(11), 1141–1154.

Cui, G., & Lui, H. (2005). Order of entry and performance of multinational corporations in an emerging market: A contingent resource perspective. *Journal of International Marketing, 13*(4), 28–56.

Day, G. S. (1994). The capabilities of market-driven organizations. *Journal of Marketing, 58* (October), 37–52.

De Oliveira, J. F. (2017). The importance of information technology as a strategic resource for hospital management process and the management skills required for managers. *International Journal of Business and Systems Research, 11*(1/2), 182–197.

Dutta, S., Narasimhan, O., & Rajiv, S. (1999). Success in high-technology markets: Is marketing capability critical? *Marketing Science, 18*(4), 547–568.

Eisenhardt, K., & Martin, J. (2000). Dynamic capabilities: What are they? *Strategic Management Journal, 21*(10/11), 1105–1121.

Evanschitzky, H. (2007). Market orientation of service networks: Direct and indirect effects on sustained competitive advantage. *Journal of Strategic Marketing, 15*(4), 349–368.

Fang, E., Palmatier, R., & Grewal, R. (2011). Effects of customer and innovation asset configuration strategies on firm performance. *Journal of Marketing Research, 48*(3), 587–602.

Forrest, J. Y.-L. (2018). *General systems theory: Foundation, intuition and applications in business decision making*. Springer.

Forrest, Y. L. J., Buttermore, J., & Wajda, T. A. (2017). At Nash equilibrium when new market competitions appear? *Kybernetes, the International of Cybernetics, Systems and Management Science, 46*(2), 256–271.

Forrest, J. Y.-L., & Orvis, B. (2016). Principles of management efficiency and organizational inefficiency. *Kybernetes: The International Journal of Cybernetics, Systems and Management Sciences, 45*(8), 1308–1322.

Forrest, J. Y.-L., Shao, L., McKinnery, R., & Shao, D. (2020). Firm performance and market expansion from the point of view of resources. *International Journal of Business and Systems Research, 15*(4), https://doi.org/10.1504/IJBSR.2021.115989

Forrest, J. Y. L., Tucker, R., Lin, C. C., & Mondal, S. (2019). The innovativeness of manufacturing firms from the angles of organizational culture, structure and leadership. *Journal of Business, Economics and Technology, 22*(1).

Fosfuri, A., Lanzolla, G., & Suarez, F. F. (2013). Entry-timing strategies: The road ahead. *Long Range Planning, 46*, 300–311.

Harmancioglu, N., Droge, C., & Calantone, R. (2009). Strategic fit to resources versus NPD execution proficiencies: What are their roles in determining success? *Journal of the Academy of Marketing Science, 37*(3), 266–282.

Helm, R., & Mark, A. (2007). Implications from cue utilisation theory and signaling theory for firm reputation and the marketing of new products. *International Journal of Product Development, 4* (3/4), 396–411.

Hooley, G., Greenley, G., Fahy, J., & Cadogan, J. (2005). Market-focused resources, competitive positioning and firm performance. *Journal of Marketing Management, 17*(5/6), 503–520.

Hult, G. T. M., & Ketchen, D. J. (2001). Does market orientation matter? A test of the relationship between positional advantage and performance. *Strategic Management Journal, 22*(9), 899–906.

Hult, G. T. M., Ketchen, D. J., & Slater, S. F. (2005). Market orientation and performance: An integration of disparate approaches. *Strategic Management Journal, 26*(12), 1173–1181.

Hunt, S. (2007). Synthesising resource-based, evolutionary and neoclassical thought. In N. Foss & P. Robertson (Eds.), *Resources, technology, and strategy* (pp. 53–79). Routledge.

Huselid, M. A., Jackson, S. E., & Schuler, R. S. (1997). Technical and strategic human resources management effectiveness as determinants of firm performance. *Academy of Management Journal, 40*(1), 171–188.

Kabue, L. W., & Kilika, J. M. (2016). Firm resources, core competencies and sustainable competitive advantage: An integrative theoretical framework. *Journal of Management and Strategy, 7* (1), 98–108.

Kozlenkova, I. V., Samaha, S. A., & Palmatier, R. W. (2014). Resource-based theory in marketing. *Journal of the Academy of Marketing Science, 42*, 1–21.

Landler, M., & Barbaro, M. (2006). International business: No, not always. *New York Times, 8*(2), A1–A3.

Lieberman, M. B., & Montgomery, D. B. (1988). First-mover advantages. *Strategic Management Journal, 9*, 41–58.

Lieberman, M. B., & Montgomery, D. B. (2013). Conundra and progress: Research on entry order and performance. *Long Range Planning, 44*, 312–324.

Lin, Y. (1999). *General systems theory: A mathematical approach*. Plenum and Kluwer Academic Publishers.

Lin, Y. (2009). *Systemic yoyos: Some impacts of the second dimension*. Auerbach Publication, an imprint of Taylor and Francis.

Makadok, R. (2001). Toward a synthesis of the resource-based and dynamic-capability views of rent creation. *Strategic Management Journal, 22*(5), 387–401.

Mamun, A. A., Mohiuddin, M., Fazal, S. A., & Ahmad, G. B. (2018). Effect of entrepreneurial and market orientation on consumer engagement and performance of manufacturing SMEs. *Management Research Review, 41*(1), 133–147.

Marakhimov, A., Joo, J., & Jaegeol Yim, J. (2018). Customer participation in business ecosystems: An integrated approach of system dynamics and fuzzy sets. *International Journal of Business and Systems Research, 12*(3), 290–316.

McGrath, R. G. (2013). *The end of competitive advantage: How to keep your strategy moving as fast as your business*. Harvard Business Review Press.

Moorman, C., & Slotegraaf, R. (1999). The contingency value of complementary capabilities in product development. *Journal of Marketing Research, 36*(2), 239–257.

Morgan, N., Slotegraaf, R., & Vorhies, D. (2009). Linking marketing capabilities with profit growth. *International Journal of Research in Marketing, 26*(4), 284–293.

Ombaka, B. E., Awino, Z. B., Machuki, V. N., & Wainaina, G. (2015). Exploring resources and performance relationships in commercial enterprises: An empirical perspective. *Journal of Management and Strategy, 6*(4), 12–28.

Penrose, E. (1959). *The theory of the growth of the firm*. Blackwell.

Peteraf, M., & Barney, J. (2003). Unraveling the resource-based tangle. *Managerial and Decision Economics, 24*(4), 309–323.

Porter, M. (1979). How competitive forces shape strategy. *Harvard Business Review, 57*(2), 137–145.

Rathod, G. H., Puranik, V. S., & Hiremath, V. (2019). Developing agility system in supply chains. *International Journal of Business and Systems Research, 13*(2), 247–258.

Richard, O. C., & Johnson, N. B. (2001). Strategic human resource management effectiveness and firm performance. *International Journal of Human Resource Management, 12*(2), 299–310.

Ruiz-Ortega, M., & García-Villaverde, P. (2008). Capabilities and competitive tactics influences on performance: Implications of the moment of entry. *Journal of Business Research, 61*(4), 332–345.

Slotegraaf, R., Moorman, C., & Inman, J. (2003). The role of firm resources in returns to market deployment. *Journal of Marketing Research, 40*(3), 295–309.

Supriyadi, S. (2017). Analysis effect of resources and dynamic capabilities to sustainable competitive advantages and its implications to the firms performance. *International Business Research, 10*(10), 209–222.

Szymanski, D. M., Troy, L. C., & Bharadwaj, S. G. (1995). Order of entry and business performance: An empirical synthesis and reexamination. *Journal of Marketing, 59*(4), 17–33.

Teece, D., Pisano, G., & Shuen, A. (1997). Dynamic capabilities and strategic management. *Strategic Management Journal, 18*(7), 509–533.

Timsit, J. P., Castiaux, A., Truong, Y., Athaide, G. A., & Klink, R. R. (2015). The effect of market-pull vs resource-push orientation on performance when entering new markets. *Journal of Business Research, 68*(9), 2005–2014.

VanderWerf, P. A., & Mahon, J. F. (1997). Meta-analysis of the impact of research methods on findings of first-mover advantage. *Management Science, 43*, 1510–1519.

Vorhies, D., & Morgan, N. (2005). Benchmarking marketing capabilities for sustainable competitive advantage. *Journal of Marketing, 69*(1), 80–94.

Wagner, J. (2012). International trade and firm performance: A survey of empirical studies since 2006. *Review of World Economics, 148*(2), 235–267.

Wagner, J. (2016). A survey of empirical studies using transaction level data on exports and imports. *Review of World Economics, 152*(1), 215–225.

Wernerfelt, B. (1984). A resource-based view of the firm. *Strategic Management Journal, 5*(2), 171–180.

Zachary, M. A., Gianiodis, P. T., Payne, G. T., & Markman, G. D. (2015). Entry timing: Enduring lessons and future directions. *Journal of Management, 41*(5), 1388–1415.

Zheng, W., Yang, B., & McLean, G. N. (2010). Linking organizational culture, structure, strategy, and organizational effectiveness: Mediating role of knowledge management. *Journal of Business Research, 63*, 763–771.

Chapter 5
Evolution of Resources: An Axiomatized Resource View

Abstract Riding on the increasing importance of the resource-based theory or the resource-based view of the firm of the past decades in the research of management and marketing, this chapter, which is mainly based on (Forrest, et al., 2019, *Proceedings of the 2019 annual conference of Pennsylvania Economic Association*), continues what has been started in the previous chapter to develop a cohesive theory of resources by employing the thinking logic of systems science. In particular, this chapter establishes definite results on the exploitation and exploration of resources; firm's technological opportunism, innovativeness, and performance; development, interactions, and decay of resources; and conditions under which a firm would fail to consistently exploit its available resources. Because all the conclusions established herein are developed by logical reasoning based on clearly true axioms, their truthfulness is not affected by any of the constraints and uncertainties of data analysis; and they are expected to be universally applicable in practice.

Keywords Capability-rigidity paradox · Firm performance · Knowledge competence · Market orientation · Organizational capability · Systems thinking · Technological opportunism

5.1 Introduction

For the purpose of predicting the future, various methods of science and mathematics have been practically tried throughout history, although not much progress has been made in terms of prediction accuracy and reliability (Lin & OuYang, 2010). Parallel to such attempts in science, in the past nearly 20 years scholars in areas of management science and marketing have been looking at the concept of resources with increasing interest and intensity, hoping that by using this concept one will be able to explain and predict the bases of a firm's competitive advantage and performance (Crook et al., 2008; Kozlenkova et al., 2014). During this time frame, scholars have piled up a good number of discoveries and theories, most of which are either explicitly or implicitly established on summarizing anecdotes or analyzing data. However, scientifically speaking, such discoveries and theories in general suffer

J. Y.-L. Forrest, Y. Liu, *Value in Business*, Contributions to Management Science,
https://doi.org/10.1007/978-3-030-82898-1_5

from severe constraints and uncertainties inherently existing in any kind of data collection, coding, and analysis. This fact explains why any business decision that involves the slightest amount of prediction of the future tends to be not reliable (Forrest et al., 2020; Lin & OuYang, 2010).

Targeting this weakness of the literature, where management and marketing issues have been carefully investigated by utilizing the resource-based theory, this chapter attempts to continue the development of a cohesive theory of resources started in the previous chapter for firms based on what has been established in the past decades since the time when Penrose (1959) initially identified the importance of organizational resources to a firm's success. In order to avoid all the uncertainties and constraints inherently existing in summaries of anecdotes and collections, coding, and analyses of data, this chapter will only apply the holistic thinking logic and methodology of systems science. Because of this reason, all the established results in this chapter are expected to be generally true and can be practically employed in real-life business decision-making.

More specifically, this chapter employs systems thinking (Lin, 1999) as the logic of reasoning and the systemic yoyo model (Lin, 2009) as the intuition for how organizations operate and how organizations interact with each other. On such a methodological basis, this chapter is able to develop general results on such important issues as when a firm is likely or certainly to develop sustainable competitive advantages, how the capability-rigidity paradox can be avoided, what could potentially improve firm performance, how relational resource impacts the innovativeness of the firm, when interactions of resources could lead to undesirable effects, when and how a capability would evolve or cease to exist, and when a firm would fail to consistently exploit its resources, among others.

The rest of this chapter is organized as follows: Section 5.2 reviews the relevant literature and explains how this chapter contributes to what is known. Section 5.3 lays down the most fundamental properties of resources. Section 5.4 looks at issues regarding the exploitation and exploration of competence. Section 5.5 addresses issues of technological opportunism and innovativeness of firms. Section 5.6 considers interactions of resources. Section 5.7 investigates the development and decay of resources. Section 5.8 analyzes when a firm would fail to exploit its resources. This chapter is concluded in Sect. 5.9.

5.2 Literature Review

It is as early as the late 1950s that Penrose (1959) first shows how important organizational resources are to the success of firms. After a quiet period of dormancy, the well-known resource-based view of the firm started to take shape in the 1980s (Porter, 1979; Wernerfelt, 1984). Then many scholars, such as Barney et al. (2011), Crook et al. (2008), Kozlenkova et al. (2014), Slotegraaf et al. (2003), Vorhies and Morgan (2005), and others, demonstrate the importance of the concept of resources and the resultant resource-based theory. Specifically, the importance of the concept of resources stems from the fact that it offers a framework for integrating multiple, dissimilar resources to explain their synergistic, differential effects on firms' performance and the contingencies associated with each effect (Fang et al., 2011), which happens to coincide with the capability of systems science and systems methodology. That is the exact background reason for why we think about using systems science and its methodology to develop the desired general theory of resources.

In terms of competitive advantages, Lages et al. (2009) consider the relationship between organizational learning and relational capabilities and product innovation and quality. They find that the learning capability boosts product innovation, while relational capability improves product quality, innovation, and relationship capabilities increase economic performance. Considering how advantages and resources interact with each other, Kaleka (2011) finds that service advantage helps to increase the performance of export venture, while product development and customer relationship capabilities enhance service advantage, that customer relationship and informational capabilities help enrich experiential resources, and that product development and informational capabilities are improved by financial resources. Vorhies and Morgan (2005) find evidence for benchmarking of marketing capabilities as a tool for enhancing capabilities that drive the development and implementation of sustainable competitive advantages. In his study of the creation of competitive advantages, Jap (2001) claims that bilateral idiosyncratic investments help the creation of competitive advantages, while ex post opportunism erodes competitive advantages; this negative effect can be lessened by idiosyncratic investments, goal congruence, and interpersonal trust. Considering how to evaluate and (re)develop strategic initiatives that put into place operational capabilities to provide new sources of firm-level competitive advantage, Liu and Liang (2015) explore the practical intersection of operations management and strategy from resource-based view. These authors develop a theoretical approach to integrate the core factors that determine the performance of operational competitiveness, followed by a case study.

As for firm performance, Li and Calantone (1998) examine firms' processes that comprise market knowledge competence, which helps increase new product advantages that in turn improve the market performance of the product. It is found (Zou et al., 2003) that export marketing capabilities affect indirectly the performance of the export venture through low-cost and positional advantages of brands, while the development, distribution, and communication of new products improve the positional advantage of the export venture, which in turn helps raise the financial performance. Ruiz-Ortega and García-Villaverde (2008) reveal that for pioneers, all capabilities improve firm performance with marketing capabilities doing so the most; for early followers, management and technical capabilities improve performance; and for late followers, technical and marketing capabilities help performance. Slotegraaf and Dickson (2004) recognize that marketing planning capability exerts a curvilinear effect on firm performance. In studying what improves the development of unique products, Ramaswami et al. (2009) find that customer management exerts the strongest impact on firm performance. Lonial and Carter (2015) suggest that each of market, entrepreneurial, and learning orientations can enhance company success, but their individual potential should not be viewed in isolation by looking at these three orientations as capabilities of small and medium-sized enterprises on the basis of the resource-based view of the firm.

In terms of resources and market deployment, Slotegraaf et al. (2003) maintain that the level of intangible marketing and technological resources is positively correlated to the effectiveness of market deployments, such as distribution and coupon activity, while the level of financial resources is negatively correlated to the effectiveness of market deployment. Hughes et al. (2010) find that investments in marketing differentiation strategy generate the necessary strategic incentive to form innovation ambidexterity, which in turn determines both marketing differentiation and cost leadership advantages, and that both marketing differentiation and cost leadership advantages improve the performance of the export venture. In terms of how an advantage could turn into a disadvantage, Boulding and Christen (2003) find that pioneering firms tend to experience an initial profit advantage for a period of time of about 12–14 years, after which early entry turns into a profit disadvantage when compared to later entrants.

Atuahene-Gima (2005) demonstrates that market orientation simultaneously empowers exploitation of existing innovation competencies and exploration of new innovation competencies. Luo et al. (2004) establish that customer relationships and business partner social capital help sales growth and return on investment with the former being the main driver behind firm performance, that governing-agency social capital helps sales growth but not return on investment. It is shown that social capital is correlated with higher economic income (Tomer, 2003). Entrepreneurs who possess stronger human capital and broader social networks are more effective networkers (Wright et al., 2009). Wright et al. (2009) and Baron and Markman (2003) indicate the importance of social capital and human capital in entrepreneurs' success. In addition, other authors connect social capital to entrepreneurship. For example, Aldrich and Zimmer (1986) suggest that social ties and social network diversity broaden the scope of opportunities for entrepreneurs and that increased connectedness increases the flow of information as well as resource availability.

Richey et al. (2010) analyze what affects the financial performance of B2B partnerships, and realize that for data storage resources but not for communication or customization resources there are resource complementarities. In their investigation of what capabilities have direct and complementary effects on revenue and margin growth, Morgan et al. (2009) find that customer relationship management capability lowers revenue growth rate while increases margin growth. Considering the fact that dynamic capabilities are sometimes inadequate for dealing with urgent and unexpected organizational and strategic problems, Ritalam et al. (2016) explore the boundary conditions of dynamic capabilities and proposes a complementary approach for dealing with unfamiliar problems: the ad hoc problem-finding and problem-solving perspective, accompanied by a series of theory-driven propositions to illustrate their arguments. Looking at how company size affects the effectiveness of investments in marketing and innovation capabilities on firm performance, Jeng and Pak (2016) analyze a sample of 692 small, medium, and large enterprises. They find that large firms prospered from building dynamic capabilities under the condition of high industry competitiveness, investments in innovation and marketing individually diminished small firms' performance, and the effect was mixed for medium-size firms.

By using a specific type of strategic resource—commitment-based human resource systems, Chadwick et al. (2015) examine the effect of CEO resource orchestration in a multi-industry sample of 190 Korean firms. They demonstrate that CEO emphasis on strategic human resource management is a significant antecedent to commitment-based human resource systems and carries primary effects on firm performance. By looking at the phenomenon of interlocking directorates, Zona et al. (2016) suggest that such phenomenon may exert either a positive or a negative effect on subsequent firm performance, depending on the firm's relative resources, power imbalance, ownership concentration, and CEO ownership. And this hypothesized effect is positively supported by a sample of 145 Italian companies. Studying the influence of the dynamic capabilities on firm's performance, Arifin and Frmanzah (2015) propose that technology adoption is a functional competence/capability that mediates the relationship between dynamic capabilities and firm's performance by identifying four determinant factors: externalities, entrepreneurial leadership, slack resources, and absorptive capability. By using a sample of 1671 firms operating in four countries, Covin et al. (2016) look at what drivers of innovativeness could be common to family and nonfamily firms. These authors reveal six configurations of behavioral proclivities and/or resources that predict radical innovativeness. Fuentes-Fuentes et al. (2017) investigate the relationship between business contacts and innovativeness in women-owned firms by using a sample of 107 women entrepreneurs in Spain. They show that maintaining close contacts with managers/entrepreneurs in different industries and with customers is significant in explaining innovativeness in women-owned firms. By considering how to theoretically incorporate factors external to the organization into the business process management lifecycle through dynamic capabilities, Bernardo et al. (2017) find that dynamic capabilities contribute to the conceptualization and understanding of business process management, assist in the process of organizational change by promoting adaptation to the environment, and enhance business process management performance.

Compared to the literature, this chapter, once again as in the previous chapter, employs the thinking logic and methodology of systems science instead of empirical means to establish a few general results that can be applied in business decision-making with reliability and certainty. In particular, some of the most important results developed herein include, among others: (1) If a resource (or a set of resources) a firm controls is simultaneously valuable, rare, and exploitable by the firm, then the firm will be able to design and implement a sustainable competitive advantage out of the resource; (2) If a resource or a set of resources of a firm is simultaneously valuable and rare, and if the managers' interests are aligned with those of shareholders, then the resource will result in a sustainable competitive advantage for the firm; (3) Market orientation of a firm and its capability to act accordingly enable the firm to resolve the capability-rigidity paradox, and consequently helps the firm to sustain its competitive advantage that in turn means superb performance; and (4) for two opposing business goals A and B, any interaction between resources that lead to the realization of goal A and those that help actualize goal B tends to produce undesirable effects.

Speaking differently, the contribution of this chapter to the literature is of the following two-folds: one is its introduction of a brand-new thinking logic and methodology into the study of various management and marketing issues based on the resource-based view; the other is the actual establishment of widely applicable results in real life.

5.3 Preliminary Properties of Resources

A firm has a competitive advantage, provided that the firm has a particular strategy or way of operation, production, and/or management with which the firm generates more economic value than other firms (Peteraf & Barney, 2003). In such a case, the firm generally has invented a special way to mobilize its resources. If, additionally, other firms in the same industry are unable to duplicate the benefits of this strategy, then the competitive advantage is known as sustained (Barney & Clark, 2007); if the advantageous situation of the strategy continues into the future, then the advantage is referred to as sustainable (McGrath, 2013). Based on the established Axioms 4.1–4.3, the following two propositions naturally follow.

Proposition 5.1 If a firm possesses such a system of available resources that few other firms have due to various reasons, such as history, relational networks, cost, and intellectual and/or physical difficulty to replicate, then the firm will have a great opportunity to develop a sustainable competitive advantage.

Barney and Hesterly (2012) initially formulated a slightly different version of this proposition, while the proposition's truthfulness is quite systemically straightforward. To this end, let us model the firm as a spinning yoyo. Then the assumption of the proposition implies that for some reason the underlying force that has the potential of making the yoyo field of the firm spin strongly is not shared by the yoyo fields of most other competing firms. Therefore, the yoyo field of this specific firm has a chance to spin strongly or differently from the ways of how the yoyo fields of other firms spin. However, the eventual appearance of a sustainable competitive advantage depends on internally how the system of the firm is composed of—or whether or not the firm is able to discipline itself to make use of its special underlying force.

The assumed scarcity of resources means directly that if the firm could make use of its scarce resources, few other firms could potentially compete with it successfully on the ground of the particular set of resources. The potential development of sustainable competitive advantages, which take advantage of the scarce resources, of course, depends on whether or not the firm could find an innovative way to mobilize its resources. For example, in their rivalry, the Eastman Kodak surely had very specific resources available to itself when compared to Fuji Films. Nonetheless, the organizational culture of Kodak prohibited itself from developing a sustainable competitive advantage through innovatively mobilizing its available resources against Fuji Films (McGrath, 2013).

A resource (or a set of resources) is valuable to the firm, provided that the resource enables the firm to strategically increase its profit by lowering costs and/or increasing revenue beyond what would have been the case without the resource. That is, a resource is valuable if the resource empowers the firm to take advantage of an external opportunity and/or to neutralize an external threat. A resource is rare, provided that a small number of competing firms possess it. A resource is exploitable by the firm that has control of the resource, provided that the firm has an appropriate organizational culture, processes, policies, and procedures to exploit the potential benefits of the resource.

Proposition 5.2 If a resource (or a set of resources) a firm has control of is simultaneously valuable, rare, and exploitable by the firm, then the firm will be able to design and implement a sustainable competitive advantage out of the resource.

A slightly different version of this result was initially formulated by Barney and Hesterly (2012), and the current version is really a corollary of Proposition 5.1. And its truthfulness is quite systemically clear. In fact, by continuing the systemic modeling given in the analysis of Proposition 5.1, the assumption of this current proposition that the firm is able to exploit the resource means that the systemic composition of the firm allows it to take advantage of the resource.

On the other hand, notice that being valuable alone is not sufficient for a resource to generate a competitive advantage for the firm. It is because other firms can very possibly have the resource too or while the firm recognizes the value of the resource, it is just neither culturally nor organizationally suitable to make use of the resource. For example, in 1979 Tom Whiteley, the head of the Emulsion Research Division of the Eastman Kodak, attended personally slide presentations with the slides produced out of Xerox Alto, an early version of a personal computer. But, with his mindset and the then-culture of Kodak, he did not see the need and urgency to take advantage of the new technology and the available human talents. In other words, the systemic structure of the Eastman Kodak was not fit at the time to monetize the just-gained new knowledge and relevant technician of the firm by introducing and mobilizing a time-changing resource. Instead, the company continued to exploit and strengthen its long-standing advantages in film. That led to the company's defeat in competing with Fuji Photo that moved into a bright future without film (McGrath, 2013).

If a resource is not rare, then exploiting such a resource, when the value is recognized, generally leads to intensified competition in the product market, because other firms will most likely have the capability to exploit the resource, too. To illustrate this, a good example is how market competition has driven the price of personal computers all the way down to the level that nearly no profit can be made. If a resource is valuable, then its rarity means that the resource is also imperfectly imitable, meaning that it is impossible for competing firms to duplicate or substitute the resource due to various constraints, such as history, prohibitive cost, intellectual limitations, physical inabilities, and organizational complexity. On the other hand, if a valuable resource is perfectly imitable, then it will no longer be rare because the recognized value definitely encourages other firms to duplicate or to substitute the resource, making the earlier rarity disappear quickly.

Lastly, for a firm to generate a sustainable or any competitive advantage out of a resource, the firm has to be able to exploit the resource, because only an organizational culture with appropriate philosophical and value system could lead to profitable exploitation of any particular resource. In other words, only when the firm is under the influence of two key organizational factors—the systemic structure and orientation of the organization— the firm is able to first recognize the potential value of a resource, second introduce a unique way to take advantage of the resource by considering the opportunities afforded by the market, and then enable the realization of the benefits embodied in the resource (Barney & Clark, 2007; McGrath, 2013).

Proposition 5.3 If a resource or a set of resources of a firm is simultaneously valuable and rare, and if the managers' interests are aligned with those of shareholders, then the resource will result in a sustainable competitive advantage for the firm.

In fact, the assumption that the managers' interests are aligned with those of shareholders implies that the particular resource or the particular set of resources is effectively exploitable by the firm. Hence, Proposition 5.2 indicates that a sustainable competitive advantage will be resulted in.

To pave the ground for the rest of this chapter, let us look at the following two game-theoretic theorems, for elementary versions of these results established for limited specific scenarios, see Forrest et al. (2017).

Theorem 5.1 If the market described below is in Nash equilibrium, then the following two conditions are equivalent:

1. The market's consumer surplus is larger than the loyal-customer base of one incumbent firm; and
2. Some firm(s) will enter the market with their versions of substitute offer with their aggregate expected profit potentially larger than that of at least one incumbent firm.

As in the statement of Theorem 3.2, the market mentioned in Theorem 5.1 satisfies the following conditions: It is served by a fixed number of firms with their horizontally differentiated offers; its operation is only affected by market forces, such as demand and supply, and consumers' forever evolving preferences and tastes; each incumbent firm has a base of loyal customers if its price is not more than their reservation price; there are switchers, collectively known as consumer surplus, who make purchases based on which firm's unit-value price is lower; and all firms' pricing strategies are known to all incumbent companies who respond by playing Nash equilibria through untainted self-analyses. For the technical proof of Theorem 5.1, go to the Appendix of this chapter.

Theorem 5.2 In the previously described market, if the consumer surplus is at least as large scale α as the loyal-customer base of some incumbent firms and there is only one new entrant, then there is a threshold magnitude α^* satisfying that when $\alpha \geq \alpha^*$, the expected price of the incumbent firms is higher than that of the entrant and the expected profit of some incumbent firms is lower than that of the entrant.

In practice, Theorem 5.1 explains why incumbent firms have to compete over the market switchers instead of just hold on to their respective bases of loyal consumers by charging them the reservation price. Especially, although the incumbent firms in theory can be risk neutral and attempt to continuously reap their respective profits from their established turfs, they still have to fight over the switchers in the market in order to reduce the size of the switcher segment. By doing so, as implied by Theorem 5.2, they can help control the size of the consumer surplus to below that of the loyal consumers' base of some incumbent firms. If not successful with this effort, a new competition will surely enter the market. And, the entering firm can also potentially make as much profit as some of the incumbent firms.

5.4 Capability Rigidity and Good Firm Performance

Such variables as marketing capabilities, technology, and R&D capabilities, innovation ambidexterity, and competencies related to sensing and responding to changes in the market environment, be they about customers or competitors or both, are considered market-based resources (Lee & Grewal, 2004; Li & Calantone, 1998; Moorman & Slotegraaf, 1999).

The well-known capability-rigidity paradox describes the following tough managerial challenge, resolution of which will help a firm sustain its competitive advantage: Many firms are adept at exploiting existing capabilities, but faltering in simultaneously exploring and developing new capabilities (Dougherty, 1992; Leonard-Barton, 1992; O'Reilly III & Tushman, 2004).

Proposition 5.4 Market orientation of a firm and its capability to act accordingly enable the firm to resolve the capability-rigidity paradox, and consequently helps the firm to sustain its competitive advantage that in turn means superb performance.

From how the paradox is formulated, it follows that the key to the resolution of this paradox is the appearance of one or several organizational factors that can ensure the firm to place simultaneous emphasis on both the exploitation of existing innovation capabilities and the exploration of new ones. The exploitation keeps the firm afloat financially for the present and the near term while the exploration, although it can be riskier and represents a more active process of discovery of new resources and technologies (Noble et al., 2002), helps the firm to pay sufficient attention to the future. Such an equal split of attention between how the business is doing now and how it will be doing in the future surely requires the firm to have sufficient knowledge resource of the current and future customers and competitors. That is, the market orientation of the firm will help resolve or lessen the severity of the capability-rigidity paradox (Atuahene-Gima, 2005; Day, 1994; Hurley & Hult, 1998).

If the firm is seen as a spinning yoyo field fighting against all other spinning pools in the ocean of many competing yoyo fields of the marketplace, it is quite clear that in order for the firm to succeed in its struggle, it has to be able to foretell how everyone spins currently and expectedly in the future. With such knowledge in place, if the firm can also adjust itself accordingly with insightful strategic actions, then of course it will be able to either resolve completely or lessen the severity of the capability-rigidity paradox and sustain its competitive advantage, which consequently leads to a superb performance.

As for why firms with analogous customer and competitor knowledge exhibit different capacities for competence exploitation and exploration, it is really rooted in differences in the interpretations of the knowledge and how the knowledge is coordinated and integrated among functional units beyond knowledge itself (Grant, 1996). That explains why the firm's capability to act accordingly enables the firm to resolve or lessen the severity of the capability-rigidity paradox in different ways, where different interpretations of the knowledge and the corresponding actions taken lead to different levels of business success and failures.

Proposition 5.5 What is most important that underlies good firm performance is the interaction of marketing, R&D, and operational capabilities.

Marketing capability of the firm bridges the firm about what it is and what it can do and the product market regarding what the current and expected focus of firms' competition is and what the consumers want (Jaworski & Kohli, 1993; Kotler & Keller, 2016). So, marketing capability both influences the innovative output of the firm and presents the firm to the market. Now, the R&D capability of the firm converts the acquired market information into conceptual products to meet consumers' evolving demand while making it extremely difficult for competitors to imitate (Irwin & Klenow, 1994). If the firm's R&D capability has a strong track record of consistent innovations that always produce externality benefits associated with its conceptualized products, then such capability will enhance the firm's image and consumers' favorable expectations. Practically, if an appropriate operational capability, the complexity of which makes it imperfectly mobile and imperfectly imitable, practically implements the knowledge gained from the marketing capability and materializes ideas and designs of newer and better products of the R&D capability into sellable goods, then good firm performance will follow (Hayes et al., 1988; McGrath, 2013). Therefore, according to Proposition 5.2, the interaction of marketing, R&D, and operational capabilities represents the most important determinant underlying the good performance of the firm.

5.5 Knowledge Competence, Technological Opportunism, and Innovativeness

By definition, a firm's capability represents a process that enables the firm to deploy its resources (inputs) to materialize its business goals (outputs). So, each firm's capability is an input-output mechanism (Dutta et al., 1999). Jointly the totality of all the firm's capabilities can be very well modeled as a systemic yoyo, where the higher the functional capability—the interaction of the marketing, R&D, and operational capabilities—the firm possesses the more efficiently it is able to deploy its productive inputs to achieve its functional objectives. In other words, the better the yoyo field of the firm is able to predict how other competing fields will behave, to design proactively appropriate ways of reaction, and to implement what is imagined necessary, the stronger (or the better performance) the yoyo field of the firm will spin (or achieve).

Proposition 5.6 Market knowledge competence increases new product advantage, which in turn increases the firm's performance in the product market.

This result is a corollary of Proposition 5.5 because market knowledge competence means such processes that generate and integrate market information and knowledge (Li & Calantone, 1998).

Proposition 5.7 The technological opportunism of a firm impacts the magnitude of radical technology adoption, while future focus and top management's advocacy of new technologies increase technological opportunism of the firm.

From the definition of technological opportunism, which is defined as the sense-and-response capability of a firm in its adoption of radical technologies (Srinivasan et al., 2002), it follows readily that the more technological opportunistic the firm is, the greater magnitude it is for the firm to adopt radical technologies. Because the top management represents the center of the organizational system of the firm, where a slight vibration of the center sends large waves of changes across the entire system (Hall & Fagen, 1956; Lin, 1999), it follows that future focus and top management's advocacy of new technologies increase technological opportunism of the firm.

Since the market invites competition and stimulates innovation (Theorems 5.1 and 5.2), networking relationships with various actors within the firm's environment are critical for information gathering, testing, and the adoption of innovations (Arnold et al., 2011; Romijn & Albaladejo, 2002), the following result reveals the influence of relational resources on innovation.

Proposition 5.8 If a firm is established for the purpose of satisfying a particular market niche through generating a positive cash flow, which can be the consequence of profits made from the marketplace, or investments from investors, or a combination of both, then the relational resource of the firm bears extremely important impact on the innovativeness of the firm.

In fact, the assumption about why the firm exists implies that the firm's yoyo field attracts through various interactions with the outside world whatever inputs necessary for the firm to adjust and to reposition itself in the marketplace in order to survive. These interactions help provide the firm with new and crucial information, knowledge, and additional resources, all of which strengthen the competitiveness and innovativeness of the firm or the intensity of spin of the yoyo structure of the firm. To such theoretical height does this systemic conclusion carry the relevant empirical discoveries (Beugelsdijk & Cornet, 2002; Freel, 2003; Kaufmann & Todtling, 2001; Landry et al., 2002; Ritter & Gemunden, 2003; Romijn & Albaladejo, 2002; Souitaris, 2002), which are either positive or nonsignificant on how relational resource and innovation are correlated. And in this context, the pattern of movement in Fig. 2.2c indicates that in general the innovation process represents a nonlinear evolution of ideas and related processes on how to implement the ideas into marketable products, as empirically confirmed by various authors such as Dosi et al. (1988), Kaufmann and Todtling (2001), Kline and Rosenberg (1986), and Malecki (1997).

5.6 Interaction and Comparison of Resources

Many resources studied in management and marketing research, such as brand, relational assets, knowledge generating capabilities, are intangible. Because intangible resources, such as these resources and similar ones, are valuable, rare, and imperfectly imitable, Proposition 5.1 implies that any firm that possesses intangible resources is likely to develop sustainable competitive advantages.

Due to systemic effects, such as the whole is greater than the sum of parts (Lin, 1999), it is reasonable and logical to imagine that when different resources are placed in service jointly, the outcome is expected to be better than that produced by only employing one of the resources (Kozlenkova et al., 2014). However, when some particular resources interact with each other, undesirable effects can appear. For example, the study of interacting capabilities related to new product development, customer management, and supply-chain management suggests that trade-offs may be involved among different resources (Ramaswami et al., 2009). In particular, capabilities of customer management interact positively with capabilities of new product development but negatively with those of supply-chain management. To this end, we have the following general result:

Proposition 5.9 For two opposing business goals A and B, any interaction between the mutually exclusive sets X and Y of resources, where resources in X lead to the realization of goal A and those in Y help actualize goal B, tends to produce undesirable effects.

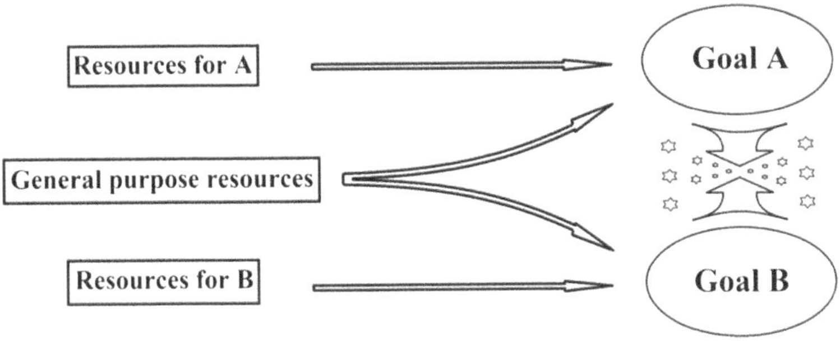

Fig. 5.1 How available resources are divided

For example, capabilities developed only purposefully for the goal of implementing the strategy of exploiting the proven successful competitive advantages generally do not work right with those capabilities introduced only for the purpose of implementing the strategy of constantly discovering new competitive advantages (McGrath, 2013). On the other hand, there are capabilities, such as strong relationships that lead to information sharing, risk taking, and adoption of innovations (Dutta et al., 1999), which can be deployed easily in multiple directions. That is, interaction of the special resources acquired only for materializing either goal A or goal B but not both generally leads to undesirable effects. Figuratively, Proposition 5.9 is depicted in Fig. 5.1, where the sets of resources developed for goal A and for goal B respectively do not overlap, while the general-purpose resources can be redeployed easily to serve different, even opposing goals.

Proposition 5.10 Market-based resources most likely exhibit stronger effects on firm's performance than nonmarket-based resources do.

By definition, market-based resources represent such assets of the firm that can be used to design and implement strategies (Harmancioglu et al., 2009) and that are closely related to the current conditions and expected trends of the market (Lee & Grewal, 2004; Li & Calantone, 1998; Moorman & Slotegraaf, 1999). For example, marketing capabilities, technology and R&D capabilities, innovation ambidexterity, and competencies related to sensing and responding to changes in the market environment, be they about customers or competitors or both, are all market-based resources. Hence, market-based resources most likely exhibit stronger effects on the firm's performance than nonmarket-based resources do because the former help bring the firm's products and services to where the market demand is, while the latter tend to play a supporting role by complementing the former. Empirically, Hooley et al. (2005) demonstrate the truthfulness of a slightly weaker version of Proposition 5.10.

5.7 Resources' Development and Decay

The anecdotes that after Walt Disney died, Disney's brand equity persevered through two decades of neglect, and that Xerox lost its innovative capability regarding photocopiers in only a few years (Collis & Montgomery, 2008) suggest that different types of resources and capabilities develop and decay at different rates. The concept of capability life cycle (Helfat & Peteraf, 2003) articulates how organizational capabilities evolve over time through stages and how heterogeneity in these capabilities appears. By organizational capability, it means the ability for an organization to perform a coordinated set of tasks, utilizing resources, for the purpose of achieving a particular end result, referred to as objective. The following result sheds light on what factors would help to lengthen or shorten an organizational capability's effective life.

Proposition 5.11 If an organizational capability A is developed by a leadership capable of coordinating joint actions to achieve a particular objective B, then capability A will evolve from the stage of founding through that of development to that of maturity until when objective B ceases to exist because it has been accomplished or is altered within the changing market environment. If objective B ceases to exist, then capability A will retire; if objective B is altered, then capability A will evolve into one of the following six branches: retirement (death), retrenchment, renovation, replication, redeployment, and recombination (pooled together with other capabilities).

In fact, when a capable person (or a team of people) identifies a market opportunity, which generally is a market's call and invitation, as shown by Theorems 5.1 and 5.2, which requires people with a special talent to identify, a group of individuals endowed with human capital, social capital, and cognition, is organized to achieve a central objective by a leadership capable of coordinating joint actions (Helfat & Lieberman, 2002; Levinthal & Myatt, 1994). In real life, such group of individuals might very well consist of total strangers. In fact, no matter whether it consists of strangers or not, the natural appearance of the leadership within the newly formed group of talents is guaranteed by Theorem 2.5 of systems science.

Around the identified central objective, a particular organizational capability is formulated or founded by holistically employing the individually endowed talents of the group (or team). When the team searches for viable ways to accomplish the end result of the central objective, an appropriately working capability appears (Winter, 2000). The choice of which viable way to pursue depends on the boundary conditions at the founding stage through learning-by-doing both individually and collectively as a team (Helfat & Peteraf, 2003). This stage of development ends when the team perceives the capability as good enough (Winter, 2000). That is when the capability enters into its maturity stage, where it is maintained and repeatedly exercised.

If at this moment of evolution, the central objective is no longer there or has been achieved successfully, then the organization, established on top of the initial group

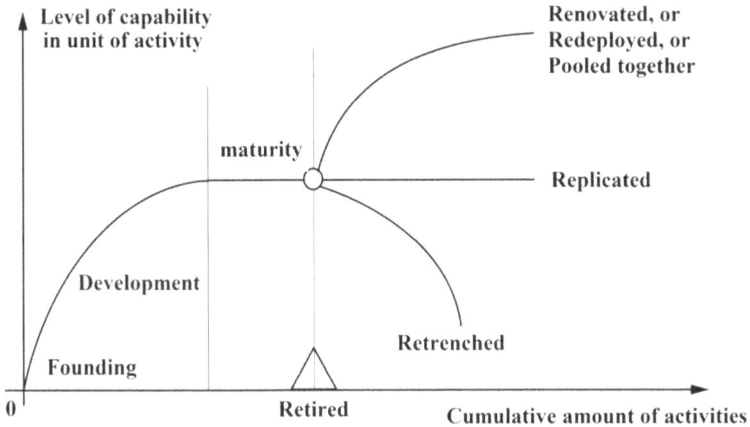

Fig. 5.2 How a capability evolves in a stylized fashion

of individuals, will dissolve or will aim at a different objective. In the former case, the established organizational capability will cease to exist, since it is no longer needed; while in the latter case, the capability will branch out in several directions, Fig. 5.2, when factors external to the capability exert a strong enough impact to alter the current development trajectory of the capability.

These factors could be either internal or external to the organization within which the capability resides. Figure 5.2 depicts the general evolution of a capability starting from the initial founding to the eventual branching. Although two particular moments are clearly labeled, one for the start of maturity and the other for that of branching, in real life, not all capabilities have clearly marked moments of separation between stages of development. As a matter of fact, branching can occur at any stage from that of development onward, depending on the initial endowments available at the founding stage and the central objective the organization aims at achieving. Additionally, the choice of which branch the capability will take on depends on the particular capability and selection effects, external to the capability, and what internal reactions the firm is taking (Helfat & Peteraf, 2003). All the selection effects are rooted in the changing nature of the central objective the capability was initially founded to address. When the central objective evolves with the changing market condition, some selection effects can appear as threats that will make the capability obsolete, while other section effects provide opportunities for the capability to grow and to change.

Proposition 5.12 Only if a firm is capable of using relevant information of the market, the firm's resources are able to increase the firm's performance.

In fact, the assumed increasing performance of the firm, as caused by the firm's resources, has to be a consequence of the fact that the firm's products and/or services are better meeting the demand of the market than those of other competing firms. To better satisfy the demand of the market, the firm's product development must have

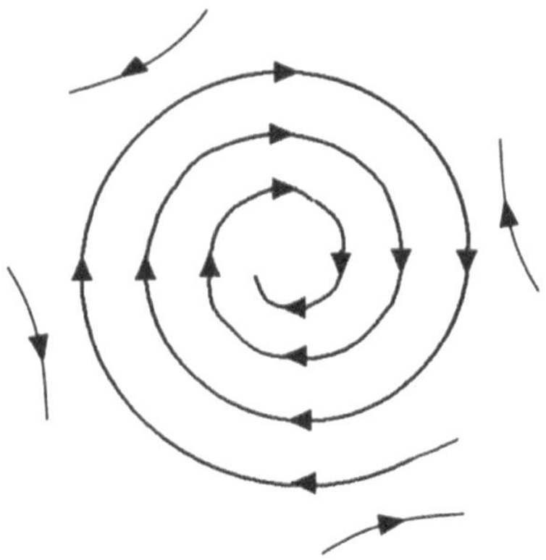

Fig. 5.3 Position its product/service in an advantageous situation

achieved better outcomes in the market competition than other firms do. That, of course, implies that the firm is able to use, either consciously or unconsciously, relevant information of the market, such as what the market is calling for (Theorems 5.1 and 5.2), how competitors are positioned in the marketplace, what the features of the competing products and services are, and what the competitors' pricing strategies are, on its product development, as confirmed by the economics of information (Stigler, 1961), a conclusion of which is that information can facilitate consumer search and increase competitive activities. Empirically, a special case of Proposition 5.12, concerning only technology and marketing resources, was shown by Moorman and Slotegraaf (1999).

Systemically speaking, the firm's increasing performance implies that the yoyo field of the firm moves in advantageous ways when compared to the movement patterns of the yoyo fields of other competing firms (Fig. 5.3).

5.8 Failures of Consistently Exploiting Resources

As for the problem of when a firm could potentially fail its exploitation of available resources, we have

Proposition 5.13 A firm fails to consistently exploit its available resources to achieve superior performance if one or more of the following conditions hold true:

1. The firm does not have any constantly reinforced policy or procedure in place to exploit its available resources;
2. The firm is not market-oriented and/or is unable to react to market invitations;
3. The firm cannot organically (or distinctively) combine marketing and R&D capabilities, supported accordingly by appropriate operations; and
4. The firm cannot sense in a timely fashion what the market is signaling when compared to competing firms.

In fact, for a firm to *consistently* exploit its available resources to achieve superior performance, the firm has to have a mechanism in place that relentlessly pushes the organization towards superb performance. Hence, Proposition 5.1 implies that the firm needs to have policies and procedures, which are constantly in effort, which support the exploitation of its resources. That is, condition 1 holds true.

When the firm is not market-oriented and is unable to react to market invitations (Theorems 5.1 and 5.2) accordingly, it means that the firm cannot consistently adjust its strategies to stay in sync with the market development (McGrath, 2013). When that is the case, the firm will not be able to consistently exploit its available resources to achieve superior performance (Proposition 5.4), even though once in a while the firm's performance could be superb by chance. That is like the situation of a broken clock that can still be correct twice a day by chance. That is, condition 2 holds true.

The reasoning for condition 3 is similar to that of condition 2 above and follows from Proposition 5.5. Condition 4 implies that the firm is mostly behind its competitors in terms of receiving and comprehending market cues; so, Proposition 5.6 implies that the firm will not be able to consistently exploit its available resources to achieve superior performance when compared to its competitors.

In terms of management, the previous proposition can be rewritten as follows:

Proposition 5.14 A firm fails to consistently exploit its available resources to achieve a competitive advantage, if one or more of the following conditions hold true:

1. The firm's leadership lacks the capability needed to maximize the potential of the available resources;
2. The self-interests of the firm's managers diverge from those of the owners; and
3. The philosophical and value systems of the managers limit the financial performance of the firm.

In fact, when condition 1 holds true, then various versions of the four conditions in Proposition 5.12 will become true. So, the conclusion that the firm fails to consistently exploit its available resources to achieve competitive advantage follows.

When either condition 2 or condition 3 is true, conditions 2 and 4 in Proposition 5.13 will likely become true. So, the conclusion of this proposition follows.

A result similar to Proposition 5.14 was also established by Hunt (1997) by using resource advantage theory (Hunt & Morgan, 1995).

Proposition 5.15 If a firm is driven to innovate by its inferior financial performance, then the drive is both endogenous and exogenous to the firm.

In fact, when the firm is driven to innovate due to its inferior financial performance, it means that the firm has employed the feedback mechanism (Lin, 1999), Fig. 5.4, of the market information: it occupies an inferior position in the marketplace relative to other competitors. That generally means that the firm possesses a comparative disadvantage when compared to its competitors. So, the drive for innovation, in this case, comes from the firm's internal analysis of its performance. In particular, by introducing desired innovation based on self-analysis of where the

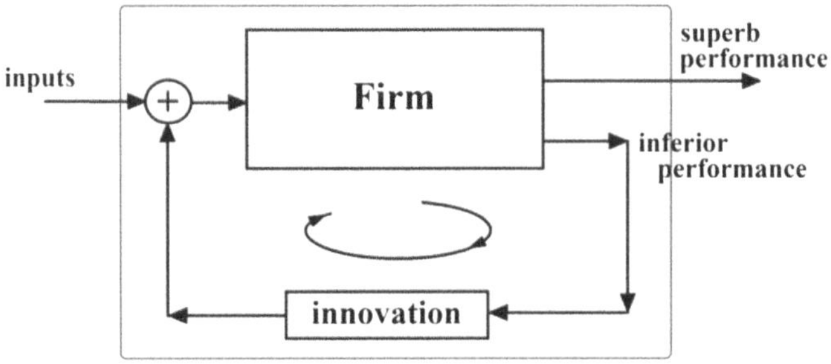

Fig. 5.4 The feedback mechanism

firm's disadvantage is located in the operation of the firm, the earlier inferior performance will improve and be soon replaced by superior performance. Here, the feedback loop represents a way for the firm to become more competitive than before.

On the other hand, if the firm takes in the market cue (Theorems 5.1 and 5.2) regarding what is demanded by the consumers, then the consequent innovation taken up by the firm will be seen as exogenous to the firm.

5.9 A Few Final Words

This chapter uses the thinking logic of systems science in general and the systemic yoyo model in particular to look at various research topics of management and marketing based on the resource-based theory. Because the issues addressed in this chapter are about characteristics and operations of organizations, we are able to beautifully employ the methodology of systems science instead of empirical analysis to establish general and practically useful results unique to domains of management and marketing from the angle of resources. In particular, this chapter develops the following results:

1. Conditions under which a firm is likely or certainly to develop sustainable competitive advantages are given;
2. How the capability-rigidity paradox can be resolved;
3. What could potentially lead to improved firm performance;
4. What impacts the magnitude of radical technology adoption;
5. How relational resource impacts the innovativeness of the firm;
6. When interactions of resources could lead to undesirable effects;
7. When a capability would evolve from the stage of founding to maturity and when it would cease to exist; and
8. When a firm would fail to consistently exploit its resources.

Other than establishing these general results, which is impossible when only empirical analysis is employed, another important contribution this chapter makes to the literature is the introduction of the thinking logic and methodology of systems science into the study of management science and marketing. When compared to methods of empirical analysis, the thinking logic and methodology of systems science avoid all limitations and uncertainties of data mining. That is the very reason why we are able in this chapter to establish general results that will be practically applicable to produce tangible and reliable outcomes.

In other words, relevant studies in the literature, be they conceptual or empirical, are generally affected adversely by the limitations and uncertainties involved in the collection and analysis of data. So, when business managers have to decide what actions to take in their daily business operations based on such studies of limited validity, they face a great deal of uncertainty in terms of what they can expect as the consequence of their chosen actions. Opposite to the theoretical and practical uncertainties that exist in the literature, this work uses logical reasoning based on Axioms 4.1–4.3 to establish results that are generally applicable in real life.

As for the limitation of this chapter, all conclusions are derived on the definitions of (respectively, valuable, rare, and exploitable) resource and competitive advantage, where the buzz words and phrases are *asset* and that the firm makes "*more economic value than other firms.*" That means that this chapter does not include situations where the firm only wants to "stay alive" instead of creating more economic value than other firms do. In real life, many firms actually fall into this category. As for the buzz word *asset*, if the firm of concern is seen as a system, then two natural questions arise: Which systemic part(s) or process(s) of the firm can be treated as assets is not clear, because the assets in one person's eyes might not amount to much or anything important at all in another person's eyes? And among all the identified assets that have contributed to the success of a particular strategy, which ones have played more important roles than others? These and other related questions need to be studied in future works, maybe first empirically and then theoretically.

Appendix: Proofs of Theorems 5.1 and 5.2

To facilitate the necessary proofs, let us assume that the given market is served by n, a fixed number, incumbent firms with their horizontally differentiated offers. Incumbent firm i has a base of loyal customers of magnitude β_i, where each loyal customer only makes purchase from her favorite firm i if its price is not more than their reservation price M_i, $i = 1, 2, \ldots, n$. Assume firm i's cost of production is C_i. Then, Propositions 3.2, 3.4, 3.6, and 3.8 indicate that in Nash equilibrium,

- The price P_i, the production costs C_i, and loyal customers' reservation price M_i can be normalized so that $0 \leq P_i \leq 1$, $C_i = 0$ and $M_i = 1$;
- The lower and upper boundary conditions L_i and M_i and the magnitude of firm i's loyal-customer base β_i can be normalized so that $L_i = 0$, $M_i = 1$, and $0 \leq \beta_i = \alpha = 1$. Then $\beta = 1 - n\alpha$ is the normalized magnitude of the consumer surplus;
- The game the competing firms play does not have any portfolio of pure strategies.

Proof of Theorem 5.1 Assume all parameters are normalized as mentioned above and that the entrant randomizes its price between cost 0 and maximum expected acceptable price 1. Now, Theorem 3.2 provides the proof for the implication of 1) → 2). In the rest of this proof, it shows the opposite implication 2) → 1).

Suppose that some new firms enter the said market occupied by n incumbent firms. Without loss of generality, assume that there is only one such entrant. Each of the n incumbents sets its price after taking into account the price of the entrant and those of all other incumbent firms. So, the equilibrium indifference condition of Firm k is

$$\alpha \times P + \beta \times P \prod\nolimits_{j \neq k}^{n} (1 - P)\left[1 - F_j(P)\right] = \alpha \times 1. \tag{5.1}$$

So, for the incumbent firm, Eq. (5.1) leads to their symmetric equilibrium pricing strategy as follows:

$$F(P) = 1 - \left(\frac{\alpha}{\beta P}\right)^{\frac{1}{n-1}}. \tag{5.2}$$

However, for $F(P)$, as defined in Eq. (5.2), to be a well-defined probability distribution, the following must be true

$$1 - \left(\frac{\alpha}{\beta P}\right)^{\frac{1}{n-1}} \geq 0, \tag{5.3}$$

which implies that $\alpha/\beta \leq P \leq 1$. That is, the consumer surplus $\beta \geq \alpha$.

Proof of Theorem 5.2 The proof of Theorem 3.2 has located one $\alpha_1^* \in (0, 1/(n+1))$ so that when $\alpha \geq \alpha_1^*$, the expected profit of the entrant is greater than at least one incumbent firm. If we can show the existence of another threshold value α_2^* such that when $\alpha \geq \alpha_2^*$, the expected price of the incumbent firms is higher than that of the entrant, we can simply set $\alpha^* = \max\left\{\alpha_1^*, \alpha_2^*\right\}$ to finish the proof of Theorem 5.2. The rest of this proof shows the existence of α_2^*. To this end, let us continue the proof of Theorem 5.1. The expected price of one incumbent firm is given as follows:

$$E_m(P) = \int_{-\infty}^{+\infty} P \times F'(P)dP = \int_{\frac{\alpha}{\beta}}^{1} P \times F'(P)dP + 1 \times \left(\frac{\alpha}{\beta}\right)^{\frac{1}{n-1}} =$$

$$= \begin{cases} \left(\dfrac{\alpha}{\beta}\right)^{\frac{1}{n-1}} + \dfrac{\left(\frac{\alpha}{\beta}\right)^{\frac{1}{n-1}} - \dfrac{\alpha}{\beta}}{n-2}, & \text{if } n \geq 3 \\ \dfrac{\alpha}{\beta}\left(1 - \ln\dfrac{\alpha}{\beta}\right), & \text{if } n = 2 \end{cases} \tag{5.4}$$

And the expected price of the entrant is $E_e(P) = 1/2$. From Eq. (5.4), it follows that

$$\frac{\partial}{\partial \alpha} E_m(P) = \begin{cases} \frac{1}{(n-2)\beta^2}\left[\left(\frac{\alpha}{\beta}\right)^{\frac{1}{n-1}-1} - 1\right], & \text{if } n \geq 3 \\ \frac{1}{\beta^2}\left(-ln\,\frac{\alpha}{\beta}\right), & \text{if } n = 2 \end{cases} \tag{5.5}$$

So, $\frac{\partial}{\partial \alpha} E_m(P) > 0$, for $0 < \alpha \leq 1/(n+1)$ because $\alpha \leq \beta = 1 - n\alpha$. That is, $E_m(P)$ is an increasing function on the interval $(0, 1/(n+1)]$. It can be readily checked that

$$E_m(P)\big|_{\alpha/\beta=1} = 1 > 1/2 = E_e(P).$$

Hence, there is $\alpha_2^* \in (0, 1/(n+1))$ such that when $\alpha \geq \alpha_2^*$, $E_m(P) > E_e(P)$.

References

Aldrich, H. E., & Zimmer, C. (1986). *Entrepreneurship through social networks*. University of Illinois at Urbana-Champaign: Academy for entrepreneurial leadership historical research reference in entrepreneurship. Retrieved from http://ssrn.com/abstract=1497761.

Arifin, Z., & Frmanzah. (2015). The effect of dynamic capability to technology adoption and its determinant factors for improving firm's performance: Toward a conceptual model. *Procedia - Social and Behavioral Sciences, 207*(October), 786–796.

Arnold, T., Fang, E., & Palmatier, R. (2011). The effects of customer acquisition and retention orientations on a firm’s radical and incremental innovation performance. *Journal of the Academy of Marketing Science, 39*(2), 234–251.

Atuahene-Gima, K. (2005). Resolving the capability-rigidity paradox in new product innovation. *Journal of Marketing, 69*(4), 61–83.

Barney, J., & Clark, D. (2007). *Resource-based theory: Creating and sustaining competitive advantage*. Oxford University Press.

Barney, J., & Hesterly, W. (2012). *Strategic management and competitive advantage: Concepts and cases* (4th ed.). Pearson.

Barney, J., Ketchen, D., & Wright, M. (2011). The future of resource-based theory: Revitalization or decline? *Journal of Management, 37*(5), 1299–1315.

Baron, R. A., & Markman, G. D. (2003). Beyond social capital: The role of entrepreneurs’ social competence in their financial success. *Journal of Business Venturing, 18*(1), 41–60.

Bernardo, R., Galina, S. V. R., & de Pádua, S. I. D. (2017). The BPM lifecycle: How to incorporate a view external to the organization through dynamic capability. *Business Process Management Journal, 23*(1), 155–175.

Beugelsdijk, S., & Cornet, M. (2002). A far friend is worth more than a good neighbour: Proximity and innovation in a small country. *Journal of Management and Governance, 6*(2), 169–188.

Boulding, W., & Christen, M. (2003). Sustainable pioneering advantage? Profit implications of market entry order. *Marketing Science, 22*(3), 371–392.

Chadwick, C., Super, J. F., & Kwon, K. (2015). Resource orchestration in practice: CEO emphasis on SHRM, commitment-based HR systems, and firm performance. *Strategic Management Journal, 36(3)*, 360–376.

Collis, D. J., & Montgomery, C. A. (2008). Competing on resources. *Harvard Business Review, 86* (7/8), 140–150.
Covin, J. G., Eggers, F., Kraus, S., Cheng, C. F., & Chang, M. L. (2016). Marketing-related resources and radical innovativeness in family and non-family firms: A configurational approach. *Journal of Business Research, 69*(12), 5620–5627.
Crook, T., Ketchen, D., Combs, J., & Todd, S. (2008). Strategic resources and performance: A meta-analysis. *Strategic Management Journal, 29*(11), 1141–1154.
Day, G. S. (1994). The capabilities of market-driven organizations. *Journal of Marketing, 58* (October), 37–52.
Dosi, G., Freeman, C., Nelson, R., Silverberg, G., & Soete, L. (Eds.). (1988). *Technical change and economic theory*. Pinter.
Dougherty, D. (1992). A practice-centered model of organizational renewal through product innovation. *Strategic Management Journal, 13*(Spring), 77–92.
Dutta, S., Narasimhan, O., & Rajiv, S. (1999). Success in high-technology markets: Is marketing capability critical? *Marketing Science, 18*(4), 547–568.
Fang, E., Palmatier, R., & Grewal, R. (2011). Effects of customer and innovation asset configuration strategies on firm performance. *Journal of Marketing Research, 48*(3), 587–602.
Forrest, Y. L. J., Buttermore, J., & Wajda, T. A. (2017). At Nash equilibrium when new market competitions appear? *Kybernetes, The International of Cybernetics, Systems and Management Science, 46*(2), 256–271.
Forrest, J. Y.-L., Nicholls, N., Schimmel, K., & Liu, S. F. (2020). *Managerial decision making: A holistic approach*. Springer.
Freel, M. S. (2003). Sectoral patterns of small firm innovation, networking and proximity. *Research Policy, 32*, 751–770.
Fuentes-Fuentes, M. D. M., Bojica, A. M., Ruiz-Arroyo, M., & Welter, F. (2017). Innovativeness and business relationships in women-owned firms: The role of gender stereotypes. *Canadian Journal of Administrative Sciences, 34*(1), 63–76.
Grant, R. M. (1996). Prospering in dynamically-competitive environments: Organizational capability and knowledge integration. *Organization Science, 7*(July), 375–387.
Hall, A. D., & Fagen, R. E. (1956). Definitions of systems. *General Systems, 1*, 18–28.
Harmancioglu, N., Droge, C., & Calantone, R. (2009). Strategic fit to resources versus NPD execution proficiencies: What are their roles in determining success? *Journal of the Academy of Marketing Science, 37*(3), 266–282.
Hayes, R. H., Wheelwright, S. C., & Clark, K. B. (1988). *Dynamic manufacturing*. The Free Press.
Helfat, C. E., & Lieberman, M. B. (2002). The birth of capabilities: Market entry and the importance of pre-history. *Industrial and Corporate Change, 11*(4), 725–760.
Helfat, C. E., & Peteraf, M. A. (2003). The dynamic resource–based view: Capability lifecycles. *Strategic Management Journal, 24*(10), 997–1010.
Hooley, G., Greenley, G., Fahy, J., & Cadogan, J. (2005). Market-focused resources, competitive positioning and firm performance. *Journal of Marketing Management, 17*(5/6), 503–520.
Hughes, M., Martin, S., Morgan, R., & Robson, M. (2010). Realizing product-market advantage in high-technology international new ventures: The mediating role of ambidextrous innovation. *Journal of International Marketing, 18*(4), 1–21.
Hunt, S. (1997). Resource-advantage theory: An evolutionary theory of competitive firm behavior? *Journal of Economic Issues, 31*(1), 59–77.
Hunt, S., & Morgan, R. (1995). The comparative advantage theory of competition. *Journal of Marketing, 59*(4), 1–15.
Hurley, R. F., & Hult, G. T. M. (1998). Innovation, market orientation, and organizational learning: An integration and empirical examination. *Journal of Marketing, 62*(July), 42–54.
Irwin, D. A., & Klenow, P. J. (1994). Learning-by-doing spillovers in the semiconductor industry. *Journal of Political Economy, 102*(6), 1200–1227.
Jap, S. (2001). Perspectives on joint competitive advantages in buyer-supplier relationships. *International Journal of Research in Marketing, 18*(1), 19–35.

Jaworski, B. J., & Kohli, A. K. (1993). Market orientation: Antecedents and consequences. *Journal of Marketing, 57*(July), 53–70.
Jeng, D. J., & Pak, A. (2016). The variable effects of dynamic capability by firm size: The interaction of innovation and marketing capabilities in competitive industries. *International Entrepreneurship and Management Journal, 12*(1), 115–130.
Kaleka, A. (2011). When exporting manufacturers compete on the basis of service: Resources and marketing capabilities driving service advantage and performance. *Journal of International Marketing, 19*(1), 40–58.
Kaufmann, A., & Todtling, F. (2001). Science-industry interaction in the process of innovation: The importance of boundary-crossing between systems. *Research Policy, 30*, 791–804.
Kline, S. J., & Rosenberg, N. (1986). An overview of innovation. In R. Landau & N. Rosenberg (Eds.), *The positive sum strategy: Harnessing technology for economic growth* (pp. 275–307). National Academy Press.
Kotler, P., & Keller, K. L. (2016). *A framework for marketing management* (6th ed.). Pearson.
Kozlenkova, I. V., Samaha, S. A., & Palmatier, R. W. (2014). Resource-based theory in marketing. *Journal of the Academy of Marketing Science, 42*, 1–21.
Lages, L. F., Silva, G., & Styles, C. (2009). Relationship capabilities, quality, and innovation as determinants of export performance. *Journal of International Marketing, 17*(4), 47–70.
Landry, R., Amara, N., & Lamari, M. (2002). Does social capital determine innovation? To what extent? *Technological Forecasting and Social Change, 69*, 681–701.
Lee, R., & Grewal, R. (2004). Strategic responses to new technologies and their impact on firm performance. *Journal of Marketing, 68*(4), 157–171.
Leonard-Barton, D. (1992). Core capabilities and core rigidities: A paradox in managing new product development. *Strategic Management Journal, 13*(Summer), 111–125.
Levinthal, D. A., & Myatt, J. (1994). Co-evolution of capabilities and industry: The evolution of mutual fund processing. *Strategic Management Journal, 15*(Winter Special Issue), 45–62.
Li, T., & Calantone, R. (1998). The impact of market knowledge competence on new product advantage: Conceptualization and empirical examination. *Journal of Marketing, 62*(4), 13–29.
Lin, Y. (1999). *General systems theory: A mathematical approach*. Plenum and Kluwer Academic Publishers.
Lin, Y. (2009). *Systemic yoyos: Some impacts of the second dimension*. Auerbach Publication, an imprint of Taylor and Francis.
Lin, Y., & OuYang, S. C. (2010). *Irregularities and prediction of major disasters*. CRC Press, an imprint of Taylor and Francis.
Liu, Y., & Liang, L. T. (2015). Evaluating and developing resource-based operations strategy for competitive advantage: An exploratory study of Finnish high-tech manufacturing industries. *International Journal of Production Research, 53*(4), 1019–1037.
Lonial, S. C., & Carter, R. E. (2015). The impact of organizational orientations on medium and small firm performance: A resource-based perspective. *Journal of Small Business Management, 53*(1), 94–113.
Luo, X., Griffith, D., Liu, S., & Shi, Y. (2004). The effects of customer relationships and social capital on firm performance: A Chinese business illustration. *Journal of International Marketing, 12*(4), 25–45.
Malecki, E. J. (1997). *Technology and economic development*. Addison-Wesley Longman.
McGrath, R. G. (2013). *The end of competitive advantage: How to keep your strategy moving as fast as your business*. Harvard Business Review Press.
Moorman, C., & Slotegraaf, R. (1999). The contingency value of complementary capabilities in product development. *Journal of Marketing Research, 36*(2), 239–257.
Morgan, N., Slotegraaf, R., & Vorhies, D. (2009). Linking marketing capabilities with profit growth. *International Journal of Research in Marketing, 26*(4), 284–293.
Noble, C. H., Sinha, R. K., & Kumar, A. (2002). Market orientation and alternative strategic orientations: A longitudinal assessment of performance implications. *Journal of Marketing, 66* (October), 25–39.

O'Reilly, C., III, & Tushman, M. L. (2004). The ambidextrous organization. *Harvard Business Review, 82*(April), 74–81.

Penrose, E. (1959). *The theory of the growth of the firm*. Blackwell.

Peteraf, M., & Barney, J. (2003). Unraveling the resource-based tangle. *Managerial and Decision Economics, 24*(4), 309–323.

Porter, M. (1979). How competitive forces shape strategy. *Harvard Business Review, 57*(2), 137–145.

Ramaswami, S., Srivastava, R., & Bhargava, M. (2009). Market-based capabilities and financial performance of firms: Insights into marketing's contribution to firm value. *Journal of the Academy of Marketing Science, 37*(2), 97–116.

Richey, G., Tokman, M., & Dalela, V. (2010). Examining collaborative supply chain service technologies: A study of intensity, relationships, and resources. *Journal of the Academy of Marketing Science, 38*(1), 71–89.

Ritalam, P., Heiman, B., & Hurmelinna-Laukkanen, P. (2016). The need for speed—Unfamiliar problems, capability rigidity, and ad hoc processes in organizations. *Industrial and Corporate Change, 25*(5), 757–777.

Ritter, T., & Gemunden, H. G. (2003). Network competence: Its impact on innovation success and its antecedents. *Journal of Business Research, 56*, 745–755.

Romijn, H., & Albaladejo, M. (2002). Determinants of innovation capability in small electronics and software firms in Southeast England. *Research Policy, 31*, 1053–1067.

Ruiz-Ortega, M., & García-Villaverde, P. (2008). Capabilities and competitive tactics influences on performance: Implications of the moment of entry. *Journal of Business Research, 61*(4), 332–345.

Slotegraaf, R., & Dickson, P. (2004). The paradox of a marketing planning capability. *Journal of the Academy of Marketing Science, 32*(4), 371–385.

Slotegraaf, R., Moorman, C., & Inman, J. (2003). The role of firm resources in returns to market deployment. *Journal of Marketing Research, 40*(3), 295–309.

Souitaris, V. (2002). Technological trajectories as moderators of firm-level determinants of innovation. *Research Policy, 31*, 877–898.

Srinivasan, R., Lilien, G., & Rangaswamy, A. (2002). Technological opportunism and radical technology adoption: An application to e-business. *Journal of Marketing, 66*(3), 47–60.

Stigler, G. (1961). The economics of information. *Journal of Political Economy, 65*(June), 213–225.

Tomer, J. (2003). Personal capital and emotional intelligence: An increasingly important intangible source of economic growth. *Eastern Economic Journal, 29*, 453.

Vorhies, D., & Morgan, N. (2005). Benchmarking marketing capabilities for sustainable competitive advantage. *Journal of Marketing, 69*(1), 80–94.

Wernerfelt, B. (1984). A resource-based view of the firm. *Strategic Management Journal, 5*(2), 171–180.

Winter, S. G. (2000). The satisficing principle in capability learning. *Strategic Management Journal, 21*(10–11), 981–996.

Wright, M., Piva, E., Mosey, S., & Lockett, A. (2009). Academic entrepreneurship and business schools. *The Journal of Technology Transfer, 34*, 560–587.

Zona, F., Gomez-Mejia, L. R., & Withers, M. C. (2016). Board interlocks and firm performance: Toward a combined agency–resource dependence perspective. *Journal of Management, 44*(2), 589–618.

Zou, S., Fang, E., & Zhao, S. (2003). The effect of export marketing capabilities on export performance: An investigation of Chinese exporters. *Journal of International Marketing, 11*(4), 32–55.

Part II
Demand- and Supply-Side Strategies

Chapter 6
The Supply-Chain Ecosystem of a Firm

Abstract This chapter, which is mainly based on (Forrest et al., 2019, *Proceedings of the 2019 annual conference of decision sciences institute*, New Orleans, LA), investigates holistically such a business system that consists of market invitations, the ecosystem of a focal firm's supply chain, the firm's innovative recognition of the market signals, and challenges consequently imposed on the ecosystem. By employing the thinking logic and methodology of systems science, we establish a series of general conclusions regarding innovative firms' internal and environmental challenges. In particular, we show that a firm's upstream component challenges and downstream complement challenges place respectively a cap on how much "value" the firm can create and capture. Thus, the more capable a focal firm is at identifying these challenges, the more adept this firm will be in creating and sustaining marketplace advantage as well as imposing greater challenges on its ecosystem. A firm's strong capability to learn and act accordingly is, in itself, an advantage. Upon the conclusion of our analysis, practical managerial significance of this chapter and open questions for future research are given.

Keywords Capability · Market invitation · Performance · Systemic yoyo · Value creation · Value capture

6.1 Introduction

Systemically speaking, when a firm recognizes a market invitation in an epoch-making way, the firm has an opportunity to develop a significant competitive advantage over its competitors. However, other than internal reasons other players in the ecosystem of the firm's supply chain greatly affect the enhanced performance the firm consequently enjoys (Adner et al., 2013). For example, the difficult daily commuting problem, confronting those individuals who live in a suburban area and commute to work inside a major US metropolitan, necessarily invites politicians and entrepreneurs alike to come up with a solution. One possible resolution to this challenge, for example, is to design, produce, and offer flying cars (Lemoussu et al., 2018) to these frustrated commuters. To manufacture realistically such cars,

J. Y.-L. Forrest, Y. Liu, *Value in Business*, Contributions to Management Science,
https://doi.org/10.1007/978-3-030-82898-1_6

a firm has to have appropriate components supplied to it in its assembling of the imagined product. However, the value created for commuters is not practically realizable until the appropriate road conditions and air traffic controls, as key complements, are constructed and installed in order for such cars to move and fly around freely and orderly. So, a natural question arises: After a firm receives a market invitation for innovation, how will the consequent success of the firm be dependent on the success of other players in its ecosystem?

This chapter seeks to address this question, which is important both theoretically and practically, by randomly selecting a firm as our focal firm and by using the thinking logic and methodology of systems science. To accomplish this end, we first distinguish suppliers and complementors so that suppliers provide their outputs for the focal firm to integrate in its effort to offer a complete product to its customers and complementors facilitate necessary conditions for customers to fully utilize the product. Then we develop the following main conclusions among others:

- The upper limits of value the focal firm can create and capture are respectively constrained by the challenges facing its upstream components and downstream complementors; and
- The more adept a focal firm is at learning and acting, the greater the challenges that will face the ecosystem of the firm's supply chain.

The remainder of this chapter is organized as follows. Section 6.2 provides a literature review while showing how this chapter contributes to the established knowledge. Section 6.3 introduces the well-known dishpan experiment and other necessary background information to make this chapter self-contained. Section 6.4 constructs the supply-chain ecosystem of an arbitrarily chosen business firm. Section 6.5 studies the relationship between the firm's capability of learning and its consequent challenges imposed on the ecosystem. Section 6.6 concludes the presentation of this chapter.

6.2 Literature Review

Due to the power of systems thinking and methodology, this chapter contributes to an array of different literatures, simultaneously covering business ecosystems, value capture and creation, learning capabilities, competitive advantage, and customer adoption of innovation. In this section, we review relevant works in each of these discrete literatures.

In the literature of ecosystems, Adner et al. (2013) publish a collection of papers addressing how system players collaborate and compete with each other. Refuting the common beliefs that knowledge ecosystems in technology hotspots lead to business ecosystems in which there are competitive advantages for each of the players in the ecosystem, Clarysse et al. (2014) find a particular case where a structured knowledge ecosystem is concentrated around only a number of central players. The result is an imagined business ecosystem that is nearly nonexistent. When a business ecosystem includes well-established companies and new ventures,

Zahra and Nambisan (2012) claim that the success of that ecosystem requires collaboration and competition that demands strategic thinking to leverage a firm's resources and capabilities. Kapoor and Agarwal (2017) study how platform firms in a business ecosystem orchestrate the ecosystem's functioning by providing platforms and setting the rules for participation by complementor firms. They develop a reason for how the ecosystem's structural and evolutionary features may shape the extent to which participating complementor firms can sustain their superior performance. Instead of using technical interdependencies, Hellström et al. (2015) explore how firms in an energy business ecosystem collaborate by using a business model lens, pointing to the importance of locating the factors that drive the business models of collaborating firms to facilitate system transitions and change in the logic of an industry. Attour and Lazaric (2018) emphasize that transformation of a knowledge ecosystem can lead to the emergence of a technological platform that provides the resources required especially for firm startup. By conceptualizing the construct of ecosystems, Adner (2017) offers an examination of the relationship between ecosystems and a host of other constructs, such as business models, platforms, coopetition, multi-sided markets, networks, technology systems, supply chains, and value networks.

In the literature on value creation and capture, Chemmanur et al. (2014) analyze how entrepreneurial firms differ when they are corporate venture capital (CVC) based or independent venture capital (IVC) based in terms of their innovativeness. What is found is that CVC-backed firms are more innovative, although they are younger, riskier, and less profitable than those backed by IVC. Kagermann (2015) discusses the impact, challenges, and opportunities of digitization and finds that the two key instruments for greater value creation in the Age of Industry 4.0 are platform-based cooperation and a dual innovation strategy. By looking at how the usage of big data analytics (BDA) affects value creation, Chen et al. (2015) conceptualize BDA use as an information processing capability that brings competitive advantage to organizations and identify paths through which factors influence the actual usage of BDA.

In terms of learning capability, Camps et al. (2016) test hypotheses that link organizational learning capability (OLC) and individual employee performance, OLC and employee flexibility, and employee flexibility and individual performance, and show that employee flexibility fully mediates the relationship between OLC and individual performance under the influence of environmental turbulence. Altinay et al. (2016) investigate the interface between OLC, entrepreneurial orientation (EO), and small business performance, and find a positive relationship between EO and sales and market share growth and OLC and EO, but not between EO and employment growth. Kalmuk and Acar (2015) emphasize the mediating role of OLC on the relationship between innovation and company's performance and attempt to demonstrate how multiple types of innovations could contribute to firm performance through OLC. Sung and Choi (2014) examine how training and development affect organizational innovation and suggest that investments in training and development do exert influence on firms' innovative performance through the promotion of various learning practices.

Regarding performance and competitive advantages, Yadav et al. (2017) look at the association between a firm's environmental efforts and the sustainability of its competitive advantage through analyzing how changes in the firm's environmental performance affect the persistence of its profitability growth. These authors find that environmental resources permit superior financial performing firms to maintain their competitive advantage, while complementing poorly performing firm's efforts to recover. This chapter empirically supports the proposition that economic value can be created by benefiting the environment. Prajogo and Oke (2016) examine the effect of human capital (HC) on service innovation advantage (SIA) and business performance (BP), and how external environmental factors play a role. These authors show that HC is positively correlated to the creation of value or SIA which in turn results in rent generation for firms, and that the effect of SIA on BP is influenced by environmental factors such that it is strengthened by dynamic environments and weakened by competitive environments. Chang et al. (2016) consider the inconsistent relationship between supply-chain integration (SCI) and firm performance by proposing that the inconsistency may be associated with selection bias, failure to consider the mediating routes of how SCI affects financial performance, and lack of investigation of moderators. Their findings confirm the early claim that each dimension of SCI indeed improves financial performance.

In terms of market adaptation of innovation, Oliveira et al. (2016) aim to identify determinants of mobile payment adoption by proposing a new model. These researchers find that compatibility, perceived technology security, performance expectations, innovativeness, and social influence have significant effects on consumer adoption. This work provides a basis for further refinement of individual models of acceptance. Bilgicer et al. (2015) study the role of social contagion in customer adoption of new sales channels by focusing on two aspects of social contagion (i.e., local contagion and homophily), as they influence a consumer's preference for either an Internet channel or a brick-and-mortar store. Bilgicer et al. find that social contagion does indeed play a major role with long-term versus short-term customers being less inclined to consider alternate sales channels. By recognizing the importance of downstream activities of the innovation process, specifically marketing and commercialization, Brem and Viardot (2015) consider the adoption of innovations by the market by providing conceptual insights and manners that appear to stimulate and facilitate the adoption of every kind of innovation. Laukkanen (2016) posits that all innovations meet consumer resistance and overcoming this unfriendliness needs to occur prior to product adoption. He examines how five adoption barriers—usage, value, risk, tradition, and image—and three consumer demographics—gender, age, and income—influence consumer adoption versus rejection decisions in Internet and mobile banking. Laukkanen finds that value is the strongest barrier, while image slows adoption; and that both gender and age significantly predict adoption and rejection decisions.

This chapter contributes to the literature in several different ways. First, it establishes general propositions that are practically applicable with reliability and certainty. Second, it introduces a new methodology for investigating important issues relating to the distribution of challenges across the ecosystem of a focal firm. Third, this chapter develops a general theory of business ecosystems without suffering from the constraints and limitations of anecdotes and data analyses (or data mining). In other words, the results established in this chapter are universally true within the system of our cohesive theory constructed on top of a few basic assumptions and logical reasoning.

6.3 Preparation

In order to understand how the yoyo model (Fig. 2.2) can be applied to the studies of evolution and development of systems, organizations, etc., this subsection will look at the well-known dishpan experiment.

Dave Fultz, University of Chicago, and his colleagues in the late 1950s (Fultz et al., 1959) constructed the following experiment: they partially filled a cylindrical vessel with water, placed it on a rotating turntable, and subjected to heating near the periphery and cooling near the center. The bottom of the container is intended to simulate one hemisphere of the earth's surface, the water the air above this hemisphere, the rotation of the turntable simulates the earth's rotation, the heating and cooling simulate the excess external heating of the atmosphere in low latitudes and the excess cooling in high latitudes.

In order to observe flow patterns at the upper surface of the water, which was intended to simulate atmospheric motion at high elevations, Fultz and his colleagues sprinkled some aluminum powder. A special camera that effectively rotated with the turntable took time exposures so that a moving aluminum particle would appear as a streak and sometimes each exposure ended with a flash, which could add an arrowhead to the forward end of each streak. The turntable generally rotated counterclockwise, as does the earth when viewed from above the North Pole.

Even though everything in the experiment was arranged with perfect symmetry about the axis of rotation, such as no impurities added in the water and the bottom of the container was kept flat, Fultz and his colleagues observed more than they had bargained for. First, both expected flow patterns (as shown in Fig. 6.1) appeared, and the choice depended on the speed of the turntable's rotation and the intensity of the heating. Briefly, with fixed heating, a transition from circular symmetry (Fig. 2.1a) would take place as the rotation increased past a critical value. With a sufficiently rapid, but fixed rate of rotation, a similar transition would occur when the heating reached a critical strength, while another transition back to symmetry would occur when the heating reached a still higher critical strength.

A few years before the previous dishpan experiment was conducted in Chicago, Raymond Hide, Cambridge University, did a similar experiment in England (Hide, 1953). Instead of a dishpan, Hide used two concentric cylinders with a fluid placed in the ring-shaped region between the cylinders. He discovered similar transitions between symmetric and asymmetric flow patterns. In this experiment, the asymmetric flow was often regular and consisted of a chain of apparently identical waves, which would travel around the ring-shaped region without changing their shapes.

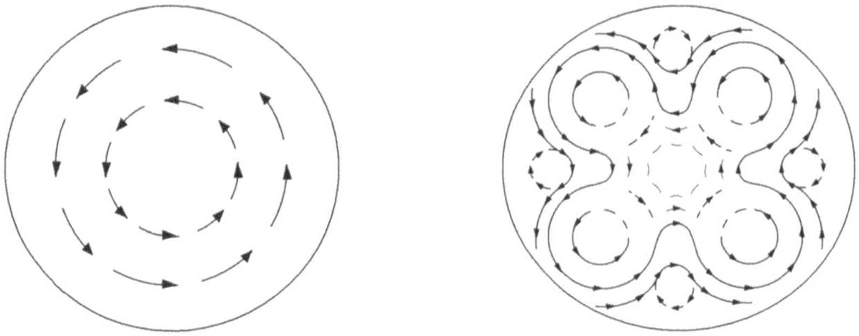

(a) Symmetric flow at the upper surface (b) Asymmetric flow at the upper surface

Fig. 6.1 Patterns observed in Fultz's dishpan experiment. (**a**) Symmetric flow at the upper surface. (**b**) Asymmetric flow at the upper surface

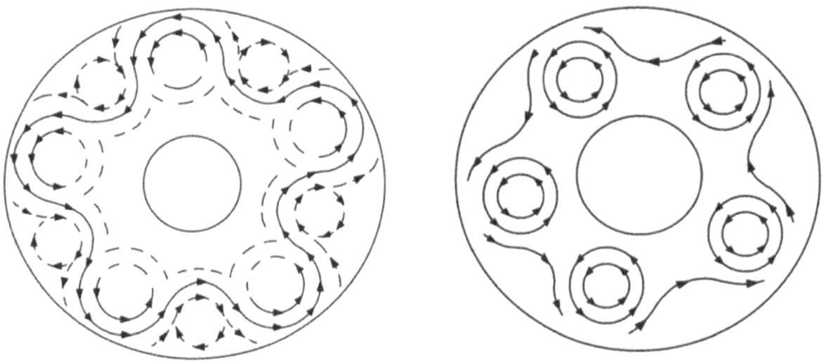

(a) One flow pattern (b) Another flow pattern

Fig. 6.2 Streak patterns of the flow at the upper surface of Hide's dishpan experiment. (**a**) Is followed by (**b**) after 8 rotations; and (**b**) is followed by (**a**) after another 8 rotations. (**a**) One flow pattern. (**b**) Another flow pattern

What is more remarkable about what Hide found is that a chain of identical waves would appear. But, as they traveled along, they would alter their shapes in unison in a regular periodic fashion and after many rotations of the turntable, they would regain their original shape and then repeat the cycle (Fig. 6.2).

What is important about these experiments is that structures, such as jet streams, traveling vortices, and fronts appear to be basic features in rotating heated fluids, and are not peculiar to atmospheres only.

To make our logical reasoning valid, we assume that each and every firm in the business world exists for the purpose of satisfying a particular market niche through generating a positive cash flow. The positive cash flow can be a result of profits from the marketplace, or investments from various investors or both.

By a *value system*, we refer (Porter, 1985) to associated activities that jointly transform raw materials into products or services for end users. These activities are usually performed by various firms with end users known as *consumers*. In contrast, intermediate, B2B purchasers in a value system are referred to as *customers* instead of consumers. When a firm delivers its products and/or services to both customers and consumers simultaneously, for the sake of convenience of communication, we will just refer to both of these users as customers.

Assume that value creation is the principal objective of any business enterprise, where the outputs are more valuable than the sum of the inputs. It consists of two components: creating value for customers as well as creating value for shareholders. The former helps sell products and services, while the latter, in the form of a rising stock price, insures the future availability of investment capital to fund the operations or to help with the positive cash flow of the enterprise. The *value creation* of a value system is determined by the expected benefits that consumers will receive from their purchases, as represented by their willingness-to-pay (WTP). Due to the lack of perfect price discrimination, consumers generally maintain a consumer surplus, because consumers' spending with payments into the value system is mostly less than the aggregate WTP.

After value has been created, the business enterprise whose efforts led to the value creation needs to capture the value in order to achieve its financial goal of first surviving and then thriving in the economic world. For example, value is created when farmers plant and grow crops; however, these farmers have to capture their created value by harvesting and selling their crops. Value capture by an individual firm is usually determined by its ownership of resources and power status within the value system, as well as by the firm's cost structure and negotiating skill in relation to other firms in the value system (Bowman & Ambrosini, 2001; Priem, 2007).

6.4 A Firm's Supply-Chain Ecosystem

This section looks at why market invitations can be at least partially answered and establishes some basic properties of an arbitrarily chosen focal firm in relation to its ecosystem.

6.4.1 The Ecosystem and Its Systemic Intuition

The success of our focal firm primarily depends on how well other firms are doing in its input-output environment because of their economic interdependence, where the inputs of one firm are the outputs of some other firms. That is, by using the flow of inputs and outputs we can talk about the ecosystem within which the focal firm resides with other firms seen as either upstream components or downstream complements of the system (Adner & Kapoor, 2010). In particular, suppliers represent some of the upstream components of the focal firm; customers, supporters, and assistants who help to make the product of the focal firm usable by consumers are the downstream complements. Supporters and assistants are called complementors

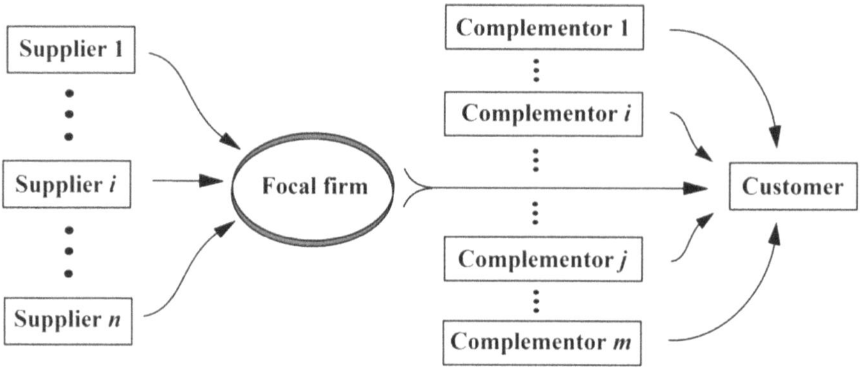

Fig. 6.3 The ecosystem of any focal firm

because although they are outside the focal firm's direct supply chain they need to invest and develop new infrastructure to make the focal firm's product practically usable by the ultimate consumer. Governments, regulators, etc., are examples of complementors that help to build and to regularly maintain, for instance, the roads necessary for transporting products, the specification of new safety procedures, etc. With this explanation in place, we can see that the set of components of the focal firm also includes upstream complementors that help to make the outputs of the suppliers usable by the focal firm.

Figure 6.3 depicts the systemic structure of this ecosystem, where the focal firm utilizes the inputs, which will be called components, from n supplies and delivers its outputs to customers with the support and assistance of m complementors, $m, n = 1, 2, 3, \ldots$ These support and assistance are called complements. As shown in Fig. 6.3, only first-tier components and complements are given, although this structure in real life should be extended forward and backward along the chain of value creation to include actors of other tiers, such as suppliers' suppliers, customers' customers, and complementors' complementors. And when the full range of all players of different tiers is included in the ecosystem, one actually develops a practical situation of linked systems (Lin & Ma, 1987). (To make the figure intuitively clear, all upstream complementors of the focal firm are omitted.) In other words, the ecosystem of the focal firm possesses the characteristics of the yoyo model in Fig. 2.2a, where the firm's internal working is around how its employees with adequate organizational supports can innovatively integrate the components from suppliers in order to deliver its functional product(s) to its customers. Note that similar systemic constructs such as this one has also attracted the attention of academics and practitioners (Adner, 2006; Iansiti & Levien, 2004; Intel Corporation, 2004; SAP Corporation, 2006).

By modeling each firm in the ecosystem of the focal firm and the market as a system, and a mapping ℓ_{ts} from an upstream supplier firm S_s to a downstream customer firm S_t as the linkage of how S_s's component m_s is applied in S_t's product m_t, we have the following result.

Theorem 6.1 Assume that T is a given partially ordered set of type α and that S is an α-type hierarchy of systems over T, satisfying that each state S_t is a nontrivial system, for $t \in T$. Then S can be made into a linked hierarchy by a family $\{\ell_{ts}: t, s \leq T, s \geq t\}$ of linkage mappings of S.

The proof of this result is very technical and is left to the Appendix.

One interpretation of this theorem of systems science can be given as follows: When the focal firm receives a market invitation for innovation, meaning that the partially ordered set T has a set of minimal elements (representing the various market demands related to the invitation), then each and every firm within the ecosystem of the focal firm will be challenged one way or another to provide a resolution to their respective challenges in order to jointly answer the market call. Although the initial market invitation might not be answered perfectly, some combinations of individual firms' resolutions to their respective challenges would provide at least a partial solution to the market demand.

6.4.2 *A Firm's Success and Challenge*

Based on the ecosystem of the focal firm constructed previously, this subsection establishes a few fundamental conclusions about the focal firm in terms of its market success, the values it creates and captures, and the magnitude of challenges it faces in the marketplace.

Proposition 6.1 The success for the focal firm to meet its challenges that appear when the firm decides to answer a market invitation is often dependent on the successes of the other firms within the ecosystem to meet their respective challenges that arise out of the focal firm's desire to answer the market invitation.

In fact, this result follows from how the ecosystem of the focal firm is constructed, as shown in Fig. 6.3. In particular, this ecosystem organically relates the firm's performance with not only the challenges the firm faces (Christensen, 1997; Henderson & Clark, 1990) but also those that confront the external partners. For instance, to meet a challenge, the focal firm may very well need to acquire certain special upstream components, which may in turn represent challenges to the relevant suppliers and their corresponding complementors. At the same time, a solution to the challenge, as identified by the focal firm, may likely require downstream complementors to successfully meet their corresponding challenges in order for the focal firm's new product to be fully usable by the customers. In other words, this ecosystem, as a systemic model of the environment of the focal firm, explicitly explains how the network of challenges centered on that particular challenge the focal firm faces needs to be successfully resolved by different actors in the ecosystem in order for the ecosystem to create value.

On the practical front, there are many real-life instances that support the conclusion of Proposition 6.1. For example, direct current (DC) generation technologies declined because of unconquerable bottlenecks in the development of DC distribution technology (Hughes, 1983), while in the semiconductor lithography industry optical lithography dominates all nonoptical approaches as a consequence of the fact that suppliers, customers, and complementors played roles in meeting their respective challenges in optical lithography technology (Henderson, 1995). On a more personal level, when a colleague had to convert her house to gas heating from its original 100% electric home, although the conversion was possible and done successfully without any trouble, the system became louder than before. Apparently, the ductwork in electrical homes is not directly (1 to 1) compatible with that of gas-heated homes. The ductwork of the former homes is narrow so that the newly installed gas system pushes more air through a smaller space, and, hence, lending itself to a noisier output.

Proposition 6.2 The upper limit of the value the focal firm can *create* when it answers a market invitation is constrained by the upstream component challenges the suppliers consequently confront, while the upper limit of the value the focal firm can *capture* from its effort to answer the market invitation is constrained by the downstream complement challenges.

In fact, the first conclusion follows from the fact that the focal firm's ability to actually make a product that is expected to be competitive in the marketplace is constrained by the challenges the upstream components face. The second conclusion is a consequence of the fact that how much of the full benefit of the focal firm's offering the customer can enjoy from consumption is influenced by the downstream complements.

By the magnitude of a challenge the focal firm faces, it means how much the challenge requires the current approach of problem-solving of the firm to change. In other words, the magnitude of a challenge to the focal firm corresponds to the degree to which this firm is charged with changing its approach to problem-solving.

Proposition 6.3 The greater the magnitude of a challenge the focal firm faces in the marketplace, the greater change the entire ecosystem of the firm will be required to undergo in order for the focal firm to resolve the challenge.

When the focal firm confronts a challenge in the marketplace, it has to make certain necessary changes in its adopted way of problem-solving in order to come up with a resolution beneficially. Now, the required changes necessary to produce the desired resolution could be in:

- The discovery of market trends, the consequent design of brand-new ideas, and the development of competitive products (Henderson & Clark, 1990; Tushman & Anderson, 1986),
- The identification of how some environmental elements can be integrated into firms' internal structures (Brusoni et al., 2001; McGrath, 2013; Takeishi, 2002), or
- The determination of how to scale up the production and delivery of identified solutions (Argote, 1999; Hatch & Mowery, 1998).

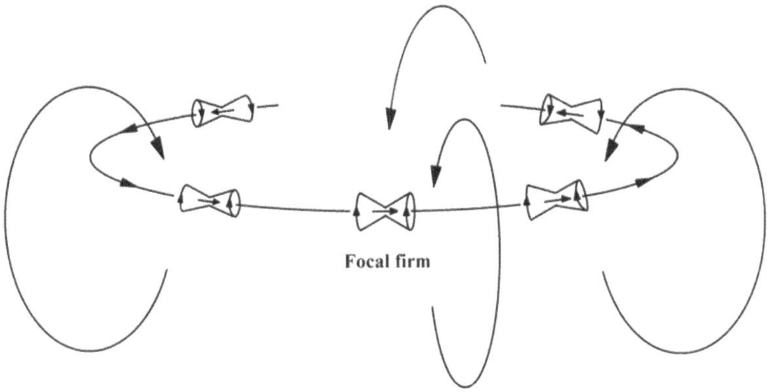

Fig. 6.4 One linear branch of the supply chain within the ecosystem of the focal firm

However, for the ecosystem's purpose of creating value, the focal firm's current approach of problem-solving has achieved a dynamic equilibrium with those of the other players in the ecosystem. In other words, in their effort of achieving the common goal of value creation, all players in the ecosystem are in a state of mutual forbearance. That is, the players, be they firms, complementors, or consumers, mitigate the overall challenge of creating and capturing value by dividing the series of necessary activities into bundles, each of which is manageable by one player according to its strength (Bernheim & Whinston, 1990). They cede efforts in less efficient areas to other more capable players, while in exchange the latter do the same in areas where the former are more efficient (Li & Greenwood, 2004). The players' codependence gradually makes the ecosystem stable in their concerted effort of value creation and value capture (Yu & Cannella, 2012). Eventually, the association of the players becomes strengthened (Fuentelsaz & Gómez, 2006) and the network of adopted approaches of problem-solving by different players reaches equilibrium (Haveman & Nonnemaker, 2000). Therefore, when the focal firm is forced to make changes in its adopted approach of problem-solving, the stability and equilibrium of this network have to be broken; and at the same time, the greater the change the focal firm has to make, the greater alteration the entire ecosystem will have to make in order for the system to reach another equilibrium.

In terms of the systemic yoyo model, the situation implied by the setup of Proposition 6.3 can be depicted in Fig. 6.4, where one linear branch of the supply chain is shown as the horizontal curve with the arrows indicating the direction of movement of inputs and outputs of the firms along the chain, and the vertical arrowed curves stand for the combined eddy field of the individual eddy fields of the firms along the supply chain. When a certain amount of value is created and captured by the firms, an equilibrium and stability between the vertical and horizontal fields are established. However, when the focal firm faces a challenge from the marketplace, it means that its outputs need to be altered, which in turn implies that components from suppliers different from before will be needed. That consequently means that the suppliers' outputs have to be different from before. At the same time, the outputs of the downstream firms will in turn be changed accordingly. Such a domino effect quickly rises upward affecting all the suppliers and the suppliers' suppliers as well as downward exerting demands for downstream firms to alter their offers accordingly. So, once again, this analysis indicates that the more a change is required of the focal firm, the greater change the entire ecosystem will have to go through.

6.5 Learning Capability and Consequent Challenges

As the title suggests, this section studies how the market signals for additional competition, new innovation, and how the focal firm's capability to learn and act accordingly leads to challenges to its ecosystem other than improved advantages. In particular, as shown in Fig. 6.5, the systemic yoyo field of the market includes inputs, which are reflected in such terms as new kinds of products and/or services supplied into the market, and outputs, which could be conceived as demands for improved products and/or services. Although various upstream supplies are available to the focal firm, when the firm learns what the market is calling for, it designs innovative products, which in turn requires the upstream suppliers to provide different kinds of innovative components. To make the rest of this chapter self-contained, let us quote Theorem 5.1 as follows.

Theorem 5.1 If the market described below is in Nash equilibrium, then the following two conditions are equivalent:

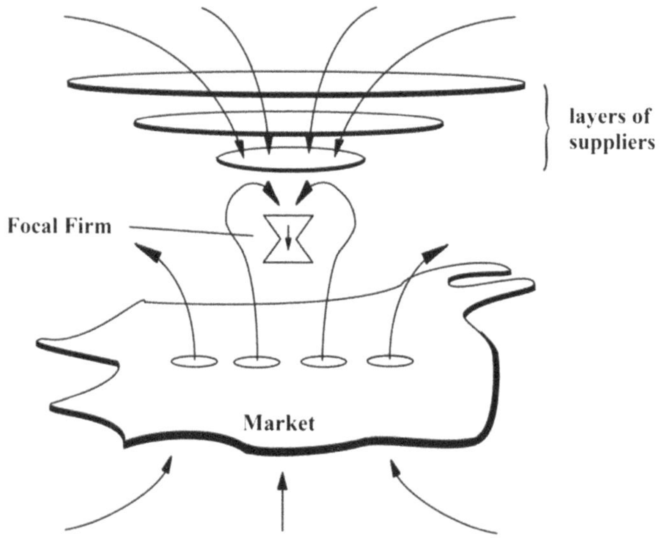

Fig. 6.5 The interaction between what the focal firm learns and what is demanded by the firm from the upstream suppliers

1. The market's consumer surplus is larger than the loyal-customer base of one incumbent firm; and
2. Some firm(s) will enter the market with their versions of substitute offer with their aggregate expected profit potentially larger than that of at least one incumbent firm.

As in the statement of Theorem 3.2, the market mentioned in Theorem 5.1 satisfies the following conditions: It is served by a fixed number of firms with their horizontally differentiated offers; its operation is only affected by market forces, such as demand and supply, and consumers' forever evolving preferences and tastes; each incumbent firm has a base of loyal customers if its price is not more than their reservation price; there are switchers, collectively known as consumer surplus, who make purchases based on which firm's unit-value price is lower; and all firms' pricing strategies are known to all incumbent companies who respond by playing Nash equilibria through untainted self-analyses.

In particular, the first statement of this theorem—the magnitude of the market segment of switchers is greater than the average market share of the incumbent firms—is such a key that it is subject to different understandings of the market players. For instance, other than frugality or other thousands of reasons, one explanation for the existence of these switchers is that they are not totally satisfied with the features of the existing products. Aiming at such an understanding of the market cue, our focal firm could potentially come up with an innovative idea on how to improve the functionalities and features of the existing products. However, such an idea could most likely post challenges to the upstream suppliers for them to design and produce relevant components for the focal firm to integrate in order to make its imagined offer to the marketplace. Additionally, at the same time, the idea could potentially post challenges to the downstream complementors, too, in order for customers to enjoy the full benefit of the revolutionary offer of the focal firm. That is, the entire ecosystem now confronts a challenge, the magnitude and location of which depends on the first innovative idea of the focal firm. At this junction, a natural question arises: Would being an early mover be advantageous for the focal firm?

Intuitively, a leader (or first mover), by virtue of having beaten all rivals in the marketplace, should be able to enjoy nonexisting or reduced competition when presenting its offer to the market, as documented by Lieberman and Montgomery (1988). That would naturally provide the leader with an advantageous exclusivity although it might be only temporary. However, from Theorem 5.1, this intuitive understanding, even so it is empirically supported, suffers from one practical problem—the existence of the desired market. In other words, if the imagined offer is too much ahead of its time, there might not be any readily available market for the product. So, first movers/leaders or pioneers in this sense will have to spend their time, efforts, and resources to first explore and define the market for their products. This end explains why studies, see (Lieberman & Montgomery, 2013) and references there, suggest that time can also benefit later entrants by helping to improve risk estimations, clear uncertainties, improve capability gaps, and facilitate learning from the costly mistakes of earlier entrants.

So, in terms of market entry timing as defined by (Zachary et al., 2015), Theorem 5.1 implies that market characteristics (or invitation) beget the entry of newcomers with the potential of making more profits than any of the earlier movers. Therefore, the general symmetry assumed in Theorem 5.1 indicates that new entrants have at least the same amounts of capabilities and resources as the earlier movers. That simply implies that established firms should not strive to be early movers in any new market until the early movers have established the market with sufficient depth. This end theoretically confirms what Golder and Tellis (1993) find empirically: pioneers enjoy an average of 10% of the market share with nearly half of them failing outright, while early followers take roughly 28% of the market share with failure rates of around 8%.

Practically speaking, Theorem 5.1 explains why large, multi-business enterprises generally exploit slack resources and complementary capabilities to scale up their operation and bypass early entrants (Markides & Geroski, 2005). In other words, size advantage, be it small or large, can make a firm more fitting for a later entry, as noted by Zachary et al. (2015).

What has just been discussed in the previous paragraphs actually explains theoretically why the key challenge underlying the early mover advantage is the opportunity for the focal firm to experiment innovative conceptual products, search for potential customers, and eventually form a market of sufficient depth. If successful, the firm would have acquired very valuable knowledge on production exploitation and market experience, while increasing its added value by improving its offer's performance and/or costs (Argote, 1999; Lieberman, 1984, 1989; Spence, 1981). At the same time, it also illustrates the reason why pioneers tend to occupy a small market share with nearly half of them failing outright (Golder & Tellis, 1993) because of the prohibitive difficulties existing first with the design and production of innovative conceptual products and then with the formation of the relevant market of sufficient depth.

Proposition 6.4 The more capable the focal firm is to learn and to discover, the greater advantage the firm is able to create and the greater the challenge the firm imposes on its ecosystem.

In fact, the previous analysis shows that the firm's capability to learn and to discover is required in the understanding of market cues, in the formation of innovative concepts, and in the development and production of competitive products to satisfy the emerging market need. Therefore, the more capable the focal firm is in terms of learning and discovery, the more likely the firm will come up with innovative concepts and the consequent design and production of epoch-making products. Now, Proposition 6.3 implies that in this case, the ecosystem will confront greater challenges in order to supply for the focal firm's design and production of new products and facilitate the delivery of the firm's products to customers.

Speaking systemically, when the focal firm is very capable of learning and discovering, it means that it is more convenient and readier for the firm to sense opportunities and emerging demands in the marketplace. That sensitivity in real life generally means that the firm has an established mechanism for learning and discovering and for how to benefit from what is learned and what is discovered, otherwise any such sensitivity would be dulled either consciously or unconsciously because resources are needed to maintain the said sensitivity. That implies that the firm's output will be replaced with improved or brand-new products based on new knowledge and discovery, which consequently imposes challenges, Fig. 6.4, on the upstream suppliers to provide different and, more likely, innovative outputs from before and on the downstream complementors to develop appropriate supports and assistance in order for customers to fully enjoy the new focal offers. Such chain effects rise quickly upward affecting all the suppliers and the suppliers' suppliers and downward disturbing all the complementors and the complementors' complementors. So, once again, this analysis indicates that the more a change is required of the focal firm by new knowledge and discovery, the greater advantage the firm is able to create and the greater change the entire ecosystem will have to go through.

Proposition 6.5 For the focal firm, its strong capability to learn and to act accordingly is an advantage of the firm over its competitors.

In fact, beyond what is given in the reasoning of Proposition 6.4, the level of learning capability depends on both the magnitude and the scope the focal firm needs to make changes. It is the emerging mastery of new routines of operation, different ways of thinking and forever changing processes of production that underpins the firm's success with its rides on the waves of transient competitive advantages in the increasingly turbulent economic market (McGrath, 2013). If little or no change from the status quo is necessary, as one of those incumbent firms described in Theorem 5.1, then there is not much for the firm to learn and to discover so that the motivation for learning and the relative advantage gained from learning and corresponding actions will be either low or nonexistent. In contrast, when the focal firm needs to bring its constantly different and supposedly better offers to the market, it has to overcome related uncertainties and complexities. That explains why a strong capability for the firm to learn and to act accordingly represents a unique advantage of the focal firm.

6.6 A Few Final Words

One important characteristic of the business landscape of the modern world is that successful firms have to ride waves of transient competitive advantages (McGrath, 2013). To do so, a firm faces challenges first in learning what the market is signaling, second in deciphering what innovative solution the firm can come up with to answer the market call by mobilizing its resources, and, third, in how to deal with considerable challenges in resolving uncertainties appearing with upstream suppliers and downstream complementors.

The literature has provided a rich collection of factors, each of which represents a challenge facing business enterprises with respect to innovativeness and good performance outcomes for a firm (Christensen, 1997; Marzi et al., 2017). However, as pointed out by Adner and Kapoor (2010) and as shown in this chapter, challenges that surround an innovative design, consequent production, and movement of the product out to the market are often situated both inside a focal firm and within the firm's ecosystem. This chapter propels us beyond the traditional analysis of the literature of innovative firms' internal and environmental challenges by theoretically establishing a series of generally true conclusions on systemic logic reasoning and relevant intuition.

By holistically looking at the ecosystem of a focal firm that recognizes an invitation from the market, this chapter provides a systemic framework by focusing on input-output threads located throughout a supply chain. Within this framework, we show that the focal firm's innovative idea introduced to answer the market invitation generally posts challenges to upstream suppliers and downstream complementors. We illustrate that suppliers' resolutions to component challenges help the focal firm to produce its ideal product, while complementors' success with their challenges facilitates the necessary conditions for the focal offer to reach customers with full potential of value creation. We examine how an epoch-making understanding of a market invitation by a focal firm can simultaneously challenge both suppliers and complementors with asymmetrical levels of effects on the performance of the focal firm. Specially, we establish the following main results among others:

- Upstream component challenges and downstream complement difficulties can respectively constrain the values a focal firm can create and capture; and
- When a focal firm is more capable to learn and to act accordingly, that firm's ecosystem will experience greater challenges.

Other than developing a series of practically useful conclusions, this chapter expands the traditional toolbox of reasoning and analysis by demonstrating the power and usefulness of systems science and methodology. By utilizing holistic thinking and logical reasoning, we are able to investigate the dynamics of the ecosystem of a focal firm, and produce different while more insightful understandings than those in the literature on the distribution of challenges, appearance of uncertainties, evolution of technologies, and performance of a focal firm, (e.g., Anderson & Tushman, 1990; Cusumano et al., 1992; Utterback & Abernathy, 1975).

The conclusions established in this chapter provide a practical managerial guideline on how to manage investments in meeting challenges originated from market calls. In particular, we have the following recommendations:

- A firm's investment in speeding up the commercialization cycles when pursuing first mover advantage has to carefully consider whether other firms in the firm's ecosystem can promptly meet their respective challenges or not.
- When introducing an innovative product to answer a market invitation, the more upstream suppliers and the less downstream complementors are challenged, the greater value the product is expected to create and capture.

- If a perceived innovative product posts challenges, downstream complementors will likely have difficulty in meeting the market's need in a timely manner. Consequently, the innovative product needs to be modified to ease the expected difficulty confronting complementors.
- Any firm that desires to succeed in its market domain needs to focus on improving its capability to learn and to discover. Potential performance advantages of the firm originate from such capability.

In other words, a firm can estimate the levels of relative challenges across its ecosystem well in advance of the commercialization of the firm's innovative idea.

As for the limitations of this chapter, the first one stems from the assumption of why a firm in the business world exists, because not all firms aim at generating positive cash flows. Firms, developed for reasons beyond generating positive cash flows, can behave in different ways than what is described in this chapter. Second, although this chapter looks at the distribution of challenges across the ecosystem of a focal firm, we did not consider the innovation challenges the focal firm faces in its core activities. These internal challenges might possibly alter some of the established results if they are somehow correlated with external challenges. Third, this chapter stays within the context of a single supply chain; however, in real-life supply chains, as systems, also interact with each other. That interaction of a higher systemic level could very possibly affect the distribution of challenges within each of the interacting supply chains. In other words, this chapter only presents a baseline structure on top of which other factors and considerations need to be overlaid.

Corresponding to these listed and other unlisted limitations, future studies should incorporate additional contingencies and antecedents in order to produce more practical useful insights than what have been established in this chapter.

Appendix: Proof of Theorem 6.1

Let each collection of elements (or things) be referred to as a set. Given two sets X and Y, f is a function or mapping from X to Y, provided that f represents a rule that spells out how each element x in X (or written as $x \in X$) is paired with an element y in Y (or $y \in Y$), written as $f: X \to Y$, and the pairing between x and y is symbolically written as $f(x) = y$.

If S is a (general) system (Lin, 1999), then S is an ordered pair (M, R) of sets, where R is a set of some relations of the elements in the set M. Each element in M is called an object of S, and M and R are called the object set and the relation set of S, respectively. A system $S = (M, R)$ is said to be trivial if $M = \varnothing$.

Given two systems $S_i = (M_i, R_i)$, $i = 1, 2$, and a mapping $h: M_1 \to M_2$, by using transfinite induction, define two classes $\widehat{M}_i$, $i = 1, 2$, and a class mapping $\widehat{h}: \widehat{M}_1 \to \widehat{M}_2$ with the following properties:

$$\widehat{M}_i = \bigcup_{n \in \mathbf{ON}} M_i^n, i = 1, 2,$$

where **ON** stands for the collection of all ordinals, and for each $x = (x_0, x_1, \ldots, x_\alpha, \ldots) \in \widehat{M}_1$,

$$\widehat{h}(x) = (h(x_0), h(x_1), \ldots, h(x_\alpha), \ldots).$$

For each relation $r \in R_1$, let $\widehat{h}(r) = \left\{\widehat{h}(x) : x \in r\right\}$ be a relation on M_2 with length $n(r)$. Without confusion, h will also be used to indicate the class mapping $\widehat{h}$ and h is now seen as a mapping from system S_1 into system S_2, denoted h: $S_1 \to S_2$.

If two systems S_1 and S_2 are the same or equal, that is, $M_1 = M_2$ and $R_1 = R_2$, and a mapping $h : M_1 \to M_2$ satisfies $h(x) = x$, for any $x \in M_1$, then we say that this mapping h is the identity mapping from S_1 to S_2, written as id_{S_1}.

Assume that $(T, \leq)$ is a partially ordered set with order type α (Lin, 1999), where the symbol $\leq$ stands for the order relation between elements of T. An α-type hierarchy S of systems over the partially ordered set $(T, \leq)$ is defined as a function defined on T such that for each $t \in T$, $S(t) = S_t = (M_t, R_t)$ is a system, called the state of the α-type hierarchy S at the moment t. Without causing confusion, we omit the words "over the partially ordered set $(T, \leq)$."

Given an α-type hierarchy S of systems, assume that ℓ_{tr}: $S_r \to S_t$ is a mapping from the state system S_r into the state system S_t, for any r, $t \in T$ with $r \geq t$, such that

$$\ell_{ts} = \ell_{tr} \circ \ell_{rs} \text{ and } \ell_{tt} = \mathrm{id}_{S_t},$$

where r, s, t are arbitrary elements in T satisfying $s \geq r \geq t$, and $\mathrm{id}_{S_t} = \mathrm{id}_{M_t}$ is the identity mapping on the set M_t. This collection $\{\ell_{ts}: t, s \in T, s \geq t\}$ is referred to as a family of linkage mappings of the α-type hierarchy S, and each ℓ_{ts} a linkage mapping from the state system S_s into the state system S_t.

An α-type hierarchy of systems S, denoted $\{S, \ell_{ts}, T\}$ or $\{S(t), \ell_{ts}, T\}$, is referred to as a linked α-type hierarchy (of systems) if a family $\{\ell_{ts}: t, s \in T, s \geq t\}$ of linkage mappings is given or can be constructed based on given conditions.

Now, for the given nontrivial α-type hierarchy S of systems in Theorem 6.1, assume that $S_t = (M_t, R_t)$, for each $t \in T$. From the Axiom of Choice (Lin, 1999), it follows that there exists a choice function C: $T \to \bigcup\{M_t : t \in T\}$ such that $C(t) \in M_t$ for each $t \in T$. A family $\{\ell_{ts}: t, s \in T, s \geq t\}$ of linkage mappings of S can now be defined as follows. For any s, $t \in T$ with $s > t$, let

$$\ell_{ts}(x) = C(t),$$

for all $x \in M_s$ and $\ell_{tt} = \mathrm{id}_{M_t}$. This ends the proof of Theorem 6.1.

References

Adner, R. (2006). Match your innovation strategy to your innovation ecosystem. *Harvard Business Review, 84*(4), 98–107.

Adner, R. (2017). Ecosystem as structure: An actionable construct for strategy. *Journal of Management, 43*(1), 39–58.

Adner, R., & Kapoor, R. (2010). Value creation in innovation ecosystems: How the structure of technological interdependence affects firm performance in new technological generations. *Strategic Management Journal, 31*(3), 306–333.

Adner, R., Oxley, J. E., & Silverman, B. S. (Eds.). (2013). *Collaboration and competition in business ecosystems*. Emerald.

Altinay, L., Madanoglu, M., Vita, G. D., Arasli, H., & Ekinci, Y. (2016). The interface between organizational learning capability, entrepreneurial orientation, and SME growth. *Journal of Small Business Management, 54*(3), 871–891.

Anderson, P. C., & Tushman, M. L. (1990). Technological discontinuities and dominant designs. *Administrative Science Quarterly, 35*(4), 604–633.

Argote, L. (1999). *Organizational learning: Creating, retaining and transferring knowledge*. In *Kluwer*. MA.

Attour, A., & Lazaric, N. (2018). From knowledge to business ecosystems: emergence of an entrepreneurial activity during knowledge replication. In *Small Business Economics*. https://doi.org/10.1007/s11187-018-0035-3

Bernheim, B. D., & Whinston, M. D. (1990). Multimarket contact and collusive behavior. *RAND Journal of Economics, 21*, 1–26.

Bilgicer, T., Jedidi, K., Lehmann, D. R., & Neslin, S. A. (2015). Social contagion and customer adoption of new sales channels. *Journal of Retailing, 91*(2), 254–271.

Bowman, C., & Ambrosini, V. (2001). "Value" in the resource-based view of the firm: A contribution to the debate. *Academy of Management Review, 26*, 501–502.

Brem, A., & Viardot, É. (2015). Adoption of innovation: Balancing internal and external stakeholders in the marketing of innovation. In A. Brem & É. Viardot (Eds.), *Adoption of innovation* (pp. 1–10). Springer.

Brusoni, S., Prencipe, A., & Pavitt, K. (2001). Knowledge specialization, organizational coupling, and the boundaries of the firm: why do firms know more than they make? *Administrative Science Quarterly, 46*, 597–621.

Camps, J., Oltra, V., Aldás-Manzano, J., Buenaventura-Vera, G., & Torres-Carballo, F. (2016). Individual performance in turbulent environments: The Role of organizational learning capability and employee flexibility. *Human Resources Management, 55*(3), 363–383.

Chang, W., Ellinger, A. E., Kim, K., & Franke, G. R. (2016). Supply chain integration and firm financial performance: A meta-analysis of positional advantage mediation and moderating factors. *European Management Journal, 34*(3), 282–295.

Chemmanur, T. J., Loutskina, E., & Tian, X. (2014). Corporate venture capital, value creation, and innovation. *The Review of Financial Studies, 27*(8), 2434–2473.

Chen, D. Q., Preston, D. S., & Swink, M. (2015). How the use of big data analytics affects value creation in supply chain management. *Journal of Management Information Systems, 32*(4), 4–39.

Christensen, C. M. (1997). *Innovator's dilemma: When new technologies cause great firms to fail*. Harvard Business School Press.

Clarysse, B., Wright, M., Bruneel, J., & Mahajan, A. (2014). Creating value in ecosystems: Crossing the chasm between knowledge and business ecosystems. *Research Policy, 43*(7), 1164–1176.

Cusumano, M., Mylonadis, Y., & Rosenbloom, R. (1992). Strategic maneuvering and mass market dynamics: The triumph of VHS over Beta. *Business History Review, 66*(1), 51–94.

Forrest, J. Y.-L., Liu, Y., Wajda, T. A., Köse, E., & Arik, O. A. (2019) How challenges and successes of a firm depend on its input-output ecosystem. *Proceedings of the 2019 annual conference of decision sciences institute*, New Orleans, LA, November 23 – 25, 2019 (pp. 2381–2402).

Fuentelsaz, L., & Gómez, J. (2006). Multipoint competition, strategic similarity and entry into geographic markets. *Strategic Management Journal, 27*(5), 477–499.

Fultz, D., Long, R.R., Owens, G.V., Bohan, W., Kaylor, R., & Weil, J. (1959). Studies of thermal convection in a rotating cylinder with some implications for large-scale atmospheric motion. Meteorol. Monographs (American Meteorological Society) 21, 4.

Golder, P. N., & Tellis, G. J. (1993). Pioneer advantage: Marketing logic or marketing legend? *Journal of Marketing Research, 30*, 158–170.

Hatch, N. W., & Mowery, D. C. (1998). Process innovation and learning by doing in semiconductor manufacturing. *Management Science, 44*(11), 1461–1477.

Haveman, H. A., & Nonnemaker, L. (2000). Competition in multiple geographic markets: The impact on growth and market entry. *Administrative Science Quarterly, 45*(2), 232–267.

Hellström, M., Tsvetkova, A., Gustafsson, M., & Wikström, K. (2015). Collaboration mechanisms for business models in distributed energy ecosystems. *Journal of Cleaner Production, 102*(1), 226–236.

Henderson, R. M. (1995). Of life cycles real and imaginary: the unexpectedly long old age of optical lithography. *Research Policy, 24*(4), 631–643.

Henderson, R. M., & Clark, K. B. (1990). Architectural innovation: the reconfiguration of existing product technologies and the failure of established firms. *Administrative Science Quarterly, 35* (1), 9–30.

Hide, R. (1953). Some experiments on thermal convection in a rotating liquid. *Quarterly Journal of the Royal Meteorological Society, 79*, 161.

Hughes, T. P. (1983). *Networks of power: Electrification in Western society 1880-1930*. Johns Hopkins University Press.

Iansiti, M., & Levien, R. (2004). *The keystone advantage: What the new dynamics of business ecosystems mean for strategy, innovation, and sustainability*. In *Harvard Business School Press*. MA.

Intel Corporation. (2004). Intel sees unified platform and ecosystem as key to enabling the digital home. Press release 17 February.

Kagermann, H. (2015). Change through digitization: Value creation in the age of industry 4.0. In H. Albach, H. Meffert, A. Pinkwart, & R. Reichwald (Eds.), *Management of permanent change* (pp. 23–48). Springer Gabler.

Kalmuk, G., & Acar, A. Z. (2015). The mediating role of organizational learning capability on the relationship between innovation and firm's performance: A conceptual framework. *Procedia - Social and Behavioral Sciences, 210*(2), 164–169.

Kapoor, R., & Agarwal, S. (2017). Sustaining Superior Performance in Business Ecosystems: Evidence from application software developers in the iOS and android smartphone ecosystems. *Organization Science, 28*(3), 379–595.

Laukkanen, T. (2016). Consumer adoption versus rejection decisions in seemingly similar service innovations: The case of the Internet and mobile banking. *Journal of Business Research, 69*(7), 2432–2439.

Lemoussu, S., Chaudemar, J. C., & Vingerhoeds, R. A. (2018). Systems engineering and project management process modeling in the aeronautics context: Case study of SMEs. *International Journal of Mechanical and Mechatronics Engineering, 12*(2), 88–96.

Li, S. X., & Greenwood, R. (2004). The effect of within industry diversification on firm performance: Synergy creation, multi-market contact and market structuration. *Strategic Management Journal, 25*(12), 1131–1153.

Lieberman, M. B. (1984). The learning curve and pricing in the chemical processing industries. *RAND Journal of Economics, 15*(2), 213–228.

Lieberman, M. B. (1989). The learning curve, technological barriers to entry, and competitive survival in the chemical processing industries. *Strategic Management Journal, 10*(5), 431–447.

Lieberman, M. B., & Montgomery, D. B. (1988). First-mover advantages. *Strategic Management Journal, 9*, 41–58.

Lieberman, M. B., & Montgomery, D. B. (2013). Conundra and progress: Research on entry order and performance. *Long Range Planning, 44*, 312–324.

Lin, Y. (1999). *General systems theory: A mathematical approach*. Plenum and Kluwer Academic Publishers.

Lin, Y., & Ma, Y. H. (1987). Some properties of linked time systems. *International Journal of General Systems, 13*(2), 125–134.

Markides, C. C., & Geroski, P. A. (2005). *Fast second: How smart companies bypass radical innovation to enter and dominate new markets*. Jossey Bass/Wiley.

Marzi, G., Dabic, M., Daim, T., & Garces, E. (2017). Product and process innovation in manufacturing firms: A 30-year bibliometric analysis. *Scientometrics*. https://doi.org/10.1007/s11192-017-2500-1

McGrath, R. G. (2013). *The end of competitive advantage: How to keep your strategy moving as fast as your business*. Harvard Business Review Press.

Oliveira, T., Thomas, M., Baptista, G., & Campos, F. (2016). Mobile payment: Understanding the determinants of customer adoption and intention to recommend the technology. *Computers in Human Behavior, 61*, 404–414.

Porter, M. E. (1985). *Competitive advantage: Creating and sustaining superior performance*. Free Press.

Prajogo, D. I., & Oke, A. (2016). Human capital, service innovation advantage, and business performance: The moderating roles of dynamic and competitive environments. *International Journal of Operations & Production Management, 36*(9), 974–994.

Priem, R. L. (2007). A consumer perspective on value creation. *Academy of Management Review, 32*, 219–235.

SAP Corporation. (2006). SAP ecosystem unites to bring innovation to chemicals industry. Press release, 8 May.

Spence, M. (1981). The learning curve and competition. *Bell Journal of Economics, 12*(1), 49–70.

Sung, S. Y., & Choi, J. N. (2014). Do organizations spend wisely on employees? Effects of training and development investments on learning and innovation in organizations. *Journal of Organizational Behavior, 35*(3), 393–412.

Takeishi, A. (2002). Knowledge partitioning in the interfirm division of labor: the case of automotive product development. *Organization Science, 13*(3), 321–338.

Tushman, M., & Anderson, P. (1986). Technological discontinuities and organizational environments. *Administrative Science Quarterly, 31*(3), 439–465.

Utterback, J. M., & Abernathy, W. J. (1975). A dynamic model of process and product innovation. *Omega, 3*(6), 639–656.

Yadav, P. L., Han, S. H., & Kim, H. (2017). Sustaining competitive advantage through corporate environmental performance. *Business Strategy and the Environment, 26*(3), 345–357.

Yu, T., & Cannella, A. A., Jr. (2012). A comprehensive review of multimarket competition research. *Journal of Management, 39*, 76–109.

Zachary, M. A., Gianiodis, P. T., Payne, G. T., & Markman, G. D. (2015). Entry timing: Enduring lessons and future directions. *Journal of Management, 41*(5), 1388–1415.

Zahra, S. A., & Nambisan, S. (2012). Entrepreneurship and strategic thinking in business ecosystems. *Business Horizons, 55*(3), 219–229.

Chapter 7
Upstream/Downstream Impacts on a Firm's Performance

Abstract This chapter, based mainly on (Forrest et al., 2019, *Proceedings of the 2019 Annual meeting of National Association of business, Economics and Technology* (pp. 83–100)), studies holistically the supply-chain ecosystem of a focal firm, when the firm innovatively deciphers a market invitation. It employs the thinking logic and methodology of systems science to establish a series of generally true conclusions regarding the challenges facing either the suppliers or the complementors. In particular, this chapter shows among results that (1) challenges upstream components face help build performance advantage for the focal firm over its competitors; however, if the firm's challenges are mostly on its upstream suppliers, then the performance advantage's lifespan will be greatly shortened. (2) An innovative firm has to consider the availability and development of appropriate complements in its introduction of innovative products. (3) In contracting with upstream suppliers, a focal firm has to deal with technological uncertainty and suppliers' behavioral uncertainty with the former affects the firm's ability to create value and the latter impacts its ability to capture value. And (4) a firm's performance advantage, as a consequence of applying vertical integration within the firm's ecosystem, increases over time within the life cycle of the technology developed to meet the firm's innovative need. At the conclusion of this chapter, practical managerial recommendations and open questions for future research are given.

Keywords Capability of learning · Innovative offer · Market invitation · Performance advantage · Systemic yoyo · Uncertainty

7.1 Introduction

When a firm comprehends a market signal innovatively, it has located a potential way to develop a significant competitive advantage over its competitors. However, to materialize such a potential, other than internal reasons the firm has to analyze how other players in its supply-chain ecosystem (Adner et al., 2013) could affect its fortune. In this regard, two natural questions arise: When a firm innovatively deciphers a market invitation, how will the consequent design of the firm's new

J. Y.-L. Forrest, Y. Liu, *Value in Business*, Contributions to Management Science,
https://doi.org/10.1007/978-3-030-82898-1_7

product(s) post challenges to other players within its ecosystem? And how can the firm systemically manage its vertical interdependence within its ecosystem?

This chapter seeks to address these two theoretically and practically important questions by randomly selecting a firm as the focal firm and by using the thinking logic and methodology of systems science. When a firm provides its outputs for the focal firm to integrate into a complete product for its customers, the firm is known as a supplier. When a firm facilitates necessary conditions for customers to utilize fully the focal firm's product, the firm is known as a complementor. With these terms defined, this chapter establishes the following main conclusions among others.

- The performance advantage of a focal firm over its competitors is positively correlated to the level of challenges the firm's suppliers face, and negatively to that of the firm's complementors;
- To successfully ride waves of transient competitive advantages, a firm has to introduce such innovative products that suppliers can possibly provide necessary components and complementors can readily facilitate needed complements; and
- When contracting with upstream suppliers, a firm has to deal with technological and behavioral uncertainty, where the former affects the firm's creation of value, and the latter impacts the firm's capture of value.

The remainder of this chapter is organized as follows. Section 7.2 provides a literature review while showing how this chapter contributes to the already established knowledge. Section 7.3 focuses on challenges facing the upstream suppliers of the ecosystem, while Sect. 7.4 focuses on those confronting the downstream players. Section 7.5 combines what is established in the previous sections to investigate the benefits of vertical integration within a supply-chain ecosystem. Section 7.6 concludes the presentation of this chapter.

7.2 Literature Review

Owing to the employment of systems thinking and methodology, this chapter contributes simultaneously to a range of different literatures, including business ecosystems, competitive advantage, customer adoption of innovation, firm uncertainties (contractual, behavioral, and technological), as well as technology life cycle. In this section, we review relevant works in each of these specific literatures.

In the literature of ecosystems, addressing how system players collaborate and compete with each other Adner et al. (2013) publish a collection of papers. Refuting the common belief that knowledge ecosystems in technology hotspots lead to business ecosystems, where each player enjoys competitive advantages, Clarysse et al. (2014) find a case where a knowledge ecosystem is only concentrated around a few central players. Zahra and Nambisan (2012) claim that the success of such an ecosystem, consisting of well-established companies and new ventures, requires collaboration and competition that demands strategic thinking to leverage a firm's resources and capabilities. Kapoor and Agarwal (2017) study how firms in a business ecosystem orchestrate the ecosystem's functioning by providing platforms and setting the rules for participation by complementor firms. By using a business model lens in an energy business ecosystem, Hellström et al. (2015) point to the importance of locating factors that drive the business models of collaborating firms in facilitating system transitions and change in the logic of an industry. Attour and Lazaric (2018) find that transformation of knowledge ecosystem(s) can lead to the emergence of a technological platform required especially for a firm start-up. Adner (2017) examines the relationship between ecosystems and a host of other constructs, such as business models, platforms, competition, multi-sided markets, networks, technology systems, supply chains, and value networks.

Regarding the literature of performance and competitive advantages, Yadav et al. (2017) look at the association between a firm's environmental efforts and the sustainability of its competitive advantage. They empirically support the proposition that benefiting the environment can lead to economic value. Prajogo and Oke (2016) show that human capital is positively correlated to the creation of value or service innovation advantage (SIA), which in turn results in rent generation, and that the effect of SIA on business performance is influenced by environmental factors. Chang et al. (2016) confirm the early claim that each dimension of supply-chain integration indeed improves financial performance.

As for the literature of market adaptation of innovation, Oliveira et al. (2016) find that compatibility, perceived technology security, performance expectations, innovativeness, and social influence have significant effects on consumer adoption. Bilgicer et al. (2015) show that social contagion does play a major role with long-term versus short-term customers being less inclined to consider alternate sales channels. Brem and Viardot (2015) consider the market adoption of innovations by providing conceptual insights and manners that appear to stimulate and facilitate the adoption of every kind of innovation. Laukkanen (2016) maintains that all innovations meet consumer resistance and overcoming this unfriendliness needs to occur prior to product adoption, and that value is the strongest barrier, while image slows adoption; and that both gender and age significantly predict adoption and rejection decisions.

As for the literature of various uncertainties, Hendrikse et al. (2015) provide empirical evidence that somewhat supports the following hypotheses: (1) general trust of the franchisor reduces the franchisor's perception of relational risk and hence the necessity to control the network relationship by more complete contract planning, and (2) knowledge-based trust increases information sharing between the partners and hence the knowledge base for specifying more detailed contracts. To better understand how contractual and relational governance interact, Cao and Lumineau (2015) conduct a qualitative review and meta-analysis of the existing literature. Some findings uncovered by these researchers include the following: (1) Contractual governance is positively related to trust and relational norms, (2) There are two sides to relational governance, (3) Contracts, trust, and relational norms jointly improve satisfaction and relationship performance while jointly reducing opportunism, and (4) Contractual and relational governance can complement or substitute for each other in certain situations. Hallberg (2015) argues that uncertain governance choices are subject to specific decision-biases and potentially corrective functions of current organization as well as asymmetries in actors' access to decision-supporting systems. Specifically, Hallberg maintains that with overestimated unbiased rationality and asymmetric access to decision-supporting systems, transaction cost economics runs the risk of underestimating the degree of vertical integration in actual firms. By looking through the lens of the relational view, Bstieler and Hemmert (2015) disentangle the effects of relational and contractual governance on collaboration outcomes by using survey data from South Korea. They find that the strength of prior business ties enhances relational governance and contributes to knowledge acquisition and collaboration satisfaction. And even though the impact of contractual governance is weaker than relational governance, when both governance mechanisms are applied simultaneously, the positive returns on collaboration satisfaction diminish.

Mondragon and Mondragon (2018) investigate the role of systems integrators in managing complex products with modular architectures under technological uncertainty in the automotive supply chain. They show that systems integrators in low technological uncertainty have to be able to redesign their architectures due to the implementation of unknown technologies in key individual components, and for complex products and architectures under high technological uncertainty, the most important source of innovation still lies in the specialization of individual activities. Addressing inconsistent empirical findings regarding the relationship between supply-chain integration (SCI) and performance, Huang et al. (2014) develop a SCI model that includes buyer-supplier-supplier relationships, and propose a contingency framework for reexamining the SCI-supplier performance relationship under demand and technological uncertainties. These scholars find that although SCI seems to have a significant positive effect on the suppliers' performance, this positive relationship can be weakened by demand uncertainty and strengthened by technological uncertainty. Considering the challenge facing the commercialization of an emerging technology that employs an immature production process, Roca et al. (2017) unpack how the characteristics of a technology may influence the options for regulatory intervention. These scholars propose a generalizable framework for regulating emerging process-based technologies in safety-critical industries. By employing the industrial economics and knowledge-based perspective, Burton and Galvin (2018) hypothesize how the combined effects of product architecture type, product complexity, and the rate of product component change may influence task, knowledge, and firm boundaries. They suggest that whether mirroring or misting is an efficient strategic choice is influenced by the characteristics of both the product architecture and the rate of technological change at the product component level, as well as changes across time as products evolve.

By drawing on effectuation theory, Jiang and Tornikoski (2018) theorize how founder teams' perceptions of uncertainty and behavioral logics develop during new venture creation processes, suggesting a possible evolution from a causal conditional relationship between perceived uncertainty and behavioral logics to an integrative relationship. Through integrating transaction cost economics and justice theory, Trada and Goyal (2017) examine the impact of perceived unfairness on distributor opportunism and uncover the dual effects of perceived unfairness on opportunism: directly enhancing opportunism and aggravating (positively moderating) the effects of economic forces on opportunism. By theoretically clarifying from a transaction cost economics point of view under what circumstances multinational companies (MNCs) should outsource their innovation functions, Yeo and Saboori-Deilami (2017) show that heterogeneity between the home and host country affects the autonomy of the innovation at the host country. This autonomy in turn leads to higher transaction cost, which turns out to be the main determinant of the decision on whether or not to outsource the innovation function. Bruneel et al. (2017) empirically show that technology complexity within new technology-based firms negatively influences the level of interorganizational trust in key partner relationships, while both relationship and partner characteristics moderate the technology complexity-trust relationship.

Regarding technology life cycle, addressing the fact that previous studies have not considered the dynamic and idiosyncratic aspects of a technology's progression, Lee et al. (2017) propose a stochastic technology life cycle analysis to trace the phases of a technology's progression based on patent citations thus leading to the identification of the patterns of technology life cycles at the individual patent level. Due to the significance of discussions on the patterns of technological innovation in terms of the efficient distribution of national R&D resources and the establishment of corporate managerial strategies, Byun et al. (2018) calculate and analyze technology cycle time (TCT) by technological area based on patent data. In contrast to the common belief that firms should match their organizational form to the prevailing nature of technology, Helfat et al. (2016) provide a new explanation for why vertically integrated and specialized firms may continue to coexist as industries evolve and why these firms may rationally choose to stay integrated and bear the sunk costs of developing integrative capabilities as well as the ongoing costs of maintaining them, even when they lose money during some time periods due to competition from lower cost specialized firms.

This chapter contributes to the literature in various ways. First, it introduces systems science into the investigation of the asymmetrical distribution of challenges across the ecosystem of a firm. Second, it establishes generally true propositions, leading to reliable applications, without suffering from the constraints and limitations of anecdotal and data analysis (or data mining). That is, the results developed herein hold true universally within the system of a few basic assumptions and logical reasoning.

To make the logical reasoning of the rest of this chapter valid, assume that each firm in our discussion exists for filling a specific market niche by creating a positive cash flow. The cash flow can be a result of profits from either the marketplace, or investments from various investors, or both.

By a *value system*, it means (Porter, 1985) associated activities, usually performed by various firms, that jointly transform raw materials into products or services for end users, known as *consumers*. And intermediate B2B purchasers are known as *customers*. When a firm delivers its products and/or services to both customers and consumers simultaneously, these users will just be referred to as customers.

Assume that value creation is the principal objective of any firm so that the outputs are more valuable than the sum of the inputs. There are two components in each value creation: creating value for either customers or shareholders. The former helps sell products and services; the latter, in rising stock prices, insures the availability of future investment capital for funding the operations or helping with the positive cash flow of the enterprise. The expected benefits consumers will receive from their purchases, represented by their willingness-to-pay (WTP), determines the *value creation* of any value system. Because perfect price discrimination is generally absent, consumers most likely maintain a consumer surplus due to the reason that consumers' spending with payments into the value system is mostly less than the aggregate WTP.

After the creation of value, the trying firm needs to capture the value in order to achieve its financial goal of first surviving and then thriving in the economic world. For an individual firm, many factors determine its value capture, such as the firm's ownership of resources and power status within the value system, as well as the firm's cost structure and negotiating skill in relation to other firms in the value system (Bowman & Ambrosini, 2001; Priem, 2007).

7.3 Challenges Facing Upstream Firms

For an arbitrarily chosen focal firm, it forms jointly with other firms within its input-output environment such an ecosystem that the inputs of one firm are the outputs of some other firms, Fig. 7.1. So, the firms in this ecosystem can be classified as either upstream components or downstream complements of the focal firm (Adner & Kapoor, 2010). Here, suppliers represent some of the upstream components of the focal firm; customers, supporters and assistants, called complementors, who help make the product of the focal firm usable by consumers are the downstream complements. Although they are outside the focal firm's direct supply chain, the complementors need to invest and develop new infrastructure to make the focal firm's product usable by the ultimate consumer. From this explanation, we see that the set of components of the focal firm also includes upstream complementors that help to make the outputs of the suppliers usable by the focal firm.

In this section, we look at the effects on the focal firm of challenges that appear in the upstream of the ecosystem if the firm aspires to be a market leader. Generally, innovative ideas of the focal firm are often objectified into actual products through necessary changes in components, while the focal firm needs to originally integrate these new components into its desired offer (Iansiti, 1998; Brusoni et al., 2001). Here, the origin of the innovative ideas really comes from an epoch-making understanding of the market cue, as what the earlier Theorem 3.1 describes.

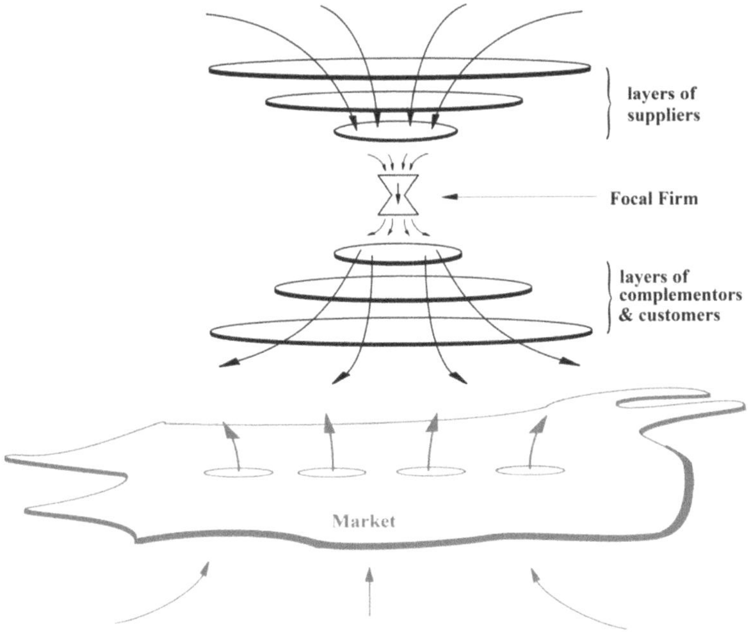

Fig. 7.1 The yoyo structure of the focal firm's ecosystem

Proposition 7.1 The greater challenges the upstream components face, the greater performance advantage the focal firm has over its competitors.

By considering a linear branch of the supply chain in the focal firm's ecosystem, it can be readily seen that when an innovation of the focal firm requires corresponding innovations in the upstream components and/or complements, the development challenges of the firm increase. In such a case, the firm has to overcome additional obstacles in specifying, sourcing, and integrating new components into its newly designed products. These obstacles generally challenge the firm to revamp its existing production as well as organizational routines or to totally replace these routines with new ones. Hence, challenges of the components increase the scale and scope of the firm's learning and doing, and consequently the performance advantage the firm gains over its rivals from progressing along its learning curve.

Additionally, the increased coordination with upstream suppliers (Dyer & Singh, 1998) and more frequent travels through the cycles of learning, design, and production that accompany challenging components (Clark & Fujimoto, 1991) benefit the focal firm through greater experiences with new knowledges, different ways of creating innovative designs, and forever improving procedures of operation and production. Such benefits enjoyed by the focal firm help to reduce the ease for later rivals to imitate this progress (Lippman & Rumelt, 1982; Rivkin, 2000) and consequently the offers. For example, such difficulty in imitating innovations, which emerged from extensive collaborations between Toyota and its suppliers, has been well documented by Dyer and Hatch (2006).

Proposition 7.2 If a challenge the focal firm brings on its ecosystem as a consequence of its particular capability of learning and practical follow-ups is mostly on the upstream suppliers, then the lifespan of the exclusive performance advantage of the firm will be greatly shortened.

This conclusion follows from the fact that the actual learning experience of the focal firm will be mainly located at the step of specifying what necessary components it needs to relatively readily assemble into its designed product(s). In other words, the key developments within the focal firm's effort to produce its new offer to the market are undertaken by the upstream suppliers. So, once the required developments are done, the components will also be available for the firm's rivals. These rivals can then free ride on the learning, discovery, and investments of the focal firm to offer their similar products. That will definitely shorten the focal firm's period of exclusivity in the market or the firm's performance advantage. This analysis also explains why Proposition 7.2 does not contradict the conclusion in Proposition 7.1.

Systemically, the input-output mechanism of the challenge the focal firm brings on its ecosystem can be depicted in Fig. 7.2a, where the amount of inputs and that of outputs need to be at equilibrium in order for the yoyo structure of the firm to stay healthy and viable for the long run. So, when the challenge places most of its burden on the mechanism of inputs while that of outputs stays unaffected, the viability of the yoyo structure of the challenge on the ecosystem cannot last too long, because the input-output mechanism will be so congested that it has to stop working (Fig. 7.2b).

As a corollary of Proposition 7.2, we have the following result conveniently.

Proposition 7.3 The more innovative the focal firm is in its effort to assemble components internally, the longer its performance advantage in the marketplace will last.

Empirically, this conclusion is supported by Brusoni and Prencipe (2001) and Hoetker (2006) when they demonstrate the increasing need in the present world of technological advances for the focal firm to possess its greater organizational and learning capabilities in its role as system integrators when specialist component

Fig. 7.2 The input-output mechanism: (**a**) in balance; (**b**) unbalanced

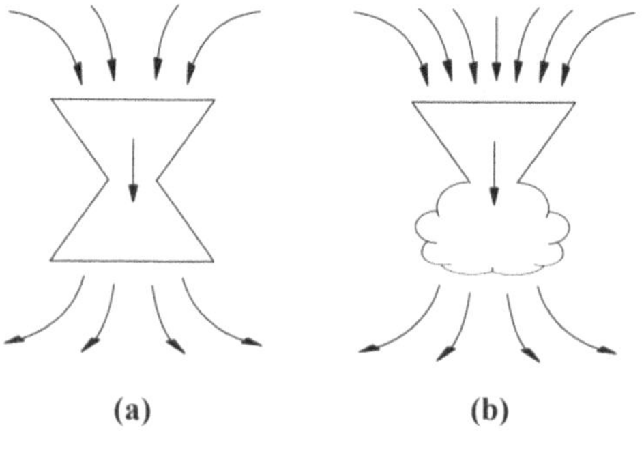

suppliers appear. At the same time, such capabilities generally take a good amount of time and great effort to acquire (McGrath, 2013).

7.4 Challenges Facing Downstream Enterprises

Compared to upstream suppliers, whose outputs enable the focal firm to practically produce its innovative offer and present it to the market, the actual value of the offer for customers is only created through the availability of appropriate complements. For example, the ability of flying cars (Lemoussu et al., 2018) to create value for users is realistically hampered until the appropriate road conditions and air traffic controls, as key complements, are constructed and installed in order for such cars to move and fly around freely. As a matter of fact, many innovative products rely on the availability of appropriate complements to expose their full value to users. So, in this section, we look at how what is learned by the focal firm and what it innovatively designs post challenges on the downstream complementors, Fig. 7.3. This figure actually also provides an explanation for why different firms receive the market invitation differently. In particular, the market calls—the upward moving arrows of the market—for additional competition and new innovation is partially shielded off by the downward moving field of the ecosystem of the firm, while different firms

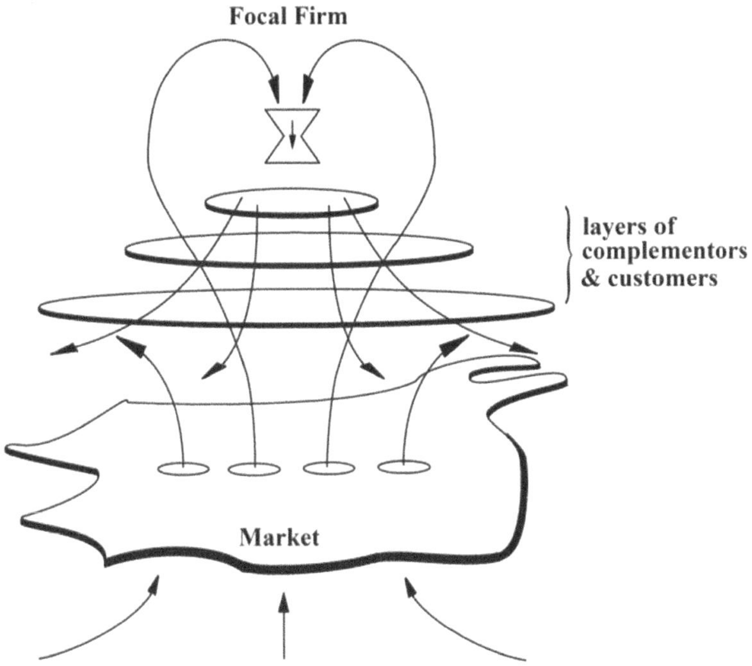

Fig. 7.3 Interaction between focal firm's innovation and challenges of complementors

have their individually different systemic yoyo fields in terms of their spinning directions, strengths, etc.

Proposition 7.4 In general, the opportunities and challenges faced by customers in adopting an innovative offer of the focal firm can be influenced by the state of development of complements. In other words, any innovative offer of the focal firm rarely constitutes a "complete innovation."

In fact, it suffices to list particular cases to support the conclusion in this proposition. To this end, the evolution of modern technology has many times experienced imbalances in the development pace of complementary innovations, which created "reverse salients" (Hughes, 1983). In the personal computer industry, for example, the value of various products developed by suppliers had been constrained by complements (Ethiraj, 2007). Similarly, over 100 years ago, the railroad and printing industries delayed their adoption of electricity due to challenges that were unconquerable at the time in complements (Goldfarb, 2005). Please note that Proposition 7.4 was initially suggested by Rosenberg (1972, p. 21).

Intuitively, this conclusion can be readily seen from Fig. 7.3. In particular, for the focal firm to deliver its offer to the market, appropriate layers of complementors and customers have to be in place first in order for the firm's outputs to reach the field of the market.

Proposition 7.5 The greater challenges complementors face, the less performance advantage the focal firm will enjoy from its innovative offer.

That is because when complementors confront great challenges, delays will appear inevitably in the availability of appropriate complements needed for customers to fully enjoy the value of the innovative offer of the focal firm. Such delays consequently reduce the timely value creation of the offer because the lack of appropriate complements slows down the adoption rate of the offer. Other than giving rivals additional time to catch up with the innovative design and production of the offer before the market takes off, the delays also prevent the focal firm from greatly expanding its production, perfecting its initial offering, and recuperating its earlier heavy investment on the design and production of the offer. Without the advantage of a sufficient period of market exclusivity, the focal firm will not be able to muster enough manpower and financial resources to embark on its next innovative frontiers. In other words, beyond the issue of imitation, later entrants will not suffer from much of the disadvantage of being late to the market. By lowering the adoption rate of the innovative offer of the focal firm and by stripping away the necessary resources for the firm to look beyond the current offer, no matter how innovative the focal offer is, greater challenges complementors face in reality not only reduce the performance advantage the focal firm from its offer, but also wear away the firm's capability to successfully ride waves of transient competitive advantages.

Hence, this analysis also implies the following result.

Proposition 7.6 For the focal firm to successfully ride waves of transient competitive advantages, other than thinking about how to introduce innovative designs of products, which generally challenge upstream suppliers, it also needs to consider whether or not appropriate complements can be readily developed by downstream complementors.

7.5 Vertical Interdependence of a Supply-Chain Ecosystem

From the discussions in the previous two sections, it can be seen quite clearly that the focal firm is closely associated with the upstream suppliers and the downstream complementors and customers. So, it is natural for us to look at how such vertical interdependence that threads through the ecosystem of the firm can possibly be managed. The yoyo structure of the focal firm's ecosystem and its relation with the market are depicted in Fig. 7.1, where the sustainability of the focal firm really depends on the outputs of the upstream suppliers and the accommodation of downstream customers and complementors.

So far in this chapter, we have looked at how an initial receipt of a market invitation for innovation by the focal firm starts a chain of associated challenges, some of which confront upstream suppliers while others confront downstream complementors. In particular, from the recognized market invitation, the focal firm first designs and then attempts to produce its innovative offer. However, to present a particular offer to answer the market invitation and a value-creating solution to customers, the firm has to combine its innovative design and effort of new production with specifically needed supplies. This will most likely post challenges to upstream suppliers, and unambiguously necessary complements, which generally presents challenges to downstream complementors, to facilitate the indispensable conditions for customers to enjoy the full potential of the focal offer. Our established propositions indicate that the distribution of the associated challenges across the ecosystem is an essential driver behind the outcomes of all firms involved. In this section, we will investigate how vertical integration can actually play the role of a governance strategy to manage such interdependence among all the parties of the ecosystem.

Proposition 7.7 When contracting with upstream suppliers, the focal firm has to deal with the following two kinds of uncertainties that affect the performance advantage of the focal firm:

- Technological uncertainty and
- Suppliers' behavioral uncertainty,

where the first uncertainty affects the focal firm's ability to create value, while the second uncertainty impacts the firm's ability to capture value.

In fact, when the focal firm contracts with upstream suppliers for special purpose components that involve high levels of development challenges, the suppliers do not really know whether or not they can discover appropriate solutions to their development challenges, and if successful with such discoveries, then when they will be able to actually deliver the contracted products (Clark, 1985). This scenario represents the technological uncertainty the focal firm has to face. Evidently, resolution of the technological uncertainty decides the focal firm's value creation: if the suppliers cannot provide the needed components, then the focal firm will not be able to bring its imagined product to the market to satisfy the demand of customers. In this case, Proposition 7.1 implies that the greater technological uncertainty the focal firm faces, the greater performance advantage it can enjoy over its competitors.

At the same time, the performance advantage of the focal firm, as a consequence of successful resolutions of technological uncertainties confronting upstream suppliers (Proposition 7.1), may realistically entice these suppliers to behave opportunistically due to suddenly increased demands from the competitors of the focal firm. That is, the focal firm has to deal with suppliers' behavioral uncertainty regarding whether or not and when (some of) the suppliers will behave opportunistically (Jap & Anderson, 2003; Sutcliffe & Zaheer, 1998). If (some of) the suppliers renegotiate the contract terms opportunistically with the focal firm because of the increasing demand for their products, then the focal firm will not be able to appropriate its originally expected rents. As implied by Proposition 7.2, suppliers' opportunistic behavior will then greatly shorten the lifespan of the exclusive performance advantage of the focal firm.

Systemically speaking, technological uncertainty appears with the situation of whether or not the desired inputs of the focal firm (Fig. 7.1) can be adequately fed by the large arrowed input curves that trespass through the layers of the upstream suppliers; and it might take a long time and a lot of effort to construct some of these needed curves. And suppliers' behavioral uncertainty exists because, in the location of the focal firm in Fig. 7.1, most of the firm's rivals can fit too.

Proposition 7.8 For each technology, the associated technological uncertainty always decreases over time within the technology's life cycle.

Generally, when a technology is in the germinating stage of its life cycle, the uncertainty about whether or not an appropriate technological solution to recognized development challenges can ever be developed is certainly high (Proposition 7.7). If the preliminary version of the technology (early stage) is supported by persistent market demand, then the development of the technology will take place. The accompanying knowledge accumulation helps to provide necessary guidance on the direction of future development, eliminating a lot of the uncertainties about *what* and *how*. That is, the forthcoming progress becomes more and more predictable than before (McGrath, 2013), which means lowering technological uncertainty. This end explains why the level of technological uncertainty decreases over time although throughout the life cycle of the technology innovation challenges are always present. For relevant empirical studies, see (Dosi, 1982; Sahal, 1981). In other words, technological uncertainty decreases with time.

This conclusion can be intuitively seen readily by using the systemic yoyo model, Fig. 7.4. In particular, at the germinating stage of a technology, it is an innovative idea that combines some different existing technologies (say, the yoyo pools M, N, and P in Fig. 7.4) together in an attempt to provide a solution to some market

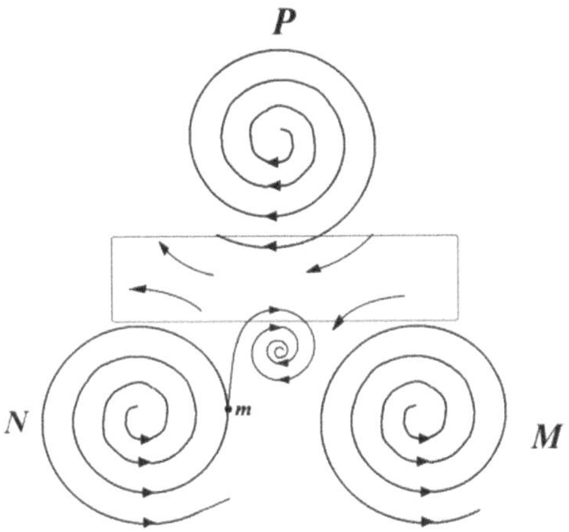

Fig. 7.4 The development of a new technology

challenges. Due to the different spinning directions of these pools, it is initially difficult for the small pool (the start of a new technology) in the center to form, because sparks of thoughts fly in all directions, as shown in the enclosed area in Fig. 7.4. As the appropriate spinning directions of *M* and *N* eventually get the little pool in the middle started, both *M* and *N* will continuously feed the fledgling technology with what they have while all irrelevant influences from *P* will be ignored. So, from such a moment on, the growth of the new technology will become more stable and predictable. That is, the relevant technological uncertainty starts to decrease with time.

Proposition 7.9 Over the life cycle of a technology developed for a need of the focal firm, the suppliers' behavioral uncertainty potentially increases except when the demand for the technology does not rise by much and the technology does not need frequent improvement over time.

In general, the assumption that each firm in the business world exists for the purpose of satisfying a particular market niche through generating a positive cash flow as a result of profits from the marketplace, or investments from various investors, or both, necessarily implies that the suppliers' behavioral uncertainty over the life cycle of a technology is a function of how much the technology is in demand and how frequently it needs to be upgraded. Hence, there are four possible scenarios for us to consider:

1. The demand for the technology does not rise much and the technology does not need frequent improvement,
2. The demand for the technology rises drastically while the technology does not need to be improved frequently,
3. The technology needs to be improved frequently while the demand for the technology does not rise much, and
4. The demand for the technology rises drastically and the technology needs to be improved frequently.

For case (1), the increased experience and repeated interactions between business partners and associates generally allow firms to specify better contracts and extend relational contracts. In such cases, behavioral uncertainty decreases over time within the life cycle of the technology, as empirically evidenced by Mayer and Argyres (2004) and Argyres et al. (2007).

For case (2), the assumptions insinuate that the particular technology is mature and that the focal firm has enjoyed good performance advantage, as a consequence of the introduction and development of the technology. That advantage in turn encourages the competitors of the focal firm to also acquire the technology. Consequently, the drastically rising demand entices (at least some of) the suppliers to renegotiate the contract terms opportunistically. In this case, the suppliers' behavioral uncertainty increases over the technology's life cycle.

For case (3), the focal firm needs the suppliers to improve their technologies frequently in order to continuously better its offer to the market. The fact that the demand for the technology does not increase much means that most of the competitors of the focal firm are not yet motivated to share the market with the focal firm. In this case, both the focal firm and its suppliers grow together in concert with increasing levels of co-specialization, creation of transaction-specific assets, etc. That generally makes any supplier switching extremely costly for the focal firm. So, for case (3), the suppliers' behavioral uncertainty could potentially increase over the technology life cycle because the frequent technology improvements place the focal firm in a relatively more vulnerable position than any of the suppliers.

For case (4), the drastically increasing demand for the technology entices (at least some of) the suppliers to behave opportunistically, while the need for continuous improvement makes the suppliers target very specific customers, such as the focal firm. So, as a consequence, at least some of the suppliers will behave opportunistically with their older versions of the technology, while the most recently improved version will be closely tied to the changing need of the focal firm. Hence, the close tie makes it extremely costly for the focal firm to switch suppliers. That increases the incentive for suppliers to behave opportunistically. That is, for case (4), the suppliers' behavioral uncertainty increases over the technology life cycle. This scenario is well illustrated by the experience between General Motors (GM) and Fisher Body. When the demand for closed metal bodies increased rapidly, GM suffered from significant contractual hazards, which led GM to eventually acquire Fisher Body (Klein, 2000).

A systemic intuitive is that when the demand for the technology rises drastically (cases 2 and 4), more of the competitors of the focal firm want to squeeze into the location of the focal firm in Fig. 7.1. That provides an increasing amount of incentive for the suppliers to behave opportunistically. When the demand for the technology does not rise by much and the technology does not need frequent improvement (case 1), the systemic structure of the ecosystem in Fig. 7.1 stays relatively stable, which means that suppliers' behavioral uncertainty decreases over time within the life cycle of the technology. And when the technology needs to be improved frequently while the demand for the technology does not rise much (case 3), it means that the meridian field of the ecosystem in Fig. 7.1 needs to be renewed frequently. That frequent renewal provides the constantly challenged suppliers with knowledge accumulations and different versions of the technology to conveniently serve other customers. So, they are in a stronger position to renegotiate with the focal firm opportunistically.

By vertical integration, we mean that the focal firm controls more than one stage of its supply chain. For example, a typical supply chain contains four stages: raw materials, manufacturing, distribution, and retail. When the focal firm controls two or more of these stages, we say that the firm integrates vertically. In terms of the systemic yoyo model (Fig. 7.1), when the focal firm integrates vertically, it means that the operation of the firm actually expands into several layers of the suppliers and complementors.

Proposition 7.10 The performance advantage of the focal firm, as a consequence of applying vertical integration within the firm's ecosystem, increases over time within the life cycle of the technology developed to meet the innovative need of the focal firm.

In fact, although it has no bearing on the technological uncertainty, vertical integration helps mitigate adverse effects of suppliers' behavioral uncertainty and provides an opportunity for the focal firm to develop a better understanding about how components and product architectures interact (Kapoor & Adner, 2012). Such firm-specific knowledge facilitates improvements of the technology. As the level of technological uncertainty decreases over time within the technology's life cycle (Proposition 7.8), the performance advantage of the focal firm gained from vertical integration within its ecosystem will become greater at later stages of the technology's life cycle.

7.6 A Few Final Words

To grow and succeed in the present world of business, a firm has to be able to ride waves of transient competitive advantages, as anecdotally confirmed by McGrath (2013). Doing so means that a firm needs to understand market invitations regularly in an epoch-making fashion, while knowing accordingly how to handle uncertainties that appear within the interactions with upstream suppliers and downstream complementors. As pointed out by Adner and Kapoor (2010) and as shown in this chapter, challenges facing an innovative design, consequent production, and eventual offer to the market are often situated inside both a focal firm and the firm's ecosystem. That makes this chapter go beyond the literature of innovative firms' internal and environmental challenges by theoretically establishing a series of generally true conclusions.

By holistically investigating the input-output ecosystem of a focal firm that is able to successfully decipher a market invitation, this chapter shows that each innovative idea of the focal firm, introduced to answer the market call, has to consider whether upstream suppliers and downstream complementors can handle their corresponding challenges or not. We demonstrate that challenges facing upstream suppliers help strengthen the focal firm's performance advantage over its competitors, while the state of development of complements insures customers' adoption of the focal offer. Specially, we establish the following main results among others:

- A focal firm's performance advantage over its competitors is positively correlated to the level of the firm's upstream component challenges and negatively correlated to that of downstream complement challenges.
- To ride waves of transient competitive advantages successfully, a firm has to introduce innovative products, to which suppliers can certainly provide necessary solutions while complementors can readily develop appropriate complements.
- The strategy of vertical integration can help increase a firm's performance advantage over the life cycle of the technology developed for that firm's innovative need.

Beyond developing a series of practically useful conclusions, this chapter grows the traditional toolbox of research by demonstrating how systems science and methodology can be employed. By exploiting logic reasoning and holistic thinking, this chapter is able to reveal some of the essential rules that govern the operation of a general input-output business ecosystem, producing more insightful knowledge than that in the literature (Utterback & Abernathy, 1975; Anderson & Tushman, 1990; Cusumano et al., 1992) on the performance of a focal firm and the distributions of challenges and uncertainties, and development of technologies within the ecosystem of the firm.

The general results developed in this chapter provide practical managerial guidelines on how to manage challenges that are originated from attempts of answering market calls in epoch-making fashions. In particular, we have the following recommendations:

- When pursuing first mover advantage, a firm has to carefully evaluate whether or not its suppliers can adequately provide what is needed and its complementors can facilitate the necessary complements.
- The more upstream suppliers and the less downstream complementors are challenged by an innovative product, the greater value the product is expected to create for customers and capture for its firm.
- Investing in internal capability to assemble components into innovative products will help lengthen a firm's performance advantage in the marketplace.
- When investing jointly with suppliers in introducing new technology, a firm needs to prepare for the increasing potential of suppliers' behavioral uncertainty. One effective method to mitigate this uncertainty is to employ the strategy of vertical integration.

Speaking differently, before a firm commercializes its innovative idea, it needs to and can well estimate the level of challenges facing its ecosystem.

To conclude this chapter, let us look at the limitations of this chapter. First, the most important limitation is that all our conclusions are derived on the assumption that each firm exists for the purpose of generating positive cash flows. This assumption is of course not universally true in the business world. Second, we need to consider the innovation challenges facing a focal firm in its core activities, especially when they are somewhat correlated with challenges external to the firm. Third, in the relationship among the players within the ecosystem of a focal firm, the history and nature of specific exchanges also need to be considered in order to reveal their possible effects on the relationship (Argyres & Liebeskind, 2002). Fourth, this chapter did not consider interactions of different ecosystems, which is in fact the norm of business and may very well affect the distribution of challenges within each of the interacting ecosystems. Fifth, this chapter assumes explicitly that all other variables, such as strategies employed and their interactions, the specific nature of innovative change, demand uncertainty, and modes of cross-firm coordination, are held constant. So, each of these other variables needs to be jointly considered either individually or collectively.

These limitations of course point out to some of the future research directions that will be necessary to make the series of conclusions established in this chapter more theoretically complete and practically useful.

References

Adner, R. (2017). Ecosystem as structure: An actionable construct for strategy. *Journal of Management, 43*(1), 39–58.

Adner, R., & Kapoor, R. (2010). Value creation in innovation ecosystems: How the structure of technological interdependence affects firm performance in new technological generations. *Strategic Management Journal, 31*(3), 306–333.

Adner, R., Oxley, J. E., & Silverman, B. S. (Eds.). (2013). *Collaboration and competition in business ecosystems*. Emerald.

Anderson, P. C., & Tushman, M. L. (1990). Technological discontinuities and dominant designs. *Administrative Science Quarterly, 35*(4), 604–633.

Argyres, N. S., Bercovitz, J., & Mayer, K. J. (2007). Complementarity and evolution of contractual provisions: An empirical study of IT services contracts. *Organization Science, 18*(1), 3–19.

Argyres, N. S., & Liebeskind, J. P. (2002). Governance inseparability and the evolution of the biotechnology industry. *Journal of Economic Behavior and Organization, 47*, 197–219.

Attour, A., & Lazaric, N. (2018). From knowledge to business ecosystems: Emergence of an entrepreneurial activity during knowledge replication. *Small Business Economics*. https://doi.org/10.1007/s11187-018-0035-3

Bilgicer, T., Jedidi, K., Lehmann, D. R., & Neslin, S. A. (2015). Social contagion and customer adoption of new sales channels. *Journal of Retailing, 91*(2), 254–271.

Bowman, C., & Ambrosini, V. (2001). "Value" in the resource-based view of the firm: A contribution to the debate. *Academy of Management Review, 26*, 501–502.

Brem, A., & Viardot, É. (2015). Adoption of innovation: Balancing internal and external stakeholders in the marketing of innovation. In A. Brem & É. Viardot (Eds.), *Adoption of innovation* (pp. 1–10). Springer.

Bruneel, J., Spithoven, A., & Clarysse, B. (2017). Interorganizational trust and technology complexity: Evidence for new technology-based firms. *Journal of Small Business Management, 55* (S1), 256–274.

Brusoni, S., & Prencipe, A. (2001). Unpacking the black box of modularity: Technologies, products and organizations. *Industrial and Corporate Change, 10*, 179–205.

Brusoni, S., Prencipe, A., & Pavitt, K. (2001). Knowledge specialization, organizational coupling, and the boundaries of the firm: Why do firms know more than they make? *Administrative Science Quarterly, 46*, 597–621.

Bstieler, L., & Hemmert, M. (2015). The effectiveness of relational and contractual governance in new product development collaborations: Evidence from Korea. *Technovation, 45–46*, 29–39.

Burton, N., & Galvin, P. (2018). When do product architectures mirror organisational architectures? The combined role of product complexity and the rate of technological change. *Technology Analysis & Strategic Management, 30*(9), 1057–1069.

Byun, J. G., Sung, T. F., & Park, H. W. (2018). Technological innovation strategy: How do technology life cycles change by technological area. *Technology Analysis & Strategic Management, 30*(1), 98–112.

Cao, Z., & Lumineau, F. (2015). Revisiting the interplay between contractual and relational governance: A qualitative and meta-analytic investigation. *Journal of Operations Management, 33–34*, 15–42.

Chang, W., Ellinger, A. E., Kim, K., & Franke, G. R. (2016). Supply chain integration and firm financial performance: A meta-analysis of positional advantage mediation and moderating factors. *European Management Journal, 34*(3), 282–295.

Clark, K. B. (1985). The interaction of design hierarchies and market concepts in technological evolution. *Research Policy, 14*, 235–251.

Clark, K. B., & Fujimoto, T. (1991). *Product development performance*. Harvard Business School Press.

Clarysse, B., Wright, M., Bruneel, J., & Mahajan, A. (2014). Creating value in ecosystems: Crossing the chasm between knowledge and business ecosystems. *Research Policy, 43*(7), 1164–1176.

Cusumano, M., Mylonadis, Y., & Rosenbloom, R. (1992). Strategic maneuvering and mass market dynamics: The triumph of VHS over Beta. *Business History Review, 66*(1), 51–94.

Dosi, G. (1982). Technological paradigms and technological trajectories. *Research Policy, 11*, 147–162.

Dyer, J. H., & Hatch, N. W. (2006). Relation-specific capabilities and barriers to knowledge transfers: Creating advantage through network relationships. *Strategic Management Journal, 27*(8), 701–719.

Dyer, J. H., & Singh, H. (1998). The relational view: Cooperative strategy and sources of inter-organizational competitive advantage. *Academy of Management Review, 23*(4), 660–679.

Ethiraj, S. K. (2007). Allocation of inventive effort in complex product systems. *Strategic Management Journal, 28*(6), 563–584.

Forrest, J. Y.-L., Liu, Y., Wajda, T. A., Köse, E., & Arik, O. A. (2019). Dependence of a firm's performance on its ecosystem's upstream/downstream challenges. *Proceedings of the 2019 Annual meeting of National Association of business, Economics and Technology* (pp. 83–100).

Goldfarb, B. (2005). Diffusion of general-purpose technologies: Understanding patterns in the electrification of US manufacturing 1880-1930. *Industrial and Corporate Change, 14*(5), 745–773.

Hallberg, N. L. (2015). Uncertainty, judgment, and the theory of the firm. *Journal of Institutional Economics, 11*(3), 623–650.

Helfat, C. E., Campo-Rembado, M. A., Helfat, C. E., & Campo-Rembado, M. A. (2016). Integrative capabilities, vertical integration, and innovation over successive technology lifecycles. *Organization Science, 27*(2), 233–504.

Hellström, M., Tsvetkova, A., Gustafsson, M., & Wikström, K. (2015). Collaboration mechanisms for business models in distributed energy ecosystems. *Journal of Cleaner Production, 102*(1), 226–236.

Hendrikse, G., Hippmann, P., & Windsperger, J. (2015). Trust, transaction costs and contractual incompleteness in franchising. *Small Business Economics, 44*(4), 867–888.

Hoetker, G. (2006). Do modular products lead to modular organizations? *Strategic Management Journal, 27*(6), 501–518.

Huang, M. C., Yen, G. F., & Liu, T. C. (2014). Reexamining supply chain integration and the supplier's performance relationships under uncertainty. *Supply Chain Management: An International Journal, 19*(1), 64–78.

Hughes, T. P. (1983). *Networks of power: Electrification in Western society 1880–1930*. Johns Hopkins University Press.

Iansiti, M. (1998). *Technology integration: Making critical choices in a dynamic world*. Harvard Business School Press.

Jap, S. D., & Anderson, E. (2003). Safeguarding interorganizational performance and continuity under expost opportunism. *Management Science, 49*(12), 1684–1701.

Jiang, Y., & Tornikoski, E. T. (2018). Perceived uncertainty and behavioral logic: Temporality and unanticipated consequences in the new venture creation process. *Journal of Business Venturing*. https://doi.org/10.1016/j.jbusvent.2018.06.002

Kapoor, R., & Adner, R. (2012). What firms make vs. what they know: How firms' production and knowledge boundaries affect competitive advantage in the face of technological change. *Organization Science, 23*(5), 1213–1522.

Kapoor, R., & Agarwal, S. (2017). Sustaining superior performance in business ecosystems: Evidence from application software developers in the iOS and android smartphone ecosystems. *Organization Science, 28*(3), 379–595.

Klein, B. (2000). Fisher-general motors and the nature of the firm. *Journal of Law and Economics, 43*(1), 105–142.

Laukkanen, T. (2016). Consumer adoption versus rejection decisions in seemingly similar service innovations: The case of the internet and mobile banking. *Journal of Business Research, 69*(7), 2432–2439.

Lee, C. Y., Kim, J., Noh, M. S., Woo, H. G., & Gang, K. W. (2017). Patterns of technology life cycles: Stochastic analysis based on patent citations. *Technology Analysis & Strategic Management, 29*(1), 53–67.

Lemoussu, S., Chaudemar, J. C., & Vingerhoeds, R. A. (2018). Systems engineering and project management process modeling in the aeronautics context: Case study of SMEs. *International Journal of Mechanical and Mechatronics Engineering, 12*(2), 88–96.

Lippman, S. A., & Rumelt, R. P. (1982). Uncertain imitability: An analysis of interfirm differences in efficiency under competition. *Bell Journal of Economics, 13*(2), 418–438.

Mayer, K. J., & Argyres, N. S. (2004). Learning to contract: Evidence from the personal computer industry. *Organization Science, 15*(4), 394–410.

McGrath, R. G. (2013). *The end of competitive advantage: How to keep your strategy moving as fast as your business*. Harvard Business Review Press.

Mondragon, A. E. C., & Mondragon, C. E. C. (2018). Managing complex, modular products: How technological uncertainty affects the role of systems integrators in the automotive supply chain. *International Journal of Production Research*. https://doi.org/10.1080/00207543.2018.1424362

Oliveira, T., Thomas, M., Baptista, G., & Campos, F. (2016). Mobile payment: Understanding the determinants of customer adoption and intention to recommend the technology. *Computers in Human Behavior, 61*, 404–414.

Porter, M. E. (1985). *Competitive advantage: Creating and sustaining superior performance*. Free Press.

Prajogo, D. I., & Oke, A. (2016). Human capital, service innovation advantage, and business performance: The moderating roles of dynamic and competitive environments. *International Journal of Operations & Production Management, 36*(9), 974–994.

Priem, R. L. (2007). A consumer perspective on value creation. *Academy of Management Review, 32*, 219–235.

Rivkin, J. W. (2000). Imitation of complex strategies. *Management Science, 46*(6), 824–844.

Roca, J. B., Vaishnav, P., Morgan, M. G., Mendonça, J., & Erica Fuchs, E. (2017). When risks cannot be seen: Regulating uncertainty in emerging technologies. *Research Policy, 46*(7), 1215–1233.

Rosenberg, N. (1972). Factors affecting the diffusion of technology. *Explorations in Economic History, 10*(1), 3–33.

Sahal, D. (1981). *Patterns of technological innovation*. Addison-Wesley.

Sutcliffe, K. M., & Zaheer, A. (1998). Uncertainty in the transaction environment: An empirical test. *Strategic Management Journal, 19*(1), 1–23.

Trada, S., & Goyal, V. (2017). The dual effects of perceived unfairness on opportunism in channel relationships. *Industrial Marketing Management, 64*(July), 135–146.

Utterback, J. M., & Abernathy, W. J. (1975). A dynamic model of process and product innovation. *Omega, 3*(6), 639–656.

Yadav, P. L., Han, S. H., & Kim, H. (2017). Sustaining competitive advantage through corporate environmental performance. *Business Strategy and the Environment, 26*(3), 345–357.

Yeo, C. S., & Saboori-Deilami, V. (2017). Strategic challenges of outsourcing innovation in global market. *Asia Pacific Journal of Innovation and Entrepreneurship, 11*(1), 5–16.

Zahra, S. A., & Nambisan, S. (2012). Entrepreneurship and strategic thinking in business ecosystems. *Business Horizons, 55*(3), 219–229.

Chapter 8
Sufficient Conditions that Lead to Synergistic Innovations

Abstract This chapter, which is mainly based on (Forrest et al., 2019, *Proceedings of the 2019 Annual Conference of Decision Sciences Institute* (pp. 2061–2080)), investigates issues related to the following questions: (1) how can producer-side synergies be created by employing the strategy of economies of scope? And (2) how can demand-side synergies be developed by making use of simultaneous consumer utilities and multi-sided markets? Because of the different approach taken, we are able to describe how resources interact with each other and how simultaneous consumer utilities, two-sided markets and consumers' willingness to pay react to each other. We first look at how such economic entities and basic concepts as business firm, resource, innovation, diversification, synergy, segment of consumers, etc., can be respectively modeled as interacting systems. We then address the previous two questions by establishing a series of 8 propositions by examining how relevant systems exert forces on each other so that some general conclusions follow. Because of the systemic certainty our discussions offer, this chapter is expected to provide practically useful guidance for managers, entrepreneurs, and retailers to create values for consumers and capture values for their companies.

Keywords Competitive advantage · Economy of scope · Resource · Simultaneous consumer utility · Two-sided market

8.1 Introduction

Recent researches in strategic management suggest and demonstrate the need for firms to develop the necessary organizational culture and capability to effectively ride waves of transient competitive advantages (e.g., McGrath, 2013). For a firm to do so successfully it can look either inwardly at its nascent, heterogeneous resources and dynamic capabilities to see what values it can create for consumers (e.g., Barney, 1991; Eisenhardt & Martin, 2000), or outwardly at product markets and consumer demands to find out where consumer synergies are located and what can be potentially developed (Drucker, 1954). No matter whether to look inwardly or outwardly, one way for a firm to possibly find potentials of different competitive advantages is

J. Y.-L. Forrest, Y. Liu, *Value in Business*, Contributions to Management Science,
https://doi.org/10.1007/978-3-030-82898-1_8

to create economies of scope by diversification at either the producer side (Porter, 1985; Santalo & Becerra, 2008) or the demand side (Ye et al., 2012) or both. If a firm adopts this approach, then the firm has to address the following natural questions: (1) how can producer-side synergies be created when using the strategy of economies of scope? (2) How can demand-side synergies be developed by making use of simultaneous consumer utilities and multi-sided markets, after consumers determine the success of companies (Penrose, 1959)?

Aiming at addressing these two questions, this chapter develops a cohesive theory on how to potentially develop synergistically innovative ideas at either the producer side or the demand side. In this theoretical development, we rely 100% on logical reasoning, systems thinking, and the intuitive yoyo model (Lin, 2009) without analyzing any set of data or anecdotes. Considering the fact that data mining and anecdotal analysis can only help uncover potential facts because of their constraints inherently existing in the available data/anecdotes and limitations existing in the analysis tools, results and conclusions established in this chapter are more reliable and practically useful than those based on data and anecdotes.

In terms of the structure of our cohesive theory, we first spell out the basic assumptions and meanings of particular terms. Then by using the reasoning of logic and the methodology of systems science we develop formal propositions to address various issues related to the two questions given above.

The rest of this chapter is organized as follows. Section 8.2 reviews the relevant literature and demonstrates the contributions of this chapter. Section 8.3 develops a few insights on synergistic innovation of the producer side. Section 8.4 looks at similar issues on the demand side. And this chapter is concluded in Sect. 8.5.

8.2 Literature Review

Because this chapter contributes simultaneously to the literatures of innovation and strategy, in this section, we will look at the relevant literatures individually and show where our contributions lie in each case.

For the literature of innovation, over 200 years ago Adam Smith (1776) acknowledges the importance of innovation. The topic of innovation of the producer side has been well investigated through the years by many scholars (Aas et al., 2015; Adner & Levinthal, 2001; Damanpour, 1991; Visnjic et al., 2016). In the business world, innovation is characterized as the introduction of original products and processes (Becheikh et al., 2006) that enable firms to competitively enter or create new markets (Smith & Tushman, 2005). And it has been treated by many as one major factor of long-term performance (Kanter, 2001). Veugelers and Cassiman (1999) find that high perceived risks and costs and low appropriation do not discourage innovation, but determines how the innovation sourcing strategy is chosen. They find that small firms are more likely to restrict their innovation strategy to an exclusive make or buy strategy, while large firms are more likely to combine both internal and external knowledge acquisition in their innovation strategy. With economic globalization and the advent of Internet-based technologies, firms experience increasing pressures to become more competitive and innovative than ever before (Buffington, 2016; Caputo et al., 2016; Zollo et al., 2016).

Contributing to this activity area of research, this chapter establishes the following general result: by adding to a successful set X of resources or replacing some resources in X by other resources, a firm can potentially create positive synergies.

For the literature of strategies, resource-based approaches tend to look inwardly at how heterogeneous resources and dynamic capabilities of firms can create value for consumers (e.g., Barney, 1991; Eisenhardt & Martin, 2000; Nason & Wiklund, 2018) through bundling internal resources in different ways (e.g., Sirmon et al., 2008). Because the concept of resources offers a framework for integrating multiple, dissimilar resources to explain their synergistic, differential effects on firms' performance and the contingencies associated with each effect (Fang et al., 2011), these approaches greatly contribute to knowledge of strategies (e.g., Lockett et al., 2009). At the same time, Barney (2001) and Priem and Butler (2001a, 2001b) recognize that when looking instead outwardly, demand issues need to be addressed equally seriously, because after all it is consumers that determine the success of firms (Drucker, 1954; Penrose, 1959).

Along this suggestion of outward looking at the demand side, scholars examine various strategic issues, such as how technological innovations and competitive advantages are affected by market demand (e.g., Adner & Zemsky, 2006; Tripsas, 2008), how value can be created and captured by employing consumer-focused strategies (e.g., Adner & Snow, 2010; Gans et al., 2008; Priem, 2007), and how entrepreneurial innovation can be influenced by consumers (Sawhney et al., 2005; Shah & Tripsas, 2007). More specifically, Ye et al. (2012) study how bundled preferences of consumers can be employed as a basis for developing firms' strategies in an effort that is parallel to the recent producer-focused synergy studies on economies of scope (e.g., Crossland & Smith, 2002; Gary, 2005; Tanriverdi & Venkatraman, 2005).

By attending to both the demand side and the producer side of the literature of strategies, this chapter establishes the following main results, among others, that are generally true: (1) if a firm makes its resources available to all business units in its attempt to produce an economy of scope, the firm will then be able to innovate synergistically. (2) Mobilizing inconsistent resources generally produce negative synergies. And (3) if a set of core resources is behind a firm's portfolio of offers, it is likely for the firm to focus on exploiting its existing competitive advantages instead of developing new advantages.

Other than contributions in the previously listed individual literatures, what is more important is that this chapter holistically incorporates the demand side simultaneously into the studies of strategic management and innovation. At the same time, by using logical reasoning, as a well-tested method in science and mathematics, this chapter integrates different streams of research into an organic whole. On top of that, a theoretical foundation is developed so that strategic relatedness in providing simultaneous consumer utilities of the demand side (Sakhartov & Folta, 2014; Ye et al., 2012) and resource-based relatedness in diversification of the supply side (Farjoun, 1998; Kim & Finkelstein, 2009; Villasalero, 2017) can be identified.

Additionally, we generally theorize some of the fundamental properties of two-sided markets uncovered previously by Rochet and Tirole (2006), Sun and Tse (2009), Ye et al. (2012), and others, either empirically or anecdotally. (For a similar but different approach, see Azevedo & Leshno, 2016).

Another important contribution this chapter makes is the introduction of systems science (especially, the systemic yoyo model) into the study of synergistic innovations. Although written statements, data analysis, and systemic expressions are essentially different forms of articulations, there are major differences. In particular, written statements, no matter how logically and precisely they are transcribed, are often inconclusive due to their nature of linearity and inability of simultaneously controlling the effect of several arguments. Although methods of data analysis overcome these weaknesses of written statements, the essence of these methods is to extrapolate the known past into the unknown future. That explains why methods of data analysis cannot be successfully employed to predict drastically different future (Lin & OuYang, 2010). In comparison, systems science is fundamentally different from written statements and data analysis. It looks into the internal structure of the organization of concern, where the structure and the organization come into being simultaneously. Hence, conclusions of investigations based on systems methodology are much closer to the real situation under investigation than those produced otherwise.

To make logical reasoning in this chapter valid, assume that each business firm exists for the purpose of satisfying a particular market niche through generating a positive cash flow, which can be profits from the marketplace, or investments from various investors, or both. In other words, the firm has to actively pursue after certain endeavors in order to meet some market demand(s).

By consumers, we mean the end users of products and/or services. By resource, it means an asset of a firm, be it tangible or not (Harmancioglu et al., 2009), which that firm can use to design and implement its strategies (Barney & Arikan, 2001). In other words, a resource stands for something, be it physical, financial, intellectual, or organizational, the firm can mobilize to accomplish its business goals. And, a resource can generally provide alternative services that differentially increase either the efficiency inside the firm or consumer value creation outside the firm.

8.3 Producer Side Synergistic Innovation

By synergistic innovation, it stands generally for such an innovative strategy of a focal firm that systemically deploys and redeploys known elements that exist either within or outside that firm to produce additionally economic benefits. In particular, known elements can be resources—either tangible or intangible, established management routines, production processes, consumer preferences, market demands, etc., and the word "systemically" means that a focal firm can develop a meaningful system or a set of systems that organically combine the known elements to produce new products with much improved functionalities, and/or new services that meet a forever evolving market demand or consumer preference. The following are some good examples of synergistic innovations in action on either producer side or consumer side.

- Different combinations of known technological processes can produce positive synergy (Crossland & Smith, 2002).
- When knowledge across business units of a focal firm is shared, benefits from transferred skills can be anticipated (Hitt et al., 2001).
- When economies of scope and skill transfers are combined, the social network of a focal firm can greatly contribute to that firm's innovativeness (Tsai, 2001).
- Evolving consumer tastes stimulate firms to embark on technological development and innovation (Adner & Zemsky, 2006).
- The benefits consumers enjoy when their knowledge learned from using one product can literally reduce or eliminate the learning time required to use another product (Priem, 2007; Tanriverdi & Lee, 2008).
- The appearance of a disruptive technology can help reveal a new dimension or confirm an existing dimension of consumer preferences for various industries (Adner & Snow, 2010).

This section establishes relevant concepts and elementary properties regarding synergistic innovations at the producer side (Priem & Butler, 2001a, 2001b). The concept of synergistic potential arising from resource relatedness has been introduced since at least the early 1970s (e.g., Palich et al., 2000; Rumelt, 1974) when Rumelt noted that firms, which diversify through employing respectively related resources, generally outclass those firms, which also diversify, but only doing so by utilizing unrelated resources.

8.3.1 Repeated Deployment of Resources

By synergistic innovation at the producer side, it means that a focal firm designs and employs an innovative strategy that systemically deploys available resources in such a way that the firm can produce an economy of scope in terms of products and/or services. By producer synergy, it means that through product diversification as a way to widen a firm's stock of assets and skills, the firm is able to expand the perimeter of its value activities (Porter, 1985, p. 380). By synergistic potential arising from resource relatedness, it means the potential of producing new products and/or services, which helps increase the economy of scope of a firm, by innovatively mobilizing related resources. When a mobilization of certain resources leads to positive economic benefits, then we say that the resources create a positive synergy. Otherwise, we say that the mobilization produces a negative synergy.

(a) Existing associations of resources (b) Additional associations are created

Fig. 8.1 Additional associations of resources are encouraged. (**a**) Existing associations of resources. (**b**) Additional associations are created

Proposition 8.1 If a firm is able to deploy and redeploy resources repeatedly in its effort to produce an economy of scope by making resources, be they tangible or intangible, available to all business units, then there is a potential for synergistic innovations to appear from the effort of diversification.

To see why this proposition holds true, let us model the firm of concern as a system $S = (M, R)$ such that the object set M consists of the resources the firm controls and each relation r in the relation set R stands for a product produced by mobilizing a subset of the resources in M. Now, the given assumption means that freedom and encouragement are given for the employees of the firm to establish possible associations among elements in any subset of M, where each element in M can be repeatedly used in as many associations as possible. That of course implies that a potential of designing and producing many additional products and/or services appear. This logical reasoning is well depicted in Figs. 8.1a and Fig. 2.2b, where the rectangle stands for the object set M, and in Fig. 8.1a A and B are the two existing products of the firm, which are produced respectively by associating resources in their enclosed areas. When employees are encouraged to associate resources any way imaginable, additional enclosed areas, such as C and D in Fig. 8.1b, can be produced.

This proposition can be shown rigorously by using the following theorem of systems science:

Theorem 8.1 If $S = (M, R)$ is a multileveled system with a finite object set M such that

- Each object system of S has a finite object set,
- Each object that appears in any relation in the system S or an object system of S only appears once in that relation, and
- S has at least one chain of object systems of more than one level and each object system on the chain contains more than one object, and $M(S)$ the set of all fundamental objects in S, then $|M| < |M(S)|$.

For a detailed, extremely technical proof of this result, see Appendix.

Now, let us model the focal firm with business units strictly defined and resources divided according to the unit divisions as a system $S = (M, R)$, where each object m in M stands for the bundle of resources available only to a particular business unit and each relation r in R as a product or service of the firm. Then the set $M(S)$ of all fundamental objects of the system S stands for the totality of resources the firm has control over. Now, the conclusion of Theorem 8.1 says that $|M| < |M(S)|$, which means that the number of available resources is greater than that of business units. Therefore, there are more ways to define relations (products) on $M(S)$ than on M. That proves Proposition 8.1.

In terms of the literature, comparing to Proposition 8.1, a more specific conclusion is given by strategic management scholars. For example, it has been argued (e.g., Porter, 1985) that positive synergies from diversification can emerge if a firm is able to share tangible functions and intangibles, such as knowledge and skills, among business units.

In terms of when mobilization of certain resources can lead to negative synergies, we have the following result.

Proposition 8.2 Assume that A and B represent two opposing business goals of a focal firm and that X and Y are mutually exclusive sets of resources that firm has control over, satisfying that resources in X support the realization of goal A and those in Y help actualize goal B. Then a mobilization of any resource in X and any resource in Y will generally produce negative synergies.

To illustrate how two business goals of a focal firm could be opposing to each other, let us consider the following situation (McGrath, 2013):

- A = the goal of implementing the strategy of exploiting the competitive advantages proven successful over time; and
- B = the goal of implementing the strategy of constantly designing and developing new competitive advantages.

Based on this illustration it can be seen that the set of resources employed to materialize goal A and the set of those used for goal B are not the same. All the involved resources can be written into three mutually disjoint sets X, Y, and Z, other than X and Y, as defined in the proposition, set Z includes all those resources that are needed for materializing both business goals A and B, Fig. 8.2. The elements in set Z could include such commonly deployable resources in multiple directions as strong relationships that lead to information sharing, risk-taking, and adoption of innovations (Dutta et al., 1999). Now, it is ready to see that a joint mobilization of the special resources in X and those in Y generally leads to undesirable effects or negative synergies.

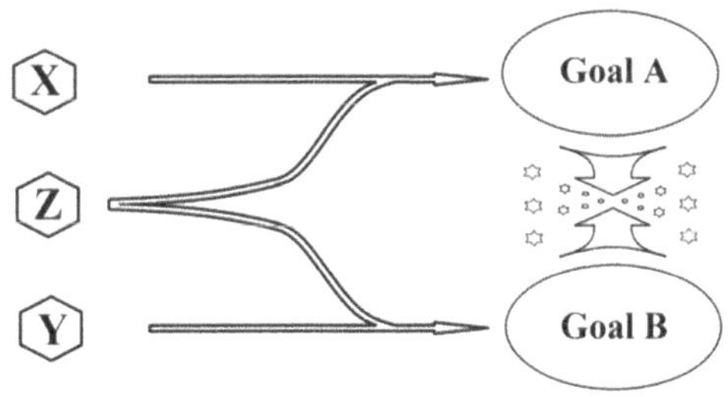

Fig. 8.2 How resources are divided corresponding to business goals A and B

8.3.2 Resources that Might Be Inconsistent

Two resources x and y are said to be inconsistent, if mobilizing these resources jointly leads to negative synergy. For example, any resource from set X and any resource from set Y in Proposition 8.2 will be inconsistent, because jointly they lead to undesirable outcome(s). Now, in terms of when a mobilization of certain resources can potentially lead to positive synergies, we have the following conclusion.

Proposition 8.3 Assume that X represents the set of resources that were used to generate a positive synergy in a focal firm's effort to create an economy of scope. Then by either adding another set of available resources into X or replacing some resources in X with another set of available resources the firm can potentially lead to positive synergies if none of the new resources under consideration is inconsistent with the other remaining resources in X.

This result holds true because altering set X is generally the way for the focal firm to produce additional versions of the product initially developed on X with varying degrees of functionality. Systemically, either adding another set of consistent resource into X or replacing some resources in X by another set of consistent resources mean that the new resources will alter the yoyo structure that underlies the original association of the resources in X in terms of spinning strength, the orientation of the spinning axis, the area the spin field covers, etc. The resultant yoyo structure of the updated set X of course means a different product or service, which is either very different from or somehow related to the original one. Depicting this situation, Fig. 8.3 also demonstrates why the condition that the set of new resources used to alter X has to be consistent with the remaining resources in X, otherwise, the resultant yoyo body can be potentially broken into pieces, as what Proposition 8.2 shows. In Fig. 8.3, the original X evolves into either X' or X", where some of the

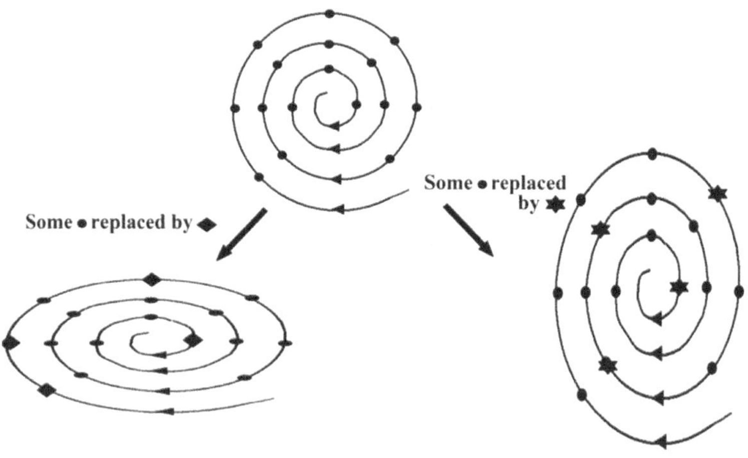

Fig. 8.3 Altering an existing systemic field by changing its composition

round dots are replaced by diamonds in the former case, while by stars in the latter case.

A set of resources is said to be relatable, provided that the association of any two resources in the set does not lead to negative synergy, although the association might not produce a positive synergy, either.

Proposition 8.4 The most natural portfolio of products and/or services of a firm is bound together by a set of related resources that generate positive synergies.

From the point of view of development (Wu & Lin, 2002), any portfolio of products and/or services must start with the very first product or service. When a firm develops its first product or service that generates a positive cash flow (seen as a positive economic benefit), either from the marketplace or investors, there must be a set X of resources the firm has control or influence over that have jointly produced positive synergies for the firm. So, Proposition 8.3 indicates that a natural way to develop different products and/or services will be finding growth points out of the set X of once-already-proven resources through modifying X by either replacing some resources from X with new and consistent ones or adding additional consistent resources into X. That end is vividly shown historically by how industrial revolutions evolve over time and how one follows another (Forrest et al., 2018); and it is partially recognized by Robins and Weirsema (2003, p. 45) when they remarked that "contemporary view(s) of related diversification might be summarized by . . . that . . . a portfolio of businesses is bound together by some shared strategic resources or capabilities."

As for why such a portfolio of products and/or services is considered most natural, it is because finding an organic association of another set of resources to produce positive economic benefits generally is much harder, which is equivalent to starting up a new business. Other than opportunities of luck, this is much more difficult to accomplish than simply finding growth points on top of the set X as described in the previous paragraph (Salamzadeh & Kawamorita, 2015).

Systemically, this result is quite intuitively straightforward: After a yoyo field emerges, as a newborn business entity, in the turbulent ocean (a model of the business world) of countless incumbent yoyo fields that intensively combat against each other first for survival and then for further development, the best and most natural way for it to survive and to grow is to make itself stronger through strengthening what the newborn has already achieved and acquiring additional elements so that new and enhanced associations of various resources will help the yoyo field of the newborn spin faster and more powerfully than ever before.

Proposition 8.5 If a firm's portfolio of products and/or services is bound together by a set of core resources, then there is a good chance for the firm to continuously exploit its established competitive advantages instead of constantly developing new advantages.

This result is a corollary of the previous propositions. In particular, if there is a set of core resources that underlies a firm's portfolio of products and/or services, the superior value of those resources will be widely acknowledged by the stakeholders of the firm if not yet the case already. So, there will be a momentum for the firm to continue its current state of affairs based on the following first law on the state of motion.

First law on the state of motion (Lin, 2009): Each imaginable and existing entity in the universe is a spinning yoyo of a certain dimension. Located on the outskirt of the yoyo is a spin field. Without being affected by another yoyo structure, each particle in the said entity's yoyo structure continues its movement in its orbital state of motion.

The scenario can be seen from a different angle: after putting in tremendous amounts of time and efforts in trying various combinations of resources to get a business operation going, the founders of the firm, as systems themselves with life cycles, can be most likely tired both mentally and physically. That explains why under such circumstances, there is a good chance for the firm to continuously exploit its established competitive advantages instead of constantly developing new advantages, unless new stimuli, either from competitors or from the changing market condition, appears to redirect the attention and energy of the founders and their successors.

8.4 Consumer Side Synergistic Innovations

Since over 50 years ago when Penrose (1959) noted that when paying attention to the needs of consumers, companies grow, studies in strategic management have advocated for paying additional attention to the demand side in decision-making (e.g., Adner & Zemsky, 2006; Priem, 2007). Riding on this trend of thoughts, this section investigates the concepts and elementary properties of synergistic innovations at the demand side (Ye et al., 2012) by placing end users of products and services in the square center.

8.4.1 When Higher Fees Can Be Collected

Similar to how the concept of producer synergy is defined earlier in the previous section we have the following (Ye et al., 2012): By consumer synergy, it stands for an increased value created for consumers and captured by a firm through employing combination(s) of products and/or services that jointly expand consumer utilities beyond the sum of the consumer utilities offered individually by the products and/or services. For illustration purpose, let us look at an excellent example constructed by Ye et al. (2012).

A state university in the northern USA has a population of over 29,000 students, most of whom are unmarried undergraduates living on tight budgets. As most of the rental properties around the campus do not offer laundry facilities, self-service, coin-operated laundromats near the university compete for the students' business.

Another successful kind of college-town business, attractive mostly to college women, is indoor tanning. Instead of offering these services separately, two entrepreneurs provide their independent, single-location businesses, each of which jointly provides coin-operated self-service laundry facility and indoor tanning service. Such business combination creates consumer synergies from diversification of commonly available assets—same or similar equipment—without much reduction in the overhead costs (from completely different sets of equipment) and labor costs (the combined business types require considerably different production and customer service skills). However, this business combination creates tremendous amount of demand-side synergies due to the following reasons:

1. Students can start washing their clothes and then tan their skin simultaneously, saving the otherwise unproductive and costly waiting time; and
2. Students are provided with a place for the potential of establishing relationships because the location of combined services provides opportunities for both males and females to meet, where women's greater participation in the tanning service attracts a larger number of men to the laundry service.

Consequently, the created consumer (or demand side) synergies help respectively support higher fees collected from the laundry service than other standalone stores and maintain competitive prices for the much-increased demand for the tanning service. In other words, a greater value is respectively created for consumers and captured by business providers because of the resultant consumer synergies from product-and-service diversification.

In general, if two or more market demands can be met simultaneously, such as the situation described in reason (1) above, while saving valuable resources for consumers, such as time and/or money, then we say that the diversification effort of meeting different market demands has created a simultaneous consumer utility effect. If two services, each of which attracts a different segment of consumers, although the segments might be overlapping, are collocated so that the increased traffic flow due to the joint offer of services from one segment of consumers attracts an increasing traffic flow from the other segment of consumers, while the latter does not have any adverse effect on the former, as described in reason (2) above, then we say that the diversification of services has created a two-sided market effect.

Proposition 8.6 When collocating diverse products and/or services leads to a simultaneous consumer utility effect, consumers will be willing to pay additionally for a product or service beyond the market norm.

First, when a simultaneous consumer utility effect is created by collocating diverse products and/or services, by definition it means that consumers have saved some of their valuable resources. That saving generally is convertible into consumers' willingness to pay more than what is expected on the average in the marketplace. When a two-sided market effect is developed by collocating diverse products and/or services, the alternatively reinforcing effects on the traffic flow of the different segments of consumers make the demand for the joint offer of products and/or services rise. That in turn supports rising prices.

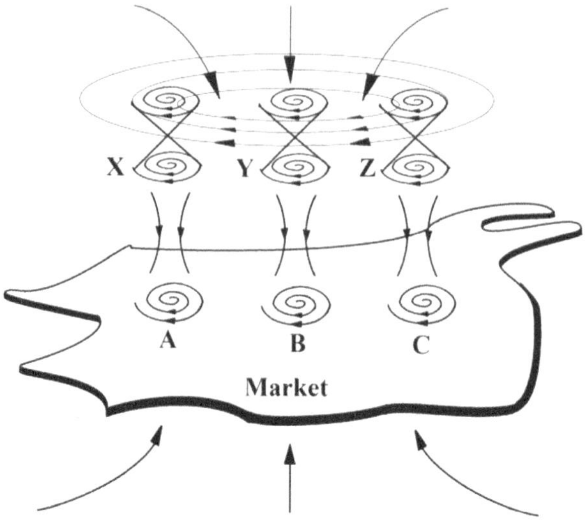

Fig. 8.4 Meeting three market demands simultaneously

Systemically speaking, each simultaneous consumer utility effect can be described by employing the yoyo model in Fig. 8.4. In this systemic depiction, three market demands A, B, and C can be satisfied by either three different business entities X, Y, and Z separately or by one business firm whose yoyo field encompasses all those of the three entities X, Y, and Z. The idea is kind like, but not exactly the same as that of one-stop shopping. The reason why consumers are willing to pay extra is demonstrated by the fact that the conflict zones existing in areas between the yoyo fields X, Y, and Z no longer exist, if these small yoyo fields can be combined into a much larger field that encompasses those of X, Y, and Z.

Proposition 8.7 When collocating diverse products and/or services leads to a two-sided market effect, consumers will be willing to pay additionally for a product or service beyond the market norm.

For two-sided market effects, the concept can be intuitively seen systemically in Fig. 8.5. In this systemic intuition two different market segments A and B of consumers are identified, where consumers in B are closely attached to those in A who at the same time are particularly attracted to the service provided by firm X while can also take advantage of the service of Y. From such a realization of potentially creating a two-sided market through diversified offers of products and/or services, the business ideas of both X and Y are combined into one business practice, whose yoyo field encompasses those of both X and Y. This new business firm will attract the consumers in A, through which those consumers in B are also pulled over from the corresponding competitors.

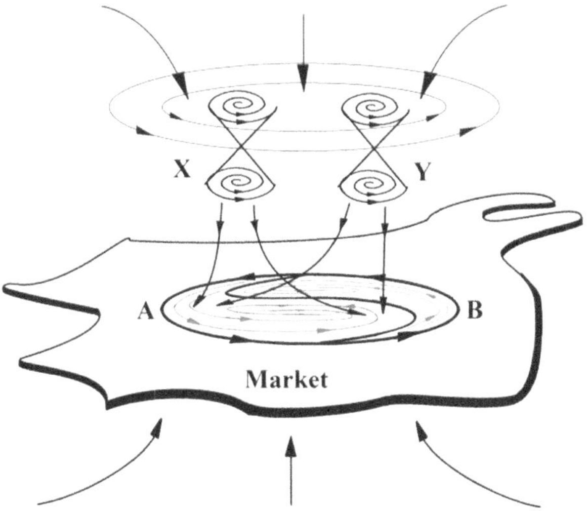

Fig. 8.5 How a two-sided market effect appears

8.4.2 Achieving Growth and Good Performance Simultaneously

When consumers are willing to pay extra beyond the expected market fees for a product or service due to the reason that diverse products and/or services are provided to consumers as bundles, we say that the diversification has produced a demand side or consumer synergy. Because at any given moment of time each human being has multiple needs from across different industries, Propositions 8.6 and 8.7 imply that demand-side synergy can be obtained via interindustry diversification of seemingly unrelated products and services from the viewpoint of producers.

Proposition 8.8 Corporate growth and improving performance can be achieved simultaneously through diversification at either the producer side, or the consumer side, or both.

At the producer side, diversification or creating an economy of scope is most likely achieved in real life by making resources available to all business units (Proposition 8.1) so that these resources can be deployed and redeployed repeatedly for different purposes, leading to a rich portfolio of products and/or services, which are more or less related due to the resources employed (Proposition 8.4). From Proposition 8.1, it follows that diversification at the producer side leads to the potential for synergistic innovations to appear, which helps to grow the business and to improve the performance. This end is like what Porter (1985) argued that firms should consider the "fit" of related diversification for better performance.

From the perspective of demand-side synergies, a synergistic fit of diversified offers can occur on the consumer side (Proposition 8.6), leading to increasing top lines of a firm. Hence, both corporate growth and improving performance can be simultaneously achieved through diversification.

By combining the previous analyses, it shows that if diversification occurs at both the producer and the consumer side, the chance for corporate growth and improving performance increases tremendously.

8.5 A Few Final Words

The present world of market economies has been turbulent. To successfully survive and grow in such a constant turbulence, a firm has to continuously look for potentials of different competitive advantages (McGrath, 2013). One way of doing so is to create economies of scope by diversification at either the producer side (Porter, 1985; Santalo & Becerra, 2008) or the demand side (Ye et al., 2012) or both. If a firm adopts this approach of diversification, then the firm faces the challenge of how to develop the necessary ideas for diversification. It is around our attempt to address this challenge that this chapter establishes a series of both theoretically and practically significant results.

By employing the thinking logic of systems science, this chapter is able to develop urgently needed conclusions generally useful for managerial problem solving and decision-making (Lockett et al., 2008) and to address some key issues in strategic management. In terms of applications, this chapter provides practically significant recommendations for entrepreneurs, managers, and retailers, respectively.

In particular, for managers, entrepreneurs, and retailers who focus on the demand side by considering collocating commonly available products and/or services, this chapter shows that other than an intimate understanding of consumers' needs and the creativity to envision a novel collocation of commonly available resources (Ye et al., 2012), superior consumer benefits can still be created by offering consumer groups utilities that they were not previously aware. In particular, this chapter shows why diversification helps firms with corporate growth and performance (Proposition 8.8); and when consumers are willing to pay extra for their needs (Propositions 8.6 and 8.7). Beyond theoretical results, this chapter also provides practically applicable recommendations. For example, when strategically planning for collocations of commonly available products and/or services, a manager/entrepreneur should always keep in mind the potential of complementary consumer benefits (Propositions 8.6 and 8.7), where consumer interactions can create another dimension of growth and satisfaction and where a two-sided or even multi-sided market can emerge.

For managers of the producer side who intend to produce an economy of scope, beyond explaining why the idea of economies of scope will help with corporate growth and performance (Proposition 8.7), this chapter also shows how to potentially create synergistic innovations by

- Making their limited resources available to all business units within their companies,
- Discovering what new ideas could be introduced through combining different resources,
- Naturally formulating a portfolio of diversified products and/or services,
- Staying alert about when their firms could likely become exploiters of existing competitive advantages instead of explorers of new advantages in the turbulent consumers' markets, and
- Avoiding negative synergies by not jointly mobilizing inconsistent resources.

To conclude this chapter, let us now look at the limitations of this chapter. One limitation stems from the assumption of why a firm actually exists. In fact, firms in the business world behave widely differently from each other depending on why and how they are established. That will explain why the results established will not be universally applicable to many actual situations. Although our attention in this chapter is on how both conceptually and practically a focal firm can synergistically collocate available resources at the producer side or commonly available products and/or services at the demand side in its effort to create economies of scope, we did not consider the limitations that firm is subject to. These listed and unlisted limitations of this chapter should help point to future research problems in order to develop additional practically useful insights beyond what have been established in this chapter.

Appendix: Proof of Theorem 8.1

Given a system $S_0 = (M_0, R_0)$, another system $S_n = (M_n, R_n)$ is said to be an nth-level object system of system S_0, provided that there are systems.

$$S_i = (M_i, R_i), \text{for } i = 1, 2, \ldots, \text{n} - 1,$$

such that S_i is an object in M_{i-1}, $1 \leq i \leq n$. In this case, each element in M_n is referred to as an nth-level object of system S_0. When the system S_0 has at least one nth-level object system, for $n = 1, 2, \ldots$, then the system is referred to as a multileveled (or multilevel) system.

A chain of object systems of S_0 is a sequence $\{S_i = (M_i, R_i) : i < \alpha\}$, for some ordinal number α, of different-level object systems of S_0, such that for each pair $i, j < \alpha$ satisfying $i<j$, there exists an integer $n = n(i, j)$, a function of i and j, such that S_j is an nth-level object system of S_i, Fig. 8.6, where each oval area stands for a system with its objects indicated by dots and relations by enclosed regions. Then the following result (Lin, 1999, p. 193) is true:

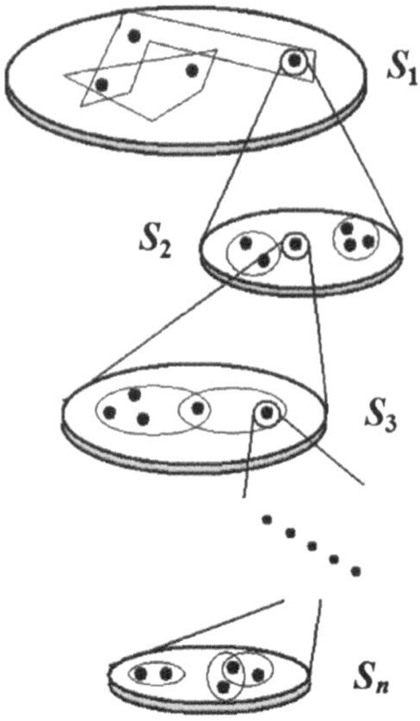

Fig. 8.6 A chain of object system

$(*)$ Each chain of object systems of S must be finite.

Now, we define the set $M(S)$ of all fundamental objects as follows: First, let us rewrite the system $S = (M, R)$ as $S = (M_0, R_0)$ for the sake of notational convenience. Define

$$^*M_0 = \{x \in M_0 \colon x \text{ is not a system}\}$$

and

$$\widetilde{M}_0 = \bigcup \{M_x \colon x = (M_x, R_x) \in M_0 - {}^*M_0\}$$

Assume that for a natural number n, two sequences $\{^*M_i : i = 0, 1, 2, \ldots, n\}$ and $\{\widetilde{M}_i \colon i = 0, 1, 2, \ldots, n\}$ have been defined, satisfying that

$$^*M_i = \left\{x \in \widetilde{M}_{i-1} : x \text{ is not a system}\right\}$$

and

$$\widetilde{M}_i = \bigcup \left\{M_x : x = (M_x, R_x) \in \widetilde{M}_{i-1} - {}^*M_i\right\}.$$

Then the following two sets can be defined for the index number $i + 1$,

$$^{*}M_{i+1} = \{x \in \widetilde{M}_i : x \text{ is not a system}\} \text{ and } \widetilde{M}_{i+1} = \bigcup \{M_x : x = (M_x, R_x) \in \widetilde{M}_i - {}^{*}M_{i+1}\}.$$

From mathematical induction, it follows that a sequence $\{^{*}M_i : i \in \omega\}$ is defined. Now, by letting $M(S) = \bigcup \{^{*}M_i : i \in \omega\}$, result (*) above guarantees that $M(S)$ consists of all fundamental objects in system S.

As for the conclusion that $|M| < |M(S)|$, it follows from the assumptions that (1) each object system of S has a finite object set, (2) each object that appears in any relation in the system S or an object system of S only appears once in that relation, and (3) S has at least one chain of object systems of more than one level and each object system on the chain contains more than one object. Here, conditions (1) and (2) are necessary for producing the resultant inequality by limiting our counting cardinalities $|M|$ and $|M(S)|$ within the realm of natural numbers. End of the proof.

References

Aas, T. H., Breunig, K. J., Hydle, K. M., & Pedersen, P. E. (2015). Innovation management practices in production-intensive service firms. *International Journal of Innovation Management, 19*(05). https://doi.org/10.1142/S1363919615500553

Adner, R., & Levinthal, D. (2001). Demand heterogeneity and technology evolution: Implications for product and process innovation. *Management Science, 47*(5), 611–628.

Adner, R., & Snow, D. (2010). Old technology responses to new technology threats: Demand heterogeneity and graceful technology retreats. *Industrial Corporate Change, 19*(5), 1655–1675.

Adner, R., & Zemsky, P. (2006). A demand-based perspective on sustainable competitive advantage. *Strategic Management Journal, 27*(3), 215–239.

Azevedo, E. M., & Leshno, J. D. (2016). A supply and demand framework for two-sided matching markets. *Journal of Political Economy, 124*(5), 1235–1268.

Barney, J. (1991). Firm resources and sustained competitive advantage. *Journal of Management, 17* (1), 99–120.

Barney, J. (2001). Is the resource-based "view" a useful perspective for strategic management research? Yes! *Academy of Management Review, 26*(1), 41–56.

Barney, J., & Arikan, A. (2001). The resource-based view: Origins and implications. In M. Hitt, R. Freeman, & J. Harrison (Eds.), *Handbook of strategic management* (pp. 124–185). Blackwell.

Becheikh, N., Landry, R., & Amara, N. (2006). Lessons from innovation empirical studies in the manufacturing sector: A systematic review of the literature from 1993 to 2003. *Technovation, 26* (5), 644–664.

Buffington, J. (2016). The future of manufacturing: An end to mass production. In J. Buffington (Ed.), *Frictionless markets* (pp. 49–65). Springer.

Caputo, A., Marzi, G., & Pellegrini, M. M. (2016). The internet of things in manufacturing innovation processes: Development and application of a conceptual framework. *Business Process Management Journal, 22*(2), 383–402.

Crossland, P., & Smith, F. I. (2002). Value creation in fine arts: A system dynamics model of converse demand and information cascades. *Strategic Management Journal, 23*(5), 417–434.

Damanpour, F. (1991). Organizational innovation: A meta-analysis of effects of determinants and moderators. *The Academy of Management Journal, 34*(3), 555–590.

Drucker, P. F. (1954). *The practice of management*. Harper & Row.
Dutta, S., Narasimhan, O., & Rajiv, S. (1999). Success in high-technology markets: Is marketing capability critical? *Marketing Science, 18*(4), 547–568.
Eisenhardt, K., & Martin, J. (2000). Dynamic capabilities: What are they? *Strategic Management Journal, 21*(10/11), 1105–1121.
Fang, E., Palmatier, R., & Grewal, R. (2011). Effects of customer and innovation asset configuration strategies on firm performance. *Journal of Marketing Research, 48*(3), 587–602.
Farjoun, M. (1998). The independent and joint effects of the skill and physical bases of relatedness in diversification. *Strategic Management Journal, 19*(7), 611–630.
Forrest, J. Y.-L., Dang, Y. G., McCarthy, L., Liu, S. F., & Yong Liu, Y. (2019). Conditions under which synergistic innovations appear. *Proceedings of the 2019 Annual Conference of Decision Sciences Institute* (pp. 2061–2080).
Forrest, J. Y.-L., Zhao, H. C., & Shao, L. (2018). Engineering rapid industrial revolutions for impoverished agrarian nations. *Theoretical Economics Letters, 8*, 2594–2640.
Gans, J. S., MacDonald, G., & Ryall, M. D. (2008). *The two sides of competition and their implications for strategy*. Working paper, Melbourne Business School. https://works.bepress.com/michael_ryall/15/
Gary, M. S. (2005). Implementation strategy and performance outcomes in related diversification. *Strategic Management Journal, 26*(7), 634–664.
Harmancioglu, N., Droge, C., & Calantone, R. (2009). Strategic fit to resources versus NPD execution proficiencies: What are their roles in determining success? *Journal of the Academy of Marketing Science, 37*(3), 266–282.
Hitt, M. A., Bierman, L., Shimizu, K., & Kochhar, R. (2001). Direct and moderating effects of human capital on strategy and performance in professional service firms: A resource-based perspective. *The Academy of Management Journal, 44*(1), 13–28.
Kanter, R. M. (2001). *Evolve! Succeeding in the digital culture of tomorrow*. Harvard Business School Press.
Kim, J. Y., & Finkelstein, S. (2009). The effects of strategic and market complementarity on acquisition performance: Evidence from the U.S. commercial banking industry, 1989–2001. *Strategic Management Journal, 30*(6), 617–646.
Lin, Y. (1999). *General systems theory: A mathematical approach*. Kluwer Academic/Plenum Publishers.
Lin, Y. (2009). *Systemic yoyos: Some impacts of the second dimension*. CRC Press (an imprint of Taylor and Francis).
Lin, Y., & OuYang, S. C. (2010). *Irregularities and prediction of major disasters*. CRC Press, an imprint of Taylor and Francis.
Lockett, A., O'Shea, R. P., & Wright, M. (2008). The development of the RBV: Reflections from Birger Wernerfelt. *Organization Studies, 29*(8–9), 1125–1141.
Lockett, A., Thompson, S., & Morgenstern, U. (2009). The development of the resource-based view of the firm: A critical appraisal. *International Journal of Management Review, 11*(1), 9–28.
McGrath, R. G. (2013). *The end of competitive advantage: How to keep your strategy moving as fast as your business*. Harvard Business Review Press.
Nason, R. S., & Wiklund, J. (2018). An assessment of resource-based theorizing on firm growth and suggestions for the future. *Journal of Management, 44*(1), 32–60.
Palich, L. E., Cardinal, L. B., & Miller, C. C. (2000). Curvilinearity in the diversification-performance linkage: An examination of over three decades of research. *Strategic Management Journal, 21*(2), 155–174.
Penrose, E. T. (1959). *The theory of the growth of the firm*. John Wiley & Sons.
Porter, M. E. (1985). *Competitive advantage: Creating and sustaining superior performance*. Free Press.
Priem, R. L. (2007). A consumer perspective on value creation. *The Academy of Management Review, 32*(1), 219–235.
Priem, R. L., & Butler, J. E. (2001a). Is the resource-based "view" a useful perspective for strategic management research? *The Academy of Management Review, 26*(1), 22–40.

Priem, R. L., & Butler, J. E. (2001b). Tautology in the resource-based view and the implications of externally determined resource value: Further comments. *The Academy of Management Review, 26*(1), 57–66.

Robins, J. A., & Wiersema, M. F. (2003). The measurement of corporate portfolio strategy: Analysis of the content validity of related diversification indexes. *Strategic Management Journal, 24*(1), 39–59.

Rochet, J. C., & Tirole, J. (2006). Two-sided markets: A progress report. *RAND Journal of Economics, 37*(3), 645–667.

Rumelt, R. P. (1974). *Strategy, structure, and economic performance*. Harvard University Press.

Sakhartov, A. V., & Folta, T. B. (2014). Getting beyond relatedness as a driver of corporate value. *Strategic Management Journal, 36*(13), 1939–1959.

Salamzadeh, A., & Kawamorita, K. H. (2015). Startup companies: Life cycle and challenges. *The 4th international conference on employment, education and entrepreneurship* (EEE), Belgrade, Serbia, 2015. Available at SSRN: https://ssrn.com/abstract=2628861 or https://doi.org/10.2139/ssrn.2628861.

Santalo, J., & Becerra, M. (2008). Competition from specialized firms and the diversification-performance linkage. *Journal of Finance, 63*(2), 851–883.

Sawhney, M., Verona, G., & Prandelli, E. (2005). Collaborating to create: The internet as a platform for customer engagement in product innovation. *Journal of Interactive Marketing, 19*(4), 4–17.

Shah, S. K., & Tripsas, M. (2007). The accidental entrepreneur: The emergent and collective process of user entrepreneurship. *Strategic Entrepreneurship Journal, 1*(1–2), 123–140.

Sirmon, D. G., Gove, S., & Hitt, M. A. (2008). Resource management in dyadic competitive rivalry: The effects of resource bundling and deployment. *The Academy of Management Journal, 51*(5), 919–935.

Smith, A. (1776). *The wealth of nations, books I-III, (1986 printing)*. Penguin Books.

Smith, W. K., & Tushman, M. L. (2005). Managing strategic contradictions: A top management model for managing innovation streams. *Organization Science, 16*(5), 522–536.

Sun, M., & Tse, E. (2009). The resource-based view of competitive advantage in two-sided markets. *Journal of Management Studies, 46*(1), 45–64.

Tanriverdi, H., & Lee, C. H. (2008). Within-industry diversification and firm performance in the presence of network externalities: Evidence for the software industry. *The Academy of Management Journal, 51*(2), 381–397.

Tanriverdi, H., & Venkatraman, N. (2005). Knowledge relatedness and the performance of multibusiness firms. *Strategic Management Journal, 26*(2), 97–119.

Tripsas, M. (2008). Customer preference discontinuities: A trigger for radical technological change. *Managerial Decision Economics, 29*(2–3), 79–97.

Tsai, W. (2001). Knowledge transfer in intraorganizational networks: Effects of network position and absorptive capacity on business unit innovation and performance. *The Academy of Management Journal, 44*(5), 996–1004.

Veugelers, R., & Cassiman, B. (1999). Make and buy in innovation strategies: Evidence from Belgian manufacturing firms. *Research Policy, 28*(1), 63–80.

Villasalero, M. (2017). A resource-based analysis of realized knowledge relatedness in diversified firms. *Journal of Business Research, 71*(Feb.), 114–124.

Visnjic, I., Wiengarten, F., & Neely, A. (2016). Only the brave: Product innovation, service business model innovation, and their impact on performance. *Journal of Product Innovation Management, 33*(1), 36–52.

Wu, Y., & Lin, Y. (2002). *Beyond nonstructural quantitative analysis: Blown-ups, spinning currents and modern science*. World Scientific.

Ye, G. L., Priem, R. L., & Alshwer, A. A. (2012). Achieving demand-side synergy from strategic diversification: How combining mundane assets can leverage consumer utilities. *Organization Science, 23*(1), 207–224.

Zollo, L., Marzi, G., Boccardi, A., & Ciappei, C. (2016). Gli effetti della Stampa 3D sulla competitivita aziendale. Il caso delle imprese orafe del distretto di Arezzo. *Piccola Impresa/Small Business, 2*(2), 80–100.

Chapter 9
Consumer Synergies: Simultaneous Utilities and Multi-Sided Markets

Abstract By employing a systemic approach, which is completely dissimilar to those used in the literature, this chapter, which is mainly based on (Forrest et al., 2019, *Proceedings of the 2019 annual meeting of national association of business, economics and technology* (pp. 101–121)), scrutinizes issues related to the following problems: (1) What are the most fundamental decisions a retailer can make in terms of its offers to consumers? (2) When can simultaneous consumer utilities be produced by collocating products and/or services? (3) When can a positively correlated multi-sided market be formed? And (4) without particular talent and luck how can a synergistic innovation be introduced? Due to the specific approach taken, we are able to describe how simultaneous consumer utilities, two-sided markets, and consumers' willingness to pay additional react with one another and how the previous problems can be addressed by establishing a series of 6 propositions. To explore the possibility of systemically generating new ideas instead of waiting for the seemingly sudden and random appearance of disruptive technologies, a mechanical procedure is developed for the potential of producing synergistic innovations on either the producer or the demand side. Because of the certainty systems science offers, the general conclusions developed in this chapter are expected to provide practically useful recommendations for entrepreneurs, managers, and retailers to create and to capture values for consumers and their companies.

Keywords Collocation of products · Market demand · Simultaneous consumer utility · Multi-sided market

9.1 Introduction

To effectively ride waves of transient competitive advantages (e.g., McGrath, 2013), a firm has to see inwardly what values it can create for consumers (e.g., Barney, 1991; Eisenhardt & Martin, 2000), and/or outwardly where consumer synergies are located so that potential products can be developed (Drucker, 1954). Either looking inwardly or outwardly, the essence is for the firm to find potentials of different competitive advantages (Porter, 1985; Santalo & Becerra, 2008; Ye et al., 2012).

J. Y.-L. Forrest, Y. Liu, *Value in Business*, Contributions to Management Science,
https://doi.org/10.1007/978-3-030-82898-1_9

More specifically, the firm needs to address the following natural questions: (1) What are the most fundamental decisions a retailer can make in terms of its offers to consumers? (2) How can collocating products and/or services lead to simultaneous consumer utilities? (3) When can the firm develop a positively correlated multi-sided market? And, (4) without particular talent and luck how can the firm introduce a synergistic innovation?

To address these questions, this chapter develops a cohesive theory of a series of formal propositions on how to produce consumer synergies by developing simultaneous consumer utilities, two-sided markets, and related issues. This theoretical development relies 100% on logical reasoning, systems thinking, and the systemic yoyo model without involving any data or anecdotes in order to derive general conclusions and widely employable recommendations. It is because our approach is not limited by any available data or anecdotes and is not constrained by the strict conditions of econometric tools.

The rest of this chapter is organized as follows. Section 9.2 reviews the relevant literature and demonstrates the contributions of this chapter. Section 9.3 addresses the concepts of simultaneous consumer utilities and consumer synergies. Section 9.4 considers the relationship between two-sided markets and consumer synergies. Section 9.5 investigates the relationship between ownership and sustainability behind the idea of collocating commonly available products and/or services. Section 9.6 develops a mechanical procedure that leads to synergistic innovations. And this chapter is concluded in Sect. 9.7.

9.2 Literature Review

This chapter contributes simultaneously to the literatures of entrepreneurship, innovation, and strategy. So, this section reviews the relevant literatures individually and demonstrates where our contributions lie in each case.

For the literature of entrepreneurship, Penrose (1959) notes that companies grow when they attend to the needs of consumers; Priem and Butler (2001a), Kor et al. (2007), and McMullen (2015) maintain that subjective entrepreneurial judgment is important to firm success and that resources gain economic value through customers' usage of them. And a resource can be utilized in multiple different ways that can both differentially increase the magnitude and scale of internal activities of a firm and create additional values for consumers externally (e.g., Augier & Teece, 2008; Kor et al., 2007). While Sirmon et al. (2008) and others investigate how internal resources can be bundled to produce efficiency inside a firm, Ye et al. (2012) consider how consumer preferences can be bundled outside a firm so that simultaneous consumer utilities and two-sided market effects will result. Their work indicates that the entrepreneurial alertness to external opportunities, the so-called pure entrepreneurial judgment (Kirzner, 1973, 2018), can come before making any investment and can exist without any intimate knowledge of available resources.

This chapter emphasizes the creation of consumer synergies through collocation of commonly available products and/or services, one aspect of pure entrepreneurial judgment. By doing so, it contributes to the entrepreneurship literature by establishing the following main results, among others, that can be generally applied in real-life scenarios: (1) A retailer's location and the depth and breadth of offers represent the most important, most fundamental, as well as mutually constraining tactical decisions of the retailing firm. (2) If two offers of a firm facilitate beneficial interactions of two consumer segments, then that firm establishes a positively correlated two-sided or multi-sided market. And (3) when a firm conducts its business within a saturated market, its demand-side advantage established through collocating commonly available goods can be sustainable, if the marginal profit from the advantage is not more than the entry cost of a new competitor.

For the literature of innovation, since when Smith (1776) acknowledges the essential significance of innovation over 200 years ago, many scholars have investigated the topic of innovation of the producer side (Aas et al., 2015; Adner & Levinthal, 2001; Damanpour, 1991; Visnjic et al., 2016). Innovation enables firms to competitively enter an existing market or create new markets (Becheikh et al., 2006; Smith & Tushman, 2005), as a major factor of long-term performance (Kanter, 2001). Veugelers and Cassiman (1999) find that high levels of perceived risks and costs and low appropriation determine the innovation sourcing strategy while not discouraging innovation from taking place. In particular, these authors find that small firms tend to restrict their innovation efforts to exclusive makes or simply buy strategies and large firms tend to acquire knowledge both internally and externally when implementing their innovation strategies. The fast development of Internet-based technologies forces firms to become forever more competitive and innovative than before (Buffington, 2016; Caputo et al., 2016; Zollo et al., 2016).

Contributing to this innovation literature, this chapter develops the following general result: Under the influence of a particular factor, such as location and circumstance, if one of a firm's collocated offers that are necessary to consumers meets a competitive market demand, while other offers provide convenience, then the firm provides simultaneous consumer utilities.

In terms of the literature of strategies focusing on looking outside into the environment, Barney (2001) and Priem and Butler (2001a, 2001b) recognize that issues of consumer demand need to be addressed seriously because consumers eventually determine the firms' successes (Drucker, 1954; Penrose, 1959). Along this line of thinking, studies consider various strategic issues, such as how market demand affects technological innovations and competitive advantages (e.g., Adner & Zemsky, 2006; Tripsas, 2008); how consumer-focused strategies can lead to value creation and capture (e.g., Adner & Snow, 2010; Gans et al., 2008; Priem, 2007); and how consumers influence entrepreneurial innovation (Sawhney et al., 2005; Shah & Tripsas, 2007). In particular, Ye et al. (2012) study how firms can employ bundled preferences of consumers to develop their strategies.

By attending to the demand side of the literature of strategies, this chapter develops the following main result, among others, that are generally true: when a firm operates a two-sided market with its collocated assortment of goods, if there are few attractors compared to attractees, then that firm can charge the attractees higher prices than the market norm.

Other than the contributions listed above, this chapter systemically incorporates the demand side simultaneously with strategic management, entrepreneurship, and innovation. Our established theory moves beyond empirically confirming and anecdotally prescribing to the fact that subjective entrepreneurial judgment or managerial ability are important variables for strategic success (e.g., Foss et al., 2007; Holcomb et al., 2009; McMullen, 2015; Ye et al., 2012). Instead, this chapter develops specific while generally applicable recommendations and a mechanical procedure for managers and entrepreneurs to explore and to identify demand-side potentials in their efforts of developing economies of scope. Additionally, by using logical reasoning, as a well-tested method in the successes of natural science, and the systemic yoyo model, this chapter develops a theoretical foundation employable to the identification of strategies that lead to simultaneous consumer utilities of the demand side (Sakhartov & Folta, 2014; Ye et al., 2012).

Another contribution this chapter makes to the literature is how it employs systems science to study consumer synergies. Because of this reason, conclusions derived in this chapter hold true in general without suffering from the weaknesses and limitations of data and anecdotal analyses, and recommendations widely applicable in real life. To this end, the mechanical procedure developed in this chapter for discovering possible collocation ideas on the demand side and possible ways of diversification on the producer side represents a vivid support.

To make this chapter self-contained, let us provide the definitions of fundamental terms and basic assumptions used in the rest of this chapter.

First, by a firm, we assume that it exists to satisfy a market niche through generating a positive cash flow and that the cash flow comes from business profits or investments or both. Speaking differently, each firm considered in this chapter needs to actively pursue after certain endeavors to meet some market demand(s). And by consumers, it means the end users of products and/or services, while customers are the intermediate users. When a product/service is used by both end and intermediate users, these users are collectively known as customers.

By resource, it means an asset of a firm, whether it is tangible or not (Harmancioglu et al., 2009), which the firm can use to design and implement its strategies (Barney & Arikan, 2001). Speaking differently, a resource is anything that can be physical, financial, intellectual, or organizational, the firm can employ to realize its business goals. Additionally, any resource provides alternative services that differentially escalate the efficiency of the firm or increase consumer value creation outside the firm.

By synergistic innovation, it represents such a strategy of a focal firm that systemically deploys and redeploys known elements, either within or outside that firm, to produce additionally economic benefits. Here, the so-called known elements can be resources, management routines, production processes, consumer preferences, market demands, etc. The word "systemically" stands for that a focal firm can develop a meaningful system or a set of systems based on the known elements to produce new products of improved functionalities and/or new services that meet a forever evolving market demand/consumer preference.

If it is possible to meet several market demands simultaneously, while saving valuable consumer resources, say, time and/or money, then the diversification effort of meeting different market demands is said to have created a simultaneous consumer utility effect. Assume that there are two such services that attract different segments of consumers, although they might be overlapping. If the joint offer of these services increases traffic flow in such a way that one segment of consumers served attracts an increasing traffic flow from the other segment, while the latter does not have any adverse effect on the former, then the diversification of services is said to have created a two-sided market effect.

9.3 Simultaneous Consumer Utilities

In this section, we look at some fundamental properties of simultaneous consumer utilities, created by a focal firm through offering a collocated assortment of products and/or services to consumers. For illustration purpose, let us look at the following example from (Ye et al., 2012).

The student population of a state university in the northern USA is over 29,000. Most of these students are undergraduates, unmarried living on tight budgets. Because most of the rental properties targeting the students do not offer laundry facilities, self-servicing coin-operated laundromats near the campus compete for the students' business. At the same time, indoor tanning is also a successful college-town business that attracts mostly college women. Seeing the potential, two entrepreneurs provide their independent, single-location businesses by jointly providing coin-operated self-service laundry facility and indoor tanning service. These business people create consumer synergies by combining commonly available assets without much reduction in their overhead costs (from different sets of equipment) and labor costs (the combined business operations require different production and customer service skills). However, these business operations create a tremendous amount of demand-side synergies due to the following reasons:

1. Clothes washing and skin tanning can be accomplished simultaneously, saving the otherwise unproductive and costly waiting time; and
2. A potential opportunity is created for students to establish relationships, because the single location of combined services helps males and females to meet, while the greater participation of women in the tanning service attracts a larger number of men to the laundry service.

The subsequent consumer (or demand side) synergies help respectively support higher fees to be collected from the laundry service than other standalone stores and sustain competitive prices for the demand for the tanning service that is much increased due to the heavy flow of male users of the laundromats. That is, the combined services create a greater value for consumers that is also captured by business providers because of the resultant consumer synergies from product-and-service diversification.

Other than the concept of simultaneous consumer utilities, there is also that of sequential consumer utilities. The former, as described in the previous example, means when the consumer saves time by performing two or more activities at the same time, while the latter describes when consumers benefit from making a series of purchases sequentially without spending much additional time from one store to another, say, in a shopping mall. This section will only focus on the concept of simultaneous consumer utility and relevant consumer synergies.

Because our focus in this section is on how a focal firm could potentially create simultaneous consumer utilities through offering a collocated assortment of products and/or services, let us first look at the strategic significance of the firm's depth and breadth of its assortment of products/services.

Proposition 9.1 For each retailer, both (1) location and (2) the depth and breadth of its assortment of products/services to be offered to consumers represent the two most important, most fundamental, as well as mutually constraining tactical decisions.

A retailer's location affects shopping time and other related costs (e.g., travel) that are generally borne by consumers when they shop for desired goods and/or services, while the depth and breadth of the assortment of products/services offered at a particular location provide the basis for consumers to evaluate the worth of their time and other related costs (Betancourt & Gautschi, 1990). In other words, if a location is not good, convenience and savings on time and other shopping-related costs will dictate consumers to shop at nearby stores for their desired goods and services; and if the depth/breadth of the product/service assortment offered is not sufficient in the nearby locations, the consumer will be happy to spend extra time and money to shop in a further off location to save time and increase their simultaneous utilities. That is, location and depth/breadth of the assortment of products/services offered are two mutually constraining factors that the value of one determines that of the other, where the values are best assessed by consumers (Moreau & Dahl, 2005).

Consumers make their patronage choices among retailers by evaluating trade-offs of retailers' depth—such as that offered by specialty stores, breadth—such as that offered by general stores and convenience—such as that offered by convenience store (Messinger & Narasimhan, 1997). Therefore, consumers' demand for benefits and maximization of their overall utilities drives the demand for particular product combinations in a retailer's assortment, which in turn determines which location is considered convenient. That explains why both location and the depth/breadth of the assortment of products/services to be offered to consumers represent the two most important, most fundamental tactical decisions of retailers.

Systemically, a retailer's choice of the depth/breadth of its assortment of products/services offered to consumers stands for the material foundation for the yoyo field of the retailer to exist, while the location represents how conveniently the yoyo field can absorb necessary inputs (e.g., consumers). Only when these two elements coexist, the yoyo field of the retailer can possibly "spin" and "spin" viciously.

Evidently, it can be seen that collocating two randomly selected products and/or services might not naturally result in simultaneous consumer utility. For example, the attempt of Sears, Roebuck and Company to provide consumers a one-stop shop of hard goods, soft goods, insurance services, and financial services was unsuccessful (Sobel, 1999). So, a natural question is: Under what condition(s) will the collocation of two or more products and/or services result in simultaneous consumer utility, which in turn leads to consumer synergies? To this end, we have the following result.

Proposition 9.2 If all collocated products and/or services offered by a focal firm are necessary to consumers under the influence of a particular factor, such as location and circumstance, and one of the products and/or services meet a competitive market demand while others provide convenience, then simultaneous consumer utilities will be produced by the collocated assortment of goods.

Because one of the collocated products and/or services is assumed to meet a competitive market demand, when a consumer decides to go with that particular focal firm to satisfy his/her need or demand, the customer's decision will be further strengthened and backed by all the other collocated products and/or services due to the convenience they provide. Hence, jointly this assortment of products and/or services results in simultaneous consumer utilities.

Next, let us use examples to illustrate how Proposition 9.2 can be employed practically, where in each case, we need to identify the products and/or services that are collocated, the group of consumers, that particular factor, a product or service that meets a competitive market demand, and what convenience other collocated products and/or services provide. First, let us look at successful collocations, where consumers receive simultaneous utilities and business firms capture values through pricing decisions designed to maximize profitability across product types collocated.

- For the previous case of collocated laundry/indoor tanning services, the tanning part serves as the service in Proposition 9.2 that meets a competitive market demand of mostly female college students, while the laundry service provides convenience—for both washing clothes and meeting opposite genders. The particular factor is the specific location—the vicinity of a large university campus of mostly unmarried undergraduate students.
- For the case of combined gasoline station and convenience store, consumers are mostly drivers of automobiles and the particular factor is the fact that all cars need to be fueled regularly. Providing gasoline satisfies a competitive market demand, and the store provides convenience items which consumers can readily obtain when their vehicles are being filled with gasoline.
- For the case of collocated coffee shop and bookstore, the group of consumers contains all coffee drinkers, the factor is the fact that a lot of people are addicted to coffee. In this case, the competitive market demand is coffee drinks, and the book store provides a convenience for consumers to avoid rough outdoor climates and a comfortable environment to enhance the enjoyment of drinking coffee, as well as staying updated with current affairs.
- For the case of major car dealers, such as General Motors (GM), each of them collocates the service of selling cars with that of purchase financing. The group of consumers includes all potential drivers, and the particular factor is the reality that

in America life can be quite difficult if a person does not have a car. The service of selling cars meets a competitive market demand, while the financing service provides convenience for car buyers and flexibility for the dealer to provide discounts on its automobiles.

- For the case of online retailers, such as amazon.com, the collocated assortment of products consists of those on sale through the Internet, the group of consumers includes everybody who has Internet access, and the particular factor is the fact that the majority of the population currently has a connection to the Internet. The competitive market demand is the retail of almost everything needed in life, and the convenience is that the Internet connection eliminates all the travel costs and consumers can purchase a variety of needed items at the same time without stepping even one step out of his/her front door. (Note: amazon.com is only a case of one-stop shopping with nearly simultaneous consumer utility because for cases of true simultaneous utilities the consumer saves time by performing two activities at the same time.)

Next, let us revisit the unsuccessful collocation attempt of Sears, Roebuck and Company, as mentioned earlier. In this case, the company collocated the retailing of hard and soft goods and the offer of insurance and financial services, while we cannot identify a clear core product(s) or service(s) that meet a competitive market demand with others providing convenience at the same time.

9.4 Markets of Multiple Sides

In this section, we look at some fundamental properties of two-sided markets, created by a focal firm through offering a collocated assortment of products and/or services to consumers, where cross-market-segment externalities give rise to complementary increases in consumer utilities.

A market is two-sided provided that there are externalities across different participating segments of consumers (cf. Sun & Tse, 2009). In other words, the participation on both sides of a two-sided market are either positively or negatively correlated to each other. For example, New York Stock Exchange (NYSE) is a positively (correlated) two-sided market, where the more stocks are listed, the more investors are drawn to the exchange; and the more investors frequent NYSE, the more companies list their stocks on the exchange. And, the interaction between consumers and advertisements might stand for a negatively (correlated) two-sided market depending on how consumers react to advertisements. On the other hand, not all markets are two-sided. For example, a typical product market is not two-sided, because increasing supply does not necessarily increase the demand, although increasing demand tends to result in rising supply.

Also, practical experiences indicate that collocating randomly selected products and/or services might not naturally result in a two-sided market effect. And many real-life cases support this conclusion. For example, pets.com tried to develop a

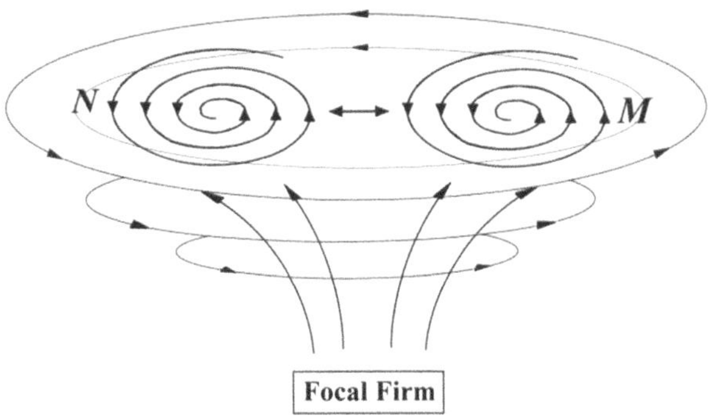

Fig. 9.1 The systemic structure of a two-sided market

two-sided market by linking pet owners and suppliers of pet products. However, its heavy investments did not create sufficient benefits for consumers to cover relevant costs. So, the expensive venture failed (Evans & Schmalensee, 2007). Hence, a natural question is: Under what condition(s) will the collocation of two or more products and/or services result in a two-sided market effect, which consequently leads to consumer synergies?

To potentially address this question, let us first look at the systemic structure of a two-sided market (Fig. 9.1), where a focal firm establishes a "platform" (the large, multilayered spin field encompassing those of N and M) by offering two (or more) market segments of consumers, as indicated by N and M, its particularly collocated assortment of products and/or services. Through that platform, value is created by that firm through facilitating interactions between mutually interdependent market segments of customers, while that firm captures this value by collecting fees from at least some of the participating consumers. If the two sides of a two-sided market are positively correlated, then participants on either side of the market receive more value when membership increases on the other side (Rochet & Tirole, 2006). Based on this systemic intuition, we have the following result.

Proposition 9.3 If two of the products and/or services collocated by a focal firm facilitate either implicitly or explicitly beneficial interactions of two market segments of consumers while other collocated products and/or services make the interactions efficient or convenient, then that firm establishes a platform for a positively correlated two-sided or multi-sided market to form.

To see why this proposition holds true, let A and B be the two assumed products and/or services in the collocated assortment of goods that focal firm offers to consumers. Assume that the particular market segments of consumers between which A and B facilitate beneficial interactions are N and M. Then, as depicted in Fig. 9.1, when all other collocated products and/or services help make these interactions efficient or convenient, a spin field (or platform) that encompasses the yoyo fields of N and M emerges. That means that the focal firm has successfully established a positively correlated two-sided market with N being one side and M the other. Note that in this systemic modeling, the fields of N and M must be

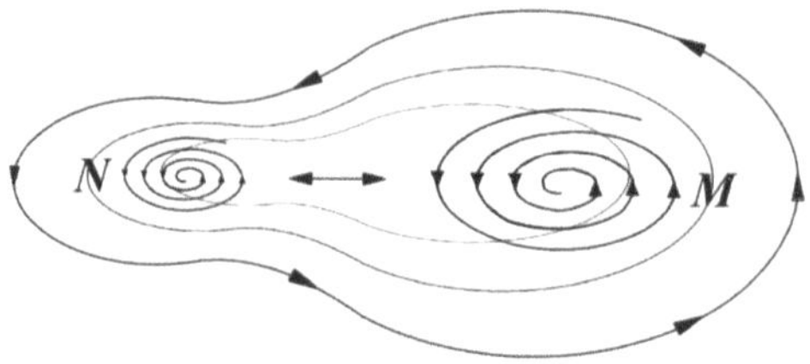

Fig. 9.2 The general systemic structure of a two-sided market

converging fields that spin in the same direction; otherwise, the platform field cannot be emerging as claimed above.

Next, let us use examples to illustrate how Proposition 9.3 can be employed practically, where in each case, we need to identify the products and/or services that are collocated, the two particular products and/or services and the corresponding market segments of consumers that will be stimulated to interact beneficially, and whether or not the rest of the collocated products and/or services are supportive.

- For the previous case of collocated laundry/indoor tanning services, evidently it is the laundry service and the indoor tanning that stimulate the following two market segments of consumers—female and male college students—to interact beneficially. In this case, the two particular services exert their functionalities only implicitly, where the focal firm only needs to attract as many female consumers to the indoor tanning service as possible and then the rest of the two-sided market will play out naturally with the least amount of effort from the focal firm.
- For the case of online retailers, such as amazon.com, the collocated assortment of products and services are those on sale through the Internet and an online platform. The online platform provides a means for sellers and buyers to interact beneficially so that the more sellers list the more buyers appear and vice versa.
- Online dating services represent another case of two-sided markets, where a focal firm operates an online platform for the male and female sides of the dating market to join.
- Organizations, be they local, national, or international, that work to develop community projects generally establish two-sided markets by offering such events as showcases, designed to join entrepreneurs and venture capitalists.

In the systemic structure of a two-sided market in Fig. 9.1, the yoyo pools of N and M do not have to be similar sizes. Instead, in real life, the general structure should look like the one depicted in Fig. 9.2, where the pool of one side of the market is smaller than the other side due to the reason that customers on different sides may enjoy different marginal utilities from participating in the platform. In other words, the presence of an additional participant, say, on side M may have a much greater positive effect on the utility of the other side N. In such a case, participants, say, on side N with lower marginal utility for participation are called attractors, and those on the other side, say, M, with higher marginal utility are attractees (Ye et al., 2012). Such designation of participants of a two-sided market naturally leads to the following pricing strategy.

Proposition 9.4 If a focal firm operates a two-sided market by offering a collocated assortment of products and/or services with the pool size of attractors a lot smaller than that of attractees, then that firm can offer a competitively discounted price offer to attractors and compensate at least some of the opportunity cost, caused by subsidizing attractors, by making attractees bear higher prices.

The marketing intuition behind this result is straightforward. When more attractors participate in a two-sided market with the said characteristics due to competitively discounted prices, a lot more attractees will be on board. That jointly generates increased demand for the platform. Therefore, more economic value for the focal firm that operates the two-sided market is captured. For related discussions, see Rochet and Tirole (2003), Armstrong (2006), and Jullien (2005).

9.5 Ownership and Sustainability

Discussions in the previous sections show that market-side strategic innovations, based on collocations of commonly available products and/or services that deliver simultaneous utilities and/or serves two-sided markets with mutual attraction of different market segments of consumers, can create value for consumers and capture value for the focal firm that offers a collocated assortment of goods. It is definitely the virtuosity of an entrepreneur who innovatively associates some seemingly unrelated assets into an assortment that produces for consumers novel synergistic benefits that accrue from consumer perceptions and needs. So, a natural question arises (Ye et al., 2012): Can the products and/or services in a collocated assortment, which provides consumer synergies through simultaneous utilities, two-sided markets, or both, be offered individually by different business firms?

Proposition 9.5 Assume that all firms exist for the purpose of satisfying a particular market niche through generating a positive cash flow from the marketplace, as assumed in Sect. 9.3. If a focal firm desires to provide simultaneous consumer utilities by operating a two-sided market through collocating commonly available products and services, then that firm needs to offer all the collocated assets by itself.

Without loss of generality, assume that only two products are collocated in this given situation. Then, the reason why this result holds true in general is the following: Let us reason by contradiction. Assume that these two collocated products are individually offered by two separate firms with N being one side of the imagined two-sided market and M the other (Fig. 9.2). Because each of the separate firms has to generate a positive cash flow from the marketplace, the stability of the resultant platform—the large field encompassing the fields of N and M in Fig. 9.2—is not guaranteed as an operating organic whole or a connected system (Lin, 1999). That means that in practice the attraction between N and M does not exist steadily. In other words, the firm that attracts side N of the imagined two-sided market can easily join hands with another firm that has been attracting a consumer segment M' that is similar to M, because the products and/or services collocated are assumed to be commonly available. The systemic significance of common availability means that in the ocean of all the yoyo fields of the commonly available products and services each local field X can readily find another local field Y to temporarily form for a platform for a short-lived two-sided market because in that ocean of yoyo fields, each individual yoyo pool is constantly fighting for survival and growth.

In terms of the literature, various studies also suggest that common ownership is efficient in different contexts. In particular, Williamson (1979) argues that collocation requires investment to cover the sunk costs so that post-contract opportunism represents a real moral hazard for separate firms. Ye et al. (2012) see the need to determine what consumer subsidies one firm must offer so that the other firm can reap the rewards of greater overall willingness to pay and then a mutually agreeable procedure for settling relevant revenues and costs must be negotiated between the separate firms. And with the changing consumer preference renegotiation will become a constant issue of concern. That will lead most likely to moral hazards and recurring contracting costs, a resolution of which, based on Williamson (1979) and Alchian and Woodward (1987), is a common ownership. Even if opportunism and contracting costs are nonexistent, common ownership may still be preferred over markets or contracting, because the acquired experiences through the sole ownership provide the knowledge for better coordination of value-creating assets (Conner & Prahalad, 1996). In short, the very concept of collocating products and/or services for the purpose of delivering consumer synergies implicitly indicates that both flexibility for what to collocate and responsiveness to changing market demands are the basics of success and are only afforded by a common ownership.

Considering the nature of the products and services—common availability—to be collocated to produce consumer synergies in a resultant two-sided market, another natural, yet also practically important question arises (Ye et al., 2012): Can a focal firm's innovative and successful collocation of commonly available products and/or services provide a sustainable advantage for that firm?

If we employ the resource-based theory (Kozlenkova et al., 2014) to look at this question, the answer does not look promising, because the following two conclusions have been developed by Barney and Hesterly (2012):

- If a firm possesses such resources that few other firms have due to various constraints, then it is likely for that firm to develop a sustainable competitive advantage, and
- If a resource(s) is simultaneously valuable, rare, and exploitable by a firm, then the resource(s) can generate a sustainable competitive advantage for that firm.

Here, the assumed common availability implies that the products and/or services to be collocated are neither rare nor difficult for any firm to acquire. So, when proven successful, imitators would possibly follow the already shown path to business success, leading to intensified competition and consequent competitive parity (Powell, 1992). However, the key technical detail in this reasoning is that the statement "if condition A holds true, then conclusion B follows" does not implies anything about what happens to A if B holds true. In the context of our current discussion, this means that although the products and/or services to be collocated are commonly available, it does not necessarily mean that the established consumer synergies and two-sided markets cannot be sustainable.

So, to address our question just given above, let us next turn our attention to the concept of managers' willingness to compete (Madhok et al., 2010) in their strategic decision-making, the essence of which is a choice about how to effectively allocate the limited resources (Grant, 2008; Lin, 2009). Indeed, in real life, beyond the ability to enter a market, no firm actually has the willingness and necessary resources to enter every market or most markets it is technically able to due to various reasons, such as that of forbearance (Bernheim & Whinston, 1990). In particular, firms yield their competition against their stronger competitors in those markets where they are less efficient, while in exchange the latter do the same in markets where the former are more efficient (Li & Greenwood, 2004). For example, Madhok et al. (2010) analyze a two-firm, two-product case and show that when one firm with limited resources has a comparative advantage for both products compared with the other firm, the former chooses to manufacture the product where it has the greater advantage while voluntarily abandoning the other market to the latter. In other words, many firms that have the ability to compete in multiple markets choose to do so in only one or a few while avoiding others so that their competitive advantages may actually be sustainable due to the choices of whom to compete with. Or speaking differently, a focal firm with limited availability of resources generally abandons less attractive markets and chooses to compete in such market(s) that allows it to capture the greatest value (Adner & Snow, 2010).

Proposition 9.6 A demand-side advantage established by a focal firm through collocating commonly available products and/or services and operating a resultant multiple-sided market can be sustainable if the firm conducts its business within a saturated market and the margin of profits due to the advantage is not more than the entry cost of a new competitor.

In fact, in a saturated market (Wagner et al., 2007), there is not much consumer movement; and incumbent firms provide an over-abundant supply of goods to a stable consumer base of a whole spectrum of heterogeneous preferences. That generally means that the margin of profit for the incumbent firms is thin (Forrest et al., 2017). Therefore, if a focal firm's collocation of commonly available products and/or services indeed provides a demand-side advantage in a saturated market, then that firm must have attracted people away from the established consumer base of the incumbent firms due to these people's simultaneous enjoyment/need of the other products and/or services within the collocated assortment. Such a group of consumers can only be a portion of the saturated market. And once the consumer segment of the focal firm, as the first mover, is well served, the cost of entering that segment is drastically increased for potential competitors up to the new entry deterrence price (Lieberman & Montgomery, 1988). That is, in this case, the first mover can obtain a better margin of profits compared to other incumbent firms. Secondly, the assumed multiple-sided market the focal firm is operating serves as an isolating mechanism that benefits that firm as the first mover. The cross-segment externalities of the multiple-sided market within the saturated market are especially difficult for a competitor to penetrate because the potential competitor must entice a critical mass of participants away from the established platform in order to be profitable.

Systemically speaking, the conditions that the firm conducts its business within a saturated market and that the margin of profits due to the advantage is not more than the entry cost of a new competitor, as assumed in Proposition 9.6, are necessary for this conclusion to hold true, as Theorem 3.2 and Theorem 4.1 indicate.

9.6 Mechanical Production of Synergistic Innovations

Continuing our discussions in the previous sections, this section focuses on introducing a mechanical procedure by following which synergistic innovations, either on the producer side or the demand side, will most likely emerge without waiting for particular talents to suddenly and randomly hit on some bright ideas (Ridley, 2016). To limit the length of our present, the following discussions will be focused on the demand side only. For the producer side, the relevant ideas are similar.

To derive new ideas, as consequences of logical reasoning and abstract computation, let us first develop a formal representation for a good, an affair (or event) or a relationship as a formal system, written as an ordered triplet $T = (X, Y, Z)$, where, respectively, T stands for the targeted good or affair or relationship, X describes what the good is or what the affair is for or what the relationship is about, Y the characteristics, and Z the value of the targeted thing. For example, for such a product P that has a market value of \$100, whose quality is ranked A, and whose brand name is excellent, we can formally describe product P as follows:

$$P = \begin{bmatrix} \text{Product P} & \text{Price} & 100 \\ & \text{Quality rank} & \text{A} \\ & \text{Brand name} & \text{Excellent} \end{bmatrix}$$

And if John needs something that can provide a source of light in his home, we can describe this affair as

$$N = \begin{bmatrix} \text{Fulfill} & \text{Need} & \text{light source} \\ & \text{Location} & \text{Home} \\ & \text{Receiver} & \text{John} \end{bmatrix}$$

And if we look at the relative positions of the parts—base, light bulb, and the brightness—of a table lamp, we can write

$$L = \begin{bmatrix} \text{up/down relationship} & \text{on top} & \text{lampholder} \\ & \text{on bottom} & \text{Light bulb} \\ & \text{brightness} & \text{100 w} \end{bmatrix}$$

When the target is an object (thing), affair, or relationship, the formal representation is respectively referred to as an objective element, a situation element, and a relation element.

Secondly, associative thinking is the key for developing new ideas (Lockwood & Papke, 2017). In other words, after having formally described products, services, demands, and relationships, we need to somehow associate these formal expressions one way or another to generate ideas for new things, be they products, services, demands, or relationships. To this end, assume that two formal elements $E_1 = (X_1, Y_1, Z_1)$ and $E_2 = (X_2, Y_2, Z_2)$ are given. Then their logic AND operation is defined as follows:

$$E_1 \wedge E_2 = (X_1 \wedge X_2, Y_1 \wedge Y_2, Z_1 \wedge Z_2)$$

$$= \begin{cases} (X, Y, Z_1 \wedge Z_2), & \text{if } X_1 = X_2 = \mathrm{X}, Y_1 = Y_2 = \mathrm{Y}, Z_1 \neq Z_2 \\ (X_1 \wedge X_2, Y, Z_1 \wedge Z_2), & \text{if } X_1 \neq X_2, Y_1 = Y_2 = \mathrm{Y}, Z_1 \neq Z_2 \\ \begin{bmatrix} X & Y_1 & Z_1 \\ & Y_2 & Z_2 \end{bmatrix}, & \text{if } X_1 = X_2 = \mathrm{X}, Y_1 \neq Y_2, Z_1 \neq Z_2 \\ \begin{bmatrix} X_1 \wedge X_2 & Y_1 & Z_1 \wedge Z_{21} \\ & Y_2 & Z_{12} \wedge Z_2 \end{bmatrix}, & \text{if } X_1 \neq X_2, Y_1 \neq Y_2, Z_1 \neq Z_2 \end{cases}$$

where Z_{21} stands for the value of X_2 with respect to Y_1 and Z_{12} the value of X_1 with respect to Y_2. And the logic OR operation of E_1 and E_2 is defined by

$$E_1 \vee E_2 = (X_1 \vee X_2, Y_1 \vee Y_2, Z_1 \vee Z_2)$$

$$
= \begin{cases}
(X, Y, Z_1 \vee Z_2), & \text{if } X_1 = X_2 = \mathrm{X}, Y_1 = Y_2 = \mathrm{Y}, Z_1 \neq Z_2 \\
(X_1 \vee X_2, Y, Z_1 \vee Z_2), & \text{if } X_1 \neq X_2, Y_1 = Y_2 = \mathrm{Y}, Z_1 \neq Z_2 \\
(X, Y_1 \vee Y_2, Z_1 \vee Z_2), & \text{if } X_1 = X_2 = \mathrm{X}, Y_1 \neq Y_2, Z_1 \neq Z_2 \\
\begin{bmatrix} X_1 \vee X_2 & Y_1 & Z_1 \vee Z_{21} \\ & Y_2 & Z_{12} \vee Z_2 \end{bmatrix}, & \text{if } X_1 \neq X_2, Y_1 \neq Y_2, Z_1 \neq Z_2
\end{cases}
$$

where both Z_{21} and Z_{12} are defined in the same way as above. Lastly, the logic NOT operation of an element $E = (X, Y, Z)$ includes either the X-negation or the Z-negation, written respectively as $\overline{E}_X = \left(\overline{X}, Y, Z\right)$ and $\overline{E}_Z = \left(X, Y, \overline{Z}\right)$. For details on relevant discussions, see (Yang & Cai, 2013).

To understand these logic operations within our current context, they mean synergistic innovations on the producer side, demand side, and market expansion. In particular, when the productions of different products are combined by using these logic operations, ideas of different products emerge; when the demands for different products and/or services are operated on by using these logic operations, potential new demands can be discovered even when consumers are still unaware of them; and when relationships between different consumer segments are jointly studied through these logic operations, two-side or even multi-sided markets can be possibly created. In the rest of this section, as an example, let us look at how to generate the synergistically innovative collocation of indoor tanning service and coin-operated laundromat as described earlier.

First, the laundry service can be formally expressed as follows:

$$
L = \begin{bmatrix}
\text{wach clothes} & \text{need} & \text{laundry machines} \\
& \text{location} & \text{near university campus} \\
& \text{who needs} & \text{students on tight budget} \\
& \text{activity} & \text{participative} \\
& \text{multi tasking} & \text{yes} \\
& \text{sequential tasking} & \text{yes}
\end{bmatrix}
$$

the indoor tanning salon can be written as follows:

$$T = \begin{bmatrix} \text{develp desirable appearance} & \text{need} & \text{indoor tanning salon} \\ & \text{location} & \text{near university campus} \\ & \text{who needs} & \text{mostly young females} \\ & \text{activity} & \text{participative} \\ & \text{multi tasking} & \text{no} \\ & \text{sequential tasking} & \text{yes} \end{bmatrix}$$

and the relationship between college males and females can be expressed formally by

$$R = \begin{bmatrix} \text{develp relationship w opposite sex} & \text{need} & \text{an indoor meeing place} \\ & \text{location} & \text{near university campus} \\ & \text{who needs} & \text{mostly males} \\ & \text{activity} & \text{participative} \\ & \text{multi tasking} & \text{yes} \\ & \text{sequential tasking} & \text{yes} \end{bmatrix}$$

Before we go on with this example, let us add some notes to explain a few details used in the previous formal expressions. First, being indoor needs to be required for everything because for this particular case, the student population is located in a northern US city, where the winter is long and cold. Second, the reason why everything has to be near the university campus is because most of the students served by the laundromat and the indoor tanning salon live on very tight budgets; distance travel will inevitably put a dent on the already tight budgets. Third, washing clothes using coin-operated machines is participative with the need of changing loads and moving washed clothes into dryers, etc., and can be done with multiple tasks simultaneously, such as chatting with others and doing homework, among many other activities. Fourth, the reason why mostly young females use the tanning service is because that way more women than men believe that suntanned skin enhances their appearance. So, living within a location and climate of long and cold winters makes female students more likely maintain their desired, tanned appearance year-round by using tanning booths or beds.

For the development of relationships, the reason why most males need an opportunity and a potential meeting place is supported by the fact that for many people, college is a time for exploring relationships and university-age males are more likely than females to seek and initiate a relationship with the opposite sex (Clark et al., 1999). Developing a relationship with the opposite sex can be worked on jointly with other tasks simultaneously as long as the latter do not demand too much attention away from the effort of the former.

Second, let us compute the following joint elements $L \wedge R$, $T \wedge R$, and $L \wedge T$, and obtain

$$L \wedge R = \begin{bmatrix} \text{wash clothes\&develop relationship} & \text{need} & \text{laundry machines\&meeting place} \\ & \text{location} & \text{near university campus} \\ & \text{who needs} & \text{mostly males on tight budget} \\ & \text{activity} & \text{participative} \\ & \text{multi tasking} & \text{yes} \\ & \text{sequential tasking} & \text{yes} \end{bmatrix}$$

$$T \wedge R = \begin{bmatrix} \text{develop appearance\&relationship} & \text{need} & \text{tanning salon\&meeting place} \\ & \text{location} & \text{near university campus} \\ & \text{who needs} & \text{mostly females on tight budget} \\ & \text{activity} & \text{participative} \\ & \text{multi tasking} & \text{no} \\ & \text{sequential tasking} & \text{yes} \end{bmatrix}$$

and

$$L \wedge T = \begin{bmatrix} \text{wash clothes\&develop appearance} & \text{need} & \text{laundry machines\&tanning salon} \\ & \text{location} & \text{near university campus} \\ & \text{who needs} & \text{mostly females on tight budget} \\ & \text{activity} & \text{participative} \\ & \text{multi tasking} & \text{no} \\ & \text{sequential tasking} & \text{yes} \end{bmatrix}$$

Now, if we compare the matching cells between $L \wedge R$ and $L \wedge T \wedge R$ $(= (T \wedge R) \wedge (L \wedge R))$, we have the following correspondences when the matching cells are different:

1. $L \wedge R(1,1) \sim L \wedge T \wedge R(1,1) \Leftrightarrow$.
   wash clothes and develop relationship
   ~ wash clothes and develop appearance and relationship
2. $L \wedge R(3,3) \sim L \wedge T \wedge R(3,3) \Leftrightarrow$
   mostly males on tight budget ~ mostly females on tight budget
3. $L \wedge R(3,5) \sim L \wedge T \wedge R(3,5) \Leftrightarrow$
   yes, multi-tasking ~ no multi-tasking
4. $L \wedge R(3,6) \sim L \wedge T \wedge R(3,6) \Leftrightarrow$
   yes, sequential tasking ~ yes, sequential tasking

All these correspondences are consistent except (3) that can be overcome by (4). So, this computational reasoning provides an explanation for why the previously given laundry/tanning collocation can successfully create simultaneous consumer utilities and a two-sided market effect.

What is discussed above is of course a 20–20 hindsight. In real-life applications, the relevant formal computations are carried out for all existing and potential business activities near the university campus. After deriving a whole collection of all possible collocation ideas, a detailed evaluation needs to be done in order to determine which possibilities represent true potentials for entrepreneurs to embark on in their efforts of creating value for consumers and capturing value for their enterprises. For the sake of communication convenience, we will call the method presented in this section for generating innovative ideas as extenic methodology (for innovation) (Yang & Cai, 2013).

9.7 A Few Final Words

To survive and grow in the turbulent world of business, each successful firm has to continuously implement different competitive advantages (McGrath, 2013). Diversification at either the producer side (Porter, 1985; Santalo & Becerra, 2008) or the demand side (Ye et al., 2012) or both represents one possible way to accomplish this desperate need. However, adopting such an approach means that a firm has to generate innovative ideas for diversification. To ease the severity of this challenge, this chapter establishes a series of 6 propositions of general conclusions that are both theoretically and practically significant. And, undoubtedly, the extenic methodology introduced in Sect. 9.7 will help reduce the practical effect of the heterogeneity existing in managers' abilities to determine how to intelligently collocate what are available to generate synergies, consumer utilities, and beneficial multi-sided markets, an issue that drastically separate managers, entrepreneurs, and retailers based on the outcomes of their decisions (e.g., Holcomb et al., 2009; Sirmon et al., 2008; Thomas, 1988).

Because of the particular approach taken here, this chapter provides respectively generally applicable recommendations for entrepreneurs, managers, and retailers. For example, for entrepreneurs, this chapter shows that other than employing pure entrepreneurial judgment (Kirzner, 1973) that is mostly based on entrepreneurs' intimate knowledge of available resources and considerable creativity, there are some quite definite factors and a mechanical procedure that can be utilized to uncover novel diversification opportunities through looking around the entire market both inside and outside their firms for complementarities of consumers' preferences.

Our results provide a completely different approach to business success from what is suggested by Baker and Nelson (2005)—apply combinations of available resources to new problems and opportunities. In particular, this chapter provides quite specific ways to organically collocate commonly available products and/or services by targeting different segments of consumers, while sufficiently consider how the served segments of consumers could potentially interact with each other to produce simultaneous consumer utilities (Proposition 9.2) and two-sided market effects (Proposition 9.3). For example, it is found that when the magnitudes of the sides of a two-sided market are significantly different, offering competitive marketing strategies on the attractors can greatly increase the overall consumer participation in the two-sided market, while the costs of marketing can be at least partially compensated by the attractees (Proposition 9.4).

For managers and entrepreneurs who focus on the demand side by considering collocating commonly available products and/or services, this chapter shows that other than an intimate understanding of consumers' needs and the creativity to envision a novel collocation of commonly available resources (Ye et al., 2012), superior consumer benefits can still be created by offering consumer groups utilities that they were not previously aware of through employing confirmed demands and systemic scanning of the market. In particular, this chapter shows why all the commonly available products and/or services in a collocated assortment need to be offered by the same ownership (Proposition 9.5).

Beyond theoretical results, this chapter provides general recommendations that are practically applicable. For example, when strategically planning for collocations of commonly available products and/or services, a manager/entrepreneur should always keep in mind the potential of complementary consumer benefits (Proposition 9.2), where consumer interactions can create another dimension of growth and satisfaction and where a two-sided or even multi-sided market can emerge (Propositions 9.3–9.5). And when a two- or multi-sided market is established, appropriate pricings can be introduced to target attractors and attractees (Proposition 9.4) in order to help with value capture. Additionally, the demand-side advantage can actually be sustained, providing a real-life support of the concept of first mover advantage (Lieberman & Montgomery, 1988). To summarize, our overall recommendation is the following.

> An intimate knowledge of available resources and consumer preferences is not always practically available, although it is important for success (e.g., Augier & Teece, 2008; Kor et al., 2007). So, to effectively generate new business ideas and desirable corporate growth and performance from the producer side, or the demand side, or both, a manager/entrepreneur needs to regularly scan the market by utilizing the extenic methodology and by going beyond the boundaries of his/her firm's current or prior activities, as opposed to what is suggested by Penrose (1959).

For retailers, this chapter shows why their locations and the depths and breadths of their assortments of products/services are important for their success. However, when determining particular locations and the depths and breadths of their assortments, these three variables need to be considered holistically, because they are also interrelated and define each other (Proposition 9.1).

And for all managers and entrepreneurs who like to promote their corporate growth and performance through economies of scope, they need to be aware of when new competitions will appear (Proposition 9.6), and how they can constantly scan the market for the appearance of new products/services, potentially different ways to collocate commonly available products and services emerging out of forever changing consumer preferences by employing the extenic methodology.

As for the limitations of this chapter, one stems from the assumption on why a firm exists. In the business world, firms conduct their business very differently from each other contingent upon why/how they are established. That means that conclusions derived in this chapter do not universally apply to every actual situation. Although our attention in this chapter is on how both conceptually and practically a focal firm can synergistically collocate commonly available products and/or services in its effort to create economies of scope, we did not consider the limitations that firm is subject to. These limitations and related issues surely help point to future research problems in order to develop additional, more in-depth practically useful conclusions and recommendations beyond what is presented here in this chapter.

References

Aas, T. H., Breunig, K. J., Hydle, K. M., & Pedersen, P. E. (2015). Innovation management practices in production-intensive service firms. *International Journal of Innovation Management, 19*(05). https://doi.org/10.1142/S1363919615500553

Adner, R., & Levinthal, D. (2001). Demand heterogeneity and technology evolution: Implications for product and process innovation. *Management Science, 47*(5), 611–628.

Adner, R., & Snow, D. (2010). Old technology responses to new technology threats: Demand heterogeneity and graceful technology retreats. *Industrial Corporate Change, 19*(5), 1655–1675.

Adner, R., & Zemsky, P. (2006). A demand-based perspective on sustainable competitive advantage. *Strategic Management Journal, 27*(3), 215–239.

Alchian, A. A., & Woodward, S. (1987). Reflections on the theory of the firm. *Journal of Institutional and Theoretical Economics, 143*(1), 110–136.

Armstrong, M. (2006). Competition in two-sided markets. *RAND Journal of Economics, 37*(3), 668–691.

Augier, M., & Teece, D. J. (2008). Strategy as evolution with design: The foundations of dynamic capabilities and the role of managers in the economic system. *Organization Studies, 29*(8–9), 1187–1208.

Baker, T., & Nelson, R. E. (2005). Creating something from nothing: Resource construction through entrepreneurial bricolage. *Administrative Science Quarterly, 50*(3), 329–366.

Barney, J. B. (1991). Firm resources and sustained competitive advantage. *Journal of Management, 17*(1), 99–120.

Barney, J. B. (2001). Is the resource-based "view" a useful perspective for strategic management research? Yes! *Academy of Management Review, 26*(1), 41–56.

Barney, J. B., & Arikan, A. (2001). The resource-based view: Origins and implications. In M. Hitt, R. Freeman, & J. Harrison (Eds.), *Handbook of strategic management* (pp. 124–185). Blackwell.

Barney, J., & Hesterly, W. (2012). *Strategic management and competitive advantage: Concepts and cases* (4th ed.). Pearson.

Becheikh, N., Landry, R., & Amara, N. (2006). Lessons from innovation empirical studies in the manufacturing sector: A systematic review of the literature from 1993 to 2003. *Technovation, 26* (5), 644–664.

Bernheim, B. D., & Whinston, M. D. (1990). Multimarket contact and collusive behavior. *RAND Journal of Economics, 21*(1), 1–26.

Betancourt, R., & Gautschi, D. (1990). Demand complementarities, household production, and retail assortments. *Marketing Science, 9*(2), 146–161.
Buffington, J. (2016). The future of manufacturing: An end to mass production. In J. Buffington (Ed.), *Frictionless markets* (pp. 49–65). Springer.
Caputo, A., Marzi, G., & Pellegrini, M. M. (2016). The internet of things in manufacturing innovation processes: Development and application of a conceptual framework. *Business Process Management Journal, 22*(2), 383–402.
Clark, C. L., Shaver, P. R., & Abrahams, M. F. (1999). Strategic behaviors in romantic relationship initiation. *Personality and Social Psychology Bulletin, 25*(6), 709–722.
Conner, K. R., & Prahalad, C. K. (1996). A resource-based theory of the firm: Knowledge versus opportunism. *Organization Science, 7*(5), 477–501.
Damanpour, F. (1991). Organizational innovation: A meta-analysis of effects of determinants and moderators. *The Academy of Management Journal, 34*(3), 555–590.
Drucker, P. F. (1954). *The practice of management*. Harper & Row.
Eisenhardt, K., & Martin, J. (2000). Dynamic capabilities: What are they? *Strategic Management Journal, 21*(10/11), 1105–1121.
Evans, D. S., & Schmalensee, R. (2007). *Catalyst code: The strategies behind the world's most dynamic companies*. Harvard Business School Press.
Forrest, J. Y. L., Buttermore, J., & Wajda, T. A. (2017). At Nash equilibrium when new market competitions appear? *Kybernetes, the International of Cybernetics, Systems and Management Science, 46*(2), 256–271.
Forrest, J. Y.-L., Dang, Y. G., McCarthy, L., Liu, S. F., & Liu, Y. (2019). Simultaneous consumer utilities, multi-sided markets and consumer synergies. *Proceedings of the 2019 annual meeting of national association of business, economics and technology* (pp. 101–121).
Foss, K., Foss, N. J., Klein, P. G., & Klein, S. K. (2007). The entrepreneurial organization of heterogeneous capital. *Journal of Management Studies, 44*(7), 1165–1186.
Gans, J. S., MacDonald, G., & Ryall, M. D. (2008). *The two sides of competition and their implications for strategy*. Working paper, Melbourne Business School. https://works.bepress.com/michael_ryall/15/
Grant, R. M. (2008). *Contemporary strategy analysis: Concepts, techniques, applications* (6th ed.). Blackwell Publishing.
Harmancioglu, N., Droge, C., & Calantone, R. (2009). Strategic fit to resources versus NPD execution proficiencies: What are their roles in determining success? *Journal of the Academy of Marketing Science, 37*(3), 266–282.
Holcomb, T. R., Holmes, R. M., Jr., & Connelly, B. L. (2009). Making the most of what you have: Managerial ability as a source of resource value creation. *Strategic Management Journal, 30*(5), 457–485.
Jullien, B. (2005). Two-sided markets and electronic intermediaries. *CESifo Economic Studies, 51* (2–3, 233), –260.
Kanter, R. M. (2001). *Evolve! Succeeding in the digital culture of tomorrow*. Harvard Business School Press.
Kirzner, I. M. (1973). *Competition and entrepreneurship*. University of Chicago Press.
Kirzner, I. M. (2018). The ethics of pure entrepreneurship: An Austrian economics perspective. *The Review of Austrian Economics*. https://doi.org/10.1007/s11138-017-0412-1
Kor, Y. Y., Mahoney, J. T., & Michael, S. C. (2007). Resources, capabilities and entrepreneurial perceptions. *Journal of Management Studies, 44*(7), 1187–1212.
Kozlenkova, I. V., Samaha, S. A., & Palmatier, R. W. (2014). Resource-based theory in marketing. *Journal of the Academy of Marketing Science, 42*, 1–21.
Li, S. X., & Greenwood, R. (2004). The effect of within industry diversification on firm performance: Synergy creation, multi-market contact and market structuration. *Strategic Management Journal, 25*(12), 1131–1153.
Lieberman, M. B., & Montgomery, D. B. (1988). First-mover advantages. *Strategic Management Journal, 9*, 41–58.

Lin, Y. (1999). *General systems theory: A mathematical approach.* Kluwer Academic/Plenum Publishers.
Lin, Y. (2009). *Systemic yoyos: Some impacts of the second dimension.* CRC Press (an imprint of Taylor and Francis).
Lockwood, T., & Papke, E. (2017). *Innovation by design: How any organization can leverage design thinking to produce change, drive new ideas, and deliver meaningful solutions.* Career Press.
Madhok, A., Li, S., & Priem, R. L. (2010). The resource-based view revisited: Comparative firm advantage, willingness-based isolating mechanisms and competitive heterogeneity. *European Management Review, 7*(2), 91–100.
McGrath, R. G. (2013). *The end of competitive advantage: How to keep your strategy moving as fast as your business.* Harvard Business Review Press.
McMullen, J. S. (2015). Entrepreneurial judgment as empathic accuracy: A sequential decision-making approach to entrepreneurial action. *Journal of Institutional Economics, 11*(3), 651–681.
Messinger, P. R., & Narasimhan, C. (1997). A model of retail formats based on consumers' economizing on shopping time. *Marketing Science, 16*(1), 1–23.
Moreau, C. P., & Dahl, D. W. (2005). Designing the solution: The impact of constraints on consumers' creativity. *Journal of Consumer Research, 32*(1), 13–23.
Penrose, E. T. (1959). *The theory of the growth of the firm.* John Wiley & Sons.
Porter, M. E. (1985). *Competitive advantage: Creating and sustaining superior performance.* Free Press.
Powell, T. C. (1992). Strategic planning as competitive advantage. *Strategic Management Journal, 13*(7), 551–558.
Priem, R. L. (2007). A consumer perspective on value creation. *The Academy of Management Review, 32*(1), 219–235.
Priem, R. L., & Butler, J. E. (2001a). Is the resource-based "view" a useful perspective for strategic management research? *The Academy of Management Review, 26*(1), 22–40.
Priem, R. L., & Butler, J. E. (2001b). Tautology in the resource-based view and the implications of externally determined resource value: Further comments. *The Academy of Management Review, 26*(1), 57–66.
Ridley, M. (2016). *The evolution of everything: How new ideas emerge.* Harper Perennial.
Rochet, J. C., & Tirole, J. (2003). Platform competition in two-sided markets. *Journal of the European Economic Association, 1*(4), 990–1029.
Rochet, J. C., & Tirole, J. (2006). Two-sided markets: A progress report. *RAND Journal of Economics, 37*(3), 645–667.
Sakhartov, A. V., & Folta, T. B. (2014). Getting beyond relatedness as a driver of corporate value. *Strategic Management Journal, 36*(13), 1939–1959.
Santalo, J., & Becerra, M. (2008). Competition from specialized firms and the diversification-performance linkage. *Journal of Finance, 63*(2), 851–883.
Sawhney, M., Verona, G., & Prandelli, E. (2005). Collaborating to create: The internet as a platform for customer engagement in product innovation. *Journal of Interactive Marketing, 19*(4), 4–17.
Shah, S. K., & Tripsas, M. (2007). The accidental entrepreneur: The emergent and collective process of user entrepreneurship. *Strategic Entrepreneurship Journal, 1*(1–2), 123–140.
Sirmon, D. G., Gove, S., & Hitt, M. A. (2008). Resource management in dyadic competitive rivalry: The effects of resource bundling and deployment. *The Academy of Management Journal, 51*(5), 919–935.
Smith, A. (1776). *The wealth of nations, books I-III, (1986 printing).* Penguin Books.
Smith, W. K., & Tushman, M. L. (2005). Managing strategic contradictions: A top management model for managing innovation streams. *Organization Science, 16*(5), 522–536.
Sobel, R. (1999). *When giants stumble: Classic business blunders and how to avoid them.* Prentice Hall.
Sun, M., & Tse, E. (2009). The resource-based view of competitive advantage in two-sided markets. *Journal of Management Studies, 46*(1), 45–64.

Thomas, A. B. (1988). Does leadership make a difference to organizational performance? *Administrative Science Quarterly, 33*(3), 388–400.

Tripsas, M. (2008). Customer preference discontinuities: A trigger for radical technological change. *Managerial Decision Economics, 29*(2–3), 79–97.

Veugelers, R., & Cassiman, B. (1999). Make and buy in innovation strategies: Evidence from Belgian manufacturing firms. *Research Policy, 28*(1), 63–80.

Visnjic, I., Wiengarten, F., & Neely, A. (2016). Only the brave: Product innovation, service business model innovation, and their impact on performance. *Journal of Product Innovation Management, 33*(1), 36–52.

Wagner, P., Olsen, J., & Thach, L. (2007). *Wine marketing & sales: Success strategies for a saturated market* (2nd ed.). Board and Bench Publishing.

Williamson, O. E. (1979). Transaction cost economics: The governance of contractual relations. *Journal of Law Economics, 22*(2), 233–261.

Yang, C. Y., & Cai, W. (2013). *Extenics: Theory, method and application*. Science Press.

Ye, G. L., Priem, R. L., & Alshwer, A. A. (2012). Achieving demand-side synergy from strategic diversification: How combining mundane assets can leverage consumer utilities. *Organization Science, 23*(1), 207–224.

Zollo, L., Marzi, G., Boccardi, A., & Ciappei, C. (2016). Gli effetti della Stampa 3D sulla competitivita aziendale. Il caso delle imprese orafe del distretto di Arezzo. *Piccola Impresa/Small Business, 2*(2), 80–100.

Part III
Value-Chain Framework

Chapter 10
The Systemic, Hierarchical Structure of the Mind

Abstract This and the next chapter, which are mainly based on Wajda et al. (to appear, The inclusiveness dimension of human cognition and categorization paradigm), form a two-part sequel of a recent study on how people dissimilarly categorize a given set of stimulus targets by mobilizing different areas of their cognitive systems. In particular, this current chapter develops a general theory on how each human being is a goal-oriented system, the mechanism underneath the operation of the four human endowments—self-awareness, imagination, conscience, and free will, and the systemic structure of the general human cognitive system. For related discussions, see Lin and Forrest (2012, *Systemic structure behind human organizations: From civilizations to individuals*. New York: Springer).

Keywords Abstraction · Cognitive system · Conscience · Consumption decisions · Free will · Human Cognition · Human endowments · Imagination · Self-awareness

Based on what is accomplished here, this chapter investigates the relationship between different layers of abstraction within human cognitive system, and explains why and how individuals demonstrate their individually varied preferences for their respectively different levels of taxonomic abstraction. Then, an empirical case study is utilized to confirm some of the main conclusions developed in these two chapters theoretically.

The motivation of this chapter and the following chapter is the wide-range use of the categorization paradigm, regarding how people receive and process market information, in explaining a host of consumer behaviors. However, there are no theoretical unsoundness and practical reliability underneath these explanations. To make up for this deficiency in knowledge, this and the next chapter develop the necessary theory on how consumers receive and process market information so that consequent marketing programs can be developed accordingly.

J. Y.-L. Forrest, Y. Liu, *Value in Business*, Contributions to Management Science,
https://doi.org/10.1007/978-3-030-82898-1_10

10.1 Introduction

Because of its ability to predict some important consumer behaviors, the categorization paradigm in recent years has received considerable attention. This paradigm employs how market information is received and processed by consumers to explain how consumers consider and compare products and/or services and then how they eventually make their consumption decisions among alternatives during choice. The main premise of this paradigm is that the way consumers categorize information in their memory influences the content and quantity of information during later retrievals. By utilizing this paradigm, scholars (e.g., Chiou et al., 2018; Chowdhury et al., 2018; Mandler, 1982; Meyers-Levy & Tybout, 1989; Moss, 2009; Nedungadi, 1990; Sahni, 2016; Schrift et al., 2018; Sujan, 1985) are able to identify various consumer-based outcome variables, including consumer attitudes towards and evaluations of products/brands, as well as consequent inferences regarding product attributes and qualities. Based on these identified variables, abundant marketing initiatives have been produced and implemented with varying degrees of success.

In order to help make the huge investment in marketing efforts more effective, this chapter attempts to develop the missing theory on the structure of human cognitive system, hoping that the theory can practically explain why different individuals employ different methods to process information. With such a theory in place, marketers will be able to produce and execute their campaigns by more reliably predicting consumer behaviors than before.

Because decisions represent consequences of mind activities, this chapter focuses on the investigation of the hierarchical structure of the mind before one can understand how and why the categorization paradigm actually works. Because each mind activity is jointly affected by self-awareness, imagination, conscience, and free will, this chapter establishes facts about these four endowments. Speaking differently, by knowing how the mind functions, marketers will have an opportunity to appropriately predict consumer behaviors under different circumstances. As for why this chapter applies logical reasoning and systems thinking as the preferred methodology is that the concept of wholeness appears within mind activities and how these activities interact with the environment. Each thought, mind activity, and environment can all be seen as a system with its individually special structure. Hence, their interactions can be more effectively examined by using concepts of systems science (Forrest, 2018) than using structure-less number-based empirical studies (Lin & OuYang, 2010; Wu & Lin, 2002). To see more about why there is a theoretical significance and practical need to go beyond the conventional methods, such as statistics-based approaches, anecdotal analyses, and calculus-based tools, see (Forrest, 2018, p. 12–16).

Because of the adopted unconventional methodology, this chapter accomplishes its goal successfully by developing the following main theoretical results of the mind.

- Each person is a system that is oriented towards the goal of being happy.
- The endowment of self-awareness is not location-specific; it helps a person examine her thoughts and respond to whatever circumstances given.
- Imagination and formation of mental images and concepts are achieved by matching what is presented with what has been experienced and learned before by mobilizing self-awareness.
- The endowment of conscience is genetically determined, although conscience is a learned two-valued $\pm$ function that is partially defined by imagination.
- The endowment of free will takes one of the following forms: promises are kept, occurrence is opposite to promised, and promises lead to no predictable outcome.
- A person's cognitive system demonstrates a clear systemic yoyo structure, consisting of hierarchical networks of her mind and environment.
- Within any taxonomy of things, events, and thoughts, no category of the highest level of abstraction can exist.

Beyond developing these generally true statements, another contribution of this and the next chapter to the literature is the introduction of systemic reasoning and logical reasoning into the study of the mind.

The rest of this chapter is organized as follows. Section 10.2 provides a review of the relevant literature and illustrates how this chapter contributes to the literature. Section 10.3 shows how each human being is a goal-oriented system towards happiness. Section 10.4 looks at the nonpositional characteristic of self-awareness. Section 10.5 explores the structure and functionality of imagination. Section 10.6 examines the composition of conscience. Section 10.7 investigates free will and how it takes one of three possible forms. Section 10.8 combines what is established in the previous sections and demonstrates the formation of the systemic field of human cognition. Then this chapter concludes in Sect. 10.9.

10.2 Literature Review

In the area of marketing, an important research topic is about how consumers gather and comprehend market information and then how they make their consumption decisions. To address the former question, one has to first understand how consumers either consciously or unconsciously mobilize their individually different cognitive systems. Speaking differently, consumers' consumption decisions are greatly affected by the composites of their cognitive systems or their natural endowments—self-awareness, imagination, conscience, and free will. Therefore, in this part of the current study, let us first focus on the development of the necessary theory on how people collect and comprehend information from the marketplace and then consequently make their decisions of consumption. To accomplish this end, this chapter will look at the tiered structure of the mind and how the four human endowments interact with each other.

The literature in this research direction is extremely large and can be traced back many hundreds of years. However, for the purpose of this chapter, let us only survey some relevant studies. By looking at sustainable happiness as a pursuit of happiness without exploiting other people, the environment, or future generations, O'Brien (2008) links sustainability and happiness together and discusses livable communities, child-friendly planning, and education. Uchida et al. (2004) examine cultural variations in terms of the meaning, motivation, and predictors of happiness by looking at the recent evidence on happiness and well-being. Shin et al. (2018) investigate words that are closely associated with happiness and find that individuals who use the crucial words frequently are much happier with their lives than others. Walsh et al. (2018) consider the association between happiness and career success. They empirically find that although these variables are correlated, happiness often precedes career success and positive emotions improve workplace performance. Enriching this line of studies, this chapter establishes the following result: Each person is a goal-oriented system, aiming either consciously or unconsciously at attaining happiness. And it provides a potential method for an individual to attain her desired happiness—constantly look for conquering further challenges and believe (either correctly or incorrectly) in her capability of success.

As for the mind, by viewing it as a dynamic system and by examining commonalities between this and other systems, Moran (2018) attempts to understand complex mental processes. By introducing a dynamic framework for how the mind wanders and how the mind relates to large-scale brain networks, Christoff et al. (2016) maintain that as a spontaneous-thought phenomenon, mind wanders with creative thinking and dreaming. By investigating how the mind processes and organizes information and how it makes decisions, De Bono (2015) proposes the reason for why the mind works only in certain ways by singly or jointly utilizing natural, logical, mathematical, and lateral thinking. By treating the mind as an intersubjective blender, Cavell (1998) believes that it is indispensable for humans to envision a shared external world in order for an individual to know her own thoughts as a subjective perspective on the world. Contributing to this line of the literature, this chapter establishes the systemic mechanism on which the mind functions.

In terms of the four human endowments, by empirically examining how self-awareness influences the perspective-taking and egocentrism of an adult, Abbate et al. (2016) are able to conceptually extend previous findings. Through designing a self-development curriculum based on Bloom's Taxonomy, Ugur et al. (2015) suggest that to unswervingly express selected values and characteristics, one needs to first internalize chosen items as guides for living, and then behave consistently, followed with a sense of well-being and personal growth. Yazdanparast and Spears (2018) look at how comparisons of the self to fashion models lead to body dissatisfaction and what can be used to potentially remedy harmful outcomes of such evaluations. They find that objective awareness can be potentially manipulated.

By referencing recent studies on how cognitive functions engross the anterior hippocampus, Zeidman and Maguire (2016) link the former to the structure of the latter. To examine how imagination creates reliable experiences, Derbaix and

Gombault (2016) find that with support of immaterial dimensions, the material dimensions can facilitate consumers' imagination. That is, by making use of imagination, invisibles can help create reliability and authenticity. McFarland et al. (2017) hypothesize that depression is often connected with over-general memory and imagination, coupled with problem-solving deficits while finding that brief interviews can temporarily enhance problem-solving capability of individuals with depression.

Based on the assumption that conscious experiences arise from one area in the brain, LeDoux and Brown (2017) derive the belief that emotional and nonemotional states are separated by the kinds of processed inputs of a general cortical network of cognition. Among others, these scholars believe that nonconscious inputs, provided by subcortical circuits, coalesce with other kinds of neural signals existing in conscious emotional experiences. Aronfreed (1968) provides a valuable survey on the literature of how behaviors and conscience interact with each other. And Langston (2001) presents a historical account on the development of the concept of conscience throughout time.

Libet (1999) experimentally addresses whether or not free will actually exists and finds that although volitional courses of taking easily volunteer acts are initiated unconsciously, conscience still controls the result. That is, free will indeed appears, even though it might not help initiate a voluntary act. Dennett (2015) argues that instead of being threatened by advances in science, variations of free will, which endorse moral and artistic responsibilities, are distinguished, explained, and justified in further detail by these advances. Kane (2005) provides a good source of information on free will for scholars who do not have much background in the subject with an extensive survey of latest views on this central area of philosophy.

By employing systems science, systems methodology, and logical reasoning, this chapter carries relevant knowledge to a much higher level by showing the following theoretical results, which can be practically applied in real-life settings:

- The endowment of self-awareness is innate and nonpositional; it assists a person to examine her thoughts and to choose appropriate courses of action;
- The endowment of imagination helps a person find parallels between what is presented and what is known from before and form her philosophical system of values and beliefs;
- Conscience is a genetically determined capability and represents a learned partial, two-valued function, defined on the reservoir of imagination. The two possible output values are + and −; and
- An individual's free will takes one of the following three possible forms—promises kept, actual occurrence is opposite to promise, and no definite connection between promise and outcome.

In short, as a consequence of the particular methodology employed, established hereafter are theoretically general and practically applicable conclusions. They provide universally applicable recommendations for marketers, managers, and entrepreneurs unless stated antecedents are violated.

10.3 Humans as Beings Oriented Towards Happiness

In this chapter, each person is modeled as a yoyo field, which evolves with the yoyo fields of many other people, things, events, thoughts, etc. As is imaginable, these fields spin differently in terms of rotational speed, direction, and strength, as well as divergence and convergence, orientation, and occupied territories of varying scales. This systemic modeling of people and their surroundings suggests that each person lives within a rough sea of thousands of thousands of spinning fields that interact with each other. Therefore, the mind of a person also interacts with other minds, be they like or dislike, where two minds are said to be alike, if they hold similar views of life and the world and similar values; in other words, their yoyo fields spin in the same fashion in terms of the listed characteristics above. Accordingly, two minds are said to be different or dislike, if one of these characteristics of their yoyo fields is different from each other. The interactions of spinning fields in the rough sea jointly make a person suffer from self-doubt (or skepticism) about the suitability and potential of her desires in life.

By happiness, it means a state of mind (or feeling), such as contentment, satisfaction, pleasure, or joy (Cambridge Advanced Learner's Dictionary, 2008). A happy person is always such a person who dreams of heights of some future achievement(s) (Hill, 1928, p. 153); and there are principles that govern enduring happiness and success (Covey, 1989, p. 23).

Proposition 10.1 The fundamental objective of all human efforts is to become a happy person. Sustaining the hope of future achievement is the only way to maintain the state of happiness. The feeling of happiness lies with the present and/or the past while sustaining the feeling depends on the future.

To show this conclusion, let us model the concept of happiness (Lin & Forrest, 2012) as follows. At a particular moment of time, a person is happy, if and only if the yoyo field of the person just achieved an upper hand in its interaction with at least one field flow in its environment. This systemic understanding of happiness agrees with Aristotle's description of the concept, as given in his Nicomachean Ethics (Aristotle, 2016)—happiness (equivalently being well and doing well) is the only thing that humans desire for their own sake. As a direct consequence of this systemic modeling of happiness, the conclusions in Proposition 10.1 follows. In particular, the constant achievement of conquering greater and more challenging difficulties over time one after another keeps a person content and continuously excited with her state of affairs and set of circumstances. And no matter what was accomplished in the past, the yoyo field of a person is always presently interacting with other fields and experiencing challenges and difficulties. Hence, the only way to maintain the state of happiness is to sustain the hope of future achievement; and the feeling of happiness lies with the present and/or the past while sustaining the feeling depends on the future.

This same analysis in fact also indicates the following result:

Proposition 10.2 For a person to possibly achieve enduring happiness, the person needs to focus on constantly conquering additional difficulties in the indefinite future and have the confidence either correctly or incorrectly that she has the ability to do so.

The discussion above also directly leads to the following result:

Proposition 10.3 No matter whether it is conscious or unconscious, each person represents a goal-oriented system.

As a goal-oriented system (Mesarovic & Takahara, 1975), each person engrosses information and knowledge in a way like a magnet from all sources in the environment through automatic use of her four human endowments: self-awareness (or self-consciousness), imagination, conscience, and free will. If we model thoughts as the input side of the person's yoyo body while her actions as the other side, then we can conclude that the person's endeavors are carried out at their individually specific moments in manners aligned with the dominating thoughts of the person's mind of the respective moments (Lin & Forrest, 2012). Therefore, if a person, for example, desires to achieve a clear goal in life and possesses the necessary determination to materially realize it, then she can intentionally fasten and embrace the goal in her mind. As time passes, the fastened and embraced goal will sooner or later infuse the self-consciousness of the mind, which will urge the person's physical body to take actions towards the ultimate realization of the goal. There are many real-life manifestations of this theoretical fact. The magnificent life of Thomas Edison demonstrated how a news butcher was eventually transformed into a leading inventor of the world.

10.4 Nonpositional Self-Awareness

By self-awareness it stands for a person's awareness that she exists as an individual and an entity that is different and separate from other people and objects with private thoughts and individual rights (Cooke, 1974, p. 106). The concept also implicitly indicates a person's realization that other people are similarly self-aware. It is because of self-awareness that people are considerate of their own attitudes and temperaments (Branden, 1969, p. 41).

Proposition 10.4 Self-awareness, as a natural existence, is not attached to any specific location. That is, the existence of self-awareness is nonpositional.

This result is a natural consequence of the underlying systemic yoyo structure of any person of concern. In particular, the said yoyo field takes in and gives out things constantly, where the things that are being taken in and given out can assume any form, such as physical, intellectual, tangible, intangible, or epistemological. Things go in and out surely from the input and output sides, as well as in all horizontal directions and elevations along the axis of spin. The situation can be vividly imagined as that of a tornado in real life. Hence, different systems constantly push against or pull towards each other. Such pushing effects between people and between people of other things make each person sense the separation of herself from other people and things, and the pulling effect creates the sense of individual rights and entitlements.

These forever existing pushing and pulling effects collectively make all involved systems or individuals aware of their own existence, the existence of others, their private thoughts, etc. Because these pushing and pulling effects universally exist no matter where a person is and guarantee the appearance of self-awareness, that explains the reason why self-awareness is not attached to any specific location.

In terms of literature, as a French existentialist philosopher and one leading figure in the twentieth century French philosophy, Jean-Paul Sartre (June 21, 1905–April 15, 1980) is the first person who conjectured the result in Proposition 10.4 (Gerassi, 1989).

Proposition 10.5 Each person relies on her self-awareness to examine her thoughts and decide on how to respond appropriately and adequately to the circumstances she is currently in, where the examination is conducted by associating the circumstances with what have been learned before.

The truthfulness of this result can be seen as follows by continuing the systemic modeling and discussion in the previous paragraphs: for any chosen person, her self-awareness emerges from the interactions of her systemic yoyo field with those of all other beings and things existing within the environment. As the person grows with increasing maturity, she experiences various dealings with others, compares her varied approaches, and remembers those most satisfactory approaches that produced desirable or optimal outcomes.

When a unique and never-seen-before situation is presented to the person, she has two options to consider: (1) experience anxiety or (2) associate the new situation with one or a set of previously seen scenarios, each of which resembles at least slightly the current situation. No matter whether the person takes option (1) or (2), she has to eventually organize her thoughts in order to be able to formulate a reasonable course of actions by jointly employing other human endowments: imagination, conscience, and free will. Specifically, when the person is presented with a particular circumstance, she will have to find a way to jump into option (2) no matter whether or not she first experiences an initial shock of anxiety if she, somehow, has to deal with the situation, where running away from the situation is surely one possible action to take. Now, the person studies related actions taken before to deal with those previously seen scenarios, each of which resembles at least slightly the current situation to come up with a possibly appropriate reaction to the current situation that might potentially lead to desirable outcomes.

Speaking differently, through various interactions with the yoyo fields of different people, events, thoughts, concepts, etc., the person has acquired an increasingly improved understanding of herself and other people, events, and things. Hence, the person amasses over time a large deposit (or reservoir) of knowledge and experiences and has advanced a capability for her to freely and frequently mobilize the elements in the reservoir in her decision-making.

10.5 Imagination and Its Functions

By imagination, it means (Egan, 1992; Norman, 2000) the capability and the action (or process) of the mind to form mental images and concepts no matter what is presented in front of the biological senses, including the case that nothing is presented; with such aptitude, a person derives meaning out of each experience encountered and understanding to whatever knowledge learned. Imagination helps each person meet and experience things throughout life. Every time when a person physically touches an object, optically sees a scene, and acoustically hears a sound, she coalesces her senses into a mental picture via imagination. Such capability to imagine is innate through which a person forms her view of life, belief of how the world operates, and philosophical system of values within the mind based on how her yoyo field interacts with the yoyo fields of other elements within the shared world, both physical and intellectual (Harris, 2000). Speaking differently, the endowment of imagination can be virtuously seen as a workplace located within the mind. In this workplace, learned concepts and established facts are innovatively associated with each other in various ways to produce new and better solutions to problems. These innovative associations are known as the creative power of the soul (Hill, 1928).

Proposition 10.6 By match-making with elements in the imagination's reservoir of experience and knowledge through using self-awareness, imagination helps a person imagine and form mental images and concepts out of what is presented either implicitly or explicitly to her sense organs.

In order to see why this result holds true, let us model imagination by going along with the systemic yoyo model of self-awareness in such a way that the human endowment of imagination becomes another natural consequence of a person's yoyo structure. With such a systemic modeling, a person's biological sense organs are simply the body parts of the three-dimensional realization of the person's yoyo field, although the underlying multidimensional spinning field of the person is very complex. Therefore, there are a great deal of the yoyo field structures and yoyo field interactions of elements in the world human sense organs cannot really detect. Speaking different, human sense organs can only collect a small portion of the extremely rich collection of experience and knowledge gained by a person through her constant interactions with the yoyo fields of other entities of the physical and intellectual world, when her yoyo field fights against others and attempts to snatch up things of common interests. In other words, a person's experience and knowledge, acquired from her interactions with others, are much richer than what her sense organs can materialistically collect.

This model of a person's imagination indicates that the endowment "imagination" can be identified with or is equal to the reservoir of all interactions between the person's underlying yoyo field and the fields of other people and things in the environment. More specifically, the reservoir contains both conscious and unconscious records of interactions, where the former records are collected by the person's sense organs, and the latter records are unknown to the person's conscious mind.

When a need arises for a person to develop images or concepts in her pursuit of happiness (Propositions 10.1 and 10.2), the person utilizes her self-awareness to reach into this reservoir. The degree of how well she can mobilize her self-awareness governs how deeply and widely (or individually and collectively) she can reach into this reservoir. This degree of utilization of her self-awareness in turn governs how thought-provoking her consequently established concepts and images will be. The common systemic structure, shared by elements of the material and intellectual world and human thoughts (Lin, 2008), implicitly implies that when called to act, imagination simply plays the role of a match-maker. It matches the situation or issue of concern with related elements within the reservoir of imagination. If a nearly perfect match can be found, then an ideal image and concept are quickly formulated, and the consequent course of action is determined naturally. If not good or no match is found at all, the reservoir of imagination will then be expanded with new elements, consisting of unprecedent experiences and/or knowledge. This end explains why training, either formal or informal, is necessary for a person to go through. It is because, with adequate training on how to commend over self-consciousness, a person will be able to mobilize a lot of information from the reservoir of her imagination. That reservoir is composed of conscious experience and knowledge learned through sense organs and those collected unconsciously without going through these sense organs.

Proposition 10.7 A person employs her imagination to form her individually specific view of life, belief of how the world operates, and philosophical system of values by associating different experiences and pieces of knowledge.

To establish this generally true result, imagine the birth of a representative person. For a period of time since this very moment of birth, this person lives within the constraints of many limiting conditions and has to acquiescently submit to what is available from the caretaker and the limited environment. These early exchanges with people and the world assist the newborn to develop its elementary beliefs, basic values, and fundamental philosophical assumptions. For example, the newborn quickly learns that "you must care for me by meeting my needs," "If not, you will have to suffer from the consequence – my crying," etc. By unconsciously using this set of elementary beliefs, values, and assumptions, the newborn commands the caretaker and the sounding world to satisfy its wants and needs. As the newborn grows, it starts to gradually use the expanding set of beliefs, values, and assumptions to explain whatever is unfathomable, construct tactics to conquer adversities, and creates ways to manage her own dealings and concerns.

With increasingly more mature mental capacity over time, the person gains more control over her self-awareness. That naturally assists the person to obtain more efficient tools and advanced knowledge from more diversified sources in the environment of an enlarging scale. Hence, her imagination's reservoir of experience and knowledge simultaneously grows with new elements constantly added, acquired either consciously and unconsciously, while the members of the reservoir start to associate with each other to form intellectual understandings of rising levels. Associations of these experiences and knowledge become more and more fortified within her self-consciousness. As soon as the strength of these associations reaches and goes beyond a threshold value, Bjerknes's (1898) Circulation Theorem, for more details, see Chap. 3, assures that abstract, multidimensional eddy motions will appear within the self-awareness and reservoir of imagination of that person. That is exactly how imagination assists a person form her view of life, belief on how the world operates, and philosophical system of values within her minds.

Proposition 10.8 Different persons possess their individually different sets of views of life, beliefs on how the world operates and philosophical systems of values. For short, each such set will be referred to as a (philosophical) system of values and beliefs.

The truthfulness of this result naturally follows from the fact that people are born into very different families, which are located within quite varied environments. Therefore, their underlying systemic yoyo fields interact with people and environments differently within their correspondingly diverse environments. These differences existing in families, environments, and interactions help lead to correspondingly different philosophical systems of values and beliefs.

10.6 Conscience, Where Innate and Acquired Capabilities Integrate

By conscience, it represents such an ability for a person to know the principles that judge whether or not her behaviors are acceptable and to sense the degree to which her thoughts are in accord with the principles. In other words, each person employs her conscience to tell if her actions are either right or wrong (Buss, 2004; Pfaff, 2007; Tinbergen, 1951). In particular, when a person acts in a way that is not in agreement with her underlying principles (or known as moral values), her conscience will tell her that a wrong has been committed, making her feel remorse. On the other hand, when a person acts in alignment with her moral values, her conscience will make her feel morality or truthfulness.

For two sets X and Y of objects, f is a partial function from X to Y, provided that there is a subset D of X, written $D \subseteq X$, such that f is a well-defined function from D to Y. It means that for any element x in D, there is an element y in Y such that f clearly rules that y is associated with x. In this case, x can be seen as an input of f and y an output of f, written as $y = f(x)$. Speaking differently, f specifies a rule that assigns each element in D with an element in Y. The set Y is known as the range of the partial function f.

Proposition 10.9 Each person is genetically endowed with her capability of conscience, while her conscience represents an acquired partial function defined on the reservoir of her imagination such that $Y = \{+, -\}$ is the range of this partial function. In other words, for any element x in the person's reservoir of imagination, the person either has a moral value assigned to x or does not have any. For the former case, the assigned moral value to x is either + or −.

To see the truthfulness of this proposition, let us build our argument on the previous systemic yoyo modeling of self-awareness and imagination. Specifically, in this modeling elements in the reservoir imagination are spin patterns of various yoyo fields and interactions of these patterns. The value of + is assigned to some of the field flow patterns and interactions, which are seen to the person as being right, acceptable, or moral; the value of - is assigned to some other field flows and interactions, which are seen as being wrong, not acceptable or immoral.

The reason why this function is only partially defined on the reservoir of the person's imagination is because there are still plenty of other field flow patterns and interactions in the reservoir that are not assigned with either + or -. For convenience of communication, let us refer this partial function as a ± function, which is defined on the domain that consists of all the elements in the reservoir with an assigned value of either + or -.

When a person grows more consciously over time towards the orientation of having a happy life (Propositions 10.1 and 10.2), her assignment of + or - to certain elements in the reservoir of her imagination starts gradually since the time when the person is still an infant; and then she continues this task throughout her entire life. In particular, in the beginning at birth, the domain of this ± function is the empty set. Along with the growth of the person, the domain of her ± function expands with her self-awareness and reinforcements of various environmental factors. If the person experiences something dramatic in her lifetime, certain elements in the domain of her ± function may switch their originally assigned values from + to − or vice versa.

For example, assume that early + values are assigned to field patterns that flow in the same direction, and - values to currents flowing in opposite directions, Fig. 10.1. One explanation for such ± sign assignments is that same directional flows help form sub-eddy pools, while other kinds of currents are not beneficial to the formation of sub-eddy pools. In the language of elementary education, initial + values (or adult approvals) are assigned to behaviors that help strengthen the well-being of all kids involved, and - values (or adult disapprovals) to those behaviors that might lead to opposite effect. For instance, most likely encouraged is the behavior of sharing, which is assigned a + value. On the other hand, violence is disapproved with an assigned − value.

As a person grows with increasing maturity, she will have her ± function well-engrained in her mind so that every time when she intends to take an action or comes up with a thought, she automatically and unconsciously compares the action or thought with relevant elements in the domain of her ± function. If the intended

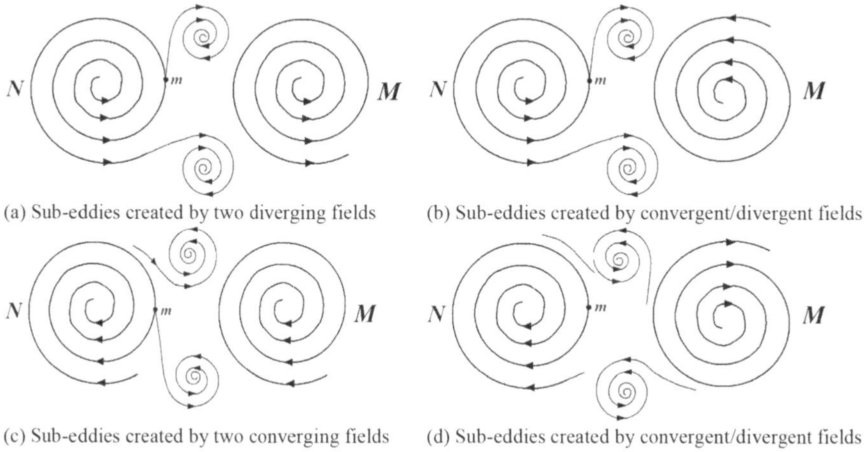

Fig. 10.1 Two adjacent fields *N* and *M* jointly produce sub-eddies in areas of their control, where *m* is a random particle in the spin field of *N*. (**a**) Sub-eddies created by two diverging fields. (**b**) Sub-eddies created by convergent/divergent fields. (**c**) Sub-eddies created by two converging fields. (**d**) Sub-eddies created by convergent/divergent fields

action or the thought in her mind has an assigned + value, the person will feel justified and moral. If the assigned value is −, emerging is a sense of remorse. And if the intended action and the thought forming in the mind does not have any + or − assignment, the person will have to examine its potential + or − value. Doing so expands the domain of her ± function.

This analysis and discussion demonstrate that each person's capacity for conscience is genetically determined, where conscience resides totally on top of imagination, while imagination develops on top of the person's self-awareness, as an innate capability of the person. At the same time, the specific contents of the person's conscience, the ± function and assignments of the ± values are learned throughout life, where some particular ± assignments are switched back and forth several times throughout the person's life.

10.7 Free Will and Its Three Different Forms

By free will, it means a person's capability to keep her promises that she makes to herself and others. That is, free will is a capability each person possesses that assists a person to make estimates for at least the near term regarding what she can do and cannot do and what choices or courses of actions are optimal for the state of affairs involved (Lin & Forrest, 2012). In other words, free will is such a human capability that helps make decisions, choices, and consequent follow-through actions that materialize those choices and decisions.

Proposition 10.10 A person's free will assumes one of the following three possible scenarios:

1. Promises are held;
2. Promises lead to opposite outcomes; and
3. Promises do not lead to any predictable outcome.

To see why this proposition is true, let us continue the discussions in the previous paragraphs based on the particular systemic yoyo modeling established earlier. Developing on her self-awareness, an innate human capability, a person's reservoir of imagination expands constantly; and the domain of her $\pm$ function of moral values enlarges with each occasion of learning and interaction with others. There are three possibilities for the specific "self," which she becomes aware of sooner or later through her increasingly awaking self-awareness in relation to the domain of her $\pm$ function.

A. The "self" is in the domain of the $\pm$ function with an assigned + value;
B. The "self" is in the domain of the $\pm$ function with an assigned - value; and
C. The "self" is not in the domain of the $\pm$ function.

If possibility A is the case, then the person is obligated to carefully investigate the situation of concern and estimate as accurately as possible what she can accomplish and what she cannot accomplish in order to actualize her promises. If she cannot assuredly foretell what she can accomplish for a presented situation, then she will promise to do her best without providing any assurance of producing the desired outcome. Therefore, in either case for possibility A, the person keeps her promises.

If possibility B is the case, then the person once again feels obligated to foretell as scrupulously and as precisely as possible about what she can achieve and what she cannot achieve. But, different from the possibility above, in this case, the person will most likely make promises that are opposite to what she is expectedly able to achieve, based on her imagination and $\pm$ function. By doing so, the negatively valued "self" will lead to an outcome of a positive value in terms of the $\pm$ function. That will make the person feel moral and truthful. That is, for possibility B, what is promised is opposite to what will actually happen.

If possibility C is the case, then the person's "self" is not in the domain of her $\pm$ function. That means that the childhood of the person has been badly spoiled so that she does not really know or care whether her "self" stands for something good or evil. In this case, no matter what the person faces, she simply makes random promises (or sounds) as long as they are pleasant to the ear without even further thinking about correspondingly taking necessary actions. As for why the person behaves this way, it is because whatever happens next, be it a promise kept or not, does not bear any conscientious impact on the person.

10.8 The Systemic Field of Human Cognition

The discussions in the previous sections jointly imply the following conclusion:

Proposition 10.11 The reservoir of a person's imagination is of the structure of a hierarchical network that extends open-endedly on the first layer of human physiological needs and the second layer of specific values of the person's philosophical system, Fig. 10.2.

This hierarchical network in Fig. 10.2 displays the general structure of the mind. The first P level stands for all physiological needs of a representative person, including such basics of life as breathing, food, water, sex, sleep, homeostasis, and excretion (Maslow, 1943, 1954). The second V level is made up of specific values of a person's philosophical system on what life means, how the world actually functions, and how she needs to behave in order to comply with her codes of morality. On top of these two basic levels, the person's mind network contains interacting blocks of various knowledge and experiences about people, things, be they physical, intellectual, conceptual, etc., events and their associations with each other.

Proposition 10.12 The seemingly endless environment of a person can be both visually and conceptually seen as a hierarchical network, within which the elements of her family and relationships among these elements constitute the bottom layer, Fig. 10.3, where by family it represents the immediate environment in which the person grows up.

This proposition illustrates that a person is initially limited to the family setting F within which her life starts. Then she ventures into an increasingly larger environment E with time and growing maturity. That is, as a person grows older and wiser, she simultaneously observes more finer details and increasingly larger wholes from the environment. She internalizes more information, more experience, and more advanced knowledge from the environment. As her vision multiplies further outward, the hierarchical network structure of her mind becomes further extended and enriched with depth. Such extensions in all reaches of vision and in the complexity of the hierarchical network structure of her mind strengthen each other; and the

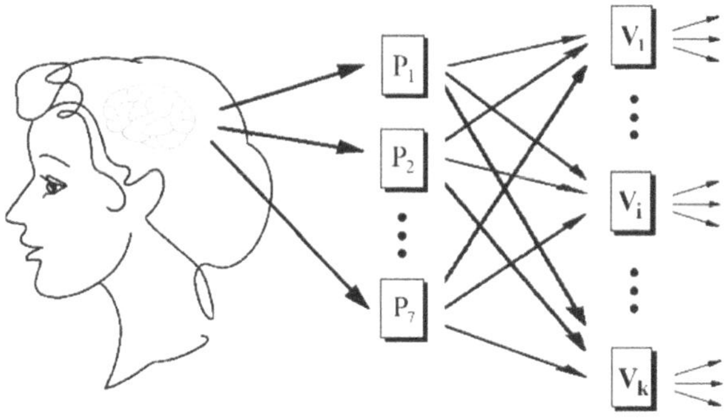

Fig. 10.2 The hierarchical structure of the networked mind

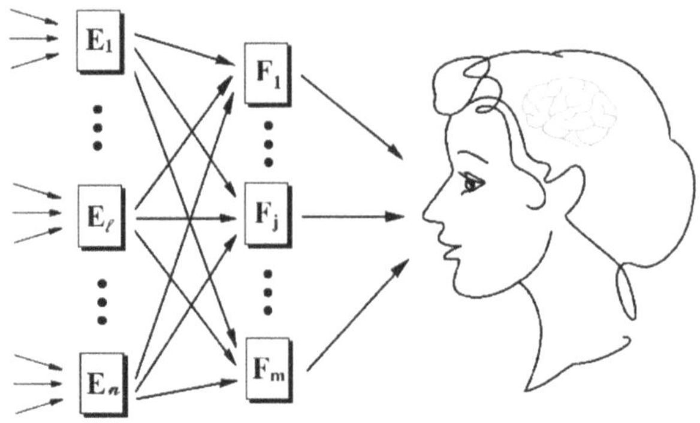

Fig. 10.3 The hierarchical network of a person's environment

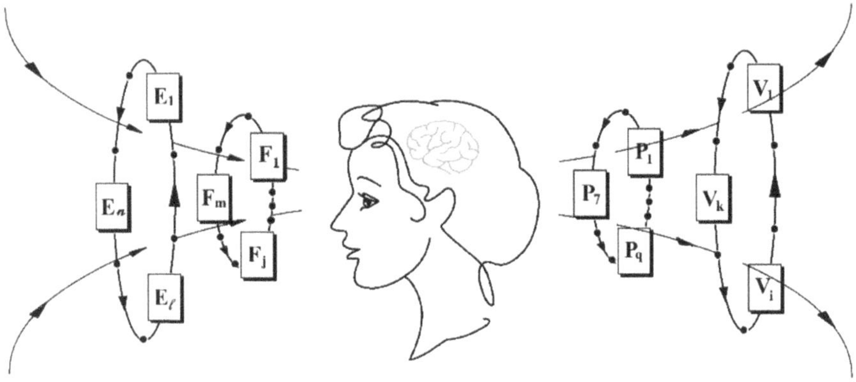

Fig. 10.4 The systemic yoyo structure of human cognitive system

further outwards her vision reaches, the wealthier her mind structure becomes. The wealthier mind structure in turn helps her see even further outwards than before. Such a reciprocating cycle of augmentation lingers until when the person is no longer exploring her sounding environment. From this discussion, the conclusion below follows:

Proposition 10.13 The hierarchically networked structure of a person's mind and that of her environment jointly form the person's cognitive system, whose systemic yoyo structure is shown in Fig. 10.4. In this figurative representation, each level, such as E, F, P, and V, is in a circular movement, which is used to model the fact that elements within each level are in fact more or less associated with each other.

This proposition illustrates the layer structure of the cognitive system within the mind pictured in both Figs. 10.2 and 10.3. Conceptually, as one moves away to the right from the shown P and V levels, units of information, experience, and knowledge become either categories (i.e., the taxonomic organization of things and events), or schemas (i.e., spatially/temporally organized structures) (Stayman et al., 1992), or product-category schemas (Meyers-Levy & Tybout, 1989). More specifically, a product category (or schema) is such a structure of knowledge residing in a person's cognitive system that encompasses a myriad of details, including complete information about the attributes of each product, values of various attributes in terms of optimizing goal-achievement, the correlation of product attributes with one another as well as the relationship between the stated product category and other categories (Stayman et al., 1992). When processing the information of a new product, the essence is how well the new information matches a product category stored in a person's cognitive system (Proposition 10.6). A product-category schema holds characteristics of a taxonomic category, and can also be perceived as a schema to the extent that the use context of a product has been specified. For the convenience of communication, the terms "category" and "schema" are used interchangeably in this chapter. In the rest of this chapter, by the cognitive system of a person, without causing confusion it means the conceptual, hierarchically networked structure in Fig. 10.2.

For the current purpose of study, let us employ Rosch's (1978) taxonomy. That is such a system that allows one category to relate to another category by means of class inclusion, denoted $\subseteq$. In particular, for two classes X and Y, $X \subseteq Y$ means that X is a subclass of Y. That is, the more members a category accommodates (i.e., is more inclusive), the higher level of abstraction the category is of.

Proposition 10.14 For any chosen taxonomy of things, events and thoughts, it is always a part of another category of a higher level of abstraction.

The argument for this result is set theoretic based on Lin (1999) and Kuratowski and Mostowski (1976). In particular, let X be an arbitrarily chosen category. There then is a condition $\varphi(x)$ that describes what this chosen category is. That is, the condition $\varphi(x)$ specifies what entity x is a member in X and what entity x is not a member in X. Assume that y is such a thing, or an event or a thought that does not satisfy the condition $\varphi(x)$ and that Y is a category that contains y as a member and X as a subclass. Then Y is a category of a higher abstraction level than X. For example, the existence of such a category Y can be readily seen as follows: let y be the chosen category X and Y be a category constructed based on this particular y and elements in X. In this construction y is simply the thought of seeing X as an intellectual existence.

Proposition 10.15 No category of the highest level of abstraction can ever exist, no matter which particular taxonomy of things, events, and thoughts is considered.

The argument for this proposition is similar to the one for the previous result. In particular, assume that there is such a category X that is of the highest level of abstraction. Then each category that contains all elements of X and the thought of X, as an existing entity represents a category of a higher level of abstraction than that of X. This end contradicts the assumption about X, as the category of the highest level of abstraction. This contradiction indicates that Proposition 10.15 holds true.

Although Proposition 10.15 holds true in theory, in practice we can still think about categories of the highest level of abstraction, because for any given practical situation, the boundary of consideration is always limited within a defined range, such as the categories of a particular group of furniture. Simultaneously, this proposition points out the fact that the clearly layered hierarchical structure of human cognitive system, as shown in Figs. 10.2 and 10.4, in theory is not quite evidently the case. In fact, the opposite is true: the reservoir of a person's imagination needs to be seen as a continuous field that does not have clearly defined hierarchies and layers that isolate one level of experience, information, and knowledge from another.

Proposition 10.16 No matter when, at least one area in the continuous field of the reservoir of a person's imagination experiences an active circulation of information, experience, and knowledge.

To see why this conclusion is true, we see from Propositions 10.1–10.3 that for the chosen person, she realistically and theoretically is a system that angles itself towards a goal, such as happiness, no matter how the person defines her happiness. Therefore, at any particular moment of time, the person can find herself focusing attention on something, such as a specific consumption or the nurturing of a particular relationship, that is more or less related to her orientation goal. This end implies that in the continuous field of the reservoir of her imagination, certain units of related experience, information, and knowledge are being connected in a special fashion, as stimulated by the puts from the current situation, in order to generate certain desired outcome. The desired outcome can be a moment of suddenly spiked utility or striking excitement. Hence, Bjerknes' Circulation Theorem, see Chap. 4 for more details, guarantees that these associated units of experience, information, and knowledge constitute an active circulation, composed of some newly or recently acquired information.

10.9 A Few Final Words

In order to improve the accuracy of marketers' prediction of consumer behaviors, this chapter develops a systemic theory about human cognition. By doing so, it attempts to introduce a new scientific methodology of logical reasoning and holistic thinking into the study of the categorization paradigm.

Jointly with all the previous results on human cognition established, we are ready to move into the next chapter on the research of the main issues of this study

References

Abbate, C. S., Boca, S., & Gendolla, G. H. E. (2016). Self-awareness, perspective-taking, and egocentrism. *Self and Identity, 15*(4), 371–380.

Aristotle. (2016). *Nicomachean Ethics* (Translated by W. D. Ross with an Introduction by R. W. Browne). Digireads.com.

Aronfreed, J. (1968). *Conduct and conscience: The socialization of internalized control over behavior*. Academic Press.

Bjerknes, V. (1898). Uber einen hydrodynamischen Fundamentalsatz und seine Anwendung besonders auf die Mechanik der Atmosphare und des Weltmeeres. *Kungl. Svenska vetenskapsakademiens handlingar, 31*, 1–35.

Branden, N. (1969). *The psychology of self-esteem*. Nash Publishing Corp.

Buss, D. (2004). *Evolutionary psychology: The new science of the mind*. Pearson Education.

Cambridge Advanced Learner's Dictionary (2008). Cambridge Press, Cambridge, UK.

Cavell, M. (1998). Triangulation, one's own mind and objectivity. *The International Journal of Psycho-Analysis, 79*(pt3), 449–467.

Chiou, J. S., Hsiao, C. C., & Chiu, T. Y. (2018). The credibility and attribution of online reviews: Differences between high and low product knowledge consumers. *Online Information Review, 42*(5), 630–646.

Chowdhury, T. G., Murshed, F., & Khare, A. (2018). Flexible flexibility! Food categorization flexibility and utilitarian preference. *Journal of Consumer Marketing, 35*(1), 1–10.

Christoff, K., Irving, Z. C., Fox, K. C. R., Spreng, R. N., & Andrews-Hanna, J. R. (2016). Mind-wandering as spontaneous thought: A dynamic framework. *Nature Reviews Neuroscience, 17*, 718–731.

Cooke, E. F. (1974). *A detailed analysis of the constitution*. Littlefield Adams & Co.

Covey, S. R. (1989). *The 7 habits of highly effective people: Powerful lessons in personal change*. Free Press.

De Bono, E. (2015). *The mechanism of mind: Understanding how your mind works to maximize memory and creative potential*. Random House.

Dennett, D. C. (2015). *Elbow room: The varieties of free will worth wanting. A Bradford Book*. MIT Press.

Derbaix, M., & Gombault, A. (2016). Selling the invisible to create an authentic experience: Imagination at work at Cézanne's studio. *Journal of Marketing Management, 32*(15–16), 1458–1477.

Egan, K. (1992). *Imagination in teaching and learning*. University of Chicago Press.

Forrest, J. Y. L. (2018). *General systems theory: Foundation, intuition and applications in business decision making*. Springer.

Gerassi, J. (1989). Jean-Paul Sartre: Hated conscience of his century. In *Volume 1: Protestant or protester?* University of Chicago Press.

Harris, P. (2000). *The work of the imagination*. Wiley-Blackwell.

Hill, N. (1928). *The law of success: In sixteen lessons (reprint of 2007)*. BN Publishing. www.bnpublishing.com

Kane, R. (2005). *A contemporary introduction to free will*. Oxford University Press.

Kuratowski, K., & Mostowski, A. (1976). *Set theory: With an introduction to descriptive set theory*. North-Holland.

Langston, D. C. (2001). *Conscience and other virtues: From Bonaventure to MacIntyre*. Penn State University Press.

LeDoux, J. E., & Brown, R. (2017). A higher-order theory of emotional consciousness. *Proceedings of the National Academy of Sciences of the United States of America, 114*(10), E2016–E2025.

Libet, B. (1999). Do we have free will? *Journal of Consciousness Studies, 6*(8–9), 47–57.

Lin, Y. (1999). *General systems theory: A mathematical approach*. Springer.

Lin, Y. (2008). Systematic studies: The infinity problem in modern mathematics. *Kybernetes: The International Journal of Cybernetics, Systems and Management Sciences, 37*(3–4), 385–542. (guest editor).

Lin, Y., & Forrest, B. (2012). *Systemic structure behind human organizations: From civilizations to individuals*. Springer.

Lin, Y., & OuYang, S. C. (2010). *Irregularities and prediction of major disasters*. Auerbach Publications, an imprint of Taylor and Francis.

Mandler, G. (1982). The integration and elaboration of memory structures. In F. Klix, J. Hoffman, & E. van der Meer (Eds.), *Cognitive research in psychology* (pp. 33–40). North-Holland.
Maslow, A. H. (1943). A theory of human motivation. *Psychological Review, 50*(4), 370–396.
Maslow, A. H. (1954). *Motivation and personality*. Harper.
McFarland, C. P., Primosch, M., Maxson, C. M., & Stewart, B. T. (2017). Enhancing memory and imagination improves problem solving among individuals with depression. *Memory & Cognition, 45*(6), 932–939.
Mesarovic, M. D., & Takahara, Y. (1975). *General systems theory: Mathematical foundations*. Academic Press.
Meyers-Levy, J., & Tybout, A. M. (1989). Schema congruity as a basis for product evaluation. *Journal of Consumer Research, 16*(1), 39–54.
Moran, G. (2018). Chaos theory and psychoanalysis: The fluidic nature of the mind. *Zeitschrift für Psychosomatische Medizin und Psychoanalyse, 40*(4), 384–403.
Moss, G. (2009). *Gender, design and marketing: How gender drives our perception of design and marketing*. Routledge.
Nedungadi, P. (1990). Recall and consumer consideration sets: Influencing choice without altering brand evaluations. *Journal of Consumer Research, 17*(3), 263–276.
Norman, R. (2000). *Cultivating imagination in adult education proceedings of the 41st annual adult education research*.
O'Brien, C. (2008). Sustainable happiness: How happiness studies can contribute to a more sustainable future. *Canadian Psychology, 49*(4), 289–295.
Pfaff, D. W. (2007). *The neuroscience of fair play: Why we (usually) follow the Golden rule*. Dana Press, The Dana Foundation.
Rosch, E. (1978). Principles of categorization. In E. Rosch & B. B. Lioyd (Eds.), *Cognition and categorization* (pp. 27–48). Lawrence Erlbaum.
Sahni, N. S. (2016). Advertising spillovers: Evidence from online field experiments and implications for returns on advertising. *Journal of Marketing Research, 53*(4), 459–478.
Schrift, R. Y., Parker, J. R., Zauberman, G., & Srna, S. (2018). Multistage decision processes: The impact of attribute order on how consumers mentally represent their choice. *Journal of Consumer Research, 44*(6), 1307–1324.
Shin, J. E., Suh, E. M., Eom, K., & Kim, H. S. (2018). What does "happiness" prompt in your mind? Culture, word choice, and experienced happiness. *Journal of Happiness Studies, 19*(3), 649–662.
Stayman, D. M., Alden, D. L., & Smith, K. H. (1992). Some effects of schematic processing on consumer expectations and disconfirmation judgments. *Journal of Consumer Research, 19*(2), 240–255. https://doi.org/10.1086/209299
Sujan, M. (1985). Consumer knowledge effects on evaluation strategies mediating consumer judgments. *Journal of Consumer Research, 12*(June), 31–46.
Tinbergen, N. (1951). *The study of instinct*. Oxford University Press.
Uchida, Y., Norasakkunkit, V., & Kitayama, S. (2004). Cultural constructions of happiness: Theory and empirical evidence. *Journal of Happiness Studies, 5*(3), 223–239.
Ugur, H., Constantinescu, P. M., & Stevens, M. J. (2015). Self-awareness and personal growth: Theory and application of Bloom's taxonomy. *Eurasian Journal of Educational Research, 60*, 89–110.
Wajda, T., Forrest, J. Y.-L., Hu, M. Y., Li, Z., & Liu, Y. (n.d.). (to appear). The inclusiveness dimension of human cognition and categorization paradigm. Submitted for publication.

Walsh, L. C., Boehm, J. K., & Lyubomirsky, S. (2018). Does happiness promote career success? Revisiting the evidence. *Journal of Career Assessment, 26*(2), 199–219.

Wu, Y., & Lin, Y. (2002). *Beyond nonstructural quantitative analysis: Blown-ups, spinning currents and the modern science*. World Scientific.

Yazdanparast, A., & Spears, N. (2018). The new me or the me I'm proud of?: Impact of objective self-awareness and standards on acceptance of cosmetic procedures. *European Journal of Marketing, 52*(1–2), 279–301.

Zeidman, P., & Maguire, E. A. (2016). Anterior hippocampus: The anatomy of perception, imagination and episodic memory. *Nature Reviews Neuroscience, 17*, 173–182.

Chapter 11
Preferred Taxonomies and Inclusive Classification of Consumers

Abstract This and the previous chapter, which are mainly based on Wajda et al. (to appear, The inclusiveness dimension of human cognition and categorization paradigm), jointly investigate how people with different focuses of attention tend to mobilize unlike areas of their cognitive systems; such different focuses naturally lead to dissimilar categorizations of given stimulus targets. Specifically, based on what have been established in the previous chapter regarding human mind and the general cognitive system of people, this chapter explains why and how people demonstrate their respectively different preferences for their correspondingly varied levels of taxonomic abstraction. Then, it uses an empirical case study to confirm some of the main conclusions developed in these two chapters theoretically.

Keywords Attribute matching · Categorization paradigm · Consumer behavior · Information processing · Processing strategy · Product category · Product information

The motivation behind the previous and this chapter is that even though a host of consumer behaviors can be explained by employing the well-known categorization paradigm about how consumers receive and process marketplace information, a very good amount of guesswork is involved in such explanations. That makes decision-making managers quite uncertain about the soundness and reliability of their logical reasoning and consequent plans of action. The objective of this chapter is to develop a practically useful theory on how consumers receive and process market information so that related marketing programs can be effectively developed accordingly.

11.1 Introduction

In recent years, the categorization paradigm has attracted significant amount of scholarly attention, because it seems to have an ability to predict many different consumer behaviors. The paradigm believes that how memory categorizes

J. Y.-L. Forrest, Y. Liu, *Value in Business*, Contributions to Management Science,
https://doi.org/10.1007/978-3-030-82898-1_11

information affects the quality and content of later information retrievals so that it ultimately affects eventual consumption decisions. Because of such practical significance, scholars, e.g., Chiou et al. (2018), Chowdhury et al. (2018), Schrift et al. (2018), Sahni (2016), Moss (2009), Nedungadi (1990), Meyers-Levy and Tybout (1989), Sujan (1985), and Mandler (1982), have identified various consumer-based outcome variables, such as consumer attitudes towards and evaluations of various products/brands, as well as inferences with respect to product attributes and qualities. These variables, possessing strategic implications, have been widely employed in marketing initiatives with varying degrees of market success. Continued from the previous chapter on how the human cognitive system is composed of, presented here aims at developing the absent theory on why different people employ different methods to process information, making predicting consumer behaviors more reliable than before. Considering how much money (in billions of US dollars in the US alone) is spent each year on marketing efforts, being able to predict consumer behaviors will no doubt make designs of marketing programs more effective.

Because of the employment of systems science and logical reasoning (as for the theoretical significance and practical need of going beyond traditional methods, such as data and anecdotal analyses and calculus-based tools, see Forrest, 2018, pp. 12–16), this chapter is able to realize its objective by establishing the following generally true results, among others, based on which recommendations instead of suggestions can be derived for decisions-makers, such as managers and entrepreneurs. Specifically, deriving recommendations instead of suggestions is part of the objective of this chapter. In terms of the categorization paradigm, this chapter establishes that

- No matter when and who is concerned with, at least one area of her imagination is occupied by an active circulation of information, experience, and knowledge.
- No matter when a person's categories of experience and knowledge are abstracted in three levels: superordinate, basic, and subordinate.
- Categorization behavior manifests itself along different levels of abstraction of a specified taxonomy in the number of subcategories generated by individual people for any given collection of product items.

Other than establishing these generally true results, another major contribution of this chapter to the literature is the introduction of systemic thinking and logical reasoning into the study of the categorization paradigm.

The rest of this chapter is organized as follows. Section 11.2 provides a review of the relevant literature and places the work presented in this chapter within the literature. Section 11.3 studies levels and individually different preferences of taxonomic abstraction. Section 11.4 presents an empirical study that confirms some of the key conclusions established earlier in this work. And then this chapter concludes in Sect. 11.5.

11.2 Literature Review

In the area of marketing, how consumers take in and deal with product information from the marketplace and how they consequently make their consumption decisions represent two of a few very important research topics. To deal with these research problems, the so-called categorization paradigm in recent years has caught considerable attention because of its capability of predicting consumer behaviors. For example, Moss (2009) and Meyers-Levy (1986) study how consumers selectively attend to various informational cues in their environments. Schrift et al. (2018), Chowdhury et al. (2018), Basu (1993), and Cohen and Basu (1987) consider how, once attended to, information is stored and categorized in memory. Sahni (2016) and Nedungadi (1990) look at how information is appropriately retrieved and utilized to process what is newly presented. These listed and other unlisted studies on information processing have helped identify a number of consumer-based outcome variables, say, consumer attitudes towards and evaluations of various products/brands, as well as consequent inferences regarding product features, functionalities, and qualities. These strategically significant variables have been employed in the design and execution of various marketing initiatives (Bilgihan, 2016; Arens & Hamilton, 2018). For instance, how information is stored or categorized in memory affects how well the information will be accessible when called for. That, of course, will influence consumers in terms of who they build consideration collections of products and how they make consumption decisions among alternatives during choice-making.

Due to the said importance, an extensive literature, pertaining to schema (or category-based) research, has been developed with considerable merit in providing insights into various processes and outcomes. By product category or schema, which will be used interchangeably in this chapter, it means (Stayman et al., 1992) such a knowledge structure within memory that contains a myriad of details, such as product attributes, values of various attributes, correlations of product attributes with one another, and relationships between product categories.

Regarding information processing, the focus of research is on the degrees of how closely a new piece of product information correlates a product category within memory. In particular, Sujan (1985) examines "expert"-consumer knowledge structures with respect to a particular product category (i.e., restaurant dining) versus those of "novice" consumers. Liu et al. (2018) investigate a product's esthetic design and its effect on consumer preference by using categorization theory. This work empirically finds that economy-segment products benefit from cross-segment imitation of luxury products, that segment prototypicality strengthens price sensitivity, and although consistency in brand stifles price sensitivity, it increases the effectiveness of advertisement.

In terms of selection of processing strategy, by employing objective knowledge Sujan (1985) assesses the structural complexity of an individual's categories. By demonstrating that moderate levels of absurdity encourage greater elaborative consumer processing of information, Mandler (1982) and Myers-Levy and Tybout

(1989) derive similar results. Based on implicit personality theory, Kim et al. (2018) demonstrate both negative, meaning dilution of the personality of a retailer, and positive, meaning expansion of the personality, effects of visual artists, associated with fashion retailers. The implication of this work is that retailers can employ visual artists as a source of value to gain additional market appeal by creating differentiated brand personalities.

As for category attribute matching in terms of product positioning and family brand extension, a moderately discrepant brand is found to be more likely to be integrated into an individual's existing product schema, while a strongly discrepant brand is subtyped into a unique category of its own (Sujan & Bettman, 1989). Evaluations of a brand extension are affected by how well the quality of the extension matches consumers' schema existing for the brand of concern (Boush & Loken, 1991). That is, the categorization paradigm is developed on the idea that the structure of all categories within the mind is hierarchical and is employed when a person needs to match an incoming piece of information with the stored knowledge (Fiske, 2018; Fiske & Pavelchak, 1986). By applying schema-triggered effect and the hypothesis of link formation, Lane and Fastoso (2016) test the prediction of how the negative spillover from low-fit extensions can be managed through proactively repeated ad exposures. By checking extension strategies, potentially effective in building brand equity, Gurhan-Canli and Maheswaran (1998) find that the effect of extensions on family brand name is determined by the typicality of extensions and consumers' motivation, while either brand-name dilution or enhancement can appear when similar extensions take place.

Comparing to the literature, this chapter aligns closely with Chowdhury et al. (2018) and Basu (1993). The approach underlying the following discussion operates like Lambert-Pandraud et al. (2017) and Rosch et al. (1976) in terms of category structure coupled with the information processing paradigm (Fiske, 2018; Fiske & Pavelchak, 1986). What follows here is built on the literature (Chiou, et al., 2018; Sujan, 1985) regarding individual differences in category structures useful in information retrieval and processing. Additionally, because of the use of systems science and methodology, this chapter is able to develop the following theoretical results that are brand new and can be practically applied in real-life settings:

- No matter when, a person's focused experience and knowledge can be classified into three descending levels of abstraction: superordinate, basic, and subordinate, where experiences and knowledge in the basic-level contain the most information and can be retrieved with minimal cognitive effort.
- A person tends to be able to easily match a given target (or a product) with a schema, if she has a broad-based imagination with loosely defined cognitive structure.
- With a marketing orientation, for a given collection of products, a person's categorization behavior manifests itself in the number of generated subcategories.

Because of the specific methodology employed, all established results in this chapter are expected to provide universally usable recommendations for marketers, managers, and entrepreneurs instead of suggestions conjected based on empirical studies that in general are constrained by various limitations.

11.3 Levels and Individual Preferences of Taxonomic Abstraction

This section investigates the hierarchical layer structure of a person's cognitive system within her mind, as pictured in both Figs. 10.2 and 10.4. As figuratively indicated, when moving away to the right from the labeled P and V levels, units of experience, information, and knowledge within the person's reservoir of imagination become either taxonomic organizations of things and events (i.e., categories), or spatially/temporally organized structures (i.e., schemas) (Stayman et al., 1992), or product-category schemas (Meyers-Levy & Tybout, 1989). Each schema (or product category) represents a block of the knowledge structure in a person's cognitive system. When the person processes new product information, she tries to match that information with a product category stored in her cognitive system (Proposition 10.6). A product-category schema holds the features of a taxonomic category, hile specifies a product use context. In this chapter, the terms "category" and "schema" are used interchangeably, and by cognitive system it means the conceptual hierarchically networked structure of a representative person shown in Fig. 10.2. In the rest of this chapter, Rosch's (1978) taxonomy will be employed, where categories are related to each other by the relation of class inclusion. Specifically, for two categories A and B, A is included in B, denoted by $A \subseteq B$, if A is a subcategory of B. That is, B contains more members than A; or B is more inclusive than A.

In terms of consumer products, let us treat the block of experience, information, and knowledge within a person's imagination regarding a category of products as the basic-level of categorization (Mervis & Rosch, 1981). This level represents the focal point of the person's attention with "within-category" similarity maximized relative to "between-category" similarity. If the category of athletic shoes is seen as basic, for instance, then the designation of this category denotes the attributes commonly shared by all members of this category and maximally different from those of any contrast category. If this category's level of abstraction is increased, then we reach a higher superordinate level of abstraction of the taxonomy. If we continue the previous example, then "footwear" stands for a superordinate category to that of "athletic shoes," because members of "footwear" share fewer attributes than those in "athletic shoes." On the other hand, if we reduce the abstraction of the basic-level, then we will reach several subordinate categories, such as those of "running shoes," "racquetball shoes," and "table tennis shoes". In this level of abstraction, categories are defined with additional attributes compared to the levels above, such as the basic and superordinate levels. In particular, the defining attributes for the membership of a subordinate category have to inform which basic-level category it is in and how it is different from another subordinate category. Because of the additional defining attributes necessary, cognitive efficiency becomes increasingly minimized as one moves downwards in levels of abstraction, where some defining attributes are shared by different categories.

Summarizing what has been discussed in the previous paragraphs leads to the following

Proposition 11.1 No matter when, the focal categories of experience, information, and knowledge in the continuous field of a person's imagination can be grouped into three descending levels—superordinate, basic, and subordinate. And, categories in the basic-level offer the most information that can be reached with minimal cognitive effort.

The undeniable expediency and significance of taxonomic categories have been well demonstrated in the literature. Specifically, the concept of such categories is closely relevant to consumer goods (Felcher et al., 2001). For example, consumer products, no matter how different they are from each other, share features that consumers utilize to profile experience, information, and knowledge, and compare alternative options. Additionally, market analysts and manufacturers of consumer products employ taxonomic categories to investigate market conditions and development trends (Felcher et al. 2001; Forrest et al., 2017).

As a conventional language (and style) in mathematics, the logical reasoning for why Proposition 10.14 holds true in general also implies the following fact. Among the three hierarchical levels—superordinate, basic, and subordinate (Proposition 11.1), one can apply the number and type of attribute of each level to determine which particular level a certain category belongs to. More specifically,

- There are fewer defining attributes for categories of the superordinate level, while these used attributes are largely general or functional in nature.
- In comparison with the superordinate level, there are more defining attributes, which are decidedly relevant, indicative, and discriminatory, for categories of the basic-level.
- The defining attributes for categories of the subordinate level are highly descriptive, maximally distinctive between objects with many common features, and mostly shared among categories in the same taxonomic hierarchy.

A more specific scenario than what is stated in Proposition 11.1 was observed by Rosch et al. (1976). And Meyers-Levy and Tybout (1989) and Sujan and Dekleva (1987) recognize that product classes, product types, and brands subsist in taxonomic hierarchies, and correspondingly represent the superordinate, basic, and subordinate levels, respectively. Among these three levels of abstraction (Proposition 11.1), it is optimal for both the basic and subordinate levels to communicate and precisely convey meaning. So, the rest of this chapter will focus more on evaluating individual differences at these levels of abstraction.

As Propositions 10.16 and 11.1 jointly indicate, people's attentions differ from each other in their individual category structures. Hence, marketers can benefit considerably if they can ascertain how the customers of a targeted population have their preferred ways to process information at respective levels of taxonomic abstraction. The general theory, developed above and in the previous chapter, and the "theory of selectivity" (Meyers-Levy, 1986) jointly imply that if a marketer first chooses a taxonomic classification of category levels, then customers will employ the same taxonomy to classify products, brands, advertisements, etc., with some at a more general (i.e., basic) level of abstraction than others. That is, no matter how the marketer predetermines product-category levels, customers individually determine their own levels of abstraction. This realization carries substantial ramifications for the positioning and promotion of new products. In particular, upon receiving the information of a new product, some people immediately relate the features of the product to their basic-level category. After that, many distinguishing features of the product will be lost as they are identified with the defining characteristics of the basic-level category. On the other hand, some people may prefer a more subordinate level of abstraction with both common category characteristics and distinguishing product features equally considered in their classifications.

If the attributes of a product cannot be easily reconciled with an existing schema in a customer's imagination, the individual will find ways to resolve the incongruity (Proposition 10.6). She will, for instance, either restructure her current schema/category configuration or create a new schema/category to accommodate the discrepant attributes, see Meyers-Levy and Tybout (1989), Stayman et al. (1992), and Mandler (1982) for relevant discussions.

The systemic structure of categories in a person's imagination and the characteristics of defining attributes of categories of different levels imply that when a customer fails to match a target with a specific schema at one level of the employed taxonomic hierarchy, she will naturally move to a lower level to potentially locate a better match. As for the scenario of moving into higher levels, it only helps the consumer find to which broad scheme the target product belongs. For a piece of related work, see Meyers-Levy and Tybout (1989).

This discussion leads to the conclusion below:

Proposition 11.2 A person with a broad-based imagination and loosely defined category structure can more easily make target-to-schema match when a stimulus target is provided than anyone with a rigidly defined category structure.

Speaking slightly differently, comparing to those with a broad-based imagination and loosely defined category structure, those people with rigidly defined category structures tend to have more difficulty to sufficiently match a stimulus target with a schema within her imagination. These people will likely move downward within their employed taxonomies to more subordinate levels to locate potentials for a more precise target-to-schema match.

As indicated by Proposition 10.15, even for the same person, given things, events or thoughts can be categorized at any level within her continuum of abstraction. That explains why different people generally provide their individually different evaluations for whatever is presented to them. This result is also obtained by Pavelchak (1989) by using a different approach. Hence, differences in people's category constructions will become evident when they face the challenge of a classification task. This discussion leads to the following.

Proposition 11.3 With a marketing orientation and a given set of product items, people display their categorization behaviors along different levels of a specified taxonomy in the number of subcategories generated by these people.

To understand this result, let us assume that a set of brands is presented to a group of people for classification into one of several product domains. Proposition 11.3 infers that if taxonomic categorization is given, some people within the group will choose aggregation at the basic-level of abstraction while some others at a subordinate level. Practically, this result implies that people, who pursue subordinate levels of inclusiveness, deal with information of attributes by focusing on specific points that distinguish one brand from another. These people can be identified by their brand groupings succinctly defined by using subtle attribute distinctions. On the other hand, people, who prefer higher levels of inclusiveness, tend to deal with information at a general abstraction level. For example, instead of attribute distinctions, these people will focus on broad category labels. Hence, these people can be identified through their brand groupings, loosely defined by using striking commonalities.

11.4 Classifying Consumer Differences Inclusively

This section provides a working classification of consumer differences by using empirical data. It segregates clusters of individuals (or processors) according to how they respond to a task of sorting bands. Different from Chiou et al. (2018), Chowdhury et al. (2018), Fiske (2018), Lambert-Pandraud et al. (2017), Basu (1993), and Sujan (1985), this chapter looks at the typology of chosen processors at the microlevel of category structure. It is expected to have ramifications for the understanding of how categories are developed in the horizontal (or breadth) dimension of category structure. These two dimensions—the vertical (inclusive) and horizontal dimensions—will jointly provide a better comprehended association between the dual nature of cognitive structure and proclivity towards preferential uses of a specific processing strategy.

Based on discussions above and in the previous chapter on people's differences in the inclusiveness dimension of their category structures, expected is the appearance of two or more distinct groups of consumers. Therefore, this attempted classification process will simultaneously consider both the number of self-directed brand groupings and details in category labels attached to each brand grouping.

11.4.1 The Methodology Used Here

This empirical study involved 119 college students, as participants, with 57 males and 62 females enrolled in business courses at a Midwestern university. These participants signed up for one of 35 experimental sessions that spanned the course of two consecutive weeks. The selection of target stimuli involved several different steps.

One first consideration was constructing the domain of products at the superordinate level, from which categories (i.e., brands) would be chosen. The constructed domain had to be one where the participants had roughly equivalent (above average) levels of knowledge and experience. To achieve this objective, a pretest based on a 7-point scale was given to all the 119 participants. The results showed that the product domain of restaurant chains among all considered, which included breakfast cereals, television shows, movies, candy bars, and computers, possessed the smallest mean difference in knowledge (and familiarity) among the participants and fairly similar and small standard deviations.

With this product domain chosen, stimuli were selected by using two criteria: brands that share commonalities and attributes that separate brands. Brands considered included Pizza Hut™, Fazolis™, Olive Garden™, etc. Shared commonalities included fast-food restaurant chains, Italian-style eateries, casual dining establishments, dine-in/carry-out restaurants, etc. And attributes included American/Italian-style cuisine, moderate prices, booth-style seating, authentic Italian-style cuisine, diverse menu offerings, self-service beverage station, etc. For example, Pizza Hut™ shares with Olive Garden™ such attributes as American/Italian-style cuisine, moderate prices, booth-style seating, etc. And it shares with Fazoli's™ such attributes as authentic Italian-style cuisine, diverse menu offerings, self-service beverage station, etc.

The objective of this chapter is to determine whether or not there exist consumer differences in ascertaining: (1) brand X is similar to brand Y or brand Z so that X and Y for the former case belong to the same category label (for the latter case X and Z belong to the same category label); (2) brand X is similar to both Y and Z equally so that X, Y, and Z all belong to the same category label; and (3) brands X, Y, and Z are not pairwise similar to each other so that each of these brands belongs to its own unique category label.

To classify the restaurant chains in the superordinate domain based on if they share important attributes with one another, while differ on a host of others, we took several steps.

1. Determine global dimensions that were generally important in influencing a consumer's decision to dine at a specific restaurant chain.

Because a consumer's belief about a brand greatly affects the underlying attribute structure that is observed in her assessment of brand typicality (Loken & Ward, 1987), this section identifies salient attributes that consumers rely on to make decisions about the product domain of a restaurant chain. To materialize this subjective, this chapter adopts the following seven attributes (from the March (2004) issue of *Restaurants and Institutions*) that were considered most important in influencing a consumer's decision to dine at particular chains: food quality, menu variety, value, service, atmosphere, cleanliness, and convenience.

To reflect these seven attributes, crafted were 12 basic-level attribute statements and 18 subordinate-level attribute statements, leading to three food quality items, four menu variety items, five value items, six service items, three cleanliness items, four atmosphere items, and five convenience items. These attribute statements were crafted to ensure that participants' genders did not affect the importance of each attribute dimension, where only the dimension of "cleanliness" revealed gender differences with $M(\text{males}) = 5.14$ versus $M(\text{females}) = 5.91$, $t[73] = -3.95$, $p < 0.01$. And they were used to ensure that the chosen attributes were intended for use in developing experimental stimuli.

Eventually, out of the 30 attribute statements, 20 were used to select 25 restaurant chains (i.e., brands), see Appendix for details, that share basic- and subordinate-level attributes with one another, while differ in subtle ways. These 20 attribute statements were used as the base from which 25 restaurant chains were selected to potentially measure individual differences in preference for a specific level of abstraction when dealing with information. Once again, these 25 restaurant chains were pretested to ensure that the participants did not significantly differ in their familiarity with specific category exemplars.

11.4.2 Measurements

After a brief explanation of the study and related procedure, the participants were provided with a survey booklet and index cards each of which was labeled with the name of a single category member (i.e., brand) from the product domain of selected restaurant chains. They then sorted all category members (the index cards) into piles based on their individual assessments of "perceived overall similarity," where most similar brands were grouped into the same pile. The key at this step was that the participants were free to determine their dimensions, such as price, cuisine, and atmosphere, as criteria for sorting. If a brand was not seen as similar to any other brand, the participants placed that brand into a group of itself. After the sorting task, participants were required to treat each pile as a category and provide a label for each of them without specific instruction on how elaborate the labels needed to be. Here, Gardner's (1953) procedural instructions were adopted.

Regarding data collection, for each participant four measures were tabulated. The first was the number of subcategories generated by each individual. The second recorded the number of groups labeled using a basic-level descriptor or a singular concept or theme. The third measured the number of groups labeled with an elaborate-level descriptor by using two or more concepts or themes. And the fourth assessed participants' knowledge or familiarity with the product domain of restaurant chains by using Johnson and Russo's (1984) two-item, seven-point scale.

11.4.3 Empirical Results

To generate a typology of category structure for the participants in the inclusiveness dimension, this work employed a nonhierarchical cluster approach, K-means, for the reason that differences in this dimension are not expected to be nested within each other. Discussions above suggest that for this current study there should be two large discrete, bipolar clusters, either basic-level and subordinate-level participants, and potentially some smaller clusters that lie somewhere between these two large clusters. To confirm this expectation, K-means cluster analysis was repeated for different number n of clusters, where $n = 2, 3, 4$.

The numbers of subcategories, basic descriptors, and elaborate descriptors were employed to determine cluster membership. Due to the method's sensitivity to outliers, four cases were removed, leaving 115 valid cases. By comparing Euclidean distances between the final cluster centers and analyzing the corresponding ANOVA table, the two-cluster solution was the optimal one. In particular, one cluster in the three-cluster solution only had 2 members. Similarly, one cluster in the four-cluster solution also had only 2 members, while two other clusters had a small distance between their centers. Additionally, the F-value on each of the independent variables was larger for the two-cluster than those of three or four-cluster solutions. Therefore, the two-cluster solution was considered as optimal.

Now, by using cluster means for independent variables and cluster sizes clusters were labeled as shown in Table 11.1.

Cluster 1 (basic-level processors), which classified 49% of the sample (57 cases), included those participants who preferred a basic-level of taxonomic abstraction. These people handled information at a general level of abstraction by giving attention to broad category labels. They loosely defined brand groupings by employing salient commonalities rather than subtle differences.

As for cluster 2 (subordinate-level processors), it classified 51% of the sample (58 cases) by including those participants who preferred a subordinate-level of taxonomic abstraction. These individuals handled information by looking at points

Table 11.1 Cluster descriptors and ANOVA values for proposed typology

Independent variable	Cluster 1 (Basic-Level Processors)	Cluster 2 (Sub-Level Processors)	Mean Square	F	Significance
Number of groups	4	6	135.82	94.10	0.000
Number of basic descriptors	3	2	41.96	28.12	0.000
Number of elaborate descriptors	1	4	328.76	352.12	0.000
Number of cases	57	58			
Percentage of participants	49	51			

of differentiation that separated one brand from another, leading to their use of subtle attribute distinctions.

Next, the soundness of a two-cluster solution is tested internally by using the labels generated above. To this end, the fundamental difference between basic- and subordinate-level processors indicates that subordinate-level processors (Cluster 2) should self-directedly divide the given set of brands into significantly more product subcategories than basic-level processors (Cluster 1) would. These subordinate-level processors should also select more elaborate-level descriptors than basic-level ones to label their self-formed product subcategories. Hence, the following hypotheses were proposed:

- H1: "Basic-level" processors are more likely to divide a given set of brands into a smaller number of self-directed product subcategories than "subordinate-level" processors.
- H2a: "Basic-level" processors are more likely to use basic-level descriptors than elaborate-level descriptors in developing category labels for self-directed product categories.
- H2b: "Subordinate-level" processors are less likely to use basic-level descriptors than elaborate-level descriptors in developing category labels for self-directed product subcategories.
- H2c: "Basic-level" processors are more likely to use a greater number of basic-level descriptors than "subordinate-level" processors in developing category labels for self-directed product subcategories.
- H2d: "Subordinate-level" processors are more likely to use a greater number of elaborate-level descriptors than "basic-level" processors in developing category labels for self-directed product subcategories.

In validating the outputs of previous cluster analysis, these hypotheses were tested with all results highly significant (Table 11.2). That is, basic-level processors

Table 11.2 Sample t-test results for numbers of self-directed product categories, basic-level descriptors and elaborate-level descriptors

Dependent measures	Test value	Mean	T	p-value
Self-directed product categories				
Basic processors ($n = 57$)	5.03	3.95	−7.98	<0.001
Subordinate processors ($n = 58$)	5.03	6.05	5.57	<0.001
B vs. S ($n = 115$)	–	–	−9.23	<0.001
Basic-level descriptors				
Basic processors ($n = 57$)	0.00	3.11	9.36	<0.001
Subordinate processors ($n = 58$)	0.00	1.83	−10.04	<0.001
B vs. S ($n = 115$)	–	–	5.63	<0.001
Elaborate-level descriptors				
Basic processors ($n = 57$)	5.35	0.84	−9.36	<0.001
Subordinate processors ($n = 58$)	5.35	4.22	10.04	<0.001
B vs. S ($n = 115$)	–	–	−18.81	<0.001

developed a significantly smaller number of brand groupings and employed more basic descriptors than subordinate-level processors did. In other words, basic-level processors were more likely to look for highly salient cues that were pertinent to all or most brands.

To understand how individual brand groupings were formed, it was necessary to find the criteria the participants used in establishing their category boundaries. Once again, the definitions of basic-level and subordinate-level processors suggest that the former aggregated brands by identifying salient attributes common to all or most brands. That is, these processors searched for common threads between brands instead of specific elements of distinction. That explained why they defined and labeled category groupings at highly macrolevel.

Opposite to the scenarios of basic-level processors, subordinate-level processors were more likely to carefully study different attribute combinations within each brand. That, of course, led to a large number of brand groupings with subtle differences, while each group contained a relatively small number of brands. Lastly, the criteria, used to develop the brand groupings, were highly detailed and multifaceted, leading naturally to extremely descriptive and succinctly defined category labels.

11.4.4 Discussion

The employed cluster analysis confirmed the existence of two bipolar groups of processors, one at the subordinate-level and the other at the basic-level. The results from relevant hypothesis testing further validated this two-cluster solution on individual proclivities for a certain level of taxonomic abstraction. More specifically, those people, who think broadly, are selective information processors and tend to advance category systems by using broad-based, highly inclusive, basic-level parameters. On the other hand, people, who are more comprehensive with whatever is presented to them, tend to develop finely discriminated and highly exclusive category systems. Hence, both this empirical study and Proposition 10.16 jointly imply that people, who regularly focus their attentions on individually very different levels of abstraction, generally have very different cognitive structures of imagination in terms of the locations of active circulations of information, experience, and knowledge.

Empirical evidence additionally confirms that people, who think in general terms, tend to apply very broadly, overarching characteristics to choose category parameters. In particular, one frequently used parameter by the basic-level participants of this chapter was the "type of cuisine." Despite the availability of other pertinent, information-laden parameters, such as price points, décor, location, and others, basic-level participants typically utilized a solo, simple parameter to divide brands. That of course had led to the situation of grouping some seemingly disparate brands together as long as a single criterion was met. For example, by using "Italian cuisine" one basic-level participant grouped Fazoli's™, Chuck E Cheese's™, Sbarros™,

Dominos™, and Olive Garden™ together in one category, even though these brands span a wide range of price points, target markets, and service offerings.

Also, it is found that not all basic-level participants chose defining parameters for the basic-level. However, a high degree of consistency existed among these participants in the context of criteria used for classification. In particular, these participants demonstrated highly congruent patterns in their classifications of brands by simultaneously provided maximal amounts of information and minimizing cognitive effort.

Opposite to the basic-level participants, the subordinate-level processors displayed a greater scale of discrimination in forming their brand divisions. In particular, they considered multiple facets of each brand before assigning a brand to a specific category. These processors used several criteria simultaneously, for instance, "Italian cuisine" jointly with some of the others, such as "type of establishment" (i.e., dine-in versus take-out), "price points" (i.e., high-priced, moderately-priced, low-priced), and "atmosphere" (i.e., formal, semi-casual, casual), in their segmentation of brands. In other words, a subordinate-level process typically created a category label involving multiple criteria, leading naturally to the formation of a large number of brand groups and the appearance of finer and highly descriptive category labels. These results collectively imply that for a brand to qualify for the membership of a particular category, the brand had to be subject to a strict set of admission standards.

To corroborate this end further, the collected data were closely examined, leading to the discovery that other than having many attributes in the labels of subordinate-level categories, the employed attributes also tended to contain multiple layers. That is, the subordinate-level processors preferred to use such verbiage that communicated very precise meanings. To illustrate, for example, the basic-level participants tended to employ such bipolar adjectives, such as "formal" or "casual," to describe the atmosphere of a dining establishment, while the subordinate-level processors resorted to adjectives, such as formal, semi-formal, and semi-casual, casual, to depict incremental variations in the atmosphere.

The descriptive data also revealed differences in preferences for different levels of taxonomic abstraction. In particular, considerably more variations, as measured by standard deviations of the dependent measures of interest, were found in the subordinate portion of this sample than in the basic portion. Here, the dependent measures of interest consist of (1) number of self-directed product categories, (2) number of basic-level descriptors, and (3) number of elaborate-level descriptors. Specifically, for (1), the variation is $SD = 1.41$ versus 1.01; for (2), $SD = 1.26$ versus 1.17; and for (3), $SD = 1.04$ versus 0.89.

One reason for such a consistent pattern of differences in sample variations is that participants at these two levels had different preferences in levels of abstraction. Because the basic-level participants likely defined their categories by using over-arching, highly salient, global cues, there would be a high level of consensus among the basic-level participants on how a particular category should be defined and on the brands that should be included in the category. At the same time, because subordinate-level processors likely preferred a finer level of abstraction, which naturally involved numerous variations on any single attribute, there would be greater differences among these individuals regarding the features of a particular category. In other words, subordinate-level processors would be less agreeable with each other, leading to greater diversity on both the boundaries and members of each category.

In short, individual differences in people's preferred levels of abstraction can be detected by using either the dependent measures of interest or the sample variance.

11.5 A Few Final Words

Other than improving marketers' quality of prediction of consumer behaviors this chapter, jointly with the previous chapter, also fill the following two important gaps in the extant literature:

- Develop a better comprehension on how individuals categorize, retrieve, and process information; and
- Introduce a novel approach of analysis into the study of categorization paradigm that is not invalidated by constraints associated with empirical methodologies.

What is demonstrated above shows that this novel approach has practically helped materialize our objectives, and produced conclusions readily employed in real-world applications in marketing.

Other than producing a list of theoretical important and practically useful results, this chapter carries the following three models, which have received the greatest attention recently (Basu, 1993), many streets forward: (1) the classical model, (2) the prototype model, and (3) the exemplar model. Specifically, this chapter aligns with model (2) and operates according to Rosch et al. (1976) regarding category structure (i.e., taxonomic categories), coupled with Fiske and Pavelchak's (1986) categorization paradigm. This and the previous chapter show that the level of the category label includes general abstraction of prototypical category-defining features. If a match can be quickly and efficiently surmised at this level, category-based information processing will follow. On the other hand, if no desirable match appears, then product attribute information will be analyzed carefully with details.

Out of such detailed analysis, one of several possible outcomes appears: the target's attributes are once again more closely compared with category-defining information. If an acceptable match is summoned this time, the target will be assimilated into the initially retrieved category. If no satisfactory match can be found at the category label level, the search for a match will move downward within the taxonomic hierarchy. In other words, before conducting a lateral search within one's reservoir of imagination for another product category, a possible match is assessed at a lower taxonomic level within the same product category. This scenario will be more likely to occur when the match sufficiency at the basic-level of product-category label is called into question and when categories at the subordinate level of a product taxonomy have a chance to make a meaningful match assessment.

Results of this chapter generalize those of Meyers-Levy (1986) to a larger population. For example, instead of claiming gender differences, it is found that both efforts spent in information processing and the number and types of informational cues attended to have important effects on people's category structures that are constructed and organized differently from one person to another. Speaking differently, this work conceptually fills the gap between processes of attended information and those related to information storage, such as category construction and organization, and to inclination to employ a particular strategy to process information.

Other than the inclusiveness dimension, this chapter did not examine the breadth dimension of a person's degree of latitude in defining categorical boundaries and measurement. Similar to what are developed in this chapter, there should be people who display varied patterns in this breadth dimension. With works in this new dimension done, one is able to holistically study how individual variations in these dimensions affect their information processing proclivities. Doing so will enable us to provide both theoretical support and empirical evidence that complete the loop of individual's categorical composition and choice of an information processing strategy.

Appendix: Survey Instruments

This work employed the following 20 basic- and subordinate-level attribute statements for individuals to select restaurant chains. Purposefully, mixed basic- and subordinate-level dimensions were included to eliminate possible response bias as caused by the appeal a particular attribute statement may have for one gender versus another. For selecting these attribute statements, a pretest was given for participants to rate the original 30 attributes. Each participant individually rated the 30 attributes on a seven-point scale, where 1 means "not important at all" in influencing a person's decision to select a particular restaurant chain to 7 as "very important."

- A "highly diverse" set of menu items.
- A "large" assortment of healthy, low-fat menu selections.
- An establishment where most of the employees have been with the organization for a considerable length of time.
- An establishment that accepts multiple forms of payments (e.g., cash, personal checks, major credit cards, etc.)
- An establishment that has locations only in the United States.
- An establishment that makes alcoholic beverages available.
- An establishment that offers booth seating (in addition to chair seating).
- An establishment that offers special menu provisions for senior citizens.
- An establishment that operates a drive-thru.
- An establishment that provides "speedy" service.
- An establishment that serves breakfast in addition to lunch and dinner.
- An establishment that serves primarily hamburgers, chicken, and French fries.

- An establishment that takes its time preparing your food.
- An establishment that tends to employ adults aged 24 years or older.
- An establishment that uses "primary" colors throughout most of its décor.
- An establishment that views children under the age of 12 as some of its best customers.
- Locations at major city intersections and in prime shopping areas.
- Menu items that are regularly customized to fit consumer preferences, and, therefore, may vary from one location to the next and from consumer to consumer.
- Menu items that carry an inexpensive price tag.
- Seating arrangement that may require you to wait for a table.

The selected restaurant chains that share important basic- and subordinate-level attributes but also differ along subtle lines of distinction are

Fazoli's	McDonalds	Olive Garden	Arby's	Blimpie
Donatos	Taco bell	Pizza hut	Shoney's	KFC
Chi Chi's	Pizzeria Uno	Captain D's	Long horn	Sharro
The cheesecake factory	Ponderosa	Houlihans	Popeyes	Rochnes
Dominos	Chuck E' cheeses	Hard rock Café	Damons	TGI Fridays

References

Arens, Z. G., & Hamilton, R. W. (2018). The substitution strategy dilemma: Substitute selection versus substitute effectiveness. *Journal of the Academy of Marketing Science, 46*(1), 130–146.

Basu, K. (1993). Consumers' categorization processes: An examination with two alternative methodological paradigms. *Journal of Consumer Psychology, 2*(2), 97–121.

Bilgihan, A. (2016). Gen Y customer loyalty in online shopping: An integrated model of trust, user experience and branding. *Computers in Human Behavior, 61*(August), 103–113.

Boush, D. M., & Loken, B. (1991). A process-tracing study of brand extension evaluation. *Journal of Marketing Research, 28*(1), 16–28.

Chiou, J. S., Hsiao, C. C., & Chiu, T. Y. (2018). The credibility and attribution of online reviews: Differences between high and low product knowledge consumers. *Online Information Review, 42*(5), 630–646.

Chowdhury, T. G., Murshed, F., & Khare, A. (2018). Flexible flexibility! Food categorization flexibility and utilitarian preference. *Journal of Consumer Marketing, 35*(1), 1–10.

Cohen, J. B., & Basu, K. (1987). Alternative models of categorization: Toward a contingent processing framework. *Journal of Consumer Research, 13*(4), 455–472.

Felcher, E. M., Malaviya, P., & McGill, A. L. (2001). The role of taxonomic and goal-derived product categorization in, within, and across category judgments. *Psychology & Marketing, 18* (8), 865–887.

Fiske, S. T. (2018). *Social cognition: Selected works of Susan T. Fiske*. Routledge.

Fiske, S. T., & Pavelchak, M. A. (1986). Category-based versus piecemeal-based affective responses: Developments in schema-triggered affect. In R. M. Sorrentino & E. T. Higgins (Eds.), *Handbook of motivation and cognition: Foundations of social behavior* (pp. 167–203). Guilford Press.

Forrest, J. Y. L. (2018). *General systems theory: Foundation, intuition and applications in business decision making*. Springer.

Forrest, J. Y. L., Buttermore, J., & Wajda, T. A. (2017). At Nash equilibrium when new market competitions appear? *Kybernetes, the International of Cybernetics, Systems and Management Science, 46*(2), 256–271.

Gardner, R. W. (1953). Cognitive styles in categorizing behavior. *Journal of Personality, 22*(2), 214–233.

Gurhan-Canli, Z., & Maheswaran, D. (1998). The effects of extensions on brand name dilution and enhancement. *Journal of Marketing Research, 35*(4), 464–473.

Johnson, E., & Russo, J. (1984). Product familiarity and learning new information. *Journal of Consumer Research, 11*(June), 542–550.

Kim, P., Vaidyanathan, R., Chang, H., & Stoel, L. (2018). Using brand alliances with artists to expand retail brand personality. *Journal of Business Research, 85*(April), 424–433.

Lambert-Pandraud, R., Laurent, G., Mullet, E., & Yoon, C. (2017). Impact of age on brand awareness sets: A turning point in consumers' early 60s. *Marketing Letters, 28*(2), 205–218.

Lane, V. R., & Fastoso, F. (2016). The impact of repeated ad exposure on spillover from low fit extensions to a global brand. *International Marketing Review, 33*(2), 298–318.

Liu, Y., Li, K. J., Chen, H. P., & Balachander, S. (2018). The effects of products' aesthetic design on demand and marketing-mix effectiveness: The role of segment prototypicality and brand consistency. *Journal of Marketing, 81*(1), 83–102.

Loken, B., & Ward, J. (1987). Measures of the attribute structure underlying product typicality. In M. Wallendorf & P. Anderson (Eds.), *NA - advances in consumer research* (Vol. 14, pp. 22–26). Provo, UT.

Mandler, G. (1982). The integration and elaboration of memory structures. In F. Klix, J. Hoffman, & E. van der Meer (Eds.), *Cognitive research in psychology* (pp. 33–40). North-Holland. maud.derbaix@kedgebs.com.

Mervis, C. B., & Rosch, E. (1981). Categorization of natural objects. *Annual Review of Psychology, 32*, 89–115.

Meyers-Levy, J. (1986). *Gender differences in information processing: A selectivity interpretation*. Northwestern University.

Meyers-Levy, J., & Tybout, A. M. (1989). Schema congruity as a basis for product evaluation. *Journal of Consumer Research, 16*(1), 39–54.

Moss, G. (2009). *Gender, design and marketing: How gender drives our perception of design and marketing*. Routledge.

Nedungadi, P. (1990). Recall and consumer consideration sets: Influencing choice without altering brand evaluations. *Journal of Consumer Research, 17*(3), 263–276.

Pavelchak, M. A. (1989). Piecemeal and category-based evaluation: An idiographic analysis. *Journal of Personality and Social Psychology, 56*(3), 354–363.

Rosch, E. (1978). Principles of categorization. In E. Rosch & B. B. Lioyd (Eds.), *Cognition and categorization* (pp. 27–48). Lawrence Erlbaum.

Rosch, E., Mervis, C. B., Gray, W. D., Johnson, D. M., & Boyes-Braem, P. (1976). Basic objects in natural categories. *Cognitive Psychology, 8*(3), 382–439.

Sahni, N. S. (2016). Advertising spillovers: Evidence from online field experiments and implications for returns on advertising. *Journal of Marketing Research, 53*(4), 459–478.

Schrift, R. Y., Parker, J. R., Zauberman, G., & Srna, S. (2018). Multistage decision processes: The impact of attribute order on how consumers mentally represent their choice. *Journal of Consumer Research, 44*(6), 1307–1324.

Stayman, D. M., Alden, D. L., & Smith, K. H. (1992). Some effects of schematic processing on consumer expectations and disconfirmation judgments. *Journal of Consumer Research, 19*(2), 240–255. https://doi.org/10.1086/209299

Sujan, M. (1985). Consumer knowledge effects on evaluation strategies mediating consumer judgments. *Journal of Consumer Research, 12*(June), 31–46.
Sujan, M., & Bettman, J. R. (1989). The effects of brand positioning strategies on consumers' brand and category perceptions: Some insights from schema research. *Journal of Marketing Research, 26*(4), 454–467.
Sujan, M., & Dekleva, C. (1987). Product categorization and inference making: Some implications for comparative advertising. *Journal of Consumer Research, 14*(3), 372–378.
Wajda, T., Forrest, J. Y.-L., Hu, M. Y., Li, Z., & Liu, Y. (n.d.). (to appear). The inclusiveness dimension of human cognition and categorization paradigm. Submitted for publication.

Chapter 12
Value Creation out of Innovation and Resources

Abstract Along with the development of modern technology, there have appeared unprecedented kinds of markets participated by traditional and nontraditional players. These new markets have simultaneously bought forward unparalleled challenges to the academic world and decision-making managers and entrepreneurs. To meet these challenges, this chapter, which is mainly based on (Forrest et al., to appear, Potentials of value creation and capture based on innovation, resources, strategic networks and blocks), explore potentials of value creation at the firm level from several diverse vantage points, and the next chapter looks at potentials of value capture at the firm level.

Keywords Market competition · Product · Profitability · Supply-chain network · System · Systemic hole

First, to possibly develop deeper insights than what has been achieved in the literature, this and the next chapter employ systems methodology and such a logical reasoning that is similar to the one commonly used in mathematics and natural science, which starts from several basic axioms with results established along with new concepts introduced. This chapter establishes a series of 8 formal propositions on value creation from the angle of market competition, innovation, and resource. Then some practically reliable general recommendations are provided to conclude this chapter along with a few topics and directions for future research.

12.1 Introduction

Globalization, be it economic or not, has created new ways for wealth creation (Forrester Research Report, 2000). Once accustomed form and rule of competition have been changing in order for businesses and individuals to adapt to emerging formats of decision-making and interorganizational interactions. These speedy changes and emergence of transient opportunities have been challenging scholars,

J. Y.-L. Forrest, Y. Liu, *Value in Business*, Contributions to Management Science,
https://doi.org/10.1007/978-3-030-82898-1_12

decision-making managers, and entrepreneurs from around the world (Hitt & Ireland, 2017), leading to enhanced information acquisition, knowledge development, and sharing of know-hows (Amit & Zott, 2001). A comparison of what is realistically happening and how the literature of scholarly works expands shows that most studies on value creation and capture provide managerial suggestions of limited validity instead of general recommendations due to their various empirical constraints. Hence, the following theoretically important question arises naturally:

> Based on original, empirical studies on issues and matters of value, which consist of a dominating part of the literature, can generally true conclusions be established so that they do not suffer from the inherent constraints of data- and anecdote-based analyses?

This question is very important epistemologically because addressing it will definitely help introduce new methodologies into studies of business-related topics and issues that can be employed widely to develop useful conclusions instead of the suggestions as currently done. At the same time, this question is also practically significant due to the fact that competitive advantages, at least some of which were once sustainable, have become mostly transient, short-lived. That is, the business world is currently in the era of transient competitive advantages; consumer preferences evolve rapidly and become less patient (Forrest & Tallapally, 2018; McGrath, 2013).

This chapter will address this important question by developing a series of 8 formal propositions on the basis of logical reasoning and systems thinking (Lin, 2009) through considering respectively innovations and resources. In particular, among others, this chapter establishes the following generally true conclusions:

- With free competition, innovation can effectively help create value if protective property rights and complementary assets are available. Within such a market of free competition, exchanges help resources reveal their dormant values.
- Mobility of resources is positively proportional to the capable value creation.

The rest of the chapter is organized as follows. Section 12.2 provides a review of related literatures, while pointing to contribution of this chapter. Section 12.3 summarizes the key concepts that will be dealt with in the rest of this chapter and the next chapter. Section 12.4 looks at value creation from the angle of innovation and resource, respectively, within a market of free competition. Section 12.5 concludes this chapter with practical recommendations for managers, entrepreneurs, and retailers.

12.2 The Literature

Innovation's importance in wealth creation has been recognized since at least the time of Adam Smith (1776). Ever since then, although the concept of innovation has been massively investigated by many scholars, relevant research has mainly focused on what factors determine the innovativeness of a firm (Ahuja et al., 2009; Becheikh et al., 2006). Within the presently phenomenal progress in scientific and technological innovations, Baizakov et al. (2019) examine the dynamics of an economic entity by exploring the connection of income and expenses. On the backdrop of this huge literature, this chapter presents a brand-new approach to reasoning and analysis. Because of this adopted new approach, this chapter is able to establish a much clearer relationship between protective property rights, complementary assets, and innovation's effectiveness in value creation.

As for resources and related studies, the role played by organizational resources in firms' success is first recognized by Penrose (1959). Gradually, scholars apply the concept of resources to logically illustrate synergistic and differential effects of various factors on firms' performance (Fang et al., 2011). So, starting in the 1980s, the so-called resource-based view of the firm begins to catch people's attention (Kozlenkova et al., 2014; Porter, 1979; Wernerfelt, 1984). For example, although nonfinancial performance of insurance companies is statistically significantly affected by resources, Ombaka et al. (2015) find that the effect of resources on firm performance indicators is mixed. By analyzing capability-performance relationship through the point of view of resources, Cacciolatti and Lee (2016) are able to elucidate how different types of marketing capabilities help improve firm performance. Contributing to this literature, this chapter shows that a firm's capabilities are also resources of the firm, how latent values of resources can be made visible, how value can be created out of resources, and under what conditions resources are unmatched and not readily substitutable.

Beyond these listed contributions, more important contribution this chapter and the next chapter make is the introduction of systems logic and methodology into the literature of value creation and capture. As detailed in Chap. 2, this logic and related methodology are generally applicable to investigating organizations, business evolutions, and interactions among organizations. More important than this, conclusions developed by employing this logic and related methodology do not suffer from limitations and constraints of data- and anecdote-based analyses. Speaking differently, conclusions, as developed in the explained manner, lead to general recommendations for practical purposes, similar to how mathematics and physics provide guidelines for engineering projects, instead of suggestions provided by most of the relevant studies.

12.3 Preparation

In order to assist in the seamless flow of logical reasoning, this chapter assumes that each business firm considered herein exists for the purpose of satisfying a particular market niche with its operation financially maintained by a positive cash flow as a consequence of its business conducts. Although in the past such positive cash flows were backed by firms' performance and profits (Sobel, 1999), that is no longer the case in recent times, where positive cash flows can be assured through investment inflows based on promises or potentials (Li & Ma, 2015). For example, within the landscape of global e-commerce, online retailers have focused on pumping up their future promises and potentials so that they can continuously attract sufficient amounts of venture capitals by emphasizing increasing their market shares instead of solely on making profits. Insomuch, some well-known retailers have been losing money 1 year after another since their inception.

By value in this chapter, it stands for the benefit offered by a product, service, or good to an economic agent, even though it is defined differently by competing schools of economics. Value in fact has been widely studied in various disciplines, such as computer science, economics, investment, marketing, mathematics, music, philosophy, and semiotics. Because the related literatures are enormous, we omit a review of the studies here. By value creation, it stands for a process a firm goes through to offer its output that is valuable to an economic agent. Such a process consists of two components: creating value either for consumers or for shareholders. When consumers are the target, value(s) are created to satisfy certain market demands. When shareholders are the end purpose, the process aims at attracting future investment capital. Opposite to the concept of value creation is that of value capture. It stands for the value a firm captures by implementing a value creation process that satisfies market demand(s).

By consumers, we mean the end users of products, services, or goods; and, by customers those firms that employ those products, services, or goods as inputs to produce their outputs. By resource, it stands for an asset of a firm that it can mobilize to introduce and implement its business strategies (Barney & Arikan, 2001; Harmancioglu et al., 2009). That is, a resource of a firm is anything that the firm can employ to realize its business objectives; it can take a physical, financial, intellectual, or organizational form. Beyond resources, it is the innovative mobilization of resources that practically provides a firm with alternative facilities that differentially improve its internal efficiency and create value externally for consumers or customers. By advantage (or competitive advantage), it represents a special way for a firm to mobilize its resources such as a strategy or particular way of operation, production, or management so that the firm is able to generate more value than other firms, considered marginal or just break even, in its market (Peteraf & Barney, 2003).

Here, the term *virtual market* means such a setting whereby Internet infrastructure helps facilitate business transactions and business deals. The Internet infrastructure is a worldwide electronic network. It helps create virtual communities and unconventional commercial arrangements (Hagel III & Armstrong, 1997); it enables knowledge and know-hows to be widely shared among firms and instantly avails product information to consumers.

Several additional characteristics of virtual markets include:

- Convenient connectivity (Dutta & Segev, 1999),
- Easy completion of transactions (Balakrishnan et al., 1999),
- Self-serving platform for cross-regional and interorganizational networks to form (Shapiro & Varian, 1999), and
- A provider of free or low-priced information (Evans & Wurster, 1999).

Entrepreneurial startups and corporate ventures have been—and still are—successfully riding the quick emergence of virtual markets while creating tremendous amount of new wealth (Amit & Zott, 2001).

12.4 Value Creation out of Innovation and Resources

This section considers potentials of value creation in two viewpoints: innovation and systems of resources, where other than isolated resources their associations are also emphasized. It consists of two subsections, one on the viewpoint of innovations and the other systems of resources.

The following game-theoretic result, which is a generalization of the results in (Forrest et al., 2017), plays the role of foundation for the discussion in the rest of this chapter. The detailed proof of this theorem is given in the appendix of this chapter.

Theorem 12.1 If the market described below is in Nash equilibrium, then the following two statements are equivalent:

1. At least one new firm enters the market competitively and profitably with their versions of substitute offer.
2. There is a consumer surplus of switchers who make purchase decisions based on which supplier's price is lower.

As in Theorems 3.2 and 5.1, the market considered here satisfies the following conditions: It is served by a fixed number of firms with their horizontally differentiated offers. Its operation is only affected by market forces, such as demand and supply, and consumers' forever evolving preferences and tastes. In other words, this market is perfectly competitive without any outside interference. Each incumbent firm has a base of loyal customers if its price is not more than their reservation price. There are switchers, collectively known as consumer surplus, who make purchases based on which firm's unit-value price is lower. And all firms' pricing strategies are known to all involved companies who respond by playing Nash equilibria through untainted self-analyses. For the technical proof of Theorem 5.1, go to Appendix of this chapter.

12.4.1 Value Potentials of Innovation

The concept of product is defined differently in management and marketing from that in economics, for related references, see discussions below, let us first define this concept for our discussions in this chapter. By product, it means such a thing or process, either tangible (physical) or intangible (none physical), which is produced by a firm to meet market demand. For example, an iPhone is a tangible and physical product that consumers can use to make their lives more enjoyable and convenient.

An annuity established with an insurance company is an intangible product through which the company makes its promises that consumers can count on when certain prescribed situations appear. Services like self-serving laundry facilities, indoor tanning salons, represent such products that consumers can utilize through using the facilities and salons. Investment recommendations of financial firms are also products, known as informational goods; they are produced by relevant firms through a "manufacturing" process, consisting of gathering, organizing, and analyzing facts and data. In general, complementary products are bundled together to promote consumer utilities, attract additional customers and improve sales.

To keep everybody on the same pace, the concept of innovation is defined as follows in a parallel fashion to that defined for the manufacturing sector (Forrest et al., 2018): By innovation, it means some activities, related to the production or offering of a product or a bundle of products, which lead to extraordinarily added value for a firm in comparison to other activities that other firms in the same economic sector simultaneously embark on. The phrase "the production or offering of a product" of course includes the related R&D efforts and actual design of one or several original products that cause extraordinarily added value for the firm of concern. In other words, each innovation consists of pertinent activities that lead to the eventual introduction of original product(s). These activities can, for example, be a redesign of the business model currently in use, works devoted to adequately comprehending market signals, efforts directed to acquiring new knowledge, energies purposefully spent in converting newly acquired know-hows to actual production, etc.

Conversely, the phrase "offering of a bundle of products" is overwhelmingly dissimilar. It means to collect several somehow relevant products together in one deliverable package to consumers. The key here is that the products that are bundled together do not need to be special or original in terms of their characteristics and functionalities. The creativity and innovative thinking behind a bundle of products, be they mundane or original, lie in the potential and/or actual developments of simultaneous consumer utilities and multi-sided markets (Ye et al., 2012). When consumers enjoy their consumptions from using several products concurrently, they experience a moment of simultaneous consumer utilities; and when different markets are served and pulled together, a so-called multi-sided market is created.

The emphasis on the comparatively added value in the previous definition of innovation implicitly assures this concept to include different aspects of applications. For instance, the emphasized added value suggests that

- Over time creativity can possibly emerge;
- Both current and predicted market demands that are comprehended internally and accepted externally will help promote creativity;
- By adopting needed business processes, expected values are created and captured;
- This particular definition of innovation implies not only inventions but also translations of them into market offers to meet identified demands;

- It maintains potentials open for newness and originality of market offers and relevant business process(es); and
- It integrates covertly potentials of technological changes, adoption of different processes, and consequent outcomes.

This definition of innovation reflects the essence that innovation is a driving force behind the creation and capture of new and exceptional value, and that it emerges in various situations and in disparate ways, such as design and production of new market offers, novel procedures of production, measures of improved efficiency, creation of new markets, discovery of new sources of supply, and reformation of industries, as claimed by Schumpeter (1934). Being able to charging high rents represents one of the benefits of the originality and scarcity brought forward by innovations. That entices and stimulates entrepreneurial spirits, even though these elevated rents gradually weaken as the initial originality turns into humdrum industry exercises and the scarcity disappears. That is why creative destruction emerges (Schumpeter, 1942): The early menaces endured by entrepreneurs pave the way for the appearance of consequent originality and scarcity, while as knowledge accumulates and information spreads over time and distance, the know-hows underlying entrepreneurial initiatives become commonly known.

Discussions above imply that each innovation is identifiable to some resolute activities formulated to satisfy a recognized market demand, no matter whether the demand is large or small. Technology plays its role in innovation by providing improvements, such as making various operations more efficient, making more relevant discoveries, and lowering the operational expenditure. Based on discretely dissimilar identifications of market demands, entrepreneurs embark their individually different innovations, leading to a spectrum of different market offers. The joint effort of these entrepreneurs transforms existing markets, creates new markets, and develops unprecedented industries. That fundamentally drives the continuous development of the economy.

Proposition 12.1 In an open, competitive economy, complementary assets and the effectiveness of protective property rights positively affect an innovation's potential of value creation.

The truthfulness of this result follows directly from the discussions above. In particular, being open and competitive encourages entrepreneurial spirit. With an emboldened entrepreneurial spirit, effective protection of property rights and mobilizability of complementary assets will jointly provoke fervent people to innovate for the purpose of satisfying market demands.

In terms of the literature, Teece (1987) provides a related, anecdote-based discussion.

Proposition 12.2 In an open, competitive economy, exchanges enabled by different markets assist the latent values of resources to be practically materialized.

This proposition is a consequence of Theorem 12.1. In particular, the assumed market in that theorem encourages consumers to develop their consumption preferences increasingly more sophisticated. Such endless progression in consumer preferences and tastes inexorably expands the consumer surplus in the marketplace. That, as indicated by Theorem 12.1, then increases the market competition among the incumbent firms, although they might be in a state of mutual forbearance. This logical reasoning illustrates how latent values of resources are developed and materialized through creative competitions of resources available to individual entrepreneurs.

See Moran and Ghoshal (1999) for associated empirical studies. For what economic variables could potentially influence a firm's innovativeness, see (Becheikh et al., 2006; Forrest et al., 2020; Hitt & Ireland, 2017).

Transaction is defined by Williamson (1983) as a trade that a product is transferred from one owner to another across a technologically distinguishable interface such that a round of production and delivery finishes and another round begins. Transaction cost is defined by Demsetz (2003) as the cost of a completed trade, where the term "trade" means the commonly known buying/selling, or emotional interactions between people, informal gift exchanges of different parties, etc. For the particular purpose of this chapter, this chapter will not establish results from Cheung's (1992) definition of transaction costs, even though Cheung's definition covers a much broader range of scenarios than Demsetz's.

As rules that are commonly accepted and followed by members of a society, institutions indirectly determine the scale of transaction costs. Various factors generally determine transaction costs (Williamson, 1979). Examples of such factors include frequency of trading, specificity, asymmetry of information, uncertainties, limited rationality, opportunistic behaviors in small-number situations, and others. That is why when companies are able to facilitate low transaction costs, they can tactfully improve their performance and growth (North, 1992). A firm is said to have transaction efficiency, if it is able to lower its transaction costs more than its competitors can. The result below follows naturally:

Proposition 12.3 Transaction efficiency is a source of value that is creatable for consumers and capturable by a firm.

Theorem 12.1 readily implies this proposition, because transaction efficiency reduces a firm's selling price. That assists the firm to keep its existent loyal consumers and capture the entire market segment of switchers. In other words, by using transaction efficiency, the firm creates value for consumers and simultaneously captures value for itself.

For empirical studies on how firms recognize the need to lower costs, risks associated with coordination and transaction, and how information technology can lessen costs and risks, see (Clemons & Row, 1992). For anecdotes on how a networked environment, such as the Internet, can help decrease transaction costs, see (Dyer, 1997).

Proposition 12.3 and discussions above jointly imply that a firm can create and capture value by reducing asymmetry of information, involved uncertainty, complexity, rationality limitations, and bargaining behavior in small-numbers situations.

That is, by economizing transaction costs with innovation, firms will be able to squeeze values out of their business transactions.

Additionally, by improving their reputation, trusts, and transactional experiences, e.g., frequent transactions, firms will be able to reduce their costs of distinctive exchanges with each other. That is, improving their reputation, trusts, and transactional experiences will help increase a firm's transaction efficiency.

12.4.2 Value Potentials of Resources

By resource, it means an asset, either tangible or intangible (Harmancioglu et al., 2009), a firm can employ to design and apply its strategies in order to realize its business objective (Barney & Arikan, 2001). In other words, a resource is such a thing, be it physical, or financial, or intellectual, or organizational, that a firm can employ to materialize its business objective. By capability, it generally represents a process, mostly information-based, that allows a firm to holistically mobilize its available resources to reach its objective (Makadok, 2001).

Proposition 12.4 Each capability specific to a firm represents a unique resource of the firm.

By combining all the discussions above, a firm can be uniquely identified with its system of firm-specific resources. Although the resources of the firm can be mostly also available to other firms, they are related to each other within the firm purposefully for particular business purposes. This identification between a firm and its unique system of resources can naturally lead to the well-known resource-based view or theory (RBV) of the firm (Kozlenkova et al., 2014) based on the recognition of the following three basic and intuitive axioms, for related details, see Forrest et al. (2019) or Chap. 4 in this book. For convenience of the reader, these axioms are cited below from Chap. 4.

Axiom 4.1 (Resource Heterogeneity): No matter which industry a firm is a part of, the firm owns its unique system of resources.

Axiom 4.2 (Resource Immobility): Due to practical difficulties of trading resources between firms and formulating similar associations among available resources within individual firms, differences in the systemness of firms' available resources endure over time.

Axiom 4.3 (Different Levels of Efficiency): Different firms enjoy their individually different levels of performance, no matter whether the firms belong to the same industry or not.

A resource is seen as valuable to a firm (Barney, 1997) if it enables the firm to introduce and to implement a particular strategy that helps increase that firm's performance beyond what would be the case without the resource. By combining the RBV and the concept of value creation (Chap. 4), the result below follows:

Proposition 12.5 If a firm can mobilize its valuable and rare resources, then the firm is able to create values.

The scarcity of the firm's valuable resources naturally leads to this proposition, if the firm is able to mobilize these resources. It is because only when the firm has the ability to employ the said resources, the scarcity and valuableness of the resources will jointly assist the firm to create certain advantages over other firms until such a time when the scarcity disappears.

Proposition 12.6 Assume that a firm has control over some valuable and rare resources. If the firm's business goal and its stakeholders do not have conflicting interests, then the firm is in a state of possibility to create value by utilizing its available resources.

In fact, the if-condition implies that the management of the firm is able to effectively exploit its valuable and rare resources. Hence, this firm is capable of creating value by mobilizing these resources, as jointly indicated by Propositions 12.4 and 12.5.

Proposition 12.7 If the management of a firm is able to fully work with its capabilities of marketing, research and development (R&D), and operation, then the firm can create value.

To see the truthfulness of this proposition, let us look at the capabilities of marketing, research and development (R&D), and operation, individually one by one. First, for each firm, it presents itself to the world by revealing what the firm is, what demands of the product market the firm can satisfy, and how the firm plans to change to accomplish its goal (Kotler & Keller, 2016). In other words, the marketing capability helps a firm acquire the knowledge and forecast about its current and future competitions (Jaworski & Kohli, 1993). Speaking differently, marketing capability helps a firm orient its effort and activities of innovation and operation, and adequately presents the firm to the market, while showing what the firm can do to satisfy market demands.

Secondly, the R&D capability of a firm translates the market information, innovatively acquired from using the marketing capability, into terms specific to the firm. Such specific terms direct the firm with its design and production of new or original products that potentially satisfy consumers' incessantly changing demands. At the same time, this capability prevents competitors from imitating the firm's new products (Irwin & Klenow, 1994). When a firm has a strong and reputable track record of R&D achievements, the firm will be able to boost its image and form favorable consumer expectations.

Thirdly, the operational capability of a firm practically realizes the market information and knowledge, acquired through marketing and R&D capability, into deliverable products.

By combining all these three capabilities together, one can see that only when the marketing capability is able to acquire novel information and knowledge from the market, it will make it possible for the R&D capability to introduce state-of-the-art designs of new and original products, which will give the operational capability a chance to adopt appropriate and accommodating procedures. Through the joint working of these three capabilities, a firm's chain of activities becomes practically mobile and imperfectly imitable. Within such a situation, the firm is perfectly positioned to create value few rivals can compete with. In other words, Proposition 12.7 follows directly from Propositions 12.4 and 12.5 jointly.

For related discussions based on anecdotes and empirical data, see McGrath (2013) and Hayes et al. (1988).

In short, the previous discussion implies from the viewpoint of the RBV that holistically organizing specialized valuable and scarce resources into a functional system is necessary to create value. In other words, only when a firm is able to identify itself with its unique system of resources, it will be able to possibly generate value(s) by systemically mobilizing its valuable and rare resources. For empirical studies that are relevant to this conclusion, please consult with Amit and Schoemaker (1993), Barney (1991), Peteraf (1993), and Wernerfelt (1984).

If the concept of time is introduced into those of resource and capability, that is, treating each resource or capability as a function of time, the resultant will naturally be a general resource-based theory of firm evolution. The scarcity of a resource is realistically time-dependent. For example, the present value of $100 million investment is a function of time. A resource, which might be initially inimitable, gradually becomes widely available as time passes. By joining what is discussed here with Proposition 12.4, it can be seen that each real-life capability (and each resource) is truly time-dependent (or a function of time). Hence, the dynamic capability, as so named by Teece et al. (1997), is a specialized scenario of the concept of capability defined in this chapter. In other words, studies of dynamic capabilities can be reframed within what is presented in this book and the concept of time does not need to be explicitly mentioned in most of the general discussions of resources and capabilities. With this understanding, the resource-based theory, as established in this book, can be employed to address many important questions, such as

- How can values be declared and necessitated?
- How can competitive advantages be made sustainable (Barney, 1991; McGrath, 2013)? and
- How are resources, which are seen valuable at certain moments of time, developed over time?

Considering each resource (or capability) as a function of time is more than a simple mind exercise. Indeed, doing so is engrained in how a firm's business operations are mostly managerial and organizational processes, targeting the formation of coordination, integration of knowledge and resources, realization of a revitalization, or completion of a transformation (Eisenhardt & Martin, 2000; Teece et al., 1997), or collective learning (Lei et al., 1996). Only such resources and capabilities that are seen as functions of time can really help firms with their value creation and capture through Schumpeterian rents (Teece et al., 1997). In particular, when resources and capabilities are time-dependent, they play their roles in strategic decision-making, knowledge acquirement, product development, how-to skill creation, social network formation, and others.

Proposition 12.8 Virtual markets make each created value short-lived and provide unprecedented opportunities for entrepreneurs to create and capture values.

This proposition follows from the fact that virtual markets reside on the Internet; and other than being widely available, the Internet also provides conveniently and readily accomplishable exchanges between buyers and sellers. Additionally, the Internet is available 24/7 for entrepreneurs to uncover new resources, to explore potential substitutes, and to exploit existing resources in inventive ways through timely information and knowledge sharing either freely or at very low prices. Such affordability and convenience quicken the circulation of knowledge and know-hows, some of which used to be difficult to acquire. That elucidates why each created value becomes short-lived.

Additionally, Proposition 12.2 also provides a reason why virtual markets offer unprecedented opportunities for value creation and value capture.

12.5 A Few Final Words

The current world of business has altered once sustainable competitive advantages to become transient and short-lived. That in turn forces firms that desire for continued successful value creation and capture to be on constant lookout for (McGrath, 2013):

- New market signals;
- Unconventional ways to comprehend market signals;
- Incremental or disruptive technological advances that are relevant; and
- Business paradigm shifts.

With worldwide widening and deepening applications of the Internet, accelerating market competitions have been taking different forms without complying with conventional rules. Over the Internet, effective platforms have been developed for buyers and sellers to interact with each other at very low costs (Amit & Zott, 2001). To stay in sync with this fast market change both academically and practically, this chapter examines a firm's potentials for value creation from several different angles—innovation and resource. The methodological novelty of this chapter assures that the derived conclusions provide dependable and reliable bases for managers and entrepreneurs to make their decisions.

Specifically, this chapter employs systems methodology and logical reasoning that is parallel to the one widely used in mathematics and physics to study the fast-changing world of business in terms of innovation and resources. Regarding innovation, this chapter pinpoints when innovation offers potentials for value creation. In the perspective of resources, it finds when latent values of resources can be practically made visible through value creation.

As for practical implications, this chapter can provide general recommendations instead of suggestions for decision-makers, such as entrepreneurs, managers, and retailers. In particular, this chapter shows the importance for entrepreneurs to innovatively understand market signals in their individually different ways. Being able to do so will lead firms to their idiosyncratic products (Theorem 12.1).

For managers, they can create additional values out of their resources by encouraging information sharing, such as exchanges of ideas, and sharing resources, and recognizing success, such as widely and openly recognizing employee contributions, offering financial incentives, (Propositions 12.1, 12.2, and 12.4). Policies and relevant procedures need to be implemented to acquire and mobilize their firm-specific resources (Proposition 12.5), while tying managers' interests with those of stakeholders (Proposition 12.6). Marketing efforts, R&D activities, and operational practicalities need to be considered jointly in managerial decision-making (Proposition 12.7). Such collective thoughtfulness will stimulate internal collaboration, leading to more reliable value creation and capture.

For retailers, incessantly try to make transactions more efficient (Proposition 12.3). Take advantage of whatever new convenience technology offers (Proposition 12.8).

To conclude this chapter, it needs to be noted that all conclusions in this chapter are established on the implicit assumption that each firm exists for the purpose of addressing a market niche by generating positive cash flows from the product market. But, in reality, firms exist for various reasons beyond what is assumed here. This end will be an important reason for why some applications of the general results developed in this chapter might not work in practice, if the firms of concern exist not for this said purpose. This limitation, along with many other ones not listed here, of this current chapter opens doors for future research.

Appendix: Proof of Theorem 12.1

What is assumed in this theorem implies that the said market is in a state of mutual forbearance (Bernheim & Whinston, 1990). The incumbent firms alleviate their rivalries by partitioning the market according to their strengths so that they surrender dominance to their stronger competitors in marketplaces where they are less efficient (Li & Greenwood, 2004). The codependence of these firms gradually decreases the rates of market entry and exit (Fuentelsaz & Gómez, 2006; Yu & Cannella Jr., 2012), while reducing interfirm hostility (Haveman & Nonnemaker, 2000). Due to this state of mutual forbearance, let us consider the aggregate of the incumbent firms as one imaginary firm with α being its market share in percentage. So, $\beta = 1 - \alpha$ denotes the percentage size of the consumer surplus.

As detailed in Chap. 3, let us assume that there are n incumbent firms and that all relevant boundary conditions are normalized such that incumbent firms' individual selling prices, production costs, loyal customers' reservation values within the said market, lower and upper boundary conditions and magnitudes of loyal-customer bases are all, respectively, normalized to P_i ($0 \leq P_i, \leq 1$), 0, 1, 0, 1, and α/n, $i = 1, 2, \ldots, n$.

Proof of 1) ⇒ 2). Assume that by randomizing its price over the interval from production cost to loyal customers' reservation value, or [0,1] after applying normalization, a firm enters into the said market of n incumbent firms, which are collectively seen as one aggregate firm. So, the consumer surplus must satisfy $\beta = 1 - \alpha > 0$.

Proof of 2) ⇒ 1). Assume that the consumer surplus satisfies $\beta = 1 - \alpha > 0$. Firstly, let α_0 be a real number so that $\beta = 1 - \alpha > \alpha_0 > 0$ and $\alpha = \ell\alpha_0$, where ℓ is a natural number. Secondly, let us imagine that the aggregate firm is divided into ℓ many identical (imaginary) "firms," named i, $i = 1, 2, \ldots, \ell$. Each of these firms provides consumers with an identical product and enjoys the market share $\alpha_0 = \alpha/\ell$ of loyal consumers. These firms compete over the switchers with adjustable prices. And thirdly, these ℓ imaginary firms do not have any symmetric pure strategy Nash equilibrium, for details see Chap. 3.

For the rest of this proof, it suffices to show that there is one firm that will be expected to profit by entering this market through uniformly randomizing its price strategy over the interval [0,1]. To this end, let $F(P)$ be the price distribution of Firm j, one of the imaginary firms of the aggregate firm. The aggregate firm or equivalently each of the ℓ imaginary forms sets its price after considering the price of the entering firm and those of all other imaginary firms. Hence, the profits for Firm j from its loyal consumers is $\alpha_0 P$ and those from its share of the switchers is $\prod_{i \neq j}^{\ell}(1 - P)[1 - F(P)]\beta P = \beta P(1 - P)[1 - F(P)]^{\ell-1}$. Hence, the profit Π Firm j generates when the firm sells its product at price P is

$$\alpha_0 P + \beta P(1 - P)[1 - F(P)]^{\ell-1}$$

and the objective of Firm j is to maximize the following by choosing its price distribution $F(P)$

$$E(\Pi) = \int_{-\infty}^{+\infty} \left\{\alpha_0 P + \beta P(1 - P)[1 - F(P)]^{\ell-1}\right\} dF(P)$$

$$= \int_{0}^{1} \left\{\alpha_0 P + \beta P(1 - P)[1 - F(P)]^{\ell-1}\right\} dF(P)$$

where $E(\Pi)$ stands for Firm j's expected profit for all possible prices. The reason why the upper and lower limits of the integral are changed respectively from $+\infty$ and $-\infty$ to 1 and 0 is because when $P < 0$ or when $P > 1$, the profits are zero.

The equilibrium indifference condition of Firm j is

$$\alpha_0 \times P + \beta \times P(1-P)[1-F(P)]^{\ell-1} = \alpha_0 \times 1. \tag{12.1}$$

So, for the ℓ imaginary firms, solving Eq. (12.1) leads to their symmetric equilibrium pricing strategy as follows:

$$F(P) = 1 - \left(\frac{\alpha_0}{\beta P}\right)^{\frac{1}{\ell-1}}. \tag{12.2}$$

From $\beta > \alpha_0$, it follows that $\alpha_0/\beta < 1$. So, for any Price P, satisfying $1 \geq P \geq \alpha_0/\beta$, Eq. (12.2) is well-defined as a probability distribution. This end implies that for the ℓ imaginary firms, or equivalently, the aggregate firm, the lowest allowed price is α_0/β.

To complete this proof, it suffices to show that the entrant actually expects to make profits in this new market. To this end because $lim_{P\to 1}\ F(P) = 1 - (\alpha_0/\beta)^{1/(\ell-1)} \neq F(1) = 1$, the cumulative price distribution function $F(P)$ has a jump discontinuity at the reservation value $P = 1$, where the amount of jump is $(\alpha_0/\beta)^{1/(\ell-1)}$. That is, $F(P)$ has a mass point of size $(\alpha_0/\beta)^{1/(\ell-1)}$ at the reservation price $P = 1$. So, the expected profit of the entrant is the following:

$$E(\Pi) = \int_0^{\alpha_0/\beta} \beta P dP + \int_{\alpha_0/\beta}^{+\infty} \beta P[1-F(P)]^{\ell} dP \tag{12.3}$$

$$= \int_0^{\alpha_0/\beta} \beta P dP + \int_{\alpha_0/\beta}^{1} \beta P[1-F(P)]^{\ell} dP + \beta\left(\frac{\alpha_0}{\beta}\right)^{\ell/(\ell-1)} \tag{12.4}$$

where the first term in the right-hand side of Eq. (12.3) stands for the expected profit of the entrant when it charges the lowest price in the marketplace and captures the entire consumer surplus, and the second term is the entrant's expected profit when it is in direct competition with the ℓ incumbent firms.

It can be readily shown that all the terms on the left-hand side of Eq. (12.4) are positive. So, it implies that if the consumer surplus $\beta = 1 - \alpha > 0$, there will be at least one entrant into the said market to compete with the incumbent firms.

References

Ahuja, G., Lampert, C. M., & Tandon, V. (2009). Moving beyond Schumpeter: Management research on the determinants of technological innovations. *The Academy of Management Annals, 2*(1), 1–98.

Amit, R., & Schoemaker, P. (1993). Strategic assets and organizational rent. *Strategic Management Journal, 14*(1), 33–46.

Amit, R., & Zott, C. (2001). Value creation in e-business. *Strategic Management Journal, 22*(6/7), 493–520.
Baizakov, S., Forest, J. Y. L., & Baizakov, N. (2019). Modeling the managements of the economies of developing countries. *Advances in Systems Science and Applications, 19*(2), 101–119. https://doi.org/10.25728/assa.2019.19.2.673
Balakrishnan, A., Kumara, S. R. T., & Sundaresan, S. (1999). Manufacturing in the digital age: Exploiting information technologies for product realization. *Information Systems Frontier, 1*, 25–50.
Barney, J. B. (1991). Firm resources and sustained competitive advantage. *Journal of Management, 17*, 99–120.
Barney, J. B. (1997). *Gaining and sustaining competitive advantage*. Addison-Wesley.
Barney, J., & Arikan, A. (2001). The resource-based view: Origins and implications. In M. Hitt, R. Freeman, & J. Harrison (Eds.), *Handbook of strategic management* (pp. 124–185). Blackwell.
Becheikh, N., Landry, R., & Amara, N. (2006). Lessons from innovation empirical studies in the manufacturing sector: A systematic review of the literature from 1993 to 2003. *Technovation, 26* (5), 644–664.
Bernheim, B. D., & Whinston, M. D. (1990). Multimarket contact and collusive behavior. *RAND Journal of Economics, 21*(1), 1–26.
Cacciolatti, L., & Lee, S. H. (2016). Revisiting the relationship between marketing capabilities and firm performance: The moderating role of market orientation, marketing strategy and organisational power. *Journal of Business Research, 69*(12), 5597–5610.
Cheung, S. N. S. (1992). On the new institutional economics. In L. Werin & H. Wijkander (Eds.), *Contract economics* (pp. 48–65). Blackwell.
Clemons, E. K., & Row, M. C. (1992). Information technology and industrial cooperation: The changing economics of coordination and ownership. *Journal of Management Information Systems, 9*, 9–28.
Demsetz, H. (2003). Ownership and the externality problem. In T. L. Anderson & F. S. McChesney (Eds.), *Property rights: Cooperation, conflict, and law*. Princeton University Press.
Dutta, S., & Segev, A. (1999). Business transformation on the internet. *European Management Journal, 17*, 466–476.
Dyer, J. H. (1997). Effective interfirm collaboration: How firms minimize transaction costs and maximize transaction value. *Strategic Management Journal, 18*(7), 535–556.
Eisenhardt, K. M., & Martin, J. A. (2000). Dynamic capabilities: What are they? *Strategic Management Journal, 21*(10–11), 1105–1121.
Evans, P. B., & Wurster, T. S. (1999). *Blown to bits: How the new economics of information transforms strategy*. Harvard Business School Press.
Fang, E., Palmatier, R., & Grewal, R. (2011). Effects of customer and innovation asset configuration strategies on firm performance. *Journal of Marketing Research, 48*(3), 587–602.
Forrest, J. Y.-L., Baizakov, S., Haile, Y. G., Gyan, A., & Liu, Y. (n.d.) (to appear). Potentials of value creation and capture based on innovation, resources, strategic networks and blocks. Submitted for publication.
Forrest, J. Y. L., Buttermore, J., & Wajda, T. (2017). At Nash equilibrium when new market competitions appear? *Kybernetes, the International of Cybernetics, Systems and Management Science, 46*(2), 256–271.
Forrest, J. Y. L., Nicholls, N., Schimmel, K., & Liu, S. F. (2020). *Managerial decision making: A holistic approach*. Springer.
Forrest, J. Y. L., Shao, L., McKinney, R., & Shao, D. (2019). Firm performance and market expansion from the point of view of resources. *International Journal of Business and Systems Research*, accepted for publication.
Forrest, J. Y.-L., & Tallapally, P. (2018). Customers are less patient, sustainable advantage becomes transient and the key for firms to succeed in fast changing markets. *Journal of Business, Economics and Technology, 21*(1), 1–14.

Forrest, J. Y. L., Tucker, R., Lin, C. C., & Mondal, S. (2018). Impacts of manufacturing firms' culture, structure and leadership on innovation: A systemic approach. *Proceedings of the 41st annual meeting of northeastern association of business, economics and technology*, State College, PA.

Forrester Research Report. (2000). Global eCommerce approaches Hypergrowth. April 2000.

Fuentelsaz, L., & Gómez, J. (2006). Multipoint competition, strategic similarity and entry into geographic markets. *Strategic Management Journal, 27*(5), 477–499.

Hagel, J., III, & Armstrong, A. G. (1997). *Net gain: Expanding markets through virtual communities*. Harvard Business School Press.

Harmancioglu, N., Droge, C., & Calantone, R. (2009). Strategic fit to resources versus NPD execution proficiencies: What are their roles in determining success? *Journal of the Academy of Marketing Science, 37*(3), 266–282.

Haveman, H. A., & Nonnemaker, L. (2000). Competition in multiple geographic markets: The impact on growth and market entry. *Administrative Science Quarterly, 45*(2), 232–267.

Hayes, R. H., Wheelwright, S. C., & Clark, K. B. (1988). *Dynamic manufacturing*. The Free Press.

Hitt, M., & Ireland, R. D. (2017). The intersection of entrepreneurship and strategic management research. In D. L. Sexton, H. Landstrom, & C. Jones (Eds.), *The Blackwell handbook of entrepreneurship* (pp. 45–63). Blackwell.

Irwin, D. A., & Klenow, P. J. (1994). Learning-by-doing spillovers in the semiconductor industry. *Journal of Political Economy, 102*(6), 1200–1227.

Jaworski, B. J., & Kohli, A. K. (1993). Market orientation: Antecedents and consequences. *Journal of Marketing, 57*(July), 53–70.

Kotler, P., & Keller, K. L. (2016). *A framework for marketing management* (6th ed.). Pearson.

Kozlenkova, I. V., Samaha, S. A., & Palmatier, R. W. (2014). Resource-based theory in marketing. *Journal of the Academy of Marketing Science, 42*, 1–21.

Lei, D., Hitt, M. A., & Bettis, R. (1996). Dynamic core competences through meta-learning and strategic context. *Journal of Management, 22*, 549–596.

Li, S. X., & Greenwood, R. (2004). The effect of within industry diversification on firm performance: Synergy creation, multi-market contact and market structuration. *Strategic Management Journal, 25*(12), 1131–1153.

Li, T., & Ma, J. H. (2015). Complexity analysis of dual-channel game model with different managers' business objectives. *Communications in Nonlinear Science and Numerical Simulation, 20*, 199–208.

Lin, Y. (2009). *Systemic yoyos: Some impacts of the second dimension*. Auerbach Publication, an imprint of Taylor and Francis.

Makadok, R. (2001). Toward a synthesis of the resource-based and dynamic-capability views of rent creation. *Strategic Management Journal, 22*(5), 387–401.

McGrath, R. G. (2013). *The end of competitive advantage: How to keep your strategy moving as fast as your business*. Harvard Business Review Press.

Moran, P., & Ghoshal, S. (1999). Markets, firms, and the process of economic development. *Academy of Management Review, 24*, 390–412.

North, D. C. (1992). *Transaction costs, institutions, and economic performance*. ICS Press.

Ombaka, B. E., Awino, Z. B., Machuki, V. N., & Wainaina, G. (2015). Exploring resources and performance relationships in commercial enterprises: An empirical perspective. *Journal of Management and Strategy, 6*(4), 12–28.

Penrose, E. T. (1959). *The theory of growth of the firm*. Blackwell.

Peteraf, M. A. (1993). The cornerstones of competitive advantage: A resource-based view. *Strategic Management Journal, 14*(3), 179–191.

Peteraf, M. A., & Barney, J. (2003). Unraveling the resource-based tangle. *Managerial and Decision Economics, 24*(4), 309–323.

Porter, M. E. (1979). How competitive forces shape strategy. *Harvard Business Review, 57*(2), 137–145.

Schumpeter, J. A. (1934). *The theory of economic development: An inquiry into profits, capital, credit, interest, and the business cycle*. Harvard University Press.
Schumpeter, J. A. (1942). *Capitalism, socialism, and democracy*. Harper.
Shapiro, C., & Varian, H. R. (1999). *Information rules: A strategic guide to the network economy*. Harvard Business School Press.
Smith, A. (1776). *The wealth of nations, books I-III, (1986 printing)*. Penguin Books.
Sobel, R. (1999). *When giants stumble: Classic business blunders and how to avoid them*. Prentice Hall.
Teece, D. J. (1987). Profiting from technological innovations: Implications for integration, collaboration, licensing, and public policy. In D. J. Teece (Ed.), *The competitive challenge: Strategies for industrial innovation and renewal* (pp. 185–219). Ballinger.
Teece, D., Pisano, G., & Shuen, A. (1997). Dynamic capabilities and strategic management. *Strategic Management Journal, 18*(7), 509–533.
Wernerfelt, B. (1984). A resource-based view of the firm. *Strategic Management Journal, 5*(2), 171–180.
Williamson, O. E. (1979). Transaction cost economics: The governance of contractual relations. *Journal of Law and Economics, 22*, 233–261.
Williamson, O. E. (1983). Organizational innovation: The transaction cost approach. In J. Ronen (Ed.), *Entrepreneurship* (pp. 101–133). Lexington Books.
Ye, G. L., Priem, R. L., & Alshwer, A. A. (2012). Achieving demand-side synergy from strategic diversification: How combining mundane assets can leverage consumer utilities. *Organization Science, 23*(1), 207–224.
Yu, T., & Cannella, A. A., Jr. (2012). A comprehensive review of multimarket competition research. *Journal of Management, 39*, 76–109.

Chapter 13
Potentials of Value Capture and General Value-Chain Framework

Abstract With the emergence of new kinds of markets, there have also appeared nontraditional players in the marketplace. These new kinds of markets and players have created unparalleled challenges to the world of learning and business decision-making, while scholars recognized that Porter's (1985, *Competitive Advantage: Creating and Sustaining Superior Performance*. New York, NY: Free Press) value-chain framework is no longer sufficient for analyzing business firms. To meet these challenges, this chapter, which is partially based on Forrest et al. (2020, *Advances in Systems Science and Applications*, *20*(3), 50–72) and Forrest et al. (2020, *Journal of Business and Economic Studies*), continues the previous chapter by exploring potentials of value capture at the firm level from several diverse angles—interorganizational network, close association between sellers and buyers, and then generalizes Porter's value-chain framework that is equally applicable to the analysis of firms that offer tangible products, services, or informational goods. Accordingly, practical and reliable general recommendations, as well as a few unsettled questions for future research, are offered.

Keywords Market invitation · Product · Resource · Value

13.1 Introduction

With the fast-changing business landscape, unprecedented ways of wealth creation have appeared recently (Forbes, 2018; Forrester Research Report, 2000). For example, virtual markets have appeared to process direct seller-and-buyer transactions remotely. That has greatly changed the conventional way of market competition and provided boundless opportunities for entrepreneurs. Such a stimulating era of business has attracted the attention of scholars, decision-making managers, and entrepreneurs (Hitt & Ireland, 2017). The emerging new avenues of value creation and capture have helped accelerate knowledge development and acquisition and promote information sharing, leading to quickened turnovers of innovative products, services, and informational goods (Amit & Zott, 2001; Priem et al., 2018). However, comparing to such exciting business development, most of the studies on value

J. Y.-L. Forrest, Y. Liu, *Value in Business*, Contributions to Management Science,
https://doi.org/10.1007/978-3-030-82898-1_13

capture have been either data-based or anecdote-based, on which managerial suggestions of limited validity were given. So, continuing what has been achieved in the previous chapter on value creation out of innovation and resources, this chapter examines value capture out of interorganizational networks and platforms that directly connect sellers and buyers.

Going along this line of thinking, this chapter also generalizes Porter's (1985) value-chain framework so that the resultant version can be appropriately applied to analyze firms that offer physical products, services, or informational goods, instead of only manufacturing firms. Such need to generalize Porter's work of over 30 years old is initially recognized by Vendrell-Herrero et al. (2017) and Stabell and Fjeldstad (1998). In particular, these scholars find that Porter's method is not quite applicable for analyzing service firms. That is, this chapter also addresses the following question: What is an adequate value-chain framework that can be applied to the present rapidly changing world of business?

In the process of accomplishing the given goals above, this chapter establishes a series of formal propositions by employing the concept of systems and logical reasoning by respectively looking at the market state of mutual forbearance, interorganizational networks, platforms that directly connect sellers and buyers, and how information flows affect the emergence of creative destruction. And, based on the previous chapter and these formal propositions, a generalized value-chain framework is developed that can be applied harmoniously to analyze firms that offer physical products, services, or informational goods.

The rest of this chapter is organized as follows: Section 13.2 provides a literature review, while pointing to what this chapter contributes to the literature. Section 13.3 looks at the market state of mutual forbearance and network structures. Section 13.4 examines the potentials of value capture out of nosiness networks and convenient platforms. Section 13.5 establishes a connection between circulation of information and the emergence of creative destructions. Section 13.6 presents the result that generalizes the earlier version of the value-chain framework initially developed by Porter (1985). Section 13.7 wraps up this chapter with practical managerial recommendations, as well as open questions for future research.

13.2 The Literature

In the business world, firms are mostly members of networks (or systems) of various business organizations and exchange associations instead of atomistic actors that compete with each other, as what the major part of the literature shows (Galaskiewicz & Zaheer, 1999; Granovetter, 1985; Gulati, 1998). Conducts and behaviors of such networked firms can naturally explain why systems methodology and logical reasoning are needed for investigating these firms' performance. More specifically, when they are individually a part of some strategic networks, other than having access to information, resources, markets, and technologies, firms also have advantages of learning, knowledge sharing, and economies of scale and scope. Such advantages collectively help lift these firms into higher levels of performance and market influence (Gulati et al., 2000). In particular, regarding networks of business entities, Harrigan (1985) and Kogut (1988) look at how interfirm partnerships are formed; Baum and Dutton (1996) investigate firms' behaviors and consequent performances; Forsgren (2016) further explores Uppsala model of the general internationalization process; and Öberg (2018) inspects the concept of innovation and a few relevant topics, among others. Furthering this literature, among other results, this chapter establishes

- Why firms in the position of systemic holes in a supply-demand network represent control and profitability;
- Why expanding market influence helps a network of business organizations improve its member firms' levels of profitability; and
- Why firms can benefit enormously from creating and maintaining platforms that directly connect sellers and buyers.

As for the relationship between information and wealth creation, literature searches did not provide much information. The small body of the relevant literature can be grouped into the following two groups—one on the relationship between an organization's interior and exterior, and the other on information and business successes. In particular, by an organization's interior, it means the knowledge on the organization's characteristics; by exterior, it stands for the state of sharing and exchange of information, knowledge, and goods with the outside world. By using the database of Standard & Poor's (S&P) 500 of the time period from 1996 to 2009, Datta (2016) suggests that at the micro-firm level there was an optimum recombination of exterior sourcing and technology distinctness. Uzuegbunam et al. (2018) investigate the association between national culture and global connectedness with the creative destruction process at the macrolevel of nations. This chapter develops hypotheses about a nation's adoption of creatively destroying technologies based on cultural dimensions of uncertainty avoidance and cultural tightness (vs. looseness), and about global connectedness dimensions of depth and breadth. These authors find significance for interaction between tightness and dimensions of global connectedness by using data of agricultural biotechnology adoptions in 58 nations.

As for the literature on information and business successes, by using the data from an in-depth interview with seven providers and consumers of electric cars, Stryja and Satzger (2018) suggest to design and use particular information systems so that potential consumers are able to test new innovations before they make their purchase decisions of the consequent products. By surveying the literature, Li and Nguyen (2017) review the related studies on when firms engage in collaboration opportunities that involve information, and what critical elements are for business success. This chapter highlights works in various angles, including economies of scale, knowledge sharing, market size and volatility, selection of strategic partners, intellectual property rights, spillover effects, collaboration costs, trust and commitment, opportunism, and overall collaboration strategies.

Enriching this literature on information and wealth creation, this chapter establishes the general fact that increasing information circulation accelerates the speed of emergence of creative destructions. By joining this result and Theorem 12.1, this chapter shows that market information and shown consumer demand jointly stimulate entrepreneurs to produce their market offers based on their individually disparate circumstances and varied cognitive aptitudes. Various incremental and disruptive innovations appear collectively on top of market information, consumer demands, and correspondingly varied entrepreneurs' efforts.

The next two sections consider a firm's potential to capture value. The first section looks at value-capture potentials that emerge from the market state of mutual forbearance, supply-chain networks, and strategic blocks. The following section examines value-capture opportunities created by interorganizational networks and platforms that directly connect sellers and buyers.

13.3 Market Forbearance and Network Structures

In the presently globalizing economy, at the time when firms compete autonomously with each other for advantages and profits, they are also players in interorganizational networks. By being part of such networks, firms team up in partnership (Galaskiewicz & Zaheer, 1999).

By a strategic network, it means a network of various business entities that are connected together by various interorganizational ties. It is strategically organized to gain certain market advantages for all partner firms (Gulati et al., 2000). For example, a firm is a system with supplies as inputs and products as outputs. So, it is naturally a part of a supply chain, known as the firm's ecosystem (or network), composed of its upstream components and downstream complements (Adner & Kapoor, 2010). In particular, a firm's suppliers represent some of its upstream components; customers, supporters, and assistants who help to make the product of the firm available to consumers are some of the firm's downstream complements. Although supporters and assistants are not exactly any part of the firm's direct supply chain, in the ecosystem of the firm, they need to help provide necessary infrastructure so that the product(s) of the firm can realistically reach their ultimate consumers.

Many firms recognize the role strategic networks play in their access to information, markets, and technologies (Gulati et al., 2000), opportunities of risk sharing, economies of scale and scope (Shapiro & Varian, 1999), knowledge sharing and advanced learning (Dyer & Nobeoka, 2000). The nature of interdependence within each network helps the member firms of the network harvest the ensuing benefits (Blankenburg et al., 1999).

Proposition 13.1 The existent state of mutual forbearance among the incumbent firms in an oligopolistic industry helps the incumbents gain improved profitability until the time when the state of mutual forbearance is disrupted.

By a state of mutual forbearance, it means such a market situation that the intensity of incumbent firms' rivalry is lessened by a market partition according to the individual strengths of the incumbents (Bernheim & Whinston, 1990). As a result, each incumbent yields stronger incumbents in those market sectors where it is less competent (Li & Greenwood, 2004) so that these incumbents become inter-reliant in their joint dominance of the market. Such market state and consequent benefits encourage the incumbents to detain their rivalries (Yu & Cannella Jr., 2012), reducing the number of entries into and exits from the market (Fuentelsaz & Gómez, 2006), while interfirm hostility declines (Haveman & Nonnemaker, 2000).

If a state of mutual forbearance exists in an oligopolistic industry, as assumed, the incumbent firms enjoy a wide-ranging influence on the industry's profitability due to their collective dominance in the marketplace relative to upstream components and downstream complements, and deepened levels of collusion, tacit, etc. Because sturdy barriers deter any firm's attempt to enter into the industry, the profitability of the incumbents improves as the state of mutual forbearance strengthens. So, such a scenario of profitability can be seen as relatively sustainable until at least some of the incumbents can no longer stay in sync with consumers' forever evolving preferences and tastes (Theorem 12.1). That is when the existing state of mutual forbearance breaks down.

Related to this proposition, Scherer and Ross (1990) and Galaskiewicz and Zaheer (1999) provide empirical evidence. They demonstrate that dense and strong ties among the incumbent firms of an industry or a market could be conducive to oligopolistic coordination, tacit or otherwise. In the banking industry, Podolny (1993) confirms that associations of top-tier banks help their member banks maintain good returns.

Assume that $S = (M, R)$ is a connected system, for details, see Chap. 2. If an object $m \in M$ satisfies

$$S' = (M - \{m\}, R|(M - \{m\})) = S_1 \oplus S_2 \oplus \ldots \oplus S_k,$$

where $S_i = (M_i, R_i)$, $i = 1, 2, \ldots, k$, is a connected system and $M_i \cap M_j = \varnothing$, for $i, j = 1, 2, \ldots, k$, satisfying $i \neq j$, for a whole number $k \geq 2$, then the object m is referred to as a (systemic) hole of the original system S. Speaking less symbolically, if removing an object from a connected system makes the system disconnect, then the particular object is a (systemic) hole of the system.

To understand this concept intuitively, assume that $S = (M, R)$ represents the systemic structure of a complete supply chain with the topmost upstream players being the providers of raw materials and the bottom-most downstream players the consumers or the end users. Specifically, in this systemic model of the supply chain, the object set M consists of all players of the supply chain, such as suppliers, producers, retailers, and consumers. And, the relation set R describes how these players are connected with each other through their inputs and outputs. For example, two players m_1 and m_2 are considered associated, provided that m_1's products are supplies of m_2 either directly or indirectly.

If a particular player $m \in M$ is a hole in this supply-chain system S, then removing m will make this supply-chain system S break up into two disjoint subsystems as follows, neither of which constitutes a complete supply chain, as follows:

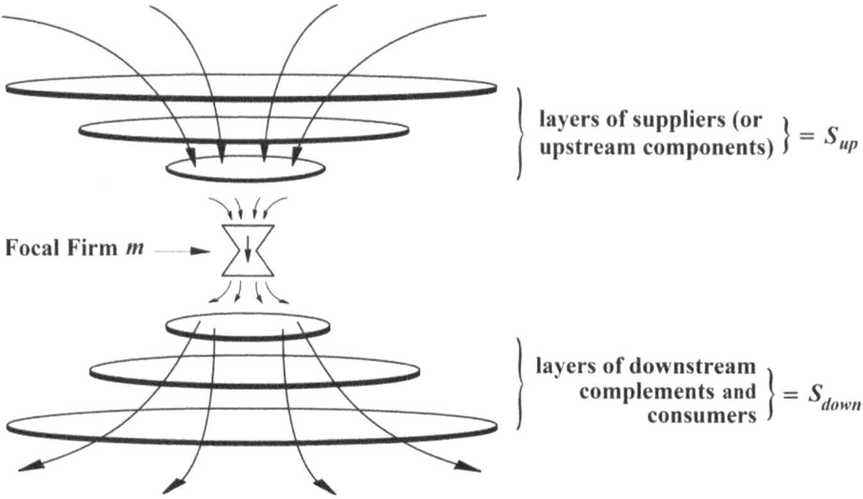

Fig. 13.1 A systemic hole in a supply-chain ecosystem

$$S_{\text{up}} = (M_{\text{up}}, R_{\text{up}}) \text{ and } S_{\text{down}} = (M_{\text{down}}, R_{\text{down}}),$$

where the object set M_{up} is composed of all players that are upstream to m, and the object set M_{down} all players that are downstream to m.

These two sub-supply chains S_{up} and S_{down} are not complete supply chains, because the outputs of S_{up} cannot satisfy the demands of S_{down}, while S_{down} cannot produce its product(s) without the input from S_{up}. In terms of business, if any two trading parties in the customer-supplier network of an industry can only close their transactions by going through a specific focal industrial sector, then this industrial sector is a systemic hole in the entire supply-demand network. Note that a supply-demand network may have several systemic holes. This concept of systemic holes is shown in Fig. 13.1, where m stands for such a hole along with the disjoint sub-systems S_{up} and S_{down}, where the large arrows indicate the direction of the input-output flows of the supply chain S.

Proposition 13.2 Each systemic hole naturally carries power and profitability within the supply-demand network which it is a part of due to its location advantage.

To show this proposition, the general supply-demand network is modeled by the connected system $S = (M, R)$. All players in the network are objects in the object set M; all supply-and-demand relationships of the players are elements in the relation set R. Let an object $m \in M$ be a systemic hole in S. Then, the locational advantage of the object m implies that removing m from the supply-and-demand system S will break S into two subsystems S_{up} and S_{down}, as defined previously, neither of which is functional. In particular, players in S_{up} cannot find the ultimate buyers of its products, while players in S_{down} cannot be served with necessary supplies. In other words, this object $m \in M$ controls the livelihood of all other players in S. Such power of life or death of the object m surely provides a solid foundation for its elevated level of profitability.

Related to this proposition, Burt (1992) empirically finds that these industrial sectors that occupy systemic hole positions are gifted with good returns by taking proportionally over-weighted large stocks of resources.

Proposition 13.3 A firm can boost its profitability through its membership in a supply-chain network of expanding consumer base by improving its market position in terms of a broad web of resource flows.

This proposition is really a corollary of Theorem 12.1. In particular, the assumed supply-chain network of expanding consumer base implies that no competitor is taking market territory away from this network (Theorem 12.1). Instead, this network is expanding its market share over time. These facts jointly mean that the market position of each member firm of the network can, at least potentially, enlarge, which in turn potentially boosts the profitability of the firm.

Conversely, for a supply-chain network of a dwindling consumer base, each of its member will unfavorably experience shrinking stock of available resources, leading potentially to a lowering level of profitability. To this end, a good example is the situation of reducing investment in a nation's military. The deteriorating capital input will make the national defense industry suffer.

Assume that the industry of concern is modeled as a symbolic system $S = (M, R)$ so that (1) each object $m \in M$ represents the unique systemic expression of a firm within the industry and each firm within the industry has a systemic expression as an object in M; and (2) each relation $r \in R$ describes how relevant firms are related to each other through business exchanges. Based on discussions above and what is studied in Chaps. 4 and 5, Such systemic modeling can be done. For example, each firm can be identified with the unique system of its resources; and the entire industry is then simply the totality of all these firm systems connected to each other by their business connections and dealings.

Assume that this industry system can be written as the following free sum of n partial systems, for details, see Chap. 2,

$$S = i = 1_n^{-} S_i = S_1 \oplus S_2 \oplus \ldots \oplus S_n$$

satisfying that $S_i = (M_i, R_i)$ is connected, $M_i \cap M_j \neq \emptyset$, when $i \neq j$, for $i, j = 1, 2, \ldots$, n. In other words, the industry is composed of n many clusters, each of which includes firms that are strategically associated to each other. In this case, each S_i is called a strategic block within the industry. The terms used here are borrowed from Nohria and Garcia-Pont (1991), who studied the global automobile industry. Speaking in business terms, each S_i represents a strategic alliance of firms, which do not have alliance relationship with other firms in the industry.

Proposition 13.4 A firm's membership in a strategic block within an industry helps the firm reach a higher level of profitability than the case that it is not a member of any strategic block. As for how high a level of profitability a firm can achieve, it varies from one strategic block to another. If a strategic block exerts an expanding market influence, then its membership brings forward a positive effect on the profitability of its member firms.

The first two conclusions are direct consequences of the real-life fact that each strategic block is organized to implement an individually unambiguous strategy with its particular accent in order to gain particular competitive advantages. Because of their individually varying strategies employed, different strategic blocks influence the market differently. That implies varying scales of profitability from one block to another, making firms in different blocks achieve correspondingly different levels of performance. Proposition 13.3 directly implies the third conclusion.

Regarding empirical supports of this proposition, Piskorski and Nohria (1999) find major variations in profitability among firms with membership in different strategic blocks within the venture capital industry. Zaheer and Zaheer (1999) look at the competitive global industry of currency trading through the interactions of banking firms. Though organizing alliances into networks, Baum et al. (2000) confirm by using data from the Canadian biotechnology industry that startups in biotechnology industry can improve their performance if they purposefully join alliances.

13.4 Business Networks and Convenient Platforms

Continuing previous discussions, this section examines the values of membership in interorganizational network(s) and in strategic blocks, and the benefits of establishing and maintaining platforms that directly connect sellers and buyers.

Proposition 13.5 A firm's membership of an interorganizational network or a strategic block generally represents an inimitable, not substitutable resource the firm can potentially mobilize.

The truthfulness of this proposition comes from the purpose of developing interorganizational networks and organizing strategic blocks and Propositions 12.8, 13.1–13.3. In particular, the main purpose for firms to join an interorganizational network includes the maintenance of their levels of profitability and the development and enhancement of their locational advantages in a broader network of resource flows. Similar goals are true for firms to join or organize strategic blocks. At the same time, interorganizational networks and strategic blocks play the role of practical barriers for other firms to enter an industry or to move across the lines of strategic groups. Hence, memberships in either an interorganizational network or a strategic block create prospects, competitive advantages as well as restrictions for member firms.

Also, this proposition follows from Theorem 12.1, which specifically states that as long as the product market possesses sufficient demand, it signals for new competitions and innovations. However, the same signal is received by firms differently due to their respectively varied capabilities of learning, histories and backgrounds, available resources, etc. These different comprehensions of the market signal naturally motivate entrepreneurial firms to develop their respectively different solutions. Therefore, these firms correspondingly demand relevant innovative supplies from their upstream suppliers and downstream complementors in order to successfully produce their market offers (Adner et al., 2013). These differences in understanding market signals, in available resources and in demands on upstream suppliers and downstream complementors all make the membership in an interorganizational network and/or strategic block an imperfectly imitable, imperfectly substitutable resource a firm is able to potentially apply.

Regarding the literature relevant to this proposition, Gulati et al. (2000) develop a related but different conclusion. As for applications, with membership in a network and strategic block, Gulati (1999) confirms a firm's access to relationships and information; Zaheer and Zaheer (1997) recognize a firm's access to knowledge, alertness, and responsiveness; Westney (1993) sees that a greater number of opportunities becomes available to a firm; and Becheikh et al. (2006) realize that additional capital, goods, services, technology, etc., become readily available to a firm. A firm can mobilize these external resources to augment its competitive position in its industry. Due to their distinctive and history-specific developments, interorganizational networks and strategic blocks represent resources that are difficult, if not impossible, for others to imitate or substitute (Gulati & Gargiulo, 1999). That once again shows why membership in an interorganizational network and association with a strategic block are a firm's resources that are inimitable or extremely difficult to imitate by competitors and are not easily substitutable with other resources. This end confirms why memberships in interorganizational networks and strategic blocks are resources a firm can enjoy and potentially mobilize.

By opportunism, it means a circumstantial advantage taking of an observant policy. Such advantage taking is primarily driven by self-interest(s) without considering any underlying principles and consequences for others. Such behaviors generally and assuredly exist with people either individually or collectively no matter what specific case and context are concerned with (Lin & Forrest, 2011). They take place within and on top of the adopted philosophical beliefs and value systems that prescribe what behaviors are considered acceptable and which are not acceptable, wrong, or immoral. This fact provides a way to detect opportunistic behaviors by comparing what is ongoing with the norms, although the standard of morality exists in a market of free competition other than laws and regulations on contracts and transaction settlements. This discussion well explains the concept of opportunism and shows no difficulty in defining this concept, as otherwise claimed by Chen et al. (2002).

Proposition 13.6 If an interorganizational network consists of firms from different industries and strategic block firms within one industry, then firm-level trust in this network and block is generally boosted with firm-level informational asymmetry reduced, and firms' opportunistic behaviors made costly.

By firm-level trust, it stands for the guarantee that no firm takes advantage of any other firm's weakness (Barney & Hansen, 1994). Business dealings generally occur within a continuously existing chain of prior associations, either directly or indirectly, and an extensive web of connections. Hence, important resources and necessary origins of referrals are commonly provided through interorganizational networks and strategic blocks with firms' characters as the basis of each referral. Because it generally takes time to have one's characters recognized while it is only a matter of quick moment to have one's recognized characters destroyed (Lin & Forrest, 2011), firms within an interorganizational network and a strategic block tend to behave professionally within the boundaries of the norms and principles of these organizations. That is the reason why interorganizational networks and strategic blocks generally increase firm-level trust and make opportunistic behaviors costly.

Through membership in such networks and blocks, member firms strengthen their business ties, interact with each other frequently to share respective capabilities and resources and to overcome vulnerabilities. The very nature of such networks and blocks empowers their member firms to acquaint with each other very well. These are the reasons why informational asymmetry of member firms within such networks and blocks is reduced.

Regarding the literature, Gulati et al. (2000) were the first people who established the first draft of this proposition, even though it was phrased very differently.

Based on the previous discussion, it can be seen that when firm-level trust and opportunism costs are elevated, the costs of business transactions get lower. And when informational asymmetry between firms increases, transaction costs also rise. Speaking differently, market-level worries decrease with any of the following: improving firm-level trust, reducing informational asymmetry, and elevating opportunism cost. Hence, the result below follows naturally:

Proposition 13.7 Assume the same as that in Proposition 13.6. Then by being a member of this network and strategic block, a firm can boost its value creation and capture.

By borrowing the ideas behind the concepts of interorganizational networks across industries and strategic blocks within an industry, the concept of platforms appears. In particular, a platform is such a network with members from the product market that connect competing sellers with contending buyers directly. Figure 13.2 graphically shows a platform within the product market with the focal firm m being a systemic hole. Then, the conclusion below follows naturally.

Proposition 13.8 If a focal firm develops and maintains a platform that simultaneously match each customer with multiple suppliers and each supplier with multiple customers, then this firm is able to capture value handsomely until similar platforms become widely available.

This proposition is a corollary of the following result developed by Forrest and Anderson (2017): In the product market, finding decided buyers represents a great challenge for sellers due to the wide availability of similar products and substitutes;

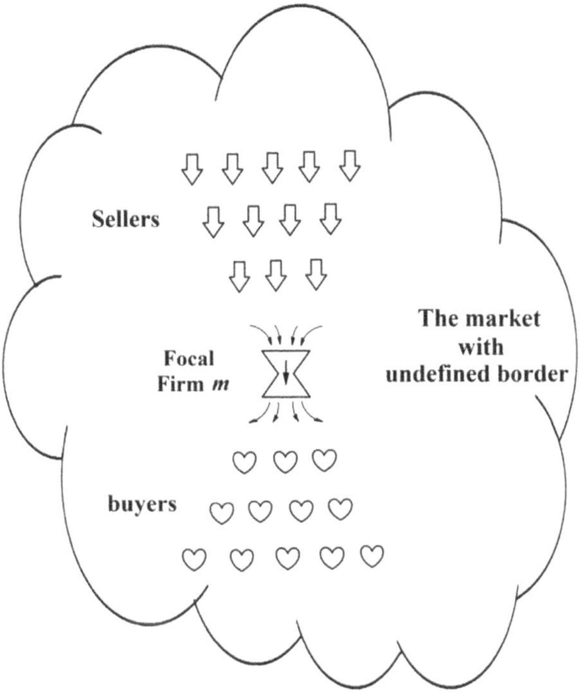

Fig. 13.2 A platform that directly connects sellers and buyers

locating willing takers of bargaining bid prices denotes a major struggle for buyers. In particular, in an afore-described platform that directly matches sellers and buyers, the provided convenience helps all involved parties save hugely in terms of information costs, bargaining costs, policing and enforcement costs, as well as time spent on searches. Hence, any firm that develops and maintains such a platform that connects each bid price to multiple sellers and each offer to multiple buyers has practically created values for both sides of the product market. Additionally, this firm will continuously capture its value until such a time when its service is no longer a systemic hole within the network of sellers and buyers (Proposition 13.2). At that time, all platform providers divide the total value potentially capturable among themselves.

Also, this proposition can follow from Proposition 12.3. Specifically, each platform, which directly connects sellers and buyers in multiplicity, greatly improves the transactions efficiency of the firm that maintains the platform.

Within the present landscape of the business world, priceline.com, among many online business operations, is such a platform as discussed above. It directly connects sellers and buyers by directly forwarding consumers' desired prices to multiple potential sellers. If a willing seller takes a bid price, then the business exchange is complete. The item of concern, such as a rental car, reservation of a resort hotel, or a plane ticket, is sold over the Internet (Amit & Zott, 2001).

13.5 Information and Emergence of Creative Destructions

To lay the necessary foundation for the introduction of the general value-chain framework, this section looks at the relationship between the availability and circulation of information and the appearance of creative destructions.

Proposition 13.9 The more available information is, the more quickly creative destructions emerge. Speaking differently, accelerating circulation of information increases the frequency for creative destructions to appear.

To see the truthfulness of this proposition, let us get a sense of how information underlies economic development. In primitive times of history, humans survived by making use of whatever was immediately available within their vicinities. At those times, each family had to construct and build what it needed to survive. Motivated by the universal desire for better living conditions beyond survival, information on various know-hows began to pass around from one family to another family, from one tribe to another tribe, and from one geographical location to another location. As flows of information were gradually accompanied by those of goods and then those of both goods and talented people, closed circulations of information, knowledge, and goods appeared over the land as the traveling talents brought these elements back to where they initially started their travels. People started to place orders over longer and longer distances. Accordingly, family-based workshops were organized to satisfy the rising demands from different families, from different tribes, then from other locations. And increasingly larger quantities of products and goods were produced with improving quality and delivered all over the land. That is, the growing demands and increasing supplies gradually led to the cultivation of the initial seeds of modern commerce, although initially only locally in today's standard.

As circulation spheres of information, knowledge, and goods increased to protracted distances, emerging demands from further away places started to appear, forcing the family-based workshops to accelerate their productions. Corresponding to the increasing number of orders to be filled and shipped across lands and waters over further and further outdistances, new and more efficient technologies for mass-production, transportation, and communication were developed. Such technological advancements connected initially isolated and separated local markets together into an economy of growing scale through long-distance shipments of goods and movements of information and knowledge. That is, the disconnected family-based workshops of the earlier times became more related through supply chains, order networks, and delivery routes.

As the circulation of information, knowledge, and goods covered an increasing size of regions, with their much-increased productivity, those small, family-based workshops began to serve their corresponding regional markets. That economic development further elevated consumer demands to a higher level, which reciprocally encouraged the small workshops to advance into large factories backed with advanced technologies. Along the path of such economic developments, industrial development appeared and started to leap forward in gradually larger scales one by one. These leaps in industrial development further stimulated the worldwide spread of information, knowledge, and products (Wen, 2016). The economic history of past several 100 years has revealed the power of accelerating circulations of information, knowledge, and goods, while the circulations induced the emergence of business enterprises of the modern form (Sobel, 1999), no matter whether they are global players or small startups. See Forrest et al. (2018) for an in-depth discussion on the evolution of manufacturing firms.

Returning to the argument for the truthfulness of Proposition 13.9, by creative destruction, it means an evolutionary process an organization goes through by unremittingly transforming its internal structure, be it organizational, economic, or cultural, into a new one (Schumpeter, 1942). From this definition and the discussion in the previous paragraphs, one can see that the forces, exerting on the transformation of an organization's internal structure, are originated from newly acquired information and knowledge and the business firm's longing for survival and growth. Therefore, the strengths of these forces rise with the speed of information circulation. That is, the speedier information circulates throughout the land, the more an organization has to transform its internal structure and the higher frequency creative destructions emerge.

E-commerce of the current business landscape represents a lively example of how Proposition 13.9 plays out in real life. The Internet, through which people from different parts of the world can communicate with each other quickly and conveniently, has merged regional economies into a giant world economy, where long-distance deliveries are facilitated by modern transportation tools without much trouble. Simultaneously, supported by the Internet, various virtual markets are developed to directly connect buyers and sellers from all over the globe. They have ingeniously reshaped and unremittingly transformed traditional commerce. With freely available (or at very low costs) information and knowledge on the Internet, entrepreneurs, be they on the supply side or the demand side, now can innovatively comprehend market signals and create wealth in various unparalleled ways (Ye et al., 2012). They find collaborators through virtual means from different supply chains and different industries, and consumers from different markets over various distances. These unprecedented opportunities provide unconventional means to create and capture values. Hence, Proposition 13.9 implies that as e-commerce further evolves, creative destructions in all areas of life will emerge in rising frequency until such a time when the Internet reaches its full potential.

13.6 The General Value-Chain Framework

By value of a product, it means either how much consumers are willing to pay for that product or the total revenue the product generates (Payne et al., 2017). By value driver for a business, it means any factor that enhances the total value of the business (Amit & Zott, 2001; Brandenburger & Stuart, 1996). To help with the value analysis of a firm, no matter whether it produces products, services, or informational goods, Porter's (1985) value-chain framework can be reformulated as follows:

1. Grasp market signal inventively,
2. Shape accordingly the strategic direction of the firm of concern,
3. Determine the key activities that will practically place the firm onto the chosen path,
4. Design the offer that is expected to satisfy the identified market demand(s), and
5. Compute the values created and captured by each activity.

The first step of this framework is about a firm's leadership that recognizes the market invitation and what it means resourcefully (Theorem 12.1). Based on this resourceful understanding of the market invitation, the leadership determines, based on what resources are available and what the firm is capable of, an appropriate strategic direction for the next stage of development (Forrest et al., 2017). The second step is for the leadership to choose a strategic direction by engaging all relevant stakeholders in the foreseeable crucial activities and decision procedures in order to later on successfully implement the adopted strategy. The third step is to design the product(s) in order to address the interpreted market invitation. And, the fourth step is for the leadership to examine the economic implications of the identified activities and procedures and evaluate the fitness of the imagined products in terms of value creation and capture. With all these due diligences done, the imagined products are produced and presented to the ultimate users in the marketplace.

When there is a market demand of sufficient depth and width, meaning that consumers are willing to pay and the number of such consumers is big, the market invites for research and development (R&D) and for the introduction of innovative products, as implied by Theorem 12.1. On the other hand, each entrepreneur or decision-making manager is bounded by her particular background and her composite of knowledge and experience, which dictates her ability to comprehend the market invitation. Hence, it is important for a firm's leadership to understand resourcefully what the market is signaling so that consequent strategic direction, plan, and actions can be determined in order for the firm to produce and deliver its imagined market offers (McGrath, 2013). In other words, the first step in this general value-chain framework is most fundamental, while the other steps follow naturally, if a firm is driven by its ambition to achieve a better and higher scale of business success.

In practice, the leadership of a firm has to carefully understand what these five steps mean before engaging in any venture. By conducting such preliminary diligence, the leadership of the firm can reconfirm that (1) its understanding of the market signal is adequate, (2) it is able to plan appropriate internal changes, (3) it configures exactly how the acknowledged market demand will be satisfied, and (4) it can pinpoint to exact offers the firm can potentially provide the market and deliver to the targeted consumers.

As for the decision-making and planning of a firm's management, the leadership can employ this value-chain framework to address: (1) what and how should its firm do to accomplish the imagined end goal? (2) What are the perceived activities that need to be carried out for the firm to add value to its catalog of market offers in order to win competitions in the product market? And (3) what can the firm do to align its strategy and operational routines to convert its recognition of the market invitation into actual market offers?

The first part of Question (i) can be split to include such more detailed questions as: what needs to be designed and produced innovatively? what method(s) of production does the firm need to adopt? And others. The second part of Question (1) focuses on operational detail, such as "how could eventual deliveries be accomplished?" "what advertisements does the firm need to employ?" "which market segments could and should the firm enter?" "how could the firm overcome the market entry barriers?" "how should the firm run its overall operation?" and others. All these and related questions need to be novelly addressed so that resultant productivity becomes efficient.

The efforts of answering Questions (2) and (3) have to holistically contemplate matters regarding complementarities of products and the lock-in effect. The former refers to the likelihood of bundling several different, but somehow related products, together for the purpose of creating simultaneous consumer utilities than the sum of selling each of the products separately (Mantovani, 2013; Sheikhzadeh & Elahi, 2013; Yan et al., 2014). The latter lock-in effect motivates consumers and customer firms to take part in repeated transactions, a major value driver along with efficiency, complementarity, and novelty, as empirically identified by Amit and Zott (2001) for an e-business to create value.

If a firm is analyzed by employing this general value-chain framework, the analyst will identify primary and supporting activities of the firm. Primary activities relate directly to the firm's creation of value and supporting activities help the analyst estimate the firm's expected value based on their effects on the performance of primary activities. In other words, primary activities directly involve the design and production of the perceived innovative products, such as internal logistics, operational routines, production procedures, outbound logistics, marketing, sales, and services.

As defined earlier in Chap. 12, a product means such a thing or process, either tangible (physical) or intangible (none physical), which is produced by a firm to meet market demand. That means that one can apply value-chain analysis to examine opportunities of value creation and capture in the manufacturing industry, various virtual markets, and the economic sector of service firms. In other words, the value-chain framework, given above, generalizes Porter's (1985) version (Stabell & Fjeldstad, 1998), and Rayport and Sviokla's (1995) virtual value chain, which consists of gathering, organizing, selecting, synthesizing, and distributing information. In terms of Rayport and Sviokla's virtual value chain, gathering information represents the first step in the virtual value chain above. It is parallel to the step of acquiring raw materials for the case of manufacturing firms. The step of organizing information is equivalent to the initial processing of raw materials in the case of manufacturing. The third and fourth steps of selecting and synthesizing information are equivalent to those of producing the eventual goods in manufacturing. The next step of distributing information is equivalent to that of offering ready-made products to end users. That is, when an analyst employs the general value-chain framework above to investigate a firm, she simply follows a sequence of steps that are parallel to those when she uses Porter's version to study a manufacturing firm.

Other than for purposes of investment, decision-making managers and entrepreneurs can surely use this general value-chain framework to find opportunities for wealth creation. For example, with the emergence of e-commerce, the majority of the traditional brick-mortar business firms also run their e-stores by appropriately managing flows of information. And, many e-enterprises develop physical stores as showcases of their products in order to make their online transactions more effective. That is, carefully understanding the general value-chain framework can help one locate value opportunities. For example, opportunities emerge when informational goods and physical products are creatively combined, when intangible and tangible products are placed side by side, when on-site and off-site services are aligned, when transactions are accommodated creatively, when available resources are tapped into innovatively, when suppliers, stakeholders, and customers are associated in revolutionary ways.

Although a holistic view is emphasized in the general value-chain framework, wealth still needs to be created little by little throughout the steps. These small bites of wealth are created by reducing costs of the firm and consumers by offering differentiated products, while raising consumers' utilities. No matter what, product differentiation is the true driver underneath the value creation (Theorem 12.1), and each differentiation stems from an entrepreneur's creative receipt of the market invitation and implementation of appropriate strategies. The receipt and strategy implementation collectively regulate what to do and how, when, where, as well as responsibility sharing among players within the supply-chain ecosystem, channels, and business divisions.

13.7 A Few Final Words

To succeed in the turbulent business world, firms have to embrace transient competitive advantages in order to create and to capture value by taking advance of unconventional opportunities (McGrath, 2013). For example, newly emerging virtual markets have provided an increasing number of firms to complete at least portions of their transactions online (Forbes, 2018; Forrester Research Report, 2000). By employing technology, direct connections of sellers and buyers have become either free or very affordable, leading to the development of convenient and efficient platforms on which buyers place orders and sellers deliver products (Amit & Zott, 2001). To understand such speedy changes in the business world, this chapter accomplishes two goals: (1) study the potentials of value capture from the respective angles of market forbearance, supply-chain network, strategic blocks, interorganizational networks, and direct seller-buyer platforms; and (2) generalize Porter's (1985) well-received value-chain framework so that the resultant framework can be equally employed to analyzing and investigating firms that offer tangible products, or useful services or informational goods. Due to the novel methodology employed here, the established conclusions in this chapter are expected to provide managers and entrepreneurs with a dependable way to make their decisions instead of suggestions of limited validity derived from anecdotes and data.

In particular, from the angle of networks and strategic blocks, this chapter demonstrates how the market state of mutual forbearance is positively related to sustainably increased profitability, why the systemic hole position of a firm provides the firm with advantages and economic benefits, and how the membership in a strategic block of increasing market influence positively affects the profitability of a firm. And, in terms of direct association of sellers and buyers, this chapter shows that each firm that creates and maintains convenient platforms will readily create and capture values.

Regarding practical applications, conclusions developed in this chapter provide general recommendations, rather than suggestions for decision-makers. Specifically, this chapter recommends managers to potentially create additional value from available resources by encouraging information sharing and idea exchanges (Proposition 13.9). Being able to do so will lead firms to their idiosyncratic products (Theorem 12.1) while locating them in systemic hole locations (Proposition 12.10). To increase the likelihood of success, join one or several networks and strategic blocks with expanding market territories (Propositions 12.11–12.15). To maintain the level of performance, help develop a market state of mutual forbearance with competitors by either joining or organizing interorganizational networks (Propositions 12.9, 12.11–12.15). Provide, manage, and recurrently improve platforms that directly connect buyers and sellers both conveniently and massively (Proposition 12.16).

All logical reasoning in this chapter implicitly assumes that a firm exists for the purpose of filling a market niche by generating positive cash flows from the product marketplace. That opens doors for future research on firms that are not established for this said purpose. At the same time, this assumption will also become a potential reason for why some applications of the conclusions developed in this chapter might not work in practice.

References

Adner, R., & Kapoor, R. (2010). Value creation in innovation ecosystems: How the structure of technological interdependence affects firm performance in new technological generations. *Strategic Management Journal, 31*(3), 306–333.

Adner, R., Oxley, J. E., & Silverman, B. S. (Eds.). (2013). *Collaboration and competition in business ecosystems*. Emerald.

Amit, R., & Zott, C. (2001). Value creation in e-business. *Strategic Management Journal, 22*(6/7), 493–520.

Barney, J. B., & Hansen, M. H. (1994). Trustworthiness as a form of competitive advantage. *Strategic Management Journal, 15*(Winter Special Issue), 175–190.

Baum, J. A. C., Calabrese, T., & Silverman, B. S. (2000). Don't go it alone: Alliance network composition and startups' performance in Canadian biotechnology. *Strategic Management Journal, 21*(3), 267–294.
Baum, J. A. C., & Dutton, J. (1996). The embeddedness of strategy. In P. Shrivastava, A. S. Huff, & J. E. Dutton (Eds.), *Advances in strategic management* (Vol. 13, pp. 3–40). JAI Press.
Becheikh, N., Landry, R., & Amara, N. (2006). Lessons from innovation empirical studies in the manufacturing sector: A systematic review of the literature from 1993 to 2003. *Technovation, 26* (5), 644–664.
Bernheim, B. D., & Whinston, M. D. (1990). Multimarket contact and collusive behavior. *RAND Journal of Economics, 21*(1), 1–26.
Blankenburg, H. D., Eriksson, K., & Johanson, J. (1999). Creating value through mutual commitment to business network relationships. *Strategic Management Journal, 20*(5), 467–486.
Brandenburger, A. M., & Stuart, H. (1996). Value-based business strategy. *Journal of Economics and Management Strategy, 5*, 5–25.
Burt, R. S. (1992). *Structural holes: The social structure of competition*. Harvard University Press.
Chen, C. C., Peng, M. W., & Saparito, P. A. (2002). Individualism, collectivism, and opportunism: A cultural perspective on transaction cost economics. *Journal of Management, 28*(4), 567–583.
Datta, A. (2016). Antecedents to radical innovations: A longitudinal look at firms in the information technology industry by aggregation of patents. *International Journal of Innovation Management, 20*(7). https://doi.org/10.1142/S1363919616500687
Dyer, J. H., & Nobeoka, K. (2000). Creating and managing a high-performance knowledge-sharing network: The Toyota case. *Strategic Management Journal, 21*(3), 345–367.
Forbes. (2018). How can businesses adapt to a rapidly changing world? https://www.forbes.com/sites/quora/2018/01/05/how-can-businesses-adapt-to-a-rapidly-changing-world/#4edbc26a5930, Accessed on Sept 17, 2019.
Forrest, J. Y.-L., & Anderson, M. (2017). Customers' indecisiveness: A pricing strategy approach. *Pennsylvania Economic Review, 24*(2), 46–64.
Forrest, J. Y.-L., Baizakov, S., Haile, Y. G., Gyan, A., & Liu, Y. (2020). Potentials of value creation and capture based on innovation, resources, strategic networks and blocks. *Advances in Systems Science and Applications, 20*(3), 50–72.
Forrest, J. Y.-L., Buttermore, J., & Wajda, T. (2017). At Nash equilibrium when new market competitions appear? *Kybernetes, the International of Cybernetics, Systems and Management Science, 46*(2), 256–271.
Forrest, J. Y.-L., Liu, Y., Golden, J., Sissoko, Y., & Li, Q. X. (2020). A reformulated value chain framework useful for different kinds of markets. *Journal of Business and Economic Studies*, accepted for publication.
Forrest, J. Y.-L., Tucker, R., Lin, C. C., & Mondal, S. (2018). Impacts of manufacturing firms' culture, structure and leadership on innovation: A systemic approach. *Proceedings of the 41st annual meeting of northeastern association of business, economics and Technology*, State College, PA.
Forrester Research Report. (2000). Global eCommerce approaches hypergrowth. April 2000.
Forsgren, M. (2016). A note on the revisited Uppsala internationalization process model – The implications of business networks and entrepreneurship. *Journal of International Business Studies, 47*(9), 1135–1144.
Fuentelsaz, L., & Gómez, J. (2006). Multipoint competition, strategic similarity and entry into geographic markets. *Strategic Management Journal, 27*(5), 477–499.
Galaskiewicz, J., & Zaheer, A. (1999). Networks of competitive advantage. In S. Andrews & D. Knoke (Eds.), *Research in the sociology of organizations* (pp. 237–261). JAI Press.
Granovetter, M. (1985). Economic action and social structure: A theory of embeddedness. *American Journal of Sociology, 91*, 481–510.
Gulati, R. (1998). Alliances and networks. *Strategic Management Journal, 19*(4), 293–317.
Gulati, R. (1999). Network location and learning: The influence of network resources and firm capabilities on alliance formation. *Strategic Management Journal, 20*(5), 397–420.

Gulati, R., & Gargiulo, M. (1999). Where do interorganizational networks come from? *American Journal of Sociology, 104*(5), 1439–1493.

Gulati, R., Nohria, N., & Zaheer, A. (2000). Strategic networks. *Strategic Management Journal, 21* (3), 203–215.

Harrigan, K. R. (1985). *Strategies for joint ventures*. Lexington Books.

Haveman, H. A., & Nonnemaker, L. (2000). Competition in multiple geographic markets: The impact on growth and market entry. *Administrative Science Quarterly, 45*(2), 232–267.

Hitt, M., & Ireland, R. D. (2017). The intersection of entrepreneurship and strategic management research. In D. L. Sexton, H. Landstrom, & C. Jones (Eds.), *The Blackwell handbook of entrepreneurship* (pp. 45–63). Blackwell.

Kogut, B. (1988). Joint ventures: Theoretical and empirical perspectives. *Strategic Management Journal, 9*(4), 319–332.

Li, M., & Nguyen, B. (2017). When will firms share information and collaborate to achieve innovation? *The Bottom Line, 30*(1), 65–86. https://doi.org/10.1108/BL-12-2016-0039

Li, S. X., & Greenwood, R. (2004). The effect of within industry diversification on firm performance: Synergy creation, multi-market contact and market structuration. *Strategic Management Journal, 25*(12), 1131–1153.

Lin, Y., & Forrest, B. (2011). *Systemic structure behind human organizations: From civilizations to individuals*. Springer.

Mantovani, A. (2013). The strategic effect of bundling: A new perspective. *Review of Industrial Organization, 42*, 25–43.

McGrath, R. G. (2013). *The end of competitive advantage: How to keep your strategy moving as fast as your business*. Harvard Business Review Press.

Nohria, N., & Garcia-Pont, C. (1991). Global strategic linkages and industry structure. *Strategic Management Journal, 12*(Summer Special Issue), 105–124.

Öberg, C. (2018). The role of business networks for innovation. *Journal of Innovation & Knowledge*. https://doi.org/10.1016/j.jik.2017.10.001

Payne, A., Frow, P., & Eggert, A. (2017). The customer value proposition: Evolution, development, and application in marketing. *Journal of the Academy of Marketing Science, 45*, 467–489.

Piskorski, M. & Nohria, N. (1999). Allocation to open and closed portfolios. Harvard Business School working paper.

Podolny, J. (1993). A status based model of market competition. *American Journal of Sociology, 98*, 829–872.

Porter, M. E. (1985). *Competitive advantage: Creating and sustaining superior performance*. Free Press.

Priem, R. L., Wenzel, M., & Koch, J. (2018). Demand-side strategy and business models: Putting value creation for consumers center stage. *Long-Range Planning, 51*(1), 22–31.

Rayport, J. F., & Sviokla, J. J. (1995). Exploiting the virtual value chain. *Harvard Business Review, 73*(6), 75–85.

Scherer, F. M., & Ross, D. (1990). *Industrial market structure and economic performance*. Houghton Mifflin.

Schumpeter, J. A. (1942). *Capitalism, socialism, and democracy*. Harper.

Shapiro, C., & Varian, H. R. (1999). *Information rules: A strategic guide to the network economy*. Harvard Business School Press.

Sheikhzadeh, M., & Elahi, E. (2013). Product bundling: Impacts of product heterogeneity and risk considerations. *International Journal of Production Economics, 144*, 209–222.

Sobel, R. (1999). *When giants stumble: Classic business blunders and how to avoid them*. Prentice Hall.

Stabell, C. B., & Fjeldstad, O. D. (1998). Configuring value for competitive advantage: On chains, shops, and networks. *Strategic Management Journal, 19*(5), 413–437.

Stryja, C., & Satzger, G. (2018). Try before you buy: How to design information systems to enhance consumer willingness to test sustainable innovations. *Journal of Technology Management & Innovation, 13*(1). https://doi.org/10.4067/S0718-27242018000100019

Uzuegbunam, I. S., Geringer, J. M., & Oberst, C. (2018). Creative destruction within and across countries: Do culture and connectedness matter. *Academy of Management Proceedings, 2018* (1). https://doi.org/10.5465/AMBPP.2018.97

Vendrell-Herrero, F., Bustinza, O. F., Parry, G., & Georgantzis, N. (2017). Servitization, digitization and supply chain interdependency. *Industrial Marketing Management, 60*(January), 69–81.

Wen, Y. (2016). *The making of an economic superpower: Unlocking China's secret of rapid industrialization*. World Scientific.

Westney, D. E. (1993). Cross-Pacific internationalization of R&D by U.S. and Japanese firms. *R&D Management, 23*(2), 171–181.

Yan, R. L., Myers, C., & Wang, J. (2014). Bundling products to success: The influence of complementarity and advertising. *Journal of Retailing and Consumer Services, 21*, 48–53.

Ye, G. L., Priem, R. L., & Alshwer, A. A. (2012). Achieving demand-side synergy from strategic diversification: How combining mundane assets can leverage consumer utilities. *Organization Science, 23*(1), 207–224.

Yu, T., & Cannella, A. A., Jr. (2012). A comprehensive review of multimarket competition research. *Journal of Management, 39*, 76–109.

Zaheer, A., & Zaheer, S. (1997). Catching the wave: Alertness, responsiveness, and market influence in global electronic networks. *Management Science, 43*(11), 1493–1509.

Zaheer, A. & Zaheer, S. (1999). The structure of global competition: A network approach. Paper presented at the SMJ Special Issue Conference, Northwestern University.

Part IV
Customer Values

Chapter 14
When Effective Consumer Value Propositions Emerge

Abstract Customer value proposition (CVP) is the key for a startup to be launched and foundation for a company to direct its business efforts. However, as of the present day, no systematic theory exists regarding how an effective CVP can be potentially formulated and strategically employed in real life. As a matter of fact, the very concept of CVP is still not properly defined; although the term has been widely used in the literature, it is only nonchalantly tossed around. Considering the importance of this concept in the area of marketing, this chapter, mainly based on (Forrest et al., to appear, Conditions under which adopted consumer value propositions are effective), attempts to lie down the first step towards the development of the badly needed theory of customer value propositions (CVP). It establishes four generally true facts on the use and effectiveness of CVPs based on Theorem 3.2 regarding how an oligopoly market signals its invitation for additional competition and new innovation.

Keywords Dissatisfied customer · Market invitation · Operating process · Supply-chain ecosystem · System

To be more specific, this chapter demonstrates that (1) an effective CVP stands for a tool of value creation and capture for a company to employ readily and strategically; (2) when an adopted CVP is effective; and (3) how a company can actually deliver promises by materializing its CVP. After the logical discussions of the basics, a real-life case is analyzed to show how the established results are underlying the business conduct of a publishing house. At the end, provided are some recommendations for practical applications and further issues for future research.

14.1 Introduction

It is on a particular customer value proposition (CVP) that a company effectively organizes itself (Webster, 2002); and it is on a specific effective CVP that an enterprise creates values for customers (Payne & Frow, 2005). Anderson et al.

J. Y.-L. Forrest, Y. Liu, *Value in Business*, Contributions to Management Science,
https://doi.org/10.1007/978-3-030-82898-1_14

(2007), for example, document how distinctive value propositions have led Sonoco Products to achieve its phenomenal performance, where its customer value propositions were formulated in the first place by methodically evaluating market demands.

Although in the area of marketing, Starch (1914) had studied the concept of propositions at least 100 some years ago, the necessity to communicate created values to customers, as a practically important way to potentially capture value for a company, is not accentuated until the 2010s (Marketing Science Institute, 2010). Accordingly, the concept of CVPs is still poorly defined, unconcernedly talked about, and discussed inconsequentially (Lanning, 2003; Payne et al., 2017). To possibly help improve this situation in the development of the theory and application of CVPs, by basing on some of the game-theoretic results developed in the previous chapters, this chapter utilizes the intuition of systems research to establish several sufficient conditions that lead to effective CVPs for a firm to acquire additional profit beyond the case without the CVP. Because the method of reasoning employed here is neither statistics-based nor anecdote-based, conclusions established in the following sections are anticipated to produce visible economic benefits for managers and entrepreneurs.

The rest of this chapter is organized as follows: Section 14.2 looks at the relevant literature and how this chapter enlarges the extant knowledge on CVPs. Section 14.3 systemically models the concept of CVPs. Section 14.4 examines when an adopted CVP is effective. Section 14.5 analyzes a real-life case to show how theoretical conclusions developed here have been consciously or unconsciously applied by business people. Section 14.6 concludes this chapter with recommendations for real-life applications and further questions for future research.

14.2 Literature Review

A literature search on the effectiveness of CVPs produced nothing except a few remotely related publications. This outcome may be well used to further support the fact that although the importance of CVPs has been recognized recently, the concept of CVPs is still not well-defined and no practically useful theory exists for detailed studies of CVPs, as mentioned earlier.

Regarding the few remotely relevant publications, Terho et al. (2012) theorize value-based selling as an operative sales tactic in product markets. Eggert et al. (2018) argue that the literature on value has advanced from an emphasis on exchanges of resources and value in exchange to that on integration of resources and value in use by clearly distinguishing value in exchange and value in use. By recognizing a CVP for the sharing economy business model, Zhang et al. (2019) are able to compare competitive advantages of different CVPs in such economies. By looking at tourist enterprises, Butler et al. (2019) suggest counterparts to the prevalent business models to realize sustainability and to create value for communities. By specifically examining hospitality companies, Shulga and Busser (2019) study in which way and why customers receive willingly these companies' value propositions.

Further developing this small literature, this chapter establishes results, based on a game-theoretic result developed in previous chapters, by employing logical reasoning on how CVPs help a firm with its value creation and capture, and under what conditions an adopted CVP becomes effective.

14.3 Modeling the General Customer Value Proposition

By a CVP, it means a business tool (Payne et al., 2017) that an organization employs to strategically facilitate communication regarding its ability to share resources and offer a superior value package to targeted customers. A CVP is said to be effective to a company, provided that by fulfilling the promises of the CVP the company captures additional net profit when compared to before the CVP is developed and implemented by the company.

To study the effectiveness of a CVP, let us first imagine that a focal company develops and implements a CVP within an oligopoly market of free competition. Assume that the market of concern is a system that is closed from or free of any interference from its environmental factors other than forces of supply and demand. That is, the market's free competition is not influenced in any way by factors of any foreign market beyond supply and demand. Assume that there are n incumbent companies that serve the focal market with their horizontally differentiated offerings, $n = 1, 2...$

To make the setting here as close to real-life situation as possible, assume that each incumbent company has successfully developed a base of loyal consumers; each of these loyal consumers makes purchases only from the company she is loyal to if the company's price is not more than her reservation price. Opposite to loyal consumers, there are also consumers in the marketplace who purchase only from one of those companies that offer the lowest price. The existence of these switchers reflects how consumer's preference and taste evolve unremittingly over time so that once satisfied consumers might very well become discontent. Collectively, this chapter will refer to these switchers as the consumer surplus in the marketplace, as in Chap. 3.

Opposite to the market side of consumers, it is the management of the incumbent companies. Assume that the managerial decision-making of these incumbents is based on their knowledge of the pricing strategies of each other and is developed by applying their best responses and by using the Nash equilibrium through pure self-analyses. Then for this said market, Chap. 3 provides the following game-theoretic result (Theorem 3.2), which is restated here in nontechnical terms for the convenience of the potential audience of this chapter.

Theorem 14.1 In Nash equilibrium, if the proportion of consumer surplus in the oligopoly market defined above is greater than the percentage magnitude of the smallest loyal-consumer base of the incumbent companies, then at least one company can profitably enter the market with its substitute offering so that the expected aggregate profit of all entrants can be potentially greater than that of the incumbent with the smallest base of loyal consumers.

In terms of CVPs, Theorem 14.1 means that seeing the existence of a good magnitude of consumer surplus, a visible market invitation for new innovations and additional competition, companies, incumbent or not, need to creatively formulate the knowledge regarding what the switchers truly look for. For example, their indeterminacy on from whom they make their purchases can be seen as a sure sign that these consumers are not pleased with the available offers in the marketplace. However, due to whatever reason, such as necessities of living, these people have to make purchases from one incumbent or another. Hence, companies in such a situation need to mobilize their learning and absorption capabilities to formulate their appropriate CVPs in order to clear their directions for the next phase of development.

The definition of CVPs, as initially phrased by Payne et al. (2017), characterizes the concept from four different angles: (1) a CVP is a communication device, (2) it promises resource sharing, (3) it identifies targeted customers, and (4) it pinpoints to the superior value of a company's offer from those of competitors. The first angle shows the functionality of a CVP that brings to the targeted customers the awareness of the value a particular offer provides, along with such a message about how the customers can help improve the economic, functional, emotional, and symbolic features of the offer. The said value that is to be communicated here is initially within the imagination of the supplier, as focused on in early studies (e.g., Kowalkowski, 2011), surely needs to be passed on to consumers, as noted by Rintamaki et al. (2007) in the retailing realm. Beyond delivering underlining value to targeted customers, this definition of CVP also stresses the involvement of customers in collaborations. Since a company's effort of creating value can be greatly enhanced by first-hand consumer knowledge, how to communicate customer value has been seen as a priority of research by marketers (Marketing Science Institute, 2010, 2014).

Let us now pay our attention to how the concept of CVPs can be naturally and intuitively seen through the lens of systems thinking and logic. The importance of such lens or intuition cannot be overlooked if we examine knowledge exploration historically. For example, all currently widely employed statistics-based and calculus-based methods in the business area are 100% developed on the intuition of the Cartesian coordinate system (Kline, 1972), although the majority of published works in business disciplines avoid mentioning the geometric intuitions underlying their methods employed. For our purposes here, the systems thinking and logic will help us reason logically based on an intuitively visible "playground," as in the cases of how statistics-based and calculus-based methods are developed in the first place (Forrest, 2018, pp. 187–191).

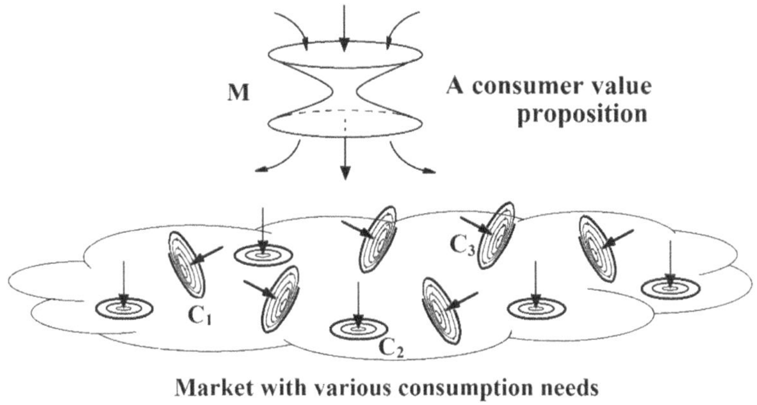

Fig. 14.1 (**a**) Systemic intuition of a general input-output system

As discussed in Chap. 2, by system, it means a whole composed of some objects and some relations that associate the objects in specific ways. In real life, such a whole emerges exactly due to specified relations (Lin, 1999). Hence, it is readily for anyone to see that business entities are systems, and that studies in economics and business are mostly about the evolutions, structures, and interactions of different economic systems (e.g., Porter, 1985). To help effectively investigate the evolution and interaction of systems (or organizations or structures), Lin (2009) introduces the yoyo model as an aid of intuition for researchers to figuratively imagine what might be happening. In particular, one can intuitively see the general input-output system as a multidimensional spinning entity. If that entity is imagined in the three-dimensional space we live in, it would possess the structure shown in Fig. 14.1a from Chap. 2, quoted here for the sake of convenience. The input side sucks in "*things,*" such as information, investment, and human talents, in order to maintain the existence of the system. After having "processed" the "*things*" through the "neck," the entity gives off outputs, such as physical offers, services, and informational goods. The name "yoyo" comes from the general shape of this intuitive structure.

Now, if the concept of CVPs is seen through the lens of this yoyo model, a systemic intuition for this concept naturally emerges as shown in Fig. 14.2. In particular, the focal company is labeled and modeled by the yoyo structure M, which is also identified with the customer value proposition the company adopts. The input-output arrowed flows indicate the communication of the company with the customers in the marketplace, shown as small yoyo fields from the marketplacc. Because of the alignment of their input-output flows with each other, the fields C_1, C_2, and C_3 are among the targeted customers of company M. The idea of resource sharing is implied by the fact that the input-output arrowed flows of company M, associated with those of targeted customers, help strengthen the flows of the yoyo fields of these customers. And the idea of superior value is depicted by the alignment of the yoyo field M with those of the targeted customers.

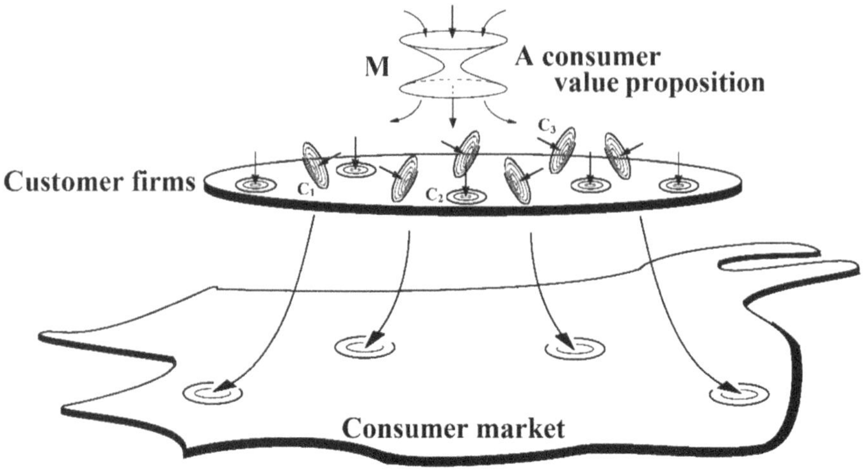

Fig. 14.2 An intuitive depiction of the concept of CVP

Other than the need for such an intuition, as discussed earlier, one benefit of exploiting such systemic modeling and intuition is that the model helps naturally combine disassociated discoveries into one organic whole of understanding. For example, the listed angles as emphasized by the concept of CVP have been in the past only occasionally while not concurrently dealt with by different studies (Ballantyne et al., 2011; Rintamaki et al., 2007). But now, with this yoyo model, they are all naturally and collectively shown in the systemic intuition given in Fig. 14.2.

The theoretical and practical importance of this systemic modeling of the concept of CVPs is the following, as shown by the discussions above:

Proposition 14.1 CVP provides not only a tool for a company to strategically create value for its targeted customers but also an approach for the company to tactically utilize to capture value.

14.4 A CVP's Effectiveness

With the concept of CVP modeled systemically, the question of whether or not a specific CVP is practically effective can be naturally addressed. Once again, the effectiveness here means that a specific CVP actually produces additional profit, after deducting all the costs of developing and implementing the CVP, for a company beyond the case without the CVP.

14.4.1 Value and Differentiation Based CVPs

In the rest of this chapter, by proposition, it stands for one of the following two unrelated situations. First, it means a statement, which is numerically labeled, such as Proposition 1 and Proposition 2, that is arguably true based on logical reasoning initiated on some preliminarily true beginnings. In such a statement, there is always an if-condition, either explicitly given or not within the proposition, followed by a conclusion. And the if-condition sometimes can also be written as an assumption. Second, a proposition stands for a particular selling point or offer as well investigated in the scholarly works of and employed practically in the context of marketing. Examples for this end include the unique selling proposition (Reeves, 1961), the emotional selling proposition (Lindstrom, 2005), customer value proposition (Payne et al., 2017), etc.

The following result outlines conditions under which an adopted CVP, which is developed to deliver superior value for customers and differentiation from competitors, is effective.

Proposition 14.2 If, as its marketing effort, a company formulates such a CVP that describes how it delivers superior value to the marketplace and how it differentiates itself from competitors, then the CVP's effectiveness is closely dependent on

- The magnitude of the market segment of customers who are discontented with the available offers in the marketplace;
- The potential for the proposed value to meet the demands of the discontented (or targeted) customers; and
- The efficiency the company communicates the particular CVP.

To see the truthfulness of this result, let us examine the listed conditions one by one. For the first condition, only with the existence of an appropriately large market segment of discontented customers, a company that aims at this market segment will have the potential of providing its offer at a high enough, as well as competitive enough, price over the cost to achieve appropriate profitability and growth (Theorem 14.1). Additionally, to catch the attention of the currently discontented customers, the adopted CVP of course has to contain the promise of delivering their expected benefits (the second condition); and only when the promised benefits of the CVP are efficiently communicated with the targeted customers (the third condition), the initial potential and promise will turn into a massive number of actual purchases. That is, only when promised benefits are well-communicated along with competitive pricing and the existence of a sufficiently large segment of discontented customers, they can produce the desired level of profit and growth for the company. Systemically speaking, an effective CVP, as depicted in Fig. 14.2 as the yoyo body M, must be able to realistically line up with a solid base of many customer yoyo fields, such as the Cs in Fig. 14.2, in the marketplace in order to sustain itself and to potentially help the company to grow its level of profits.

To see how promises and relevant communications help a company succeed in the real-life business world, one only needs to look at IBM's past dominance in the area of general business computers until the Digital Equipment Corporation (DEC) appeared (Lanning & Michaels, 1988). In particular, IBM's computers were not superior products when compared to those offered by some other companies. At the same time, IBM's price was not competitive, either. However, IBM's earlier dominance relied on its reliability (promised value) and clear communication (the third condition) of this benefit to a targeted large clientele of customers (the first condition). However, when the promised benefits dissipated with the advancement of technology, in particular, when reliability became a nonissue, IBM's dominance in general business computers ended due to the reason that the company could not provide the emerging market demand for interconnectivity among offices and production facilities.

Next, let us look at an application of Proposition 14.2 to such an economic sector that its customers and market demands are more or less well-defined. Examples of such economic sectors include those that deliver necessities to satisfy the very basics of human living. For a company within such an economic sector, if it desires to generate extraordinary profits, it needs to develop and implement such a CVP that focuses on the delivery and efficient communication of distinguishingly superior value to customers, such as organic foods and green products. To practically achieve this, the company needs to understand and stay in sync with customers' perspectives (Berry et al., 2002; Molineux, 2002; Smith & Wheeler, 2002) and use experiences (Lanning, 1998). To illustrate this end, let us, for example, consider a New Jersey-based commerce bank that attempted to make banking experiences fun so that customers would like to visit the bank's branch locations repeatedly. When a new branch just opened, fun-filled events, highlighting comedians, acrobats, foods, caricaturists, and others, were organized for an entire week as a way to advertise itself to the community and to attract potential customers. While both adults and children were attracted to the events, the bank was able to effectively communicate its superior value to customers—opening 7 days a week and most branches operating until 8 pm each weekday, very different from other banks. On the other hand, such efforts may not work out successfully. For example, Tandy Corp. tried in the 1990s to increase the profitability of its superstores in the area of consumer electronics by increasing the interaction between staff and customers. Although an approach similar to Disney's was employed, the effort did not work out successfully. See Morgan and Rao (2003) for the relevant details of these two examples.

From these two examples of very different outcomes, one has to wonder what is making the difference. Do the examples imply that Proposition 14.2 does not hold true generally? The answer to this question is NO. One important reason for the second example to not work out as expected is that there are not a segment of well-defined customers and clear demands in the economic sector of consumer electronics. In other words, consumer electronics is not among the necessities of human living. More generally, this economic sector does not have a well-defined market segment of discontented customers, condition 1 in Proposition 14.2.

14.4.2 CVPs that Are Jointly Created

A CVP is said to be cocreated by a company (or supplier) and customers, if both the company and customers, at least some of them, are involved in the development of the CVP so that the reciprocal benefits, potentially offered to and from the company and some customers, are known to the company and the customers. A cocreated CVP generally addresses concerns of different stakeholders from both supply and demand sides, such as a company' resources, capabilities and investors, and consumers' specific desires—social and environmental preferences. Such approach to developing CVPs, known as an outside-in approach (Day, 2011), can have not only the potential of developing an effective CVP, but also practically establish and fortify an intimate relationship between a company and its major customers (Ballantyne, 2003). A good example of such an intimate relationship is the important associations between upstream suppliers and downstream complements in a supply-chain ecosystem (Adner et al., 2013).

Proposition 14.3 If a company (as a supplier) cocreates a CVP with some of its downstream customer firms so that the company's outputs are supplies of the latter firms, then the CVP's effectiveness to the supplier closely depends on one of the following two conditions:

- Collectively, the customer firms occupy a sufficiently large magnitude of dissatisfied customers with the offers currently available in the marketplace;
- The developed CVP is also attractive to other customer firms so that all interested customer firms collectively occupy a sufficiently large magnitude of dissatisfied customers with offers currently available in the marketplace.

Theorem 14.1 quite readily implies this proposition. Specifically, the supplier's economic fortune is clearly and closely dependent on how well its downstream customer firms do, which is also indicated by the definition of cocreated CVPs. And, how well the downstream customer firms do can only be assured by the existence of a sufficiently large market segment of discontented customers with the offers available in the market (Proposition 14.2). Likewise, this same logic also holds true for the second condition with the following key point: the market's depth or the purchasing power is sufficient to support the supply-chain ecosystem, within which the particular supplier is in some sense the uppermost player, Fig. 14.3. In particular, the firms the particular CVP M attracts, such as those labeled as C_1, C_2, and C_3, have to be well supported by a large number of consumers in the ultimate product market. Without such a sufficiently large number of such consumers, the focal supplier firm cannot be financially sustained. This recognition of the market condition of course has to be considered naturally in the development of the cocreated CVP M.

In terms of practical applications, collectively Propositions 14.1 and 14.2 emphasize that if a company wants its adopted CVP to be effective (or brings in the desired, elevated level of profit), it has to clearly communicate to the targeted customers its differentially promised delivery of superior value. Other than producing strong reverberation with the customers, the communication also needs to stress what

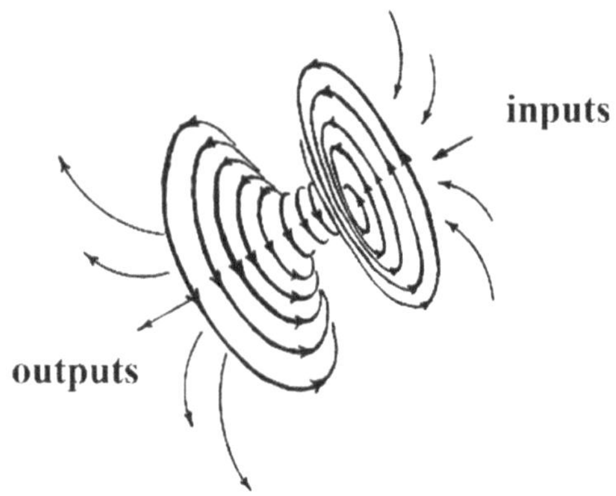

Fig. 14.3 Consumer market depth of a CVP

benefits the targeted customers can expect to receive along with their related costs (Anderson et al., 2006; Rintamaki et al., 2007; Ulaga & Eggert, 2006). As the very principle underneath the concept of cocreated CVPs, the company needs to distribute the created value appropriately across customer relations (Payne et al., 2017).

Another key consideration for the development of a CVP is that the underlying supply-chain ecosystem has to have the necessary context and capability to implement the eventually developed CVP. Without such consideration, it will be extremely difficult for a focal company to deliver the promised superior value, because the delivery of value depends on how well the entire supply-chain ecosystem can meet challenges brought on by the focal company and how the ecosystem's complementors can adequately afford the necessary infrastructure (Adner et al., 2013). For example, Theranos, a now-defunct health technology company, offers an excellent illustration of what is just discussed (Carreyrou, 2018). In particular, as the founder and CEO of the company since its inception in 2003, Elizabeth Holmes formulated her CVP for the company and the targeted consumers: with small amounts of blood pricked from fingertips, the company's small and automated devices can perform blood testing rapidly. The very idea Theranos effectively communicated was that the company had revolutionized blood testing that needed surprisingly small amounts of blood to produce a large number of test results. Since the promises, as described in the CVP and communicated to customers, were not truly delivered, the company started to fall in 2015. Various investigations into the company and its conducts exposed qualms about the company's CVP and whether or not Holmes misled investors and the government.

The US SEC in 2018 charged the company and Elizabeth Holmes of using false or exaggerated claims to mislead investors. And, Holmes was barred from serving any public company for 10 years as an officer or director. The company's communication was extremely effective and successful during its rise from 2003 to 2014. It obtained the support of such influential people as Henry Kissinger, Hillary Clinton, George Shultz, and many others. However, the delivery of promised superior value did not materialize due to the reason that the imagined small, automated device was never successfully developed.

Other than pointing to where Theranos failed, this example also illustrates the importance of cocreated CVPs with upstream suppliers in order to successfully build the vitally important device, for example. In other words, in the B2B market, the competitive environment really forces companies to develop their effective CVPs jointly with their upstream suppliers and downstream complementors. The reason is, as revealed by the case of Theranos, a creative understanding of a market invitation generally challenges suppliers technically to provide innovative supplies, and complementors to design and construct infrastructures that never existed before.

14.4.3 Making an Adopted CVP Effective Internally

As indicated above, such as the definition of CVPs, one reason a company desires to have an effective CVP is that it can attract prospects and retain customers by means of sharing resources and of delivering better value than any of the current and potential rivals. To actually accomplish this goal in real life, the company must introduce nontraditional operating processes (Davenport, 1993; McGrath, 2013) beyond established practice within the organization's domain and control. These new processes will need reconfiguration and rearrangement of related tasks and resources to deliver the promises contained in the adopted CVP. Naturally, the actual operation of these new processes will require the investment of resources, such as personnel, knowledge, and networks. Without such investment, the implemented new processes will not successfully commingle as new capabilities for the organization (Forrest et al., 2019).

For a CVP to be effective for a company, the company has to be capable of inculcating a market invitation into its organization and culture for the purpose of eventual introduction of new and innovative offers, such as products, services, or informational goods (Forrest et al., 2017). That is, an effective CVP directly reconfigures a company's business processes, organizational structure, and culture that in turn contribute overtly to resource sharing and to the creation of promised superior value.

Each company represents a live system with different operational processes working simultaneously and interactively. An effective CVP assists a company with its design and introduction of innovatively new operational processes. To see how such an assistance practically works out, let us concentrate on a few macrolevel processes underneath business tasks, which are basic, common as well as critical for achieving the goal of sharing resources and delivering superior value. To this end, the discussion above, the value-chain framework, Chap. 13, and the dependence of a firm's success on its supply-chain ecosystem (Adner et al., 2013), jointly imply that a company needs to successfully accomplish the following tasks:

1. Specify customer solution(s), either new or revamped, to meet the signaled market demand(s);
2. Acquire adequate supplies and make the desired offers; and

3. Fortify current networks of various stakeholders, such as customer firms and end users, and develop new associations.

Congruently, the company needs to create and manage the following three essential macrolevel operating processes:

- management of customer solutions,
- management of the supply-chain ecosystem, and
- management of customer relations.

The first operating process emphasizes on innovatively interpreting the consumer surplus existing in the product marketplace and inventively advance answers to meet the market demand (Theorem 14.1). It is in this process, a potentially effective CVP is developed and adopted. Each advanced answer can be realistically challenging downstream complementors in terms of materialistically providing necessary infrastructures to help push the solution to reach the end users. The second operating process deals with the concrete production of the desired offer(s). It involves the effort of acquiring necessary supplies and that of bringing together supplies into imagined customer solution(s). The former step here can potentially represent technological challenges to upstream suppliers, while the latter step can be relatively easier because the relevant how-to information should have been considered in the step of advancing inventive answer(s). The third operating process handles various aspects of communication with customers identified when the CVP is initially developed and adopted. It also addresses potential prospects, such as the creation of market knowledge, maintenance, and further development of customer relations, shaping market perception of the organization and its offers.

In real life, each of these macrolevel operating processes consists of several microlevel processes (Day, 1994). In particular, each of these macrolevel processes contains such next-level subordinate processes as human resource management, technology development, and financial supervision. The reason why these latter processes are seen as subordinate to the three macrolevel ones is because when seen as independent processes, they only play their respective supporting roles for the macrolevel processes.

Proposition 14.4 If a company desires to make its adopted CVP effective, it needs to saturate its managers' minds with the promises of the CVP, as well as to mobilize their supports towards adopting the relevant operating processes, necessary for the ultimate delivery of promises, no matter how unusual these processes might be.

Based on the definition of a CVP's effectiveness, for an adopted CVP to be effective to a company, the CVP needs to have a capability to help the company improve its efficiency with some parts of the afore-described macrolevel operating processes. Specifically, the improvement needs to be visible with a reduction in costs, or an increase in productivity, or the production of new/original market offers, or a combination of these enhancements. However, as an input-output system, the very being and operation of a company tend to stay within the influence of the inertia of its past trajectory of development, as shown by the First Law on State of Motion, see Chap. 8 or Lin (2009). That is, for a meaningful improvement to appear in any part of the company's operating processes, the management of the company has to be acted upon by external force(s). To this end, the given setting of the proposition—a CVP has been adopted and the company wants to make it effective—means that there are indeed some market forces that have been acting on the managers of the company. These forces can be on developing market trend in terms of consumer preferences, a new advancement in technology, competitive pressures from rival companies, etc. Therefore, to materially promote necessary changes in the operation of the company, as indicated by the company's desire of making its CVP effective, managers at all levels have to both support and participate in the development and actual introduction of changes to the existent operating processes or of brand-new ones. That sanctions this proposition's conclusion.

14.5 Effectiveness Analysis of a Real-Life Case

This section analyzes the great success of a real-life business case. After sensing a market invitation, the company of concern creatively construes the essence of the invitation. With its specific understanding of the market demand, the company develops its original answer to the market call and then produces the imagined product. As the last step, the company effectively pushes the product into the marketplace to satisfy the creatively comprehended market demand. To help protect the underlying trade secret, if there is any, the name of the company and related details of operation are omitted. For the convenience of communication, this particular company is referred to as Company Z.

To stay current, demand for the offering of new courses appears frequently in higher education. One recent situation is closely related to the concept of big data, emerging since the 1990s (Mashey, 1998). As big data become an important way for businesses to discover next profit opportunities, business schools around the globe have begun to provide related courses to students at undergraduate and graduate levels, such as business analytics, accounting data analytics, and many others. A course along this line generally focuses on topics of how deluges of available data can be processed and relevant decisions can be made regarding trends in consumer needs and market development.

A quickly mushrooming consumer surplus of instructors, seeking suitable textbooks for their designed courses, indicates the presence of a profit opportunity of great magnitude in higher education. Sensing the arrival of such a magnificent market demand, textbook publishers, either large or small, take their forays into this market territory. They develop their relevant CVPs by efficiently mobilizing their respectively available resources to aim at the particular market demand of drastically increasing magnitude.

The focus of this case analysis is the subject of business analytics. This subject covers areas of knowledge that help future managers gain enhanced awareness of business operations and assist them to make better decisions based on facts (Evans,

2020). Such a course joins different conventional disciplines of knowledge into an organic whole, covering statistics, information technologies, principles and methods of management science, mathematical and computer modeling, optimization, among others. To potentially provide their offers to the marketplace of increasing scale in higher education, textbook publishers actively seek potential authors, reviewers, and instructors. They find enthusiastic authors to suggest new textbooks on business analytics and instructors for information on what is considered important for such a course.

A commonly used approach by publishers is to develop a potentially effective CVP in the current content by introducing the best possible solution that will satisfy the market demand. To accomplish this, Company Z put in its diligence by investigating

1. How the ideal textbook should look like in this emerging area of learning,
2. What challenges from the side of instructors and that of students need to be met, and
3. What the best practices there are in related areas of teaching and learning.

Enthusiastic potential authors, either with or without successful track records in textbook writing, are invited to draw up proposals of their envisioned textbooks. Then, the proposals that address most of the comments and suggestions from those involved in the study of the previous three questions are invited to enrich their proposals by including missing suggestions one way or another. This step then leads to several acceptable proposals targeting different teaching methodologies and learning styles. Because a published textbook on business analytics will have to meet multifaceted demands of instructors and students, each accepted proposal is further developed in order to produce an "ideal" final product. In particular, when such a textbook is offered to the marketplace of higher education, if successful, it will serve large bodies of students through instructor adoptions of the textbook for their courses. So, for instructors, the final produced textbook needs to possess the following features and functionalities as much as possible:

- All necessary materials are covered in a conventionally adopted fashion;
- Examples, which are thought-provoking, short, and straight-to-point, are included;
- Real-life applications without unnecessary messy details are provided to demonstrate the relevance of learning to real life;
- Learning objectives are given to entail what knowledge and skills students will acquire from each chapter;
- Relevant data files and software, such as Excel or R commands, are provided; and
- A test bank, power-point presentations, and relevant supports, such as online homework system and online testing and grading facilities, are made available.

From the angle of students, the produced textbook needs to have as many of the following components as possible:

- Indications of relevance to students' chosen business majors;
- Knowledge and skills students will gain from the course;
- Precisely stated and easy-to-understand concepts and illustrative examples;
- Highlighted key concepts and terms to assist students with their test-taking; and
- Online helps, such as information on various supports, help with homework problems, etc.

Consequent to all the market-oriented activities, as described above, Company Z eventually settled down and produced at least one textbook on business analytics with necessary supporting auxiliaries. For the convenience of communication, the rest of this section assumes that Company Z produced only one such textbook. With great efforts placed in making the textbook as ideal as possible, the produced book also provided various ways to address challenges facing instructors and students of the subject matter or at least lessen the severity of these challenges.

As the next step of business operation, Company Z is to offer its textbook to as many potential instructors as possible for their adoptions. Generally, massive sales of the textbook will follow from student purchases.

Due to the fact that Company Z's production of its textbook was entirely based on such a CVP that was cocreated by the publisher and potential customers—instructors and students, most of the customer concerns, if not all, had been taken care of more or less appropriately. Hence, the next step of offering the produced textbook to the entire market of higher education is simply about attracting and convincing as many instructors to adopt the textbook as possible. And, this stage of the business operation—selling the product—is managed in a manner that is parallel to that of the development of initial textbook proposals and the subsequent publication of the textbook. Specifically, groups of instructors that are currently teaching or will be potentially teaching the course of business analytics are first identified by Company Z's sales representatives, who generally have first-hand, face-to-face contracts with college instructors all over the nation. Then these instructors are invited to participate in open, interactive discussions on the following issues, as given in the following order:

1. Goals of teaching in classrooms, challenges instructors face, and what might help with achieving the goals and with overcoming the challenges in today's environment of learning.

   Highlights of the focal textbook and how the textbook will assist instructors to accomplish their goals of teaching and students to achieve academic successes in their learning.
2. What challenges students in such a course face, and what experiences instructors have can be used to improve the solutions already designed to handle the challenges.
3. Assessments and what better tools are available.
4. The increasing market demand for accessibility, mobility, and personalization.
5. Available technologies and how they help improve instruction and engage students.

6. Academic integrity and test proctoring.
7. What can be done to materialize ideals in terms of the instruction and learning of business analytics. In answering this question, each discussant thinks as a budding entrepreneur.

Five representatives: an event planner, a marketer, a product developer, a technician, and a supervisor. From Company Z, participate in activities of this marketing stage along with the invited instructors. Specifically, for discussions to take place smoothly the event planner deals with all the detailed logics, including travels, lodging, foods, conference facilities, entertainments, etc. The marketer heads all sessions of interactive discussions and harmonizes group activities. The product developer familiarizes the discussants with available products and technology and picks up ideas about what can be either improved or introduced. The technician confirms readily accessible technical supports and what additional supports instructors might be looking for. Finally, the supervisor holistically directs the progression of the afore-described marketing activities and looks for ways to improve the effectiveness of the event.

Here is a one-line summary of the event's outcome: Most participating instructors feel that the very textbook presented to them by Company Z is the ideal book for them to use in their classroom instructions because all concerns and issues with regard to such a textbook have been addressed.

14.6 A Few Final Words

Although the importance of CVPs for a company has been well recognized (Marketing Science Institute, 2010), the concept of CVPs is only seriously defined recently by Payne et al. (2017) without the needed theory of CVPs developed to offer practically applicable results. To fill this needed gap in the literature, this chapter attempts to be the first step in the effort of developing such a theory of CVPs, as called for by Payne et al. (2017). One major hurdle in doing so successfully is the adoption of a novel methodology that has the potential of providing the power of logical reasoning while avoiding the constraints of statistics-based and calculus-based methods. Here, statistics-based and calculus-based approaches have been widely employed in areas of marketing and management. The adopted methodology in this research is systems science, the systemic yoyo model, and the kind of logical reasoning as that commonly used in natural science and mathematics (Kline, 1972), very unconventional to the literature relevant to this work. Specifically, for this chapter, the rigor of game theory is mobilized, while phrased in the language familiar to the targeted audience of this work, as the basis of logical reasoning, and systemic modeling as the foundation of intuition. Because of the adoption of such an unconventional methodology, this chapter establishes generally true facts instead of hypotheses, including:

- CVPs place crucial roles in a company's effort to capture value (Proposition 14.1).
- If a CVP is developed and implemented for marketing purposes, the CVP's effectiveness is then closely dependent on the underlying market depth and the efficiency of relevant communication (Proposition 14.2).
- The effectiveness of a cocreated CVP depends on the width of the ultimate market (Proposition 14.3).
- With sufficient internal managerial support, a company can practically deliver the promises of its adopted CVP (Proposition 14.4).

The generally true conclusions established above provide the following managerial recommendations (other than mere suggestions of limited validity): each ambitious company that wishes to flourish in the marketplace needs to bring its adopted CVP into line with what the market surplus signals (Theorem 14.1 and Proposition 14.1). To successfully achieve this alignment, the company's management needs to scrupulously evaluate the depth and width of the identified market segment (Proposition 14.2 and 14.3). And, it needs to beseech full-hearted managerial backing of all levels (Proposition 14.4).

The proof of Theorem 14.1 (or Theorem 3.2), for details, see Chap. 3, suggests that all established conclusions in this chapter hold true only if the existence of all companies of concern is for the reason of filling a market position so that each of the companies is financially sustained by positive cash flows from the product marketplace. In real life, this assumption is not universally true. In other words, there are companies in the business world that exist for reasons beyond what is just given here. Hence, two good questions for future thinking are:

1. For companies that are not established for the said purpose above, how would the previous conclusions established in this chapter look like? and
2. What will the resultant conclusions recommend for front-line, decision-making managers and entrepreneurs?

References

Adner, R., Oxley, J. E., & Silverman, B. S. (Eds.). (2013). *Collaboration and competition in business ecosystems*. Emerald.

Anderson, J., Narus, J., & Van Rossum, W. (2006). Customer value propositions in business markets. *Harvard Business Review, 84*(3), 91–99.

Anderson, J. C., Kumar, N., & Narus, J. A. (2007). *Value merchants: Demonstrating and documenting superior value in business markets*. Harvard Business Press.

Ballantyne, D. (2003). A relationship-mediated theory of internal marketing. *European Journal of Marketing, 37*(9), 1242–1260.

Ballantyne, D., Frow, P., Varey, R. J., & Payne, A. (2011). Value propositions as communication practice. *Industrial Marketing Management, 40*(2), 202–210.

Berry, L. L., Carbone, L. P., & Haeckel, S. H. (2002). Managing the total customer experience. *Sloan Management Review, 43*(3), 85–89.

Butler, R. W., Szromek, A. R., & A.R. (2019). Incorporating the value proposition for society with business models of health tourism enterprises. *Sustainability, 11*(23), 6711. https://doi.org/10.3390/su11236711

Carreyrou, J. (2018). *Bad blood: Secrets and lies in a Silicon Valley startup*. Alfred A. Knopf.

Davenport, T. H. (1993). *Process innovation: Reengineering work through information technology*. Harvard Business School Press.

Day, G. S. (1994). The capabilities of market-driven organizations. *Journal of Marketing, 58*(4), 37–52.

Day, G. S. (2011). Closing the marketing capabilities gap. *Journal of Marketing, 75*(4), 183–195.

Eggert, A., Ulaga, W., Frow, P., & Payne, A. (2018). Conceptualizing and communicating value in business markets: From value in exchange to value in use. *Industrial Marketing Management, 69*(February), 80–90.

Evans, J. R. (2020). *Business analytics: Methods, models, and decisions* (3rd ed.). Pearson.

Forrest, J. Y.-L. (2018). *General systems theory: Foundation, intuition and applications in business decision making*. Springer.

Forrest, J. Y.-L., Ashimov, A., Ren, L., Gias, S., & Borovskiy, Y. (n.d.). (to appear). Conditions under which adopted consumer value propositions are effective. Submitted for publication.

Forrest, J. Y.-L., Buttermore, J., & Wajda, T. A. (2017). At Nash equilibrium when new market competitions appear? *Kybernetes, 46*(2), 256–271.

Forrest, J. Y.-L., Liu, Y., Swift, E., & Karimi, T. (2019). Exploitation, exploration, development and decay of resources: A systemic view. *Proceedings of the 2019 annual conference of Pennsylvania Economic Association* (pp. 139–165).

Kline, M. (1972). *Mathematical thought from ancient to modern times*. Oxford University Press.

Kowalkowski, C. (2011). Dynamics of value propositions: Insights from service-dominant logic. *European Journal of Marketing, 45*(1/2), 277–294.

Lanning, M. (1998). *Delivering profitable value*. Perseus Publishing.

Lanning, M. (2003). An introduction to the market-focused philosophy, framework and methodology called delivering profitable value. Retrieved February 24, 2012 from http://www.exubrio.com/whitepapers/DPVIntro-eXubrio.pdf.

Lanning, M., & Michaels, E. (1988). A business is a value delivery system. *McKinsey staff paper* no. 41, July.

Lin, Y. (1999). *General systems theory: A mathematical approach*. Plenum and Kluwer Academic Publishers.

Lin, Y. (2009). *Systemic yoyos: Some impacts of the second dimension*. Auerbach Publication, an imprint of Taylor and Francis.

Lindstrom, M. (2005). *Brand sense: Build powerful brands through touch, taste, smell, sight, and sound*. Free Press.

Marketing Science Institute. (2010). *MSI research priorities 2010–12*. MSI.

Marketing Science Institute. (2014). *MSI research priorities 2014–16*. MSI.

Mashey, J. R. (1998). Big data ... and the next wave of infrastress. *Slides from invited talk*. https://static.usenix.org/event/usenix99/invited_talks/mashey.pdf. Accessed on February 12, 2020.

McGrath, R. G. (2013). *The end of competitive advantage: How to keep your strategy moving as fast as your business*. Harvard Business Review Press.

Molineux, P. (2002). *Exploiting CRM: Connecting with customers*. Hodder & Stoughton.

Morgan, I., & Rao, J. (2003). Making routine customer experiences fun. *Sloan Management Review, 45*(1), 93–95.

Payne, A., & Frow, P. (2005). A strategic framework for customer relationship management. *Journal of Marketing, 69*(4), 167–176.

Payne, A., Frow, P., & Eggert, A. (2017). The customer value proposition: Evolution, development, and application in marketing. *Journal of the Academy of Marketing Science, 45*, 467–489.

Porter, M. E. (1985). *Competitive advantage: Creating and sustaining superior performance*. Free Press.

Reeves, R. (1961). *Reality in advertising*. Alfred A. Knopf.

Rintamaki, T., Kuusela, H., & Mitronen, L. (2007). Identifying competitive customer value propositions in retailing. *Managing Service Quality, 17*(6), 621–634.

Shulga, L. V., & Busser, J. A. (2019). Customer acceptance of four types of hospitality value propositions. *International Journal of Hospitality Management*. https://doi.org/10.1016/j.ijhm.2019.102388

Smith, S., & Wheeler, J. (2002). *Managing the customer experience: Turning customers into advocates*. Financial Times Prentice Hall.

Starch, D. (1914). *Advertising: Its principles, practice, and technique*. Scott, Foresman.

Terho, H., Haas, H., Eggert, A., & Ulaga, W. (2012). 'It's almost like taking the sales out of selling'—Towards a conceptualization of value-based selling in business markets. *Industrial Marketing Management., 41*(1), 174–185.

Ulaga, W., & Eggert, A. (2006). Value-based differentiation in business relationships: Gaining and sustaining key supplier status. *Journal of Marketing, 70*(1), 119–136.

Webster, F. E. (2002). *Market-driven management: How to define, develop and deliver customer value* (2nd ed.). Wiley.

Zhang, T. C., Gu, H. M., & Jahromi, M. F. (2019). What makes the sharing economy successful? An empirical examination of competitive customer value propositions. *Computers in Human Behavior, 95*(June), 275–283.

Chapter 15
Values and Competitive Advantages Based on Customer Value Propositions

Abstract Continuing the effort of the previous chapter to develop a theoretically rigorous and practically useful theory of customer value propositions (CVP), as called for repeatedly by the extant literature, this chapter, mainly based on (Forrest et al., 2021, *Studies in Business and Economics Journal, 16*(2)), establishes eight generally true facts that do not suffer from the constraints of data-based and anecdote-based approaches, as widely used in the relevant literature.

Keywords Ambition · Cash flow · Market invitation · Mission · Operating process · Value added

Specifically, these results developed in this chapter disclose how a newly espoused CVP is allied with a company's operation through its three essential underlying processes, how it will be fundamental for a company to acquire competitive advantages over its rivals, how the value, either created or captured or both, as a result of adopting a particular CVPs can be estimated, etc. The conclusion of this chapter provides recommendations for decision-making managers and entrepreneurs and potential questions for future research.

15.1 Introduction

As a vital process of communication with upstream suppliers, downstream complementors, ultimate consumers, and other players within a company's supply-chain ecosystem, the significance of customer value propositions (CVPs) has been broadly documented (Webster, 2002). For example, in the research and application of marketing, proposition has been conceptualized in various forms; and related studies can be traced back to at least 100 years ago (Starch, 1914). Complementing value creation, the necessity of making customers aware of a created value has been greatly emphasized (Marketing Science Institute, 2010) within the past 10 years.

J. Y.-L. Forrest, Y. Liu, *Value in Business*, Contributions to Management Science,
https://doi.org/10.1007/978-3-030-82898-1_15

Although the concept of CVPs has such a long history, its importance has been recognized and its complementarity to value creation has been seriously emphasized, this concept remains poorly defined until Payne et al. (2017) introduce a workable definition. Consequently, the concept of CVPs is still casually tossed around and applied trivially due to the lack of a rigorously developed and practically useful theory. Speaking differently, the world of learning and business operations need a theoretically developed theory of CVPs that can practically reveal how CVPs play their parts in general aspects of business operation and how CVPs assist the creation and capture of values for individual companies.

Aiming at achieving this end while adding to what has already been developed in the previous chapter, this chapter investigates how CVPs affect the development of competitive advantages for a company, how adopted CVPs could increase values for shareholder, and how a CVP's impacts can be materialized.

The rest of this chapter is organized as follows: Section 15.2 reviews the relevant literature and demonstrates the contribution of this chapter. Section 15.3 explores the association between CVPs and the development of competitive advantages of a company. Section 15.4 considers the values added specifically for shareholder through adopting effective CVPs. Section 15.5 examines relevant signs, risks, and values of an implemented CVP. Section 15.6 concludes this chapter with managerial recommendations and questions for future research.

15.2 Literature Review

This chapter touches on three areas of literature: competitive advantages, value addition based on CVPs, and CVPs' impacts. This section describes how these areas of the literature are enriched by this work.

Regarding CVP-based competitive advantages, Rintamaki et al. (2007) study CVPs in retailing and advance a method for recognizing competitive CVPs. Frow and Payne (2011) examine the role the concept of CVPs plays in generating insights on value creation within a network of values. Through using a service-dominant logic lens and position reciprocal CVPs, Ballantyne et al. (2011) join the following three aspects of business together: exchange activities, relationship development, and knowledge renewal by examining the concept of CVPs and how it functions. By looking at the influence of rivals, Hsieh and Vermeulen (2014) study how a company is inclined to follow its rivals into new markets. To find out how a company is performing relative to its rivals, Sheehan and Bruni-Bossio (2015) outline a useful managerial test. In order to find which products and services are valuable for customers, Wouters and Kirchberger (2015) explore in which way interorganizational managerial accounting could assist companies to collaborate with each other. Based on e-enterprises, Zheng et al. (2017) examine the effect of inter-user relationship on user's perceptions of an adopted CVP. By looking at two important steps of value creation, Carlson et al. (2019) identify an unmet opportunity and show how to develop a relevant CVP. By focusing on the sharing economy,

Zhang et al. (2019) classify specific CVPs and compare their competitive advantages. Carrying these listed and unlisted studies to a general scenario, this chapter develops generally true results on when a market invites for additional competition.

A literature search on CVP-based value additions only leads to a few relevant studies. In particular, by focusing on the development of new technologies of a small entrepreneurial high-tech firm, which is located in the upstream of the ecosystem of a large company, Wouters (2010) explores under what conditions real options can be employed to develop value propositions. To go beyond Western markets in the research of customer values (CVs), Blocker (2011) looks at five culturally diverse countries in his study of the measurement and modeling of CVPs from the prospect of business management. In comparison, this chapter contributes to this stream of literature by providing insights on when a CVP helps a firm augment its competitive advantages and capabilities, and how to analytically calculate particular business value produced by developing and implementing a CVP.

As for the impacts of CVPs, Parasuraman (1997) investigates the monitoring and leveraging of CVs by proposing a detailed framework. Randall et al. (2010) compared 100 US motor carriers in terms of their CVPs. Clarke III (2001) examines such CVPs that might help with the development of productive mobile e-commerce strategies. By examining organizational misuses of customer resources, Smith (2013) studies the process of value co-destruction. Through inspecting how values are experienced and evaluated in practices by consumers, Holttinen (2014) attempts to contextualize CVPs. From the perspective of retail executives, Yrjölä (2015) ranks CV dimensions. By developing and validating a scale for measuring CV, Rintamäki and Kirves (2017) illustrate how the contextual perspective can help with the evaluation of CVPs. Contributing to this stream of research, this chapter uncovers the following results, among others: to implement a CVP, a company needs to (1) focus more on earlier cash flows than later ones (2) maximizing related revenues while minimizing relevant investment; and (3) supplement the firm's long-term overall business value.

With this quick literature review in place, the objective of this chapter is once again to greatly improve the current landscape of CVP-related research and to carry Payne et al.'s (2017) initiative to the next level so that a more rigorous theory of CVPs can be developed so that established theoretical results can be beneficially applied in practice.

15.3 Competitive Advantages Attained from CVPs

In this chapter, end users will be known as consumers and B2B (business to business) buyers as customers. When both customers and consumers simultaneously use a product, these users will be jointly known as customers.

15.3.1 Association between CVPs and Macrolevel Operating Processes

In business, customer value (CV) of a product is defined as the worth in monetary terms of technical, economic, service, and social benefits a customer or a consumer receives in exchange for the price it pays for the product (Anderson & Narus, 1998). The said benefits that really matter are those seen as important in the customer's eyes. The price stands for the thing, whatever it is, the customer perceives as paid for the product (Lanning & Michaels, 1988). In other words, the value of a product is equal to the benefits acquired minus the price paid out. Each product has its value to customers and to supplier(s) and its price is the cost of a customer paid to acquire the product.

By customer value proposition (CVP), Payne et al. (2017) define it as a strategic tool through which a company communicates its offer of resource sharing and delivery of particular values to customers. This definition of CVP includes several key elements, such as a device for communication, the sharing of resources, the existence of targeted customers, and specific value that differentiates a company's product from those of its rivals. By employing its adopted such a CVP, a company makes its identified customers aware of the value its product brings and how these customers are able to reciprocally help improve the features and functionalities of the product, see Rintamaki et al. (2007) for a special discussion on the retail industry. Other than sending the imagined value of a supplier (e.g., Kowalkowski, 2011), an implemented CVP additionally stresses the involvement of customers. Such involvement is very important and seen as a research priority by marketers (Marketing Science Institute, 2010).

In order to make an adopted CVP effective, a company must be able to internally translate market invitations for new competition (Chap. 3; Forrest et al., 2017) into innovative design(s) and efficient production of original product(s). That is, the effectiveness of an adopted CVP directly disturbs the company's processes of business operation in such a way that the relevant processes are altered to accommodate the delivery of promises. Because each business enterprise is a functional system, it naturally contains many processes that play out their roles concurrently and interactively. For the purpose of investigating how an adopted CVP might transform a company's existing business processes, among all the concurrently and interactively working processes, let us concentrate on the following essential macrolevel processes. As noted by Day (1994), these essential processes address essential, while common, tasks that are crucial for the eventual successful delivery of the promises given in the CVP:

- Management of customer solution (process 1),
- Management of supply-chain ecosystem (process 2), and
- Management of customer relations (process 3).

Process 1 identifies customer solution(s) to meet the current and near-future market demand(s), which evolve constantly, by carefully analyzing market signals.

In this process, considered includes potential challenges to upstream suppliers and downstream complementors in order to first produce imagined market offer(s) and second deliver the offer(s) to the ultimate users (Adner et al., 2013). Process 2 considers the actual production of the imagined product(s) by first acquiring necessary supplies and second assembling parts into the eventual market product, the imagined solution(s). Process 3 focuses on communications with the market and the targeted customers, as acknowledged initially with the development and implementation of the CVP. This process includes the creation of market knowledge, the development of customer relations, and the determination of market perception of the company, and eventual products. All microlevel processes of a company will not be examined in this chapter because of their supporting roles to the previous three processes.

Proposition 15.1 Suppose that the company of concern has acquired its market position as a steady player and good performer in its product market through pursuing after its clearly stated mission and a long-term, unwavering ambition. If the company desires to further its success in the marketplace and financial performance by adopting a particular CVP, then this CVP has to associate with the three essential operating processes in the following two fashions:

1. It has to be closely associated with the company's stated mission and the long-term, unwavering ambition; and
2. Tasks required by the CVP have to be coordinated, integrated, and streamlined through subprocesses across the essential processes.

To see the reason behind Fashion 1, let us first look at why a company must have a clearly stated mission and a long-term unwavering ambition for it to establish itself in the marketplace as a steady player and good performer. Inside a company, there are naturally employees who have conflicting ideas about how the organization should be run, how operations should be carried out, and how labor should be allocated. Such conflicting ideas stem from individually different philosophical assumptions and values of the employees (Forrest & Orvis, 2016). Speaking differently, any two employees generally understand their company and its operations differently, as an organizational system. Such differences in opinion are rooted in their individually varied personal philosophical assumptions and values. So, they naturally lead to conflicting ideas and inconsistent behavioral patterns in the employees and the outcomes of their works (Forrest et al., 2018). What this fact implies is that for a business organization to be steadily functional with good performance, it has to plainly express and firmly exercise its mission and ambition. It is the mission and ambition that help the company avoid being torn apart by conflicting ideas and inconsistent philosophical assumptions and value systems of individual employees.

This fact explains why thc CVP adopted for furthering the company's success in the product market and financial performance has to be closely associated with the company's mission and ambition. Specifically, each process, irrespective of its level, within a company is generally developed, maintained, and operated by a team of individuals. So, all the teams that respectively maintain and operate the essential processes must be devoted to the eventual delivery of the promises of the CVP. This end helps bring corresponding promised solutions to every subprocess for the purpose of collectively meeting a specific market invitation that is innovatively comprehended. Otherwise, each process or subprocess is really pointless when seen in separation.

Next, let us see the reason behind Fashion 2. Based on the discussion above for Fashion 1, the adopted CVP, as an outcome of cooperative effort of the essential processes, customer solution, supply-chain, and customer-relation management, permeates a customer orientation into subprocesses of these essential processes through the CVP-demanded tasks as the medium. With such special concentration on market demand and customers and resources, including capabilities as particular resources, that are presently available or potentially available to a company, some critical tasks as demanded by CVP include

- Create the knowledge of customers' current and future needs, carefully group customers into segments according to their characteristics so that appropriate customer solutions can be developed;
- Generate in-depth knowledge about the capabilities of players within the focal company's supply-chain ecosystem and identify other potential companies that might become part of the ecosystem; and
- Strengthen communication and relationship with customers by learning about customers' uses and applications of offered products, and by assessing customers' responses to rivals' current and potential substitutes, etc.

Permeating the perspective of the CVP-based customer orientation into subprocesses are generally achieved by employing the following two approaches:

1. Bolster subprocesses of process 3 often, and
2. Associate the CVP-specific tasks inherent in subprocesses of process 3 with those in both processes 1 and 2.

Intensifying market competition and quickly evolving customer preferences and tastes have placed the present business world squarely in the era of transient competitive advantages (McGrath, 2013). Hence, if a company desires to uphold its competitiveness in the marketplace, the first approach is a must. That has been typified some time ago since marketing was clearly distinguished from sales orientations (Kotler, 1977). The execution of subprocesses generally falls into task routines or easy-to-follow task procedures, frequently bolstering subprocesses of process 3 is very important for a newly adopted CVP to eventually deliver its promises in the marketplace.

Next, let us see the reason behind approach 2. To this end, let us examine how feedback from customers on product use and application and consequent new demands can result in data and information, and how such data and information need to be practically considered in the design and execution of pertinent subprocesses within the two essential processes 1 and 2. The first process deals with the generation of innovative ideas, while the second process considers the capabilities of upstream suppliers and downstream complementors in terms of what they can supply and what assistance they can provide to eventually make imagined products reach the targeted customers. Therefore, this discussion demonstrates that the CVP-demanded tasks must be coordinated, integrated, and streamlined into the works inherent in subprocesses across the essential processes. That is, Fashion 2 in Proposition 15.1 holds true.

15.3.2 How a Company's Adopted CVP Affects Its Competitive Advantages

Before anything, let us examine how companies, either incumbent or not in a market, are driven by forever-evolving customer preferences and tastes of the market. To do this, we repeat relevant contents from Chaps. 3–5 in order to make the reading of this chapter as smooth and as self-contained as possible.

Assume that the operation of the market of concern is totally under the functions of free competition, driven solely by forces of demand and supplies. Assume that there are a fixed number of incumbent companies that serve this market with their horizontally differentiated market offers. Each of these incumbent companies enjoys the support of its base of loyal consumers, each of whom makes purchases only from her particular company, provided the price is not more than her reservation price. Consequent to forever shifting preference and taste, some consumers in the marketplace are not entirely pleased with the available offers in the product market. For one reason or another, these consumers make their purchases only from the company that offers the lowest price per unit value. Jointly, they are known as switchers, and the totality of them as the consumer surplus in the marketplace. Hence, all the incumbent companies compete with one another over these switchers by using pricing strategies.

Assume that all the companies, which are currently incumbent or thinking about, even though only remotely, entering this market, are well aware of each other's pricing strategies. To develop their best responses to market competition, they play the Nash equilibrium through applying pure self-analyses. Then the following theorem, which is equivalent to Theorems 3.2, 4.1, and 5.1, describes the dynamics of competition of this market. For relevant discussions and technical proofs, see Chaps. 3–5.

Theorem 15.1 Assume that the afore-defined market is in Nash equilibrium and that the consumer surplus is of a magnitude greater than that of the loyal-consumer base of some incumbent company. Then the statements below follow:

- There are rooms for new entrants to join the market with their substitute products; and
- The aggregate of all entrants' expected profits can be potentially greater than that of the incumbent with the smallest base of loyal consumers when the entrants

randomize their prices uniformly between their costs of production and maximum expected selling price.

Directly from Theorem 15.1, the result below follows.

Proposition 15.2 If an adopted CVP is established on the basis of a company's original interpretation of a market invitation supported by a sufficiently large segment of customers, then the company can potentially acquire a competitive advantage by its effort to deliver the promises of the CVP.

To see why this proposition holds true, let y be the magnitude of consumer surplus in the marketplace and P_j the price of Company j. If this company can capture the entire market surplus, then its profit will be yP_j, where the cost of producing and delivering the said product and the expected reservation price of the loyal consumers are assumed to have been normalized to 0 and 1, respectively. For details, see Chap. 3.

For a specific Company j to fully capture this expected amount of profit, the adopted CVP has to be sufficiently attractive to the entire consumer surplus. This end is equivalent to that Company j can potentially acquire a competitive advantage by its effort to deliver the promises of the CVP.

In terms of the literature, this proposition provides conditions under which Srivastava et al.'s (1999) prima facie case actually holds true in general.

15.4 Association between CVPs and Shareholder Values

The previous discussion looked at how evolving consumer preferences and market orientation affect the three essential processes of business operation. As the title suggests, this section examines how CVPs can affect shareholder values.

15.4.1 How CVPs Drive Shareholder Values

By extending earlier discussions on mission and ambition, the proposition below follows:

Proposition 15.3 Assume that the performance of a focal company is progressively good, as guided by its mission and a long-term, unwavering ambition. If the company implements such a CVP with accompanying investment in the three essential macrolevel operating processes that are expected to further its market and economic successes, then the adopted CVP will drive shareholder value higher through producing competitive advantages.

From Proposition 15.1, the given assumption implies that the accompanying investment helps connect tasks demanded by the CVP with the essential operating processes so that they are practically accomplishable, at least to a degree. The CVP and investment jointly play the role of coordinating, integrating, and streamlining the details of the essential processes.

The adoption of the said CVP functions as an effort to reorient the three existing essential processes: customer solutions, supply-chain partner associations, and customer relationships. The CVP in reality is composed based on constant market sensing and interpretation of market signals. It then helps publicize the relevant information and knowledge within the organization, then within the company's supply-chain ecosystem and in the marketplace. The effort of sensing and interpreting market signals leads to a good formulation of an appropriate CVP that reflects what a particular market signal means in comparison with available resources and background knowledge of the company. It helps develop customer solutions that are expected to meet the sensed market demand. If technicians and engineers are the dominating force behind the development of customer solutions and the management of the supply-chain ecosystem, then the tasks, as demanded by the adopted CVP, will inspire necessary cultural changes within the company towards satisfying customer needs.

Specifically, the focal company can empower process 1 in different ways such as invest in research and development (R&D), develop platforms that produce multi-sided markets while directly bridge sellers and buyers (Adner, 2017), form networks for sharing knowledge, resources, and know-hows (Fuentelsaz & Gómez, 2006; Hendrikse et al., 2015) so that their members can expand their market exposures (Nohria & Garcia-Pont, 1991). By empowering process 1, the company produces additional assets of extraordinary value, both tangible and intangible, such as superior products, innovative services and informational goods, patents, proprietary business secrets, relationships with other members of the supply-chain ecosystem.

Likewise, the company can enhance its process 2 by building additional infrastructures, such as capital assets (Robins, 1992) and intelligent manufacturing (Acemoglu & Restrepo, 2016). Investments in additional infrastructures surely lead to the creation of tangible assets that underlie further growth of the company and more efficient operation of its supply-chain ecosystem.

Investment in process 3 strengthens the development of brand(s) and brand equity and customer-installed base. It augments the company's customer-based territory in the marketplace. Jointly, the investment produces market-based assets (Khantimirov, 2017; Pyper et al., 2019; Williams & Naumann, 2011), including brand reputation, customer loyalty, and market share.

When it is more capable than its competitors to (1) sense and interpret market signals, (2) synchronize, assimilate, and restructure works both internally and externally in its development of customer solutions, and (3) publicize information internally, within its supply-chain ecosystem and throughout the market, the focal company has already acquired such competitive advantages that it can employ to drive its shareholder value.

As spelled in Proposition 15.2, an adopted CVP could represent a critical tool for a company to produce a competitive advantage. And it could give rise to the appearance of competitive advantages and company-specific capabilities if it is appropriately funded (Proposition 15.3). Hence, the following natural question arises: how can the value added by implementing a particular CVP be evaluated? The following section attempts to address this question analytically.

15.4.2 Evaluating a CVP-Based Value Addition

Implementing an adopted CVP tends to affect a company over at least several accounting periods starting from the inception of the relevant works. Hence, the present value of adopting and implementing the CVP is given in the following result. In this proposition, E_t represents the mathematical expectation of period t, δ is equal to the (physical) depreciation rate of the capital invested in implementing a CVP, D_{t+1} and d_{t+1}, respectively, stand for the relevant stochastic dividends and gross discount rate at period t for dividends received in period $(t + 1)$, and V_{t+1} the value added for period $(t + 1)$.

Proposition 15.4 The value V_t added for period t after initially implementing a particular CVP is equal to the following:

$$V_t = E_t\left(\frac{D_{t+1} + (1-\delta)V_{t+1}}{d_{t+1}}\right) \tag{15.1}$$

$$= E_t\left(\frac{D_{t+1}}{d_{t+1}} + (1-\delta)\frac{D_{t+2}}{d_{t+1}d_{t+2}} + (1-\delta)^2\frac{V_{t+2}}{d_{t+1}d_{t+2}}\right) \tag{15.2}$$

$$= E_t\left(\frac{D_{t+1}}{d_{t+1}} + (1-\delta)\frac{D_{t+2}}{\prod_{i=1}^{2}d_{t+i}} + (1-\delta)^3\frac{D_{t+3}}{\prod_{i=1}^{3}d_{t+i}} + (1-\delta)^3\frac{V_{t+3}}{\prod_{i=1}^{3}d_{t+i}}\right)$$

$$= \cdots, \tag{15.3}$$

where the values of D_{t+i} and d_{t+i} are assumed to be known in period t, for $i = 1$ in Eq. (15.1), for $i = 1, 2$, in Eq. (15.2), and $i = 1, 2, 3$, in Eq. (15.3).

One can calculate the value, as added by implementing a CVP, in many different ways, such as, for example, price/earnings multiples, market/book-value ratio, economic value added (EVA), market value added (MVA), and shareholder value (SHV). Because implementing an adopted CVP tends to affect a company for longer than simply one or two time periods, only those methods that are based on cash flows, such as EVA and SHV (Fernando et al., 2017; Grant, 2003), can be applied here for the current problem of estimation. Specifically, the EVA method jointly considers the systems of compensation and incentive to those factors that contribute to economic value added, when the method is combined with approaches of associated value-based management. However, this method mostly emphasizes the near term without attending to future growth potential and relevant intangible assets

(Grant, 2003). In contrast, the SHV method calculates the net present value of all projected cash flows of the future, created by implementing the particular CVP.

The market value of a depreciable capital, such as an investment or a project or a business initiative, is echoed in the net present value of all cash flows of the future, as expected to ensue over the indefinite future or until the conclusion of the investment (Bernanke & Gertler, 1999; Forrest et al. (2013)). Symbolically, this recognition can be written as follows:

$$V_t = E_t\left(\sum_{i=0}^{\infty}\left[\frac{(1-\delta)^i D_{t+1+i}}{\prod_{j=0}^{i} d_{t+1+j}}\right]\right)$$
$$= E_t\left(\frac{D_{t+1}}{d_{t+1}} + \frac{(1-\delta)^1 D_{t+2}}{d_{t+1}d_{t+2}} + \frac{(1-\delta)^2 D_{t+3}}{d_{t+1}d_{t+2}d_{t+3}} + \ldots \frac{(1-\delta)^{k-1} D_{t+k}}{\prod_{i=1}^{k} d_{t+i}} + \ldots\right) \quad (15.4)$$

where V_t is the market value in period t, and D_{t+i} and d_{t+i} stand respectively for the relevant stochastic dividends and gross discount rate at t for dividends received in period $(t + i)$.

Rewriting Eq. (15.4) produces:

$$V_t = E_t\left[\frac{D_{t+1}}{d_{t+1}} + \frac{(1-\delta)}{d_{t+1}}\left(\frac{D_{t+2}}{d_{t+2}} + \frac{(1-\delta)^1 D_{t+3}}{d_{t+2}d_{t+3}} + \ldots \frac{(1-\delta)^{k-2} D_{t+k}}{\prod_i}\right.\right.$$
$$= 2^k d_{t+i} + \ldots)]$$
$$= E_t\left(\frac{D_{t+1}}{d_{t+1}} + \frac{(1-\delta)}{d_{t+1}} V_{t+1}\right), \text{if values of } D_{t+1} \text{ and } d_{t+1} \text{ are known}$$
$$= E_t\left(\frac{D_{t+1} + (1-\delta)V_{t+1}}{d_{t+1}}\right), \text{if values of } D_{t+1} \text{ and } d_{t+1} \text{ are known.}$$

This equation directly implies Eq. (15.1). To derive Eqs. (15.2) and (15.3), …, simply apply Eq. (15.1) repeatedly on itself by consecutively substituting the following, as obtained from Eq. (15.1), into Eq. (15.1) with the proper assumptions on D_{t+i} and d_{t+i}:

$$V_{t+1} = E_t\left(\frac{D_{t+2} + (1-\delta)V_{t+2}}{d_{t+2}}\right) \text{ and } V_{t+2} = E_t\left(\frac{D_{t+3} + (1-\delta)V_{t+3}}{d_{t+3}}\right).$$

This concludes the argument of Proposition 15.4.

15.5 Signs, Risks, and Values of an Implemented CVP

In the value estimation in Eq. (15.1), the stochastic dividends D_{t+i} and stochastic gross discount rate d_{t+i}, for $i = 1, 2, 3, \ldots$, reflect perceived growth potentials and associated risks of the invested depreciable capital. Hence, the practical significance of Proposition 15.4 is highlighted in the proportions, such as V_{t+1}, $V_{t+2}, \ldots$, of future cash flows, as estimated by using perceived future growth potentials and associated risks. Now, a practical concern is the demonstration and measurement of the expected value V_t, created by implementing a specifically adopted CVP in terms of current outcomes and perceptions of future financial performance.

15.5.1 Signs of an Implemented CVP

Equations (15.1)–(15.3) collectively provide a theoretical means to measure the impact of an adopted CVP on value creation and capture. The following Propositions 15.5 through 15.8 point to relevant emphases of implementing a particular CVP.

Proposition 15.5 When a particular CVP is implemented, an important focus needs to be on the accelerated earlier cash flows over later ones.

Because the factor $(1 - \delta)$ represents a real number between 0 and 1, the sequence $(1 - \delta)$, $(1 - \delta)^2$, $(1 - \delta)^3, \ldots$ consists of positive numbers of drastically decreasing magnitudes. This fact indicates that further-out future contributions in terms of cash flows of the adopted CVP play decreasing and ignorable roles in the calculation of the net present value. Besides, the globalizing economy has made once sustainable competitive advantages transient and short-lived (McGrath, 2013), which in turn accelerates changes in the marketplace. Such increasingly accelerated evolution of the market makes expectations of long-term values more uncertain and much riskier than ever before. Hence, all these reasons jointly diminish the magnitude of value contribution of further-out future cash flows.

Regarding the three essential macrolevel operating processes, defined earlier, a company's cash flows of the future can be materially accelerated by

- Enhancing learning capabilities, keeping pace with technology development, and timely translating market invitations into market offers;
- Connecting efficiently and effectively with supply-chain partners such that upstream suppliers can produce adequate components and downstream complementors can construct appropriate infrastructure, needed for customers to make their practical adoptions; and
- Reducing the time needed for achieving market acceptance through effective marketing and involving customers in the development of market solutions.

Proposition 15.6 When a particular CVP is implemented, the relevant investment needs to be minimized and the relevant revenues need to be increased.

This conclusion follows directly from Eqs. (15.1)–(15.3). In particular, let us identify "revenues" with reflected in the periodic "dividends" D_{t+1}, D_{t+2}, D_{t+2}, . . ., and "investment" with the "rate δ of physical depreciation" of the invested capital. That is, the net present value V_t of implementing an adopted CVP will be a greater value if the numerators in Eqs. (15.1)–(15.3) are larger, and at the same time, the factors $(1 - \delta)$, $(1 - \delta)^2$, $(1 - \delta)^3$, . . ., are closer to 1.

Regarding the three essential macrolevel operating processes, as focused on in this chapter, the investment in implementing an adopted CVP can be practically minimized and consequent revenue increased if a company

- Distinguishes its market offers from those of rivals through;
- Originating new offers while cannibalizing existing ones;
- Genuinely reflecting customer inputs in new offers;
- Making resources and knowledge available to all departments internally;
- Employing most advanced technologies; and
- Minimizing costs;
- Reengineers supply-chain processes and communicates efficiently with its ecosystem partners in order to;
- Reduce the number of troublesome incidences;
- Acquire ideal supplies from upstream suppliers;
- Accelerate the speed for new offers to reach the consumer market by joining forces with downstream complementors;
- Eliminate needless expenses, while lowering capital investments;
- Cut inventories and boost capacity use;
- Optimize consumer values and company revenue through;
- Combining customer needs into market offers by providing competitive products and services;
- Establishing branded offers and superior service;
- Bundling separate offers into packages;
- Building and reaching out to customer bases;
- Minimizing expenditures, such as those used for product launches, sales, and services.

15.5.2 Risks and Values of an Implemented CVP

To maintain the expected future as assuredly as possible, one needs to reduce the uncertainties associated with the stochastic quantities D_{t+i} and d_{t+i}, $i = 1, 2, 3, \ldots$ That is, later terms in Eqs. (15.1)–(15.3) need to be somehow kept around their expected values. Practically speaking, that means that future periodic cash flows need to involve as little volatilities and vulnerabilities, such as the cost of capital, as possible. Hence, the proposition below follows:

Proposition 15.7 When a particular CVP is implemented, a company needs to target at lowering all risks associated with future cash flows.

In terms of the three essential macrolevel operating processes, the goal of reducing all risks associated with future cash flows can be materially accomplished through

- Developing such an organizational culture that allows agility, flexibility, and transformation needed for riding the wave of transient competitive advantages so that
- Activities of R&D can be harmonized with the fast-changing consumer demands so that original offers, driven by the market, can be timely designed, produced, and provided to the market;
- Interorganizational and strategic networks and alliances are established for the purpose of sharing resources, information and knowledge, and developing market-driven standards and leading market positions;
- Other than exploiting the residual value of current offers, the business will focus on creating extraordinary differentiations and original offers or unique bundles of products and services that are hard to imitate;
- Dynamic market demands and forever changing consumer preferences are met by developing and maximizing synergies among offers and between production and consumption;
- Suppliers can provide desired components and originally produced offers can successfully reach the end users through clearly communicating and closely associating with supply-chain partners. Specifically, a company needs to;
- Work closely with upstream suppliers in their development and production of desired components by considering their capabilities;
- Entangle associations with distributors by employing such programs as services, incentives, and loyalties in order to make their switching costs high and to proactively manage competitions among and within delivery channels;
- Design and configure processes of supply and delivery so that rivals will have difficulty to imitate;
- Integrate production and order delivery systems for flexibility so that they can naturally incorporate constantly changing market opportunities into production-related decisions;
- Customer value is dynamically optimized in sync with customers' forever-evolving preferences and tastes. To this end, for example, any of the following strategies can be adopted;
- Feature customer needs within developed market offers;
- Bundle products and services together to raise customer switching cost;
- Design and deliver intangibles and experiential attributes to enhance customer satisfaction;
- Optimally leverage market-based assets, such as value networks and online blogs, by providing customer education and training programs;
- Apply flexible selling methods, such as leasing options, cross-sales or cross-leasing of bundled parts, consumables, and complementary services.

Proposition 15.8 When a particular CVP is implemented, a company needs to focus on improving its long-term overall value of the business.

This proposition is a direct consequence of the fact that the presently globalizing world economy embraces transient competitive advantages (McGrath, 2013). In this current business era, today's adequate CVPs become obsolete quickly due to the quickly changing market demands and forever-evolving consumer preferences. In such a fast-moving business environment, companies that desire to survive and then succeed have to learn about how to ride waves of short-lived competitive advantages. To this end, when a company implements a well-crafted CVP, it needs to consider the growth of the underlying business in order for it to gain sufficient strength and enough capabilities to surf the next tidal wave of competitive advantages. It has to develop agile and flexible, as well as powerful, operating processes in order to build necessary assets, either tangible or intangible.

15.6 A Few Final Words

The adopted CVPs signify how companies apply their particular strategic tools to create and capture value. However, the concept of CVPs remains poorly defined and casually tossed about, while the related theory still does not exist (Payne et al., 2017). To drastically change the situation on the research of CVPs, this chapter attempts to develop a theoretically rigorous and practically useful theory of CVPs for scholars and front-line decision-makers, where the rigor is required so that the consequent theory can enjoy a wide range of applications. By doing so, this work helps grow a systematic knowledge of CVPs; and at the same time, it lifts the scattered studies, most of which are empirical with conclusions of limited validity, on related topics to a more or less unified theoretical framework.

To accomplish such a goal, a calculus-based analytical approach is utilized here as the method of reasoning without referencing to any statistics-based logic. That explains why all conclusions established in this chapter are free from the limitations of data and anecdotal analyses, which are commonly used in the literature. Specifically, this chapter develops the generally true theoretical results below among others:

- When a company implements a particular CVP to further its market success, the adopted CVP needs to be closely associated with the company's stated mission and long-term, unwavering ambition; it has to demand coordination, integration, and streamlining of relevant works within the essential processes 1, 2, and 3 (Proposition 15.1).
- If implementing a particular CVP helps add value for a company, then the value V_t added for period t is approximately equal to the expected sum of the present values of periodic future returns (Proposition 15.4).
- When a company implements a CVP, it prefers accelerated earlier cash flows (Proposition 15.5), minimized investments, and increased revenues (Proposition 15.6).

15.6.1 Recommendations for Decision-Makers

Implementing best practices and making data-driven decisions have been commonly employed in the present landscape of managerial decision-making (Duan et al., 2019; McGrath, 2013). However, the result of implementing best practices has been fluctuating widely; and the logic underlying data-driven decision-making is generally not correct for any situation that is different from the place and time where and when the original data were from. On top of such backdrops, this chapter's practical significance demonstrates itself vividly through the fact that this chapter develops general recommendations instead of suggestions for decision-making managers and entrepreneurs. In particular, what is developed in this chapter naturally produces the following recommendations for general managerial purpose.

- Each company needs to have its clearly stated mission and a long-term, unwavering ambition before it develops its specific CVP based on an innovative comprehension of a market invitation with sufficient market depth and width (Propositions 1.51 and 15.2);
- When a company implements a CVP, it needs to estimate expected returns and focus on the goal of developing new competitive advantages through investing in its essential operating processes 1–3 (Propositions 15.3 and 15.4); and
- When a company plans to adopt a CVP, it needs to focus on augmenting its long-term business value through accelerating periodic cash flows, especially earlier ones, minimizing the investment in the CVP adoption and related risks (Propositions 15.5–15.8).

15.6.2 What Opens for Future Research

The basic assumption underneath all established results in this chapter is that the economic soundness of each considered company is supported by positive cash flows from product market exchanges. However, in the landscape of the present business world, such assumption does not hold true generally (Li & Ma, 2015). So, for future research, a series of important questions will include: If a company does not exist for this assumed purpose, how can then one establish generally true results that are parallel to the ones derived above? What kinds of useful recommendations will these new results produce for decision-making managers and entrepreneurs?

As for the concept of big data, although various companies have been using empirical approaches to make strategic decisions with different levels of success (Duan et al., 2019; McGrath, 2013), the logic behind such decision tactic is not scientifically sound (Lin & OuYang, 2010). Therefore, instead of implementing recognized best practices and making decisions based on discovered data patterns, one needs to demonstrate the generality of the practices and the common existence of the patterns.

References

Acemoglu, D., & Restrepo, P. (2016). The race between man and machine: Implications of technology for growth, factor shares, and employment. NBER Working Papers No.22252.

Adner, R. (2017). Ecosystem as structure: An actionable construct for strategy. *Journal of Management, 43*(1), 39–58.

Adner, R., Oxley, J. E., & Silverman, B. S. (Eds.). (2013). *Collaboration and competition in business ecosystems*. Emerald.

Anderson, J. C., & Narus, J. A. (1998). Business marketing: Understand what customers value. *Harvard Business Review, 76*(6), 53–65.

Ballantyne, D., Frow, P., Varey, J. R., & Payne, A. (2011). Value propositions as communication practice: Taking a wider view. *Industrial Marketing Management, 40*(2), 202–210.

Bernanke, B.S., & Gertler, M. (1999). Monetary policy and asset price volatility. *Economic Review, Federal Reserve Bank of Kansas City*, fourth quarter (pp. 17–51).

Blocker, C. P. (2011). Modeling customer value perceptions in cross-cultural business markets. *Journal of Business Research, 64*(5), 533–540.

Carlson, C., Polizzotto, L., & Gaudette, G. R. (2019). The "NABC's" of value propositions. *IEEE Engineering Management Review, 47*(3), 15–20.

Clarke, I., III. (2001). Emerging value propositions for M-commerce. *Journal of Business Strategies, 18*(2), 133–148.

Day, G. S. (1994). The capabilities of market-driven organizations. *Journal of Marketing, 58*(4), 37–52.

Duan, Y. Q., Edwards, J. S., & Dwivedi, Y. K. (2019). Artificial intelligence for decision making in the era of big data – Evolution, challenges and research agenda. *International Journal of Information Management, 48*(October), 63–71.

Fernando, C. S., Sharfman, M. P., & Uysal, V. B. (2017). Corporate environmental policy and shareholder value: Following the smart money. *Journal of Financial and Quantitative Analysis, 52*(5), 2023–2051.

Forrest, J. Y.-L., Buttermore, J., & Theresa Wajda, T. (2017). At Nash equilibrium when new market competitions appear? *Kybernetes: The International Journal of Cybernetics, Systems and Management Sciences, 46*(2), 256–271.

Forrest, J. Y.-L., Hopkins, Z., & Liu, S. F. (2013). Currency wars and a possible self-defense (I): How currency wars take place. *Advances in Systems Science and Application, 13*, 198–217.

Forrest, J. Y.-L., Novikov, D. A., Larson, S., Wang, F., & Yang, J. (2021). Competitive advantages and values created and attained out of well-crafted customer value propositions, *Studies in Business and Economics Journal, 16*(2), 53–73.

Forrest, J. Y.-L., & Orvis, B. (2016). Principles of management efficiency and organizational inefficiency. *Kybernetes: The International Journal of Cybernetics, Systems and Management Sciences, 45*(8), 1308–1322.

Forrest, J. Y.-L., Tucker, R., Lin, C. C., & Mondal, S. (2018). Impacts of manufacturing firms' culture, structure and leadership on innovation: A systemic approach. *Proceedings of the 2018 annual conference of NABET* (pp. 95–120).

Frow, P., & Payne, A. (2011). A stakeholder perspective of the value proposition concept. *European Journal of Marketing, 45*(1–2), 223–240.

Fuentelsaz, L., & Gómez, J. (2006). Multipoint competition, strategic similarity and entry into geographic markets. *Strategic Management Journal, 27*(5), 477–499.

Grant, J. I. (2003). *Foundations of economic value added*. Wiley.

Hendrikse, G., Hippmann, P., & Windsperger, J. (2015). Trust, transaction costs and contractual incompleteness in franchising. *Small Business Economics, 44*(4), 867–888.

Holttinen, H. (2014). Contextualizing value propositions: Examining how consumers experience value propositions in their practices. *Australian Marketing Journal, 22*(2), 103–110.

Hsieh, K. Y., & Vermeulen, F. (2014). The structure of competition: How competition between one's rivals influences imitative market entry. *Organization Science, 25*(1), 299–319.

Khantimirov, D. (2017). Market share as a measure of performance: Conceptual issues and financial accountability for marketing activities within a firm. *Journal of Research in Marketing, 7*(3), 587–592.

Kotler, P. (1977). From sales obsession to marketing effectiveness. *Harvard Business Review, 55* (6), 67–75.

Kowalkowski, C. (2011). Dynamics of value propositions: Insights from service-dominant logic. *European Journal of Marketing, 45*(1/2), 277–294.

Lanning, M., & Michaels, E. (1988). A business is a value delivery system. *McKinsey staff paper* no. 41, July.

Li, T., & Ma, J. H. (2015). Complexity analysis of dual-channel game model with different managers' business objectives. *Communications in Nonlinear Science and Numerical Simulation, 20*, 199–208.

Lin, Y., & OuYang, S. C. (2010). *Irregularities and prediction of major disasters*. CRC Press.

Marketing Science Institute. (2010). *MSI research priorities 2010–12*. MSI.

McGrath, R. G. (2013). *The end of competitive advantage: How to keep your strategy moving as fast as your business*. Harvard Business Review Press.

Nohria, N., & Garcia-Pont, C. (1991). Global strategic linkages and industry structure. *Strategic Management Journal, 12*(Summer Special Issue), 105–124.

Parasuraman, A. (1997). Reflections on gaining competitive advantage through customer value. *Journal of the Academy of Marketing Science, 25*, 154. https://doi.org/10.1007/BF02894351

Payne, A., Frow, P., & Eggert, A. (2017). The customer value proposition: Evolution, development, and application in marketing. *Journal of the Academy of Marketing Science, 45*, 467–489.

Pyper, K., Doherty, A. M., Gounaris, S., & Wilson, A. (2019). Investigating international strategic brand management and export performance outcomes in the B2B context. *International Marketing Review, 37*(1), 98–129.

Randall, W. S., Defee, C. C., & Brady, S. P. (2010). Value propositions of the U.S. trucking industry. *Transportation Journal, 49*(3), 5–23.

Rintamäki, T., & Kirves, K. (2017). From perceptions to propositions: Profiling customer value across retail contexts. *Journal of Retailing and Consumer Services, 37*(July), 159–167.

Rintamaki, T., Kuusela, H., & Mitronen, L. (2007). Identifying competitive customer value propositions in retailing. *Managing Service Quality, 17*(6), 621–634.

Robins, J. A. (1992). Organizational considerations in the evaluation of capital assets: Toward a resource-based view of strategic investment by firms. *Organization Science, 3*(4), 442–562.

Sheehan, N. T., & Bruni-Bossio, V. (2015). Strategic value curve analysis: Diagnosing and improving customer value propositions. *Business Horizons, 58*(3), 317–324.

Smith, A. M. (2013). The value co-destruction process: A customer resource perspective. *European Journal of Marketing, 47*(11/12), 1889–1909.

Srivastava, R. K., Shervani, T. A., & Fahey, L. (1999). Marketing, business processes, and shareholder value: An organizationally embedded view of marketing activities and the discipline of marketing. *Journal of Marketing, 63*(4), 168–179.

Starch, D. (1914). *Advertising: Its principles, practice, and technique*. Scott, Foresman.

Webster, F. E. (2002). *Market-driven management: How to define, develop and deliver customer value* (2nd ed.). Wiley.

Williams, P., & Naumann, E. (2011). Customer satisfaction and business performance: A firm-level analysis. *Journal of Services Marketing, 25*(1), 20–32.

Wouters, M. (2010). Customer value propositions in the context of technology commercialization. *International Journal of Innovation Management, 14*(6), 2099–1127.

Wouters, M., & Kirchberger, M. A. (2015). Customer value propositions as interorganizational management accounting to support customer collaboration. *Industrial Marketing Management, 46*(April), 54–67.

Yrjölä, M. (2015). Uncovering executive prioritization: Evaluating customer value propositions with the pairwise comparison method. *Journal of Service Science and Management, 8*(1). https://doi.org/10.4236/jssm.2015.81001

Zhang, T. C., Gu, H. M., & Jahromi, M. F. (2019). What makes the sharing economy successful? An empirical examination of competitive customer value propositions. *Computers in Human Behavior, 95*(June), 275–283.

Zheng, C. D., Yu, X. F., & Jin, Q. M. (2017). How user relationships affect user perceived value propositions of enterprises on social commerce platforms. *Information Systems Frontiers, 19*, 1261–1271.

Chapter 16
Market-Sensing Capabilities and Their Vital Importance in Firm Performance

Abstract This chapter, which is mainly based on (Forrest et al., to appear, Market-sensing capabilities, profitability within stagnant industries and crafting of customer value propositions), studies the concept of market-sensing capability and its impact on a company from the perspective of market competition. Then it looks at, as a case analysis, sources of profits within a slow growth industry. As a second case analysis, results, developed in this chapter, are employed to reveal how a company can formulate effective customer value propositions (CVPs). This chapter shows, among others, that (1) the capability of market-sensing permits a company to classify customers into various classes, (2) it lowers operational costs while increasing expected profits, (3) the profit of a company within such an industry that experiences slow growth can grow only through a combination of raising unit price and lowering unit cost, and (4) how market knowledge and its innovative understanding can potentially lead to the formulation of effective CVPs. Other than employing any approach that is either statistics- or anecdote-based, this chapter makes use of the rigor of game theory to establish its results. Therefore, compared to empirical studies, results established in this chapter are generally true unless given conditions are violated. At the conclusion, managerial recommendations are listed for decision-makers and potential questions for future research are given.

Keywords Customer relation · Market invitation · Market knowledge · Market-level growth · Nash equilibrium · Profit portfolio

16.1 Introduction

It has been widely recognized that market-sensing is an important capability for a company to survive and succeed in increasingly fast-evolving markets (Day, 1994; Forrest et al., 2017; McGrath, 2013). Even so, the existent literature points to inconsistent findings. For example, Ardyan (2016) finds that such capability does not have any positive effect on companies' profitability; and Lindblom et al. (2008) reveal some effect but not significant. In terms of its impacts on the quality of market entry and the creation of knowledge, scholars find the capability has either some

J. Y.-L. Forrest, Y. Liu, *Value in Business*, Contributions to Management Science,
https://doi.org/10.1007/978-3-030-82898-1_16

positive or quite significant impacts (Alshanty & Emeagwali, 2019; Sugiyarti & Ardyan, 2017). Hence, a natural question arises as follows: what do these inconsistent empirical results mean in theory and in practice?

As briefly discussed in the previous paragraph, the effort of addressing this question is both theoretically and practically important. For example, for either a market leader or a follower, knowing the future direction of consumer demands is essential for a firm to plan itself strategically. It is especially so for the current landscape of business where once sustainable competitive advantages have become transient (McGrath, 2013). To produce an answer to the question above, this chapter attempts to go methodologically beyond statistics- and anecdote-based approaches. The logic behind this attempt is that the existent literature is frequently empirical, and in many circumstances, inconsistencies and contradictory discoveries are simply results of the tools of analysis employed in relevant studies (Einstein, 1997). Therefore, replacing any statistics- or anecdote-based approach, this chapter employs the rigor of game theory and the logical reasoning that is parallel to the one widely used in mathematics (Kline, 1972).

Beyond attempting to answer the question posed above, this chapter also enables us, as case studies, to look at profit opportunities in stagnant industries without much market growth and show how market knowledge and relevant innovative understanding of the knowledge can assist a company to construct effective CVPs.

The organization of this chapter is as follows: Section 16.2 looks at the relevant literature. Section 16.3 demonstrates the dynamic significance of market-sensing capabilities for companies to achieve success in the marketplace. Section 16.4 examines, as a case study, the potential of profitability in stagnant industries. And as applications of established results, Sect. 16.5 investigates the connection between the managerial capability of customer relationships and the potential of profitability, and when a company can draft effective CVPs. Section 16.6 concludes this chapter by providing relevant managerial recommendations and related questions for future research.

16.2 Literature Review

This chapter touches on and contributes to three literature areas: the concept of market-sensing capabilities and related studies, profit opportunities within a stagnant industry, and the development of CVPs. First, investigations on market-sensing capabilities have revealed inconsistent findings due to the empirical nature of these works. In particular, it is found (Bharadwaj & Dong, 2014) that for a company to synchronize with market change and deliver superior value to consumers, its learning of the market signals should cover all customers, which is a statistically and practically impossible task, while market-sensing capability is shown (Ardyan, 2016; Lindblom et al., 2008) to have neither positive nor significant effect on the profitability of firms. To link market-sensing capability and firm performance,

Bayighomog Likoum et al. (2018) propose an empirically research agenda; and it is confirmed (Alshanty & Emeagwali, 2019) that market-sensing capability does have a positive effect on knowledge creation for small- and medium-sized enterprises. Contributing to this literature, at the same time when it avoids all constraints of statistics- and anecdote-based approaches, this chapter demonstrates that market-sensing capability actually empowers a company to identify and better understand target customers; it helps lower the average costs of business operation; and it increases expected profits.

As for the studies on the profitability in stagnant industries, our literature search points to only limited number of works. Even so, all of these works examine a particular industry or market. In particular, Ferber and Schlappa (2016) show how brownfield land can be managed more effectively and sustainably by using good practices and a model for the purpose of facing the challenge of stagnant European urban land markets. Tournois (2016) examines how a breakthrough product in the cosmetics market, the BB cream, led to fruitful economic growth. By overviewing the geography of the seaweed hydrocolloid industry, Porse and Rudolph (2017) provide several industry-wide suggestions. By looking at challenges facing small-scale furniture productions in Jepara, Indonesia, Clements et al. (2019) are able to evaluate the association of regional producers. By focusing on stagnant or low-productivity-growth US industries, Le et al. (2020) inspect the financial development and productivity growth of member firms in these industries. Different from these existing works on specific industries, this chapter is able to establish results with broad-range applicability to many industries that lack market-level growth.

Our literature search found that existing studies on the formulation of CVPs are mostly oriented towards applications. For example, processes and tools are developed (Osterwalder et al., 2014) to help design and produce products that actually sell and generate sales for companies. Applying such processes and tools to financial planning leads to a framework of CVP development (Lecours, 2017). It is recommended (Wouters et al., 2018) that startups construct two sequential value propositions so that the first one demonstrates superior values created for customer firms and the second one on values that are captured by customer firms. In particular, the first proposition details the innovative offering, and the second one leveraging assistance. By employing the so-called "service-dominant logic," four principles are developed (Kowalkowski, 2011) based on the literature for the purpose of guiding a company to uncover market opportunities for itself. Enriching this literature area, this chapter develops knowledge on how market knowledge can assist the creation of effective CVPs and why a creative comprehension of market invitations is fundamentally important in the creation of effective CVPs.

To summarize, this chapter augments the literature in the following three ways: (1) it clears inconsistencies in relevant empirical findings, (2) it offers conclusions generally applicable to various markets when stated conditions are met, and (3) it points to the importance of market knowledge and its innovative understanding in the formulation of effective CVPs.

16.3 Why Is Market-Sensing Capability Important?

To theoretically answer this question, consider an oligopoly market of free competition, served by n companies, where $n = 1, 2, 3, \ldots$ is a natural number. Assume that this market is only operationally affected by the forces of demand and supply, while the demand and supply are functions of consumers' forever-evolving preferences and tastes. Because technology influences the equilibrium between market demand and supply, this assumption indirectly implies that technological advances do impact market competition.

For the sake of convenience of communication, assume that in the said marketplace the incumbent companies provide their horizontally differentiated offers such that each of them offers only one product. And each incumbent company is fully supported by its loyal customers as long as the company's price is not more than their reservation price. To this end, consumers' forever-evolving preferences and tastes may practically turn once loyal customers into switchers who purchase from the company which offers the best value. The very presence of switchers indicates that some customers in the marketplace are not totally satisfied with any of the available products.

For switchers, the assumption that they make purchases only from the company that offers the best price means the best price per unit value, because in real life no two products in the marketplace are perfectly substitutable for each other. So, to compare the prices of two non-substitutable products, one can simply compare their prices of corresponding unit values. In the rest of this chapter, collectively, the totality of all switchers is referred to as the consumer surplus in the marketplace. So, other than trying to keep their corresponding loyal customers, the incumbent companies compete with each other over the consumer surplus by using pricing strategies.

In terms of the incumbent companies' decision-making, assume that the managers know how others price their products so that they all set their prices to achieve Nash equilibrium. To do this, the managers provide their best responses through pure self-analysis.

The competitive dynamics of this market is well illustrated by Theorem 3.2. For the sake of convenience and self-containment of this chapter, let us restate that theorem as follows.

Theorem 16.1 In Nash equilibrium, if the proportion of consumer surplus is greater than that of the loyal-customer base of at least one incumbent company, then the afore-described market invites the entry of other companies along with their versions of substitute offers, satisfying that the total of all entrants' expected profits can be potentially greater than the expected profit of at least one incumbent company.

By market-sensing capability, it means such a capability with which a company can acquire an innovative understanding of market signals. Speaking differently, market-sensing capability assists a company to acquire its particular understanding of customers, competitors, the supply-chain ecosystem, and a much broader market environment. This definition of market-sensing capability generalizes that given by Day (1994). It is also different from that given by Lindstrom (2005) so that other than sensing the market, this definition is not concerned with employing any innovative understanding of the market to plan for subsequent actions.

Examining this concept of market-sensing capability in light of Theorem 16.1 produces the following result. [Note: In this chapter, the word "proposition" means one of two unrelated concepts: (1) a statement that is in some way arguably true; and (2) a particular selling point, as studied in the discipline of marketing, such as the unique selling proposition (Reeves, 1961), the emotional selling proposition (Lindstrom, 2005), or customer value proposition (Payne et al., 2017). A "proposition" in sense (1) reveals the methodological novelty of this chapter: hypothesis testing, as widely employed in the literature, is a statistics-based approach designed to uncover potential facts, but not any arguably true fact.]

Proposition 16.1 If a company possesses a superb market-sensing capability, then it is capable of

1. Locating market segments of either not adequately served or unsatisfied customers with currently existing market offers; and
2. Capturing market insights that help lower average business costs by adopting appropriate technologies, acquiring and deploying resources, etc., to satisfy evolving demands and to take advantage of prospective opportunities.

Conclusion (1) follows directly from Theorem 16.1, where the market surplus is made up of both not adequately served and unsatisfied customers. Because customers' preferences and needs constantly evolve, presently happy customers can become either unsatisfied or not adequately served so that they become switchers. So, if a company possesses a superb market-sensing capability, then it can identify existing and potential switchers. Now, Theorem 16.1 indicates that if switchers can be identified as the next target, then new competitions will germinate accordingly.

For conclusion (2), it means particularly (Theorem 16.1) that if the magnitude of the market surplus is greater than that of an incumbent company's loyal-customer base, then, jointly, the new entrants can potentially make more profits than some of the incumbent companies. For this situation to actually happen in real life, the entrants must have acquired something new, such as knowledge, technology, or managerial routine, that helps greatly lower their average costs of production. Instances of this "something" can be a new technology, a recently managerial breakthrough that drastically increases management efficiency, or a novel capability that helps improve the level of production. That is, with a superb market-sensing capability, a company is able to attain some of these listed resources to lower the average unit cost of its product.

In terms of relevant studies, Slater and Narver (2000) provide a specific case of conclusion (1); and Hult (1998) and Morgan, Slotegraaf, and Vorhies (2009a) derive particular cases of conclusion (2).

Proposition 16.2 The strength of a company's market-sensing capability is positively correlated to the profit the company can expect to make.

To see the logical reasoning behind this general result, let us examine the following two scenarios:

1. The focal company conducts business in a magnanimous environment, where the supply cannot satisfy the demand; and
2. The company is situated in such an industry or market where market-level growth is absent. That is, in the industry or market, the demand is ≤the supply

Assume scenario (1). The unsatisfied market demand offers the potential for concurrent growth in sales and margins. Hence, additional profits can be expectedly made in these two dimensions. Within such circumstances, the company's market-sensing capability assists it to find where high demand is located and how to tailor its product to satisfy particular consumer needs not fulfilled by rivals' offers. In particular, areas of high demand can be located within underserved or unsatisfied market segments. Such market knowledge can effectively enable the company to conduct logistically effective sales (Proposition 16.1 (1)) and reach a much broader variety of customers. This end can be accomplished either by employing more targeted CVPs or by altering the functionalities of its offers (Proposition 16.1 (2)).

If scenario (2) is the case, then there are two ways (for more details, see Proposition 16.3 below) for the company to grow its profits.

1. Enlarge its market share by selling more to current customers, or to new customers, or a combination of these two methods; and
2. Elevate margins of each unit sold by raising the unit price for each output sold, or lowering the cost of each unit output, or a combination of both.

For this scenario, the strength of the company's market-sensing capability can expectedly help the company realistically know the environment in which it operates—market-level growth is absent. With the equipment of this knowledge, the company can expectedly take appropriate actions to remedy the environment by adjusting its products so that it can appeal to additional customers from both the current market and other markets, or by lifting its production efficiency (Morgan, Vorhies, & Mason, 2009b). That is, the strength of a company's market-sensing capability is positively correlated to profit potential.

16.4 Markets without Market-Level Growth: A Case Analysis

The purpose of growing profit can generally be achieved by increasing sales revenue or margins or both. Other than the scenario (1), which is very rare in life (Dickson, 1992; Keats & Hitt, 1988), and happens when a market is expanding, for example, the market of personal computers in the 1970s (Sobel, 1999), the following proposition portrays the general profit picture of a company:

Proposition 16.3 When the focal company conducts business in an industry that has no market-level growth, its growth in profit can only be materialized by lifting margins, as actualized by a combination of either raising the unit price or lowering the unit cost.

To demonstrate this proposition, it suffices to show that the focal company cannot successfully improve its sales revenue. In general, sales revenue can be increased either by raising the sales of additional units to current customers or by obtaining new customers or both.

Assume that n companies, $n = 1, 2\ldots$, jointly serve the industry (or market) in which the focal company conducts business with their mutually substitutable offers to their respective bases of loyal customers, if the prices are not more than their reservation prices. For these loyal customers, their demands are generally satisfied by their respective companies. Thus, it is nearly impossible to sell additional units of offered products to these customers except when the price per unit value is lowered substantially and/or the demand increases from these loyal customers.

As for selling additional units, enticing a loyal customer to purchase one additional unit of product is equivalent to acquiring a new customer. To obtain new customers, the incumbent companies need to compete by adjusting their per unit-value prices, because assumed is that they all serve the market with their mutually substitutable offers, including features and functionalities of products and follow-up services.

As noted by various scholars, the present world of business is commonly embraced with transient competitive advantages (McGrath, 2013) and customers' patience become much less than before (Forrest & Tallapally, 2018). Therefore, the product market contains an increasing number of discontent and underserved customers, partially due to their quickly changing preferences and tastes. No matter whether these customers were once loyal to their particular companies or not, they become potential new customers of the focal company. And, as adjustable per unit-value prices are used to entice new customers, they also need to be available to loyal customers in order to sustain their loyalty by making them feel they are treated fairly. Assume that the forever advancing technology makes all companies' pricing strategies known to each other. Hence, to possibly attract new customers, all competing companies play the Nash equilibrium through pure self-analysis.

By continuing the proof of Theorem 3.2, the following game-theoretical analysis shows that each incumbent company expects to make the same amount of profit as if it only sells to its loyal customers with the price at their reservation value, although it tries to attract as many new customers as possible. Therefore, this conclusion and the analysis above jointly imply that the focal company cannot expect to raise its sales revenue by increasing unit sales to either current customers, or acquiring new customers, or a combination of both. For establishing this conclusion, as part of the argument for Proposition 16.3, the following looks at game-theoretical details.

In the previous set-up, all competing companies are identical, because their individually different boundary conditions can be normalized as in the proof of Theorem 3.2. Therefore, there is no pure strategy Nash equilibrium (Forrest et al.,

2017; Narasimhan, 1988), for details see Chap. 3 and all the symbols used below are the same as those in the proof of Theorem 3.2. In order to compute for a symmetric mixed strategy Nash equilibrium, let $F_i(P)$ be the price distribution of company i, $i \in \{1, 2, \ldots, n\}$. Then, company i's objective is to maximize the following function by choosing appropriate F_i:

$$\begin{aligned} E(\Pi_i) &= \int_{-\infty}^{\infty} \left\{ uP + \prod\nolimits_{j \neq i}^{m} [1 - F_k(P)] vP \right\} dF_i(P) \\ &= \int_0^1 \left\{ uP + \prod\nolimits_{j \neq i}^{m} [1 - F_k(P)] vP \right\} dF_i(P) \end{aligned} \tag{16.1}$$

where $E(\Pi_i)$ stands for company i's expected profit for all possible pricing strategies, and all other symbols are the same as those defined and normalized in the proof of Theorem 3.2. The reason why the upper and lower limits of the integral are changed respectively from the initial $+\infty$ and $-\infty$ to 1 and 0 is that when $P < 0$ (=the cost of production) or when $P > 1$ (=reservation price), the profits are zero.

Company i can readily earn its expected amount of profit u by charging its loyal customers the reservation price, which is normalized to 1. However, Theorem 16.1 implies that it cannot simply do so; otherwise, new competitions might invade its market territory. That is, to potentially maximize its profit and to protect its market territory, each incumbent company has to adjust price P to absorb as many switchers as possible as new customers so that the consumer surplus can be potentially exhausted. Also, no company is motivated and incentivized to price its product below $u/(u + v)$, because doing so will yield less profit than u despite of attracting all market surplus, because $uP + vP \geq u \to P \geq u/(u + v)$. Solving the following equilibrium indifference condition for company i

$$u \times P + \prod\nolimits_{j \neq i}^{n} [1 - F_k(P)] v \times P = u \times 1, \; \frac{u}{u+v} \leq P \leq 1, i = 1, 2, \ldots, n, \tag{16.2}$$

produces the following symmetric equilibrium price distribution

$$F(P) = F_i(P) = F_j(P) = 1 - \left[\frac{(1 - P)u}{P(1 - nu)} \right]^{\frac{1}{n-1}}, \; \frac{u}{u+v} \leq P \leq 1, \tag{16.3}$$

So that $F(u/(u + v)) = 0$ and $F(1) = 1$. So, the expected profit of each incumbent company is equal to

$$E(\Pi) = \int_{-\infty}^{+\infty} \left\{ uP + \prod_{j \neq i}^{m} [1 - F_k(P)] vP \right\} dF(P)$$
$$= \int_{\frac{x}{x+y}}^{1} u dF(P) = uF(P)|_{\frac{u}{u+v}}^{1} = u. \tag{16.4}$$

In other words, in Nash equilibrium of the symmetric mixed strategy, the expected profit of each incumbent company does not vary and is equal to that it can readily earn from its loyal customers, although these incumbents try to obtain as many new customers as possible.

Proposition 16.4 When a company does business in such an industry that experiences no market-level growth, its growth in profit from expanding sales revenue and that from margins, two basic components that drive a company's profit, do not ordinarily appear simultaneously.

To see the argument for this proposition, first examine what resources stand for. Specifically, it represents an asset a company possesses, which is either tangible or not (Harmancioglu et al., 2009) and can be employed to carry out the strategies adopted by the company (Barney & Arikan, 2001). In other words, a resource is a thing that a company can mobilize to achieve its corporate objectives through implementing its adopted strategies. It can take different forms, which can be, for example, physical, financial, informational, intellectual, or organizational.

Without loss of generality, each company can be identified with its unique system of resources (Barney & Hesterly, 2012; Peteraf & Barney, 2003), consisting of set resources and a set of relations that associate the resources in ways that are unique to the company. Such identification is unique, because, for instance, among other factors, each company has its own unique set of human resources and particular ways to utilize these human resources. Therefore, the assumption given in the proposition indicates that

1. If a company desires to either maintain the current level of or improve its sales revenue, it has to increase the number of units of product sold to existing customers and/or to acquire new customers. For details, see the argument and analysis of Proposition 16.3 and
2. If the company plans to improve its margins, it has to raise the price materialized for each unit sold, lower unit cost, or use a combination of these options.

For the accomplishment of item (1), Proposition 16.3 implies that other than lifting sales revenue, the situation will be more about maintaining the level of the sales revenue. So, the company has to assign specific resources to the exploitation of the market value of their existing offers. As for the accomplishment of item (2), the company's resources need to be classified into two classes: (i) one useful for the exploration of incremental improvements of the existing market offers and (ii) the other for the production of completely new offers that feature more use values to customers than the currently available ones. Note that resources tenaciously adopted

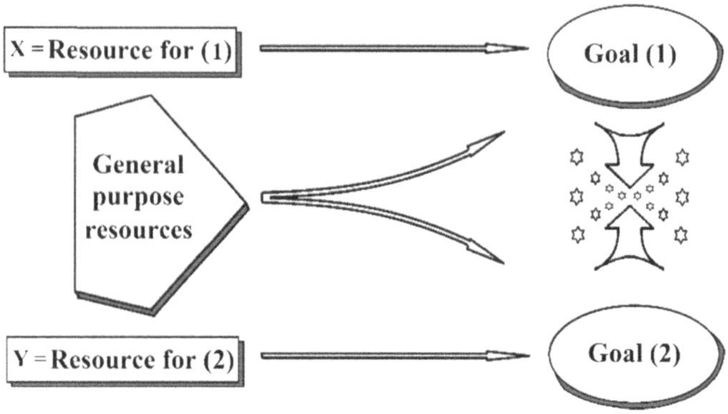

Fig. 16.1 Separation of available resources

to accomplish (1) and those for (2) do not generally work well together (McGrath, 2013), since they aim at achieving inconsistent goals. At the same time, there are common resources that can be either deployed easily or needed to realize different business goals. Examples of such common resources include, but not limited to, networks that lead to information sharing, risk-taking, and adoption of innovations (Dutta et al., 1999).

Let the set of resources that lead to the realization of goal (1) and the set of those that help with that of goal (2) be denoted as X and Y, respectively. Assume that the sets X and Y do not include any general-purpose resources so that X and Y are mutually exclusive. Then when resources from X and those from Y interact, they will likely produce unwanted outcomes. In other words, the profit generated from sales revenue and that from margins do not generally increase simultaneously, because resources in X and those in Y tend to produce neutralizing or adverse effect(s). Figure 16.1 depicts the situation discussed here, where the collection of resources exclusively developed for business goal (1) and that for (2) do not overlap. Additionally, resources available for general purposes can be deployed repeatedly to attend varying, even inconsistent purposes.

In terms of the literature, various scholars, such as Markman and Gartner (2002) and Steffens et al. (2009), have confirmed different specific versions of the general result stated in Proposition 16.4.

What is stated in Proposition 16.4 raises the following practical challenges to decision-making managers:

- What could and should they do when it becomes clear that the current competitive advantage is eroding, especially, when the situation is only clear to themselves?
- How could and should their organizations be restructured in such a way that they can concurrently uncouple the current advantage and mobilize resources, either existing or newly acquired, into the development of the next perceived advantage?

To illustrate these challenges, let us examine the well-known American media-services provider and production company, named Netflix. As a cofounder, chairman, and CEO of Netflix during the mid-2000s, Wilmot Reed Hastings recognized from the success of YouTube that the preferred method for users to access contents would be streaming so that Netflix's DVD business would soon be obsolete and no longer be the core of the company's future.

Based on this creative realization of the market trend, which was most advanced in time among all rivals, Qwickster, a spinoff of Netflix, was formed to continuously exploit the company's successful DVD services by maximizing the underlying cash flow for the duration of the expected decline of these services. At the same time, the leadership of Netflix looked to the future. Its job was to manage rapid development of and access to digital content. To minimize potentially adverse effects between the exploitation of the continuing DVD services and the exploration of newly discovered opportunities, Netflix operated intelligently into two independent business directions.

In terms of Proposition 16.4 and business operating processes, Netflix's decision makes practical sense. However, because no one else at the time in the marketplace saw the coming of the expected new era, many customers became enraged by the switch. They were infuriated by the potential that they might have to double their effort when they desired to access contents in both formats and in two different places. Additionally, cable television, satellite dish services, and other content providers experienced growing wary of Netflix's disruptive nature. See McGrath (2013) for a more in-depth discussion of this case study.

16.5 Customer Relations and Crafts of CVPs: Additional Cases

All propositions in the previous sections are derived on the bases of the rigor of game theory and the logical reasoning that is parallel to the one widely used in natural science and mathematics (Kline, 1972). The rigor plays the role of a solid foundation for the construction of the propositions, while the particular logical reasoning helps avoid limitations of data- and anecdote-based approaches. That explains the reason why these established propositions represent generally true conclusions unless the correspondingly stated assumptions are not satisfied. Contrary to this end, because the technique of hypothesis testing is widely employed in the literature, relevant empirical studies can provide only managerial suggestions. As a matter of fact, hypothesis testing stands for a statistics-based approach, very useful in terms of uncovering potential facts but not proving any fact (Forrest et al., 2020).

This section, as an application, examines how the propositions developed earlier in this chapter can help establish results on how customer relationship management capability and profitability and the formulation of effective CVPs are related to one another. By the capability of customer relationship management (CRM) or CRM capability for short, it represents a company's ability that it enables the company to influence and control customer-level profits through

1. Clearly classifying customers and prospects according to their profit potentials,
2. Developing individually appropriate relationship with each echelon of customers and prospects, and
3. Sustaining developed customer relations so that they are optimal based on the assessed profit potentials of the respective customer echelons.

This definition of the concept of CRM capabilities expresses the realization that beyond individual business transactions, it is these relationships that lead to profits for companies and satisfaction for customers (Verhoef, 2003), and that customers and potentials are not similarly attractive for a company to meet their needs profitably (Niraj et al., 2001).

Proposition 16.5 A company's CRM capability is positively correlated to the company's ability to generate profits from the growth of both sales' revenue and margins.

Let us see why this proposition is true from two angles: (1) sales revenue growth and (2) margin growth. As for (1), a company's CRM capability helps orient the company's effort towards the understanding of how customers and prospects can contribute to the overall profitability picture differently (Reinartz et al., 2005) before any of these people is converted into a switcher (Theorem 16.1). Hence, when a company is more capable of working with customer relationships, it will more adequately classify customers and prospects into multilayered echelons of attractiveness so that it will be able to develop and manage relationships with the echelons corresponding to their individually varied levels of attractiveness. Specifically, at the high extreme on the attractiveness spectrum, a company will like to devote its valuable resources to the instigation and conservation of relationships with high-potential prospects and high-revenue-producing customers. And at the low extreme of the spectrum, without putting in any additional effort a company simply employs prices to enlarge its market territory by targeting switchers of the market (Theorem 16.1). Hence, one can expect that companies with stronger CRM capabilities generate more profits or at least maintain their profitability from customers of all different echelons of attractiveness.

At this junction, it is worth pointing out that there are such companies (Ryals, 2005) that only pay attention to and devote their efforts to high-potential prospects and highly profitable customers. For these companies, they tend to have fewer customers and generate less profits (Reinartz et al., 2005). To this end, one very good case that vibrantly demonstrates how such narrow-minded attempts can and do lead to disastrous outcomes is the story of Pan American World Airways (Sobel, 1999). In particular, the airline company mostly concentrated on lucrative international air traffic, while ignoring the domestic market. Hence, when it suffered from skyrocketing energy prices and terrorist attacks, the company was wiped out as a major player in the air travel industry. The very reason why such efforts of concentrating on high profile prospects and customers led to disastrous failures is the difficulty or impossibility of maintaining such relationships over relatively long periods of time, during which customers' needs and preferences evolve constantly.

Regarding (2)—margin growth, a strong CRM capability assists a company to fine-tune its effort on how to position resources on the instigation and upkeep of appropriate relationships with prospects and customers of different echelons of attractiveness. When its knowledge and experience on prospects and customers of different attractiveness increase, a company is able to lower its service costs gradually over time (Reinartz & Kumar, 2000), while obtaining a rising level of customer retention and more experienced users of its market offers. All these improved operational efficiencies further lower the company's service cost over time (Ryals, 2005). Therefore, it can be seen that strongly CRM capable companies are generally able to increase their margins by continually lowering their service costs, as a result of appropriately allocating resources to prospects and customers of different attractiveness levels. Such allocation of resources consequently raises the level of satisfaction of customers in different echelons of attractiveness. When suitable relationships are upheld with those customers whose purchases are not affected merely by prices, companies with strong CRM capabilities are able to charge higher prices for their products (Cao & Gruca, 2005). So, Proposition 16.5 is supported.

In terms of the literature, this result is inconsistent with Morgan, Vorhies, and Mason (2009b). Specifically, these scholars hypothesized and empirically confirmed that a company's revenue growth rate is a decreasing function of the company's CRM capability, while its margin growth rate is an increasing function.

Market knowledge is a concept closely related to that of CRM capabilities. It refers to knowledge about market demand in terms of who customers and prospects are, what their needs in life are, what problems they face, whom competitors are, and what features their offers possess. By an effective CVP (to a company), it means that the CVP can materialistically help the company to acquire additional profit beyond the case when the CVP is absent, after subtracting all the costs of formulating and adopting the CVP.

Proposition 16.6 As a market-based resource, market knowledge can be used to craft effective CVPs.

Indeed, market knowledge is a resource based on market(s), because the very existence of such knowledge depends on that of some market(s). To see why such knowledge can be employed to formulate effective CVPs, let us respectively treat a focal company and its customers and prospects as input-output systems in such a way that the mission of the company represents its CVP and market demands the desired inputs of the customers and prospects. Then when the CVP is communicated to the market, the communication helps associate the company's input-output flows with those of targeted customers and prospects in the market. In this systemic modeling, resource sharing is echoed in affiliated input-output flows of the company and the targeted customers and prospects. The concept of superior value is reflected in the alignment of these harmonious input-output flows, for details, see Fig. 16.2.

This systemic intuition demonstrates that if an adopted CVP is effective to a company, then the input-output flows of the company need to align with those of a

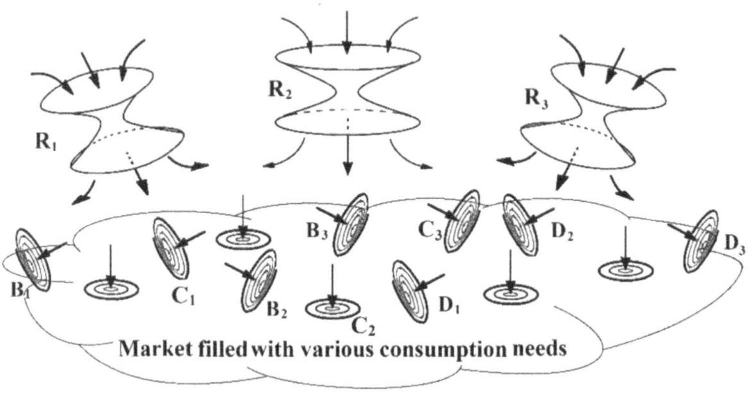

Fig. 16.2 Rivals and their market territories

sufficient number of customers and prospects. And, the company's market knowledge is intuitively seen as how the company understands the patterns of its rivals' input-output flows relative to the market and how the input-output flows of customers and prospects respond to the marketing efforts of these companies.

Intuitively, in Fig. 16.2 three of the rivals of the focal company are shown as the input-output flows R_1, R_2, and R_3, respectively. The market territory of R_1 includes B_1, B_2, and B_3; that of R_2 contains C_1, C_2, and C_3; and that of R_3 has D_1, D_2, and D_3. In such setting, the focal company's market knowledge covers how R_1 and Bs associate with each other, that between R_2 and Cs, that between R_3 and Ds, and how these semi-separate territories interact with one another. When we say that the focal company possesses such market knowledge, it represents that this company knows how its rivals interact with their associates and customers and how these interactions do not meet the expectations of customers. Such knowledge can readily assist the focal company to discover market niche(s) that it can possibly fill based on its available resources and capabilities; and the niche(s) demonstrate which customers are not adequately satisfied by its rivals, how large such consumer surplus (Theorem 16.1) is, and what these discontent customers are longing for.

Discussions above collectively elucidate the reason why market knowledge is such a resource that it is market-based and that the focal company can utilize to formulate its effective CVP(s).

In terms of the relevant literature, Proposition 16.6 is different from what is confirmed by Kozlenkova et al. (2014), and also dissimilar to the statement that market knowledge is a resource for crafting CVPs (Payne et al., 2017). As a matter of fact, if one does not consider the effectiveness of CVPs, he/she can formulate CVPs without using any market information. Although Proposition 16.6 is different from these mentioned works, this chapter still likes to use this opportunity to honor these scholars: It is their pioneering work that eventually led to the development of this proposition.

Proposition 16.7 In terms of the formulation of effective CVPs, creative comprehension of market invitations is the key element.

Theorem 16.1 readily implies this conclusion. In particular, the market normally gestures the call for additional competition and innovation by showing a large consumer surplus. For the case when $n = 1$, that is when the market is monopolistic, rather than switchers in Theorem 16.1, the level of customer dissatisfaction and the magnitude of discontent customers are employed as signs of market invitation for new competition and innovation. In this case, a large host of producers potentially appear to answer the market call, depending on the relevant profit potential. Specifically, after sensing the market call and great profit potential, entrepreneurs with different background knowledge and experience proactively introduce their individually different versions of market offers to possibly satisfy the recognized customer demand. Hence, a different way to uncover a market opportunity is to discover where a producer surplus appears. In other words, for a CVP to be effective, it has to be an offspring of pioneering and wide-ranging understanding of market signals.

To help further comprehend this proposition, a real-life case of business from Ye et al. (2012) comes to mind.

In the northern part of the United States, there is a state university of over 29,000 students. They are mostly unmarried undergraduates and live on tight budgets. Because these students are young and physically active, most rental properties do not offer easily damaged laundry facilities to their student tenants. To fill this market need, many standalone laundromats appeared near the campus and compete for students' business by offering self-servicing, coin-operated laundry machines. And because of the limited amount of outdoor sunlight, as caused by the long and cold winters of the region, another successful business, targeting appearance-conscious college women, is indoor tanning salons.

Different from the laundromat services, which do not really differentiate themselves from each other than some minor variations, such as locations, indoor tanning salons compete for business by providing competitive prices and indoor experiences. That is, these two kinds of businesses respectively serve their individually different market needs, while customers move freely from one service location to another or depending on who is more competitive.

Recognizing innovatively the large magnitude of a consumer surplus (Theorem 16.1), two entrepreneurs creatively launched their independent coin-operated self-service laundry facility combined with indoor tanning service at one single location under the same roof. The innovativeness of this new business can be seen in the following two ways:

1. It saves students their otherwise unproductive and costly waiting time by concurrently completing two tasks—clothes washing and skin tanning, and
2. It provides a natural opportunity and location for male and female students to develop relationships. Specifically, the apparent savings of time from doing two unrelated errands simultaneously attract greater participation of women in the business installment in general and the tanning service in particular. Such a greater number of women in the facility entices to the laundry service a much

larger number of men who do not have any preference as to which laundromat to use.

As a consequence of this creative understanding of the market signal, the joined business is able to amass higher fees from the laundry service than other standalone laundromats. And, it endures the market competition of the tanning service. Specifically, the female demand for the tanning service rises as a consequence of the heavy flow of male users of the laundry facility.

16.6 A Few Final Words

Consequent to the recognition of the importance of market-sensing in terms of company performance (Day, 1994; Forrest et al., 2017; McGrath, 2013), a number of scholars have contributed to the study of the concept of market-sensing. However, their works have led to inconsistent conclusions, which points to the theoretical and practical need to clarify these inconsistencies. To this end, this chapter takes the approach of going beyond the methodology commonly used in the literature, and uses game theory and the logical reasoning that is widely used in mathematics (Kline, 1972). After successfully accomplishing this objective through developing a series of generally true propositions, this chapter then looks at two closely related topics, as case studies to demonstrate how developed propositions can be employed to produce useful conclusions. One topic is on where profit sources are located within a stagnant industry without much market-level growth; and the other on the formulation of effective CVPs. Consequent to the specific methodology chosen here, developed conclusions in this chapter are free from the constraints of data-based and anecdote-based analyses, as widely and commonly employed in the literature. Specifically, among others this chapter establishes the following generally true conclusions:

- A superb market-sensing capability implies an ability to create vitally important market knowledge and increase its expected profits. Such consequences are generally achieved by adopting appropriate technologies, by acquiring and deploying resources (Propositions 16.1 and 16.3).
- Within a stagnant industry that lacks market-level growth, a superb market-sensing capability points to potential profit growth through raising margins (Proposition 16.3). and
- An effective CVP can be materialistically drafted out of market knowledge and innovative understanding of the knowledge (Propositions 16.6 and 16.7).

As commonly practiced in business operations (Duan et al., 2019; McGrath, 2013), implementing best practices and data-driven decision-makings have been widely employed. However, no matter how good the practices had been recorded and how effectively decisions have been data-driven, the consequences of implementing the practices and decisions can be and have been often widely different from what was expected (Forrest et al., 2020). This business reality richly reveals the practical significance of this chapter, because the developed conclusions produce general recommendations instead of suggestions for decision-making managers and entrepreneurs. Specifically, the recommendations below for general managerial purposes can follow readily from the established results in this chapter.

- To achieve better performance, a company has to attain and advance its market-sensing capability. With such capability, the company is able to know underserved customers, dissatisfied ones with currently available market offers, and appropriate technology that can potentially improve productivity (Propositions 16.1 and 16.2);
- When the potential of market-level growth is absent, a company needs to either raise the price and/or lower the cost of each unit sold (Proposition 16.3) while not giving much effort to improving the sales revenue (Proposition 16.4);
- To succeed in the marketplace, a company needs to manage customer relationships in such a way that resources are deployed corresponding to the levels of attractiveness of customers and prospects (Proposition 16.5); and.
- To formulate effective CVPs, companies need to make joint use of their market knowledge and innovatively understanding of that knowledge (Propositions 16.6 and 16.7).

Behind all results established in this chapter is the following implicit assumption: each company, considered here, is financially solvent, because of the positive cash flows from the product market as a consequence of its meeting market demand (s) with its particular offer(s). However, this assumption in today's world of business is no longer generally true. So, as future research attempts, results parallel to those developed in this chapter need to be considered for companies that exist for other purposes, such as producing promising futures, expanding market territories, or others (Li & Ma, 2015).

In terms of methodologies of research, it is well-known that implementing best practices and making data-driven decisions cannot definitely lead to expected outcomes (Lin & OuYang, 2010). However, contemporary business decisions have been increasingly data-driven (Duan et al., 2019; McGrath, 2013). Therefore, there is a theoretical and practical need to develop procedures that can help turn recognized best practices and statistics or anecdote patterns of limited validity into recommendations for applications beyond the origins from which the practices and patterns are initially revealed.

References

Alshanty, A. M., & Emeagwali, O. L. (2019). Market-sensing capability, knowledge creation and innovation. *Journal of Innovation & Knowledge, 4*(3), 171–178.

Ardyan, E. (2016). Market sensing capability and SMEs performance. *DLSU Business & Economics Review, 25*(2), 79–97.

Barney, J., & Arikan, A. (2001). The resource-based view: Origins and implications. In M. Hitt, R. Freeman, & J. Harrison (Eds.), *Handbook of strategic management* (pp. 124–185). Blackwell.

Barney, J., & Hesterly, W. (2012). *Strategic management and competitive advantage: Concepts and cases* (4th ed.). Pearson.

Bayighomog Likoum, S. W., Shamout, M. D., Harazneh, I., et al. (2018). Market-sensing capability, innovativeness, brand management systems, market dynamism, competitive intensity, and performance. *Journal of the Knowledge Economy*. https://doi.org/10.1007/s13132-018-0561-x

Bharadwaj, N., & Dong, Y. X. (2014). Toward further understanding the market-sensing capability–value creation relationship. *Journal of Product Innovation Management, 31*(4), 799–813.

Cao, Y., & Gruca, T. S. (2005). Reducing adverse selection through customer relationship management. *Journal of Marketing, 69*(4), 219–229.

Clements, C., Alwang, J., & Achdiawan, R. (2019). Value chain approaches in a stagnant industry: The case of furniture production in Jepara, Indonesia. *Bulletin of Indonesian Economic Studies, 55*(3), 341–365.

Day, G. S. (1994). The capabilities of market-driven organizations. *Journal of Marketing, 58*(4), 37–52.

Dickson, P. R. (1992). Toward a theory of competitive rationality. *Journal of Marketing, 56*(1), 69–84.

Duan, Y. Q., Edwards, J. S., & Dwivedi, Y. K. (2019). Artificial intelligence for decision making in the era of big data – Evolution, challenges and research agenda. *International Journal of Information Management, 48*(October), 63–71.

Dutta, S., Narasimhan, O., & Rajiv, S. (1999). Success in high-technology markets: Is marketing capability critical? *Marketing Science, 18*(4), 547–568.

Einstein, A. (1997). In R. Schulmann et al. (Eds.), *The collected papers of Albert Einstein: The Berlin years* (Vol. 8). Princeton University Press.

Ferber, U., & Schlappa, H. (2016). Managing brownfield land in stagnant land markets. In W. J. V. Neil & H. Schlappa (Eds.), *Future directions for the European shrinking City* (pp. 138–154). Routledge.

Forrest, J. Y.-L., Buttermore, J., & Theresa Wajda, T. (2017). At Nash equilibrium when new market competitions appear? *Kybernetes: The International Journal of Cybernetics Systems and Management Sciences, 46*(2), 256–271.

Forrest, J. Y.-L., Gong, Z. W., Scott, E., & Soto Ruiz, G. E. (n.d.). (to appear). Market-sensing capabilities, profitability within stagnant industries and crafting of customer value propositions. Submitted for possible publication.

Forrest, J. Y.-L., Nicholls, J., Schimmel, K., & Liu, S. F. (2020). *Managerial decision making: A holistic approach*. Springer.

Forrest, J. Y.-L., & Tallapally, P. (2018). Customers are less patient, sustainable advantage becomes transient and the key for firms to succeed in fast changing markets. *Journal of Business, Economics and Technology, 21*(1), 1–14.

Harmancioglu, N., Droge, C., & Calantone, R. (2009). Strategic fit to resources versus NPD execution proficiencies: What are their roles in determining success? *Journal of the Academy of Marketing Science, 37*(3), 266–282.

Hult, G. T. (1998). Managing the international strategic sourcing process as a market-driven organizational learning system. *Decision Sciences, 29*(winter), 193–216.

Keats, B. W., & Hitt, M. A. (1988). A causal model of linkages among environmental dimensions, macro organizational characteristics, and performance. *Academy of Management Journal, 31* (3), 570–598.

Kline, M. (1972). *Mathematical thought from ancient to modern times*. Oxford University Press.

Kowalkowski, C. (2011). Dynamics of value propositions: Insights from service-dominant logic. *European Journal of Marketing, 45*(1/2), 277–294.

Kozlenkova, I., Samaha, S., & Palmatier, R. (2014). Resource-based theory in marketing. *Journal of the Academy of Marketing Science, 42*(1), 1–21.

Le, N., Nguyen, H. M., & Schultz, E. (2020). Financial development and productivity growth in stagnant industries. SSRN: https://ssrn.com/abstract=3275778, Accessed May 4, 2020.

Lecours, M. (2017). Developing a framework to craft a value proposition. *Journal of Financial Planning, 30*(3), 23–25.

Li, T., & Ma, J. H. (2015). Complexity analysis of dual-channel game model with different managers' business objectives. *Communications in Nonlinear Science and Numerical Simulation, 20*, 199–208.

Lin, Y., & OuYang, S. C. (2010). *Irregularities and prediction of major disasters*. CRC Press.

Lindblom, A., Olkkonen, R., Mitronen, L., & Kajalo, S. (2008). Market-sensing capability and business performance of retail entrepreneurs. *Contemporary Management Research, 4*(3), 219–236.

Lindstrom, M. (2005). *Brand sense: Build powerful brands through touch, taste, smell, sight, and sound*. Free Press.

Markman, G. D., & Gartner, W. B. (2002). Is extraordinary growth profitable? A study of Inc. 500 high-growth companies. *Entrepreneurship: Theory and Practice, 27*(1), 65–75.

McGrath, R. G. (2013). *The end of competitive advantage: How to keep your strategy moving as fast as your business*. Harvard Business Review Press.

Morgan, N., Slotegraaf, R., & Vorhies, D. (2009a). Linking marketing capabilities with profit growth. *International Journal of Research in Marketing, 26*(4), 284–293.

Morgan, N. A., Vorhies, D. W., & Mason, C. (2009b). Market orientation, marketing capabilities, and firm performance. *Strategic Management Journal, 30*(8), 909–920.

Narasimhan, C. (1988). Competitive promotional strategies. *Journal of Business, 61*(4), 427–429.

Niraj, R., Gupta, M., & Narasimhan, C. (2001). Customer profitability in a supply chain. *Journal of Marketing, 65*(3), 1–16.

Osterwalder, A., Pigneur, Y., Bernarda, G., & Smith, A. (2014). *Value proposition design: How to create products and services customers want*. Wiley.

Payne, A., Frow, P., & Eggert, A. (2017). The customer value proposition: Evolution, development, and application in marketing. *Journal of the Academy of Marketing Science, 45*, 467–489.

Peteraf, M., & Barney, J. (2003). Unraveling the resource-based tangle. *Managerial and Decision Economics, 24*(4), 309–323.

Porse, H., & Rudolph, B. (2017). The seaweed hydrocolloid industry: 2016 updates, requirements, and outlook. *Journal of Applied Phycology, 29*, 2187–2200.

Reeves, R. (1961). *Reality in advertising*. Alfred A. Knopf.

Reinartz, W., & Kumar, V. (2000). The CRM process: Its measurement and impact on performance. *Journal of Marketing Research, 64*(4), 17–35.

Reinartz, W., Thomas, J., & Kumar, V. (2005). Balancing customer acquisition and retention resources to maximize customer profitability. *Journal of Marketing, 69*(1), 63–79.

Ryals, L. (2005). Making customer relationship management work: The measurement and profitable management of customer relationships. *Journal of Marketing, 69*(4), 252–261.

Slater, S. F., & Narver, J. C. (2000). Intelligence generation and superior customer value. *Journal of the Academy of Marketing Science, 28*(1), 120–127.

Sobel, R. (1999). *When giants stumble: Classic business blunders and how to avoid them*. Prentice Hall.

Steffens, P., Davidsson, P., & Fitzsimmons, J. (2009). Performance configurations over time: Implications for growth- and profit-oriented strategies. *Entrepreneurship Theory and Practice, 33*(1), 125–148.

Sugiyarti, G., & Ardyan, E. (2017). Market sensing capability and product innovation advantages in emerging markets: The case of market entry quality and marketing performance of Batick industry in Indonesia. *DLSU Business & Economics Review, 27*(1), 175–189.

Tournois, L. (2016). When markets stagnate: Finding new territory through reverse innovation. *Journal of Business Strategy, 37*(6), 18–27.

Verhoef, P. C. (2003). Understanding the effect of customer relationship management efforts on customer retention and customer share development. *Journal of Marketing, 67*(4), 30–45.

Wouters, M., Anderson, J. C., & Kirchberger, M. (2018). New technology startups seeking pilot customers: Crafting a pair of value propositions. *California Management Review, 60*(4), 101–124.

Ye, G. L., Priem, R. L., & Alshwer, A. A. (2012). Achieving demand-side synergy from strategic diversification: How combining mundane assets can leverage consumer utilities. *Organization Science, 23*(1), 207–224.

Part V
Some Roles of Manufacturing and Artificial Intelligence

Chapter 17
Manufacturing in Industrial Transformations

Abstract At the present time, the world is expecting the arrival of another new era of industrial development. To position themselves appropriately in order to either maintain or acquire their leadership in the world, some nations, either developed or developing, have introduced and adopted their individually different plans on how to transform their current states of industrial manufacturing by focusing on innovations.

Keywords Economic growth · Innovation · Manufacturing · Market · System

To understand this phenomenon theoretically and to see whether such efforts might practically succeed, this chapter, which is mainly based on (Forrest et al., 2021, Role of manufacturing sector in the forthcoming national industrial transformations. Accepted for publication by *International Journal of Sociotechnology and Knowledge Development*, to appear), addresses the question of how important the manufacturing sector is in the overall development of a nation's economic strength.

In order to derive generally true conclusions, this chapter employs systems thinking and a logical reasoning that is parallel to the one commonly used in mathematics and natural science. By doing so, this chapter is able to establish three formal propositions on related issues without suffering from limitations of statistics- and anecdote-based approaches. At the end, policy recommendations are provided along with some open problems for future research.

17.1 Introduction

Riding on the maturing development of the Internet and the fast emergence of artificial intelligence, a number of nations around the globe have been engaging in the next round of industrial transformation. They desire to transform their current states of increasingly automated manufacturing to a potential future state of intelligent manufacturing. These pioneering nations include the Great Britain (Hall &

J. Y.-L. Forrest, Y. Liu, *Value in Business*, Contributions to Management Science,
https://doi.org/10.1007/978-3-030-82898-1_17

Pesenti, 2017), the USA (OWH, 2016), Germany (Industry 4.0), Japan (Government of Japan, 2015), China (State Council of the PRC, 2015), and France (Marc et al., 2018). Such forward-looking efforts are theoretically planned and implemented based on a domineering lesson of the past rounds of industrial revolutions (Wen, 2016). These proactive nations have been vigorously pushing for their desired transformations to happen in numerous business activities in the manufacturing sector. Such international scene of business efforts of leading powerful nations naturally raises the following question:

> What is manufacturing's role in a nation's effort to generate and advance its self-sustained moment of economic growth?

This is a very important question for developed nations as well as developing nations. For those developed nations that desire to enjoy their dominant positions in the increasingly globalized world economy, they surely have to examine if their adopted policies and provided government supports are sufficient in their efforts. For those developing nations that have been trying to modernize themselves (Wen, 2016), they surely have to face the likelihood of trailing further behind developed nations economically and politically. Hence, these developing nations need to consider (Forrest et al., 2018) how to actually modernize their societies and how to possibly provide their citizens with their longed prosperity.

Due to its academic and practical importance, this question and related issues have been considered by many scholars from different angles. However, all of these studies relied on either data- or anecdotes-based approaches, producing inconclusive or inconsistent results (Wen, 2016), because these employed approaches suffer from many visible constraints. Because of this problem, Rostow (1960) loudly declared the need for the introduction of new methods of analysis and different thinking logics for studies of such questions and issues.

On top of this 60-year-old call and the necessity to theoretically confirm the rationale behind various nations' attempts to continue their leading positions in the world economy, this chapter introduces systems thinking and methodology, and the logical reasoning that has won victories one after another for mathematics and natural science (Kline, 1972). Because of such novelty in terms of methodology, this chapter is able to break through and avoid the constraints of conventional data- and anecdote-based approaches when it develops sharp theoretically important and practically meaningful conclusions and insights.

In particular, some of the visible constraints of data- and anecdote-based approaches are listed as follows: The desired quality of data cannot generally be guaranteed; those data that are crucial for obtaining the correct conclusions might be missing; when a business success is eventually recognized, there is normally no way to go back in time to collect the relevant data for analysis. Additionally, no matter whether it is the employment of a statistics-based approach or a calculus-based method, the consequent results represent the discovery of some potential facts, where the former approach extrapolates the past into the future, while the latter method extends the present into the future. Therefore, results derived out of such approach and method tend to have limited validity; and if such validity exists, it is always conditional. That is the reason why conclusions, produced by analyzing data or anecdotes, lead to suggestions instead of recommendations. See Chap. 1 of Forrest et al. (2019) for an in-depth discussion on this matter.

Before concluding this section, let us briefly discuss the methodology—logical reasoning and systems thinking—employed in this chapter. Although relevant studies in the literature all employ certain kinds of logical reasoning, the one used throughout this chapter is fundamentally different. It parallels the one well employed in Euclidean geometry and most parts of mathematics (Kline, 1972). It starts with a set of obviously true postulates (or axioms) and is supported by an intuition that assists advanced thinking and reasoning in the consequent theoretical development and practical applications. Then, generally true propositions or conclusions are established with new terms and concepts introduced one by one. Scientific knowledge generation, as of this writing, has universally utilized these kinds of logical reasoning and intuitive visualization, and they have helped such well-established disciplines as physics and chemistry to win many successes (Kuhn, 1962). One key difference between the scientific theories derived this way and those theories developed in the literature this chapter and this book contribute to is that conclusions of the former are mostly definite, while those of the latter claims with their validity not assured. In other words, from the theoretical conclusions established in the following sections, one can derive general and reliable policy recommendations instead of suggestions of limited validity out of data- and/or anecdote-based theories. Speaking differently, the main difference between conclusions established in this chapter (and this book) and the related ones developed empirically in the literature is the following: empirical studies provide pioneering surveys that reveal potential facts, and offers anecdotal reconfirmation of theoretical conclusions such as the ones established in this chapter.

Regarding systems thinking and methodology, their concepts and approaches have been progressively and assuredly employed in studies of business-related topics by different scholars (Forrest, 2018; Mazzei et al., 2017; Porter, 1985). Because of the novelty of such a way of thinking, its unconventional perspective naturally leads to drastically enriched conclusions and insights.

Discussions above point to the contribution this chapter makes to the extant literature. First, conclusions established here help fill a gap existing between works that are either in support of or against the belief that manufacturing positively assists a nation in developing its self-sustained momentum of economic growth. Second, this chapter plays the role of specifically showcasing how systems science can benefit the investigation of economic development.

This chapter is organized as follows. Section 17.2 reviews the relevant literatures. Section 17.3 examines how the feedback mechanism helps closely and systemically connect market exchange and manufacturing production. Section 17.4 studies such a feedback mechanism assists the kick-start of a self-sustaining momentum of economic growth, and when disruptive innovations positively improve the performance of manufacturing firms. This chapter concludes in Sect. 17.5, where recommendations for national governments and open questions for future research are provided.

17.2 Literature Review

The literature this chapter is based on is closely related to that of industrialization. In particular, by clearly spelling out the concept of industrialization, it is confirmed (Hopkins, 2000) that industrialization or industrial advancement appears once manufacturing plays a leading role in economic development. The European Commission has published since 2010 communication papers to encourage EU countries to understand the fundamental importance of manufacturing for advancing the region's competitiveness and sustainable growth (EC, 2014).

Specific economic challenges Africa faces and how industrialization has historically propelled economic growth, as anecdotally seen, jointly point to the characteristics and benefits of the manufacturing sector (Scholz, 2018). However, the African story seems to disagree with such an understanding. In particular, Africa experienced positive economic growth for a few decades when it deindustrialized itself, although the recent drop in commodity prices has caused a deceleration of growth. So, it is natural to question (Moyo & Jeke, 2019) whether promoting manufacturing would actually bring about higher and sustainable economic growth, while lowering unemployment. By suggesting a yes answer to this question, Moyo and Jeke then recommend policymakers adopt measures to boost manufacturing output. It is argued (Naudé & Szirmai, 2012) that for developing countries, when their primary activities evolve into those of modern production, a manufacturing sector is needed to provide high-quality employment; and it aligns with the international division of labor and does not reverse back to the earlier primary state. In terms of particular manufacturing systems, an algorithm is developed by Merati et al. (2019) to conduct evaluations and optimization.

Opposite to the previously discussed literature, there are also studies that disagree with the belief that manufacturing had historically played an important role in industrialization. For instance, recent data from both developed and developing countries indicate that services account for more than 50% to over 80% of the economy; that naturally challenges the long-held belief of positive correlations between economic growth and the development of manufacturing (Szirmai & Verspagen, 2015). The current weight of services in the economy makes it increasingly difficult for developing economies to industrialize through purposefully growing their manufacturing because that sector steadily provides fewer employment opportunities.

Such inconsistencies in the literature surely make policy decision-making difficult for developing countries. To provide practical benefits, this chapter first breaks out from the limitations of the conventional methods, as commonly used in the literature. And then it demonstrates why among all five economic sectors, the most important and most fundamental is generally manufacturing, which underlies the economic stability and strength of a nation that longs to develop its self-sustained momentum of growth.

Before the conclusion of this section, let us look at the basic terms and assumptions needed for the rest of this chapter. Specifically, assume that each business firm, considered in this chapter, is established to satisfy a particular market niche so that its operation is financially maintained by positive cash flows from the product market. To be in alignment with theories of economics, a product stands for a physical thing, such as a television and iPhone, or a service, such as a coin-operated laundry facility and indoor tanning salon, or informational goods, such as a packaged wealth-management advice and live financial market information. Consumers represent the end users of a product, and customers the intermediate firms that use the products of upstream suppliers to produce their offers for downstream complementors or end users in the product market. For situations when both consumers and customers coexist, they will be jointly known as customers. The term "proposition" means a statement that is arguably true based on logical reasoning starting from a rigorous beginning. To avoid confusion between growth and development, "growth" will be used to mean moving towards a higher level of economic prosperity.

17.3 Feedback between Markets and Manufacturing

For the purpose of uncovering how market exchange and manufacturing associate with each other, this chapter also splits the economy of a focal nation into five sectors as Selstad (1990) and Kellerman (1985) do: the primary, secondary, tertiary, quaternary, and quinary sectors. In particular, the primary sector is made up of those industries that are involved in the mining and gathering of raw materials or natural resources. Examples of such industries include farming, forestry, hunting, fishing, mining, and others. The secondary (or manufacturing) sector is composed of those industries that produce finished product(s) either as supplies for downstream businesses to assemble their products or as offers for end consumers in the ultimate product market. Industries in this secondary sector employ the outputs of the primary sector as supplies. The tertiary (or service) sector consists of such industries that serve both businesses and end consumers. Accounting, trades (e.g., mechanic or plumber services), computer services, restaurants, and tourism are some examples of this tertiary sector. The quaternary sector contains the part of the economy that is knowledge-based. Consultation, design, education, financial planning, information generation and sharing, media, and research and development are some of the representative businesses in this economic sector. Lastly, the quinary sector includes all those social activities that involve management and control. Examples of such businesses include governments, information services, and new technologies.

In order to understand the interaction and synergy of these five sectors, let us examine the crucial steps for the focal nation to go through, if the nation desires to kick-start and maintain its self-sustaining momentum of growth. By "self-sustained momentum of growth," it means a self-sustained societal development that penetrates all economic, social, and political aspects of the society; with the development brand new, while more advanced, technologies are introduced and alter the way how people live and that how the government operates; and supported by the development, a population explosion follows, and the methods of production and service are revolutionized (Heaton, 1933).

In the past 100 plus years, many developing countries tried to create such a momentum of economic growth. However, they did not succeed due to the reason that even the world of economists does not know exactly what is behind the desired economic takeoff. Along with the economic development from the earlier primary states to the present one, instead of locating true antecedents, scholars confuse consequences, identified individually throughout the evolution of time, as necessary antecedents (Wen, 2016). As realized by Forrest et al. (2018), one possible way for related studies to avoid repeating such mistakes is to employ systems science and systems methodology instead of anecdote- and statistics-based approaches. Consequently, Forrest et al. (2018) are able to provide the following detailed procedure for an impoverished agrarian nation to engineer its self-sustained momentum of economic growth.

- Step 1: Formulate a national goal to reach social stability and prosperity;
- Step 2: Implement particular ways to improve the standards of moderate living;
- Step 3: Engineer the fermentation of various market exchanges;
- Step 4: Purposefully promote targeted primary industries; and
- Step 5: Effectively round off the first successes of the developed self-sustaining momentum of economic growth.

Numerous studies of micro-firm level successes and large-scale national developments, both successful and unsuccessful, suggest (e.g., McGrath, 2013; Wen, 2016) that not adopting a long-term, unwavering development goal for how to create wealth and assure security for the entire economic entity of concern, be it a firm or a nation, is a recipe for failure. Such an overarching goal helps enable the focal economic entity to mobilize its capable citizens to accomplish targeted ends, such as job creation, stable social order, and political harmony. It motivates government officers to resolve crises and to play the necessary roles of visionaries, planners, and pursuers of a better future for their people. Such an overarching goal encourages nationwide coordination and joint efforts so that the nation will be able to develop and maintain its economic and political stability on which the desired economic prosperity becomes a possibility. Such nationwide coordination and joint efforts will help the nation establish its military security and strength, needed to protect its interests against foreign aggressions and to compete with these aggressors. The necessity of these economic and political stability and military security and strength has been well supported by the despondent modern history of many developing nations, such as that of India (Stein & Arnold, 2010).

To further our analysis along the line above, and to make our communication convenient, let us quote Theorem 3.1. Assume that the oligopoly market of our concern satisfies the following conditions:

- There are m firms, $m = 1, 2, \ldots$, that serve the market with mutually substitutable products;

- Each of these firms is loyally supported by those customers who make purchases only from their favorite firms if the price is no more than their reservation value;
- These firms compete over those customers who make purchases from the firm that offers the lowest price; and
- The management of the firms know the pricing strategies of their competitors and respond to the competition by playing the Nash equilibrium through pure self-analysis.

Theorem 17.1 When the afore-described market is in Nash equilibrium, if the number of switchers is greater than zero, then at least one new firm will possibly enter the market with its version of substitute product, as a competitor of the m incumbent firms.

In order to better illustrate this theorem, let us define innovation as an assortment of happenings that underlie the offering of one product or a bundle of products that provides remarkably added value to a firm when compared to other happenings at the same time in the same economic sector. The generality of this definition of innovation is well embedded in *the offering of a product* and *the offering of a bundle of products*. In particular, the former includes the creation of original product(s) that create remarkably added value for the firm. In this sense, the word "innovation" includes all relevant details, such as, for example, product design, production and marketing, and business model, necessary for the introduction and production of the product. As for the latter, the situation is profoundly different. It represents bundling (or packaging) a few products into one deliverable package, where the products can be special or original or none of these. The underlying innovativeness here is found in the potential creation of simultaneous consumer utilities and/or multi-sided markets; such potential makes consumers willing to bear higher prices than when the bundled products are purchased individually. The focus of this definition of innovation on the remarkably added value is employed to reflect many aspects of applications, including, for example, extraordinary level and quality of creativity, both internal and external acceptance, appropriate operating processes adopted for value creation and capture, inventions and relevant conversion to consumable market offers.

Riding on this understanding of innovation, Theorem 17.1 implies that rather than generating demand by communicating to consumers about what they want, market demand directs the level of supply and necessitates consequent innovations. Here, the level of supply and relevant sophistication of innovations are determined by sufficient market depth (or purchasing power). In other words, if there is not enough market depth for a particular product, no manufacturer will be enticed to produce much or any at all. Speaking differently, beyond existent demand in the market place, its massiveness in terms of either quantities or dollar values can practically stimulate mass and/or advanced production and the appearance of related innovations. For an economic entity to develop the needed market depth, a portion of available labors need to be freed from manufacturing the necessities of accustomed living. At the same time when the freed labors enter into the manufacturing of luxurious goods, where technologies are employed to create extravagantly added values, the remaining labors need to be sufficient so that the supply of necessities is not jeopardized.

It is at this stage of labor division that government policies play their roles in handling details, such as land use or ownership, freedom and mobility of people, taxes, among others. Beyond promoting industrial and technological advances by using policies, the government can simultaneously, while indirectly, take part in economic activities by making certain resources available to the public. As indicated by Theorem 17.1, the development of technologies is a consequence of market exchanges. In particular, market competition constantly calls for new market offers. That in turn stimulates the furtherment of both incremental and disruptive technologies that correspond adequately to the prevalent skills and purchasing power of the labor force.

As a result of the production of luxurious goods, the reallocated labors are able to earn much-increased incomes than before. That in turn deepens the depth of the product market with quickly rising purchasing power. Similar to the situation of a rolling snowball, market competition brings forward new technologies and then causes industries to advance to higher levels with much-improved management routines and increased energy efficiencies. All these developments simultaneously make various markets much more expanded and filled with great varieties of products and services beyond basic needs of living. With the market expansion, distribution networks have become continuously sophisticated.

Let X and Y be two linear spaces and $S : X \to Y$ and $S_f : Y \to X$ linear functions respectively from X to Y and from Y to X. The so-called feedback system (Lin, 1999) of the functional system S with a feedback component S_f is such an input-output system $S' \subseteq X \times Y$ that satisfies

$$(x, y) \in S' \subseteq X \times Y \leftrightarrow (\exists z \in X)\big((x + z, y) \in S \text{ and } (y, z) \in S_f\big).$$

To relate this concept of feedback systems to the current discussion over here, let us visualize the input x as a particular government policy and y the corresponding economic output produced by adopting the policy. It is intuitive to see that although the output y is a consequence of the policy x, it represents in reality the combined effect of the policy and the market reaction, denoted by $S_f(S(x))$, to the policy. In this explanation, the symbol $S(x)$ stands for the expected market output of the policymakers. This symbolic expression is similar to the idea of Soros' (2003) reflexivity.

Summarizing the previous discussion produces the following conclusion:

Proposition 17.1 When a horse race appears between market exchange and manufacturing production, where these players contest against each other, a feedback system emerges in the marketplace, satisfying that:

- Market demands motivate and stimulate producers to produce more products and entrepreneurs to introduce better products;

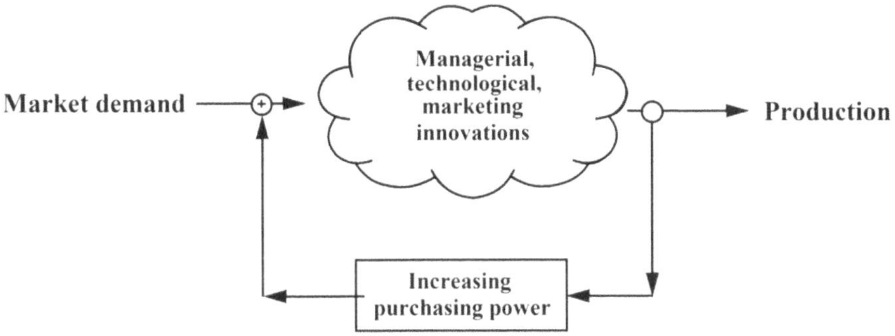

Fig. 17.1 The feedback system of the horse race between demand and production

- Firms are forced to hire additional employees with rising salaries in order to satisfy the increasing need for production; and
- This mutual reinforcement simultaneously further deepens the purchasing power of the market and the demand of the market.

Graphically, Fig. 17.1 demonstrates the mechanism underneath the working of the feedback system outlined in Proposition 17.1. If anecdotes are employed to help with the understanding of this proposition, one can readily pick the first and the second industrial revolution from around the world that had occurred one after another in the past several 100 years. In these circumstances, every time when a horse race between market exchange and manufacturing production appeared, the situations, as described in this proposition, followed.

Relating what is stated in Proposition 17.1 with the current economic state of the United States raises the following natural question:

> What is really in the play with the United States presently? Specifically, the third industrial revolution has not helped the U.S. to raise employee salaries, even though the economy is nearly at its full employment as of February 2020.

By comparing the specific situation of the United States with what is stated in Proposition 17.1, one can readily see that in the present state of the globalizing world economy, the known picture of the US economy merely describes the economic state of a local region. On the other hand, Proposition 17.1 expresses what will hold true only when the entire economic system, within which the said horse race occurs, is concerned with. To illustrate this end, let us look at the production and sales of iPhone. Specifically, with the increasing demand of the world market for iPhone Apple elevates its magnitude of production, while a large portion of this production is located outside the US. The elevated scale of production in turn requires the firm to hire additional employees with increasing salaries. However, most of the new hires happened in foreign countries, while the increasing salaries mostly benefited the domestic economies of those hiring foreign countries. This end actually explains at least partially the reason why foreign demand for iPhones has been rising in recent years.

Beyond the previous question, another question also naturally arises at this junction:

> What will happen in the economy of concern if the condition in Proposition 17.1 does not emerge?

In other words, this question can be rephrased as follows: what would occur to the economy if no horse race appears between market exchange and manufacturing production? In theory, this end can realistically occur in two possible ways:

1. Market exchange cannot keep up with the manufacturing production; or
2. Manufacturing production cannot satisfy the call of the market exchange.

When case (1) appears, the economy dwindles. Manufacturers will be forced to cut down their productions by the insufficient purchasing power of the population or consumer's desire to spend (or market depth). Their funding and effort devoted to research and development will have to be eliminated, causing economic decline. This end is precisely what is occurring presently in Japan's economy, where the young generation has not been in the mood of extravagant consumption as their parent and grandparent generations used to. When the situation in case (2) occurs, employees' wages will stay stagnant or be falling. That weakens the purchasing power of the market exchange and causes society-wide apprehension, leading naturally to challenges to the underlying political system. That is precisely what is happening in the United State currently.

17.4 The Launch of a Self-Sustaining Growth

All the five sectors of the economy, as described earlier, are enthused to function individually and cooperatively by the feedback system given in Proposition 17.1. Generally, technologies that are improved or newly developed for increasing production efficiency require unconventional resources and raw materials that were never known existent before. Increasingly advanced services, such as bookkeeping, finance, and decision support, are needed by the accelerating evolving exchanges of information, knowledge, and goods, and are demanded by the constantly rising efficiencies of production. In order to rapidly translate signals of market demand into relevant offers (Theorem 17.1), encouraged are efforts and activities in the quaternary sector, such as consultations, generation and sharing of knowledge, increased investment in and strengthened facilities of R&D, among others.

To safeguard the integrity of the nation and provide a stable and secure business environment, entities of the quinary sector have to play their roles to uphold a healthy balance of accelerating economic development. Generally, these roles are played through introducing both industry-specific and general policies to help promote further economic growth.

The previous discussion naturally leads to the following fact, while it is also supported by the empirical literature. In particular, the social and political effects of immersive wearables are examined (Briggs, 2018); how green information technology plays its role in higher education is explored by Hernandez (2017); the catch-up process of the aircraft industry in a developing country is considered by Manteghi et al. (2018). And robotics and labor-intensive jobs are investigated by McLeod (2019); it is demonstrated by Rudinsky and Hvannberg (2017) that voice communication in games can be used to train for skill and readiness of crisis management; and it is confirmed by Sivathanu and Pillai (2019) that technology could improve talent management.

Proposition 17.2 If a nation desires to kick-start a self-sustaining momentum of economic growth, it needs to purposefully create a feedback mechanism within the economy that supports the competitive race between market demands and manufacturing production.

Additionally, the previous discussion demonstrates the importance of manufacturing for a nation that desires to either maintain or acquire its world leadership through a well-planned industrial transformation from automated manufacturing, the current state, to intelligent manufacturing, an imagined future state. It also illuminates the logic of why historically manufacturing has been the path that led to major leaps in economic development (Reinert, 2019). This logic has been confirmed again and again by the rise of England in the nineteenth century, that of the US, Germany, Japan, and the Soviet Union of the twentieth century, and newly industrialized nations like South Korea and China.

By disruptive innovation, it stands for such an innovation that creates a new market and new value network that disruptively alters the existent ones so that established business entities, market offers, and organizational alliances are displaced with new ones (Christensen, 1997). In terms of where disruptive innovations tend to appear, Theorem 17.1 implies that they are generally produced by business enterprises that are not a leader in an existing market. That is because such innovations generally involve high risk and require a long time to achieve profitability, which discourages leaders of lucrative markets from pursuing activities related to such innovations. Due to the very nature of such innovations, they mostly involve unconventional R&D works to become deliverable and consumable products to consumers; they generally need new infrastructures to be constructed by downstream complementors in order for the imagined products to ultimately reach end users; and most critically, during a long time of R&D works a relevant product market has to be formed so that it has the depth and width sufficient to financially support the life cycle of the product(s) out of the disruptive innovation. That is exactly why leading companies in lucrative markets cannot afford to spend their valuable and limited resources away from sustaining their already established market positions, where they have to constantly fight off the forever prevailing forces against their current market positions (Christensen, 1997).

The previous discussion also implies that when empirically checking whether an adoption of a disruptive innovation can improve the performance of the manufacturing sector, one needs to assess if the said innovation offers the following incentive effects for manufacturing enterprises.

1. Managerial efficiency and efficient use of resources are improved for individual manufacturing enterprises;
2. Manufacturers are motivated to use advanced, never-before-existed equipment to improve productivity;
3. Collaborations both internal and external are promoted for information and knowledge exchange and sharing;
4. Carious deficiencies are continuously overcome that exist in managerial routines, in the equipment in use, and inter-unit and inter-firm manufacturing collaborations; and
5. The modes of production and management are unremittingly expanded, reflecting the ambitions of firms, availability of resources, technologies employed.

Proposition 17.3 If a disruptive innovation helps a manufacturer sharpen its market awareness and creatively comprehend market cues, then it will positively influence the manufacturer's productivity, managerial efficiency, and organizational structure.

This conclusion is a corollary of Theorem 17.1. In particular, the theorem implies that if a manufacturer can creatively spot market invitations for new products from adopting a disruptive innovation, then it will be in an advantaged position to positively and profitably respond to the market call. To illustrate this end, let us use the concept of environmental sustainability as an example. Presently, consumers and market exchanges have been becoming very conscious about this concept. So, relevant environmentally friendly products, such as Tovato's Pruis and Tesla's Model 3, have been designed and produced, creating such a phenomenon that if a firm is able to accomplish one of the 17 United Nations' Sustainable Development Goals, it can potentially serve the world with profit. That means that disruptive innovations positively affect manufacturers in terms of potentially improving their productivity, managerial efficiency, and its organizational structure. However, whether such an innovation can realize material improvement for a firm certainly depends on the firm's capability to move along its spotted direction; that direction is determined by many factors, such as organizational culture (McGrath, 2013), the CEO's ability (Papadakis & Bourantas, 1998), and others.

When a specific disruptive innovation is under the protection of a patent, the following question naturally arises. In such a case, will the business that owns the patent right of the disruptive innovation be less incentivized to optimize its productivity, managerial efficiency, and organizational structure? For the general case, the present answer to this question is NO. With the globalization of the world economy, industries are no longer as clearly separated from each other as before. Newly developed products have found their ways to penetrate several conventionally different industries simultaneously (McGrath, 2013). Such a trend of blurring industry boundaries poses threads to business entities that intend to limit themselves to their well-defined industries and market territories. Therefore, the business that owns the patent right of a disruptive innovation in general is still motivated to constantly make improvements on what is in use, what is practiced, and consider to optimize various aspects of its business.

Additionally, the approaching fourth industrial revolution or the Industry 4.0 demonstrate how the conclusion in Proposition 17.3 is presently unfolding in the world arena of business. In this upcoming new era, the Internet of Things is expected to be an emerging disruptive technology. It will systemically assimilate the knowledge, relevant methodologies and technologies of big data to digitally develop the current manufacturing based on automation into one based on intelligence. Within the new era of manufacturing, machines will be augmented smartly with wireless devices. They will be centrally controlled by a system that also administers production and provides unmanned fact-based decisions (Hermann et al., 2016).

17.5 A Few Final Words

This chapter studies the importance of the manufacturing sector in a nation's effort of transforming the current state of automated manufacturing to an imagined future state of intelligent manufacturing. To do so effectively, it employs systems thinking, systemic intuition, and logical reasoning that is parallel to the one successfully employed in mathematics and natural science.

This chapter contributes to the literature by presenting a novel approach of reasoning and method of analysis so that it can establish clear results on manufacturing's role in the forthcoming industrial transformation for nations to maintain or to acquire their leadership in the world. It establishes the main results listed below, even though certain particular cases of the results were previously conjectured by various scholars through empirical analysis. And, Fig. 17.2 figuratively summarizes how different factors strengthen each other.

- A horse race between market demands and manufacturing production is needed for a nation to launch its self-sustaining momentum of economic growth.
- For a nation to maintain its launched self-sustaining momentum of economic growth, it needs to purposefully create a feedback mechanism that strengthens the horse race between market demands and manufacturing production.
- If a manufacturing enterprise is assisted by a disruptive innovation to sharpen its ability to creatively comprehend market cues, then its productivity, managerial efficiency, and organizational structures will be likely optimized through monetizing that innovation.

By employing a novel methodology that has never been considered before in relevant studies, this chapter is able to discover some of such insights whose validity is not constrained by the limitations of data- and anecdote-based approaches. Due to this reason, the theoretical conclusions established in this chapter naturally provide dependable bases for government officers and policymakers to make their decisions.

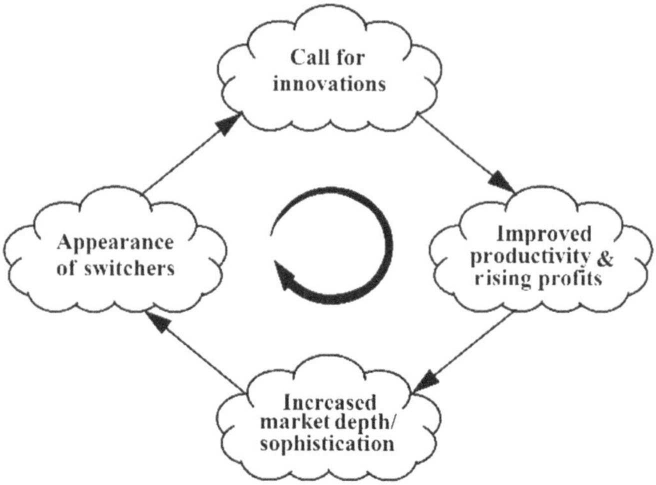

Fig. 17.2 Feedback forces that strengthen each other

Specifically, the following general recommendations for the government of a nation can be derived from the established conclusions in the previous sections:

- The government needs to inspire innovations in the manufacturing sector, be they incremental or disruptive, for the goal of orienting manufacturers towards creatively interpreting market signals (Theorem 17.1). Such innovations will benefit the resolution of the current phenomenon of stagnating entrepreneurship (Naudé & Nagler, 2018) and enable sustaining human development (Arocena & Sutz, 2018). Examples of such developments include, but are not limited to, unceasing learning of constantly renewing technologies and knowledge, and effort to create efficiencies in management and production, and creative usage of resources and present development of industrialization (Proposition 17.3).
- It needs to encourage and endorse entrepreneurial spirits so that activities of creative nature will strengthen the feedback mechanism between market exchanges and manufacturing production. Such reinforced race between market exchanges and manufacturing is the basis for a nation to rise as an economic power (Propositions 17.1 and 17.2). A predictable outcome of such encouragement and endorsement will be one of the solutions to the challenge of economic development that is similar to the African challenge (Scholz, 2018).

At the conclusion of this chapter, let us look at some of the limitations this chapter suffers from. First, beyond what is employed here there are still many tools and methods in systems science advanced unambiguously for studying organizations, their evolutions and interactions (Forrest et al., 2013). Hence, there is a need for scholars to apply these tools and methods in their investigations of economics and business issues and problems. From what are established in this chapter, it can be expected that by employing these tools and methods finer conclusions with reliable applicability in real life can be developed.

Second, this chapter assumes that each firm considered exists to satisfy a specific market niche with its operations financed by positive cash flows from the product market. However, this assumption is not 100% true in real life. It can go wrong in at least two possible ways:

1. Positive cash flows from product-market transactions are not necessary; and
2. Firms in real life also exist for some other reasons beyond the one given above.

As for possibility 1, cash flows were needed to back by a firm's performance (Sobel, 1999). However, it is no longer the case in today's world of business. In particular, for example, some major online retailers have been pumping up their future promises and potentials in order to continuously attract sufficient venture capitals. And some of these retailers have been losing money each year since their inception (Li & Ma, 2015). Hence, when the conclusions developed in this chapter are applied to the cases of such business entities, it is very likely that these conclusions will not hold true. The situation described in possibility 2 is well-known in accounting area. For instance, there are nonprofit organizations and even such firms that are established for tax purposes.

These two listed and other unlisted limitations, in fact, open doors for future research.

References

Arocena, R., & Sutz, J. (2018). Re-thinking innovation as a lever for development: Taking into account inequality. In B. Laspra & J. A. L. Cerezo (Eds.), *Spanish philosophy of technology* (pp. 125–138). Springer.

Briggs, J. R. (2018). Immersive wearables: Their political and social effects and what both mean for western liberal democracy. *International Journal of Sociotechnology and Knowledge Development, 10*(3), 54–71.

Christensen, C. M. (1997). *The innovator's dilemma: When new technologies cause great firms to fail*. Harvard Business School Press.

European Commission (EC). (2014). *For a European industrial renaissance: Communication from the commission to the European Parliament, the council, the European economic and social committee and the committee of the regions*. European Commission. Eurostat, & European Commission.

Forrest, J. Y.-L. (2018). *General systems theory: Foundation, intuition and applications in business decision making*. Springer.

Forrest, J. Y.-L., Duan, X. J., Zhao, C. L., & Xu, L. D. (2013). *Systems science: Methodological approaches*. CRC Press, an imprint of Taylor and Francis.

Forrest, J. Y.-L., Liu, J., Martorell, G., Xu, L., Liu, Y., & Darvishi, D. (2021). Role of manufacturing sector in the forthcoming national industrial transformations. Accepted for publication by *International Journal of Sociotechnology and Knowledge Development*, to appear.

Forrest, J. Y.-L., Nicholls, J., Schimmel, K., & Liu, S. F. (2019). *Managerial decision making: A holistic approach*. Springer.

Forrest, J. Y.-L., Zhao, H. C., & Shao, L. (2018). Engineering rapid industrial revolutions for impoverished agrarian nations. *Theoretical Economics Letters, 8*, 2594–2630.

Government of Japan. (2015). https://www8.cao.go.jp/cstp/kihonkeikaku/5basicplan_en.pdf. Accessed on April 16, 2019.

Hall, W., & Pesenti, J. (2017). Growing the artificial intelligence industry in the UK. https://assets.publishing.service.gov.uk/government/uploads/system/uploads/attachment_data/file/652097/Growing_the_artificial_intelligence_industry_in_the_UK.pdf, Accessed on October 15, 2017.

Heaton, H. (1933). Industrial revolution. In: *Encyclopeadia of the social sciences*; Hartwell, R. M. (ed.), *The causes of the industrial revolution in England*, VIII (1033) 30–52. : Routledge.

Hermann, M., Pentek, T., & Otto, B. (2016). Design principles for industry 4.0 scenarios. *The 49th Hawaii international conference on system sciences (HICSS)*, Koloa, HI (pp. 3928–3937).

Hernandez, A. A. (2017). Green IT adoption practices in education sector: A developing country perspective. *International Journal of Sociotechnology and Knowledge Development, 9*(3), 1–16.

Hopkins, E. (2000). *Industrialisation and society a social history, 1830–1951*. Routledge.

Kellerman, A. (1985). The evolution of service economies: A geographical perspective. *The Professional Geographer, 37*(2), 133–143.

Kline, M. (1972). *Mathematical thought from ancient to modern times*. Oxford University Press.

Kuhn, T. (1962). *The structure of scientific revolutions*. University of Chicago Press.

Li, T., & Ma, J. H. (2015). Complexity analysis of dual-channel game model with different managers' business objectives. *Communications in Nonlinear Science and Numerical Simulation, 20*, 199–208.

Lin, Y. (1999). *General systems theory: A mathematical approach*. Kluwer Academic/Plenum Publishers.

Manteghi, A., Manteghi, M., Aslani, A., & Mohaghar, A. (2018). Catch-up process in aircraft industry: A model based on experiences of six latecomer countries. *International Journal of Sociotechnology and Knowledge Development, 10*(1), 37–56.

Marc, S., Yann, B., Charly, B, et al. (2018). Mission Villani: for a meaningful artificial intelligence towards a French and European strategy. https://www.aiforhumanity.fr/pdfs/MissionVillani_Report_ENG-VF.pdf, Accessed on March 8, 2018.

Mazzei, M. J., Ketchen, D. J., Jr., & Shook, C. L. (2017). Understanding strategic entrepreneurship: A "theoretical toolbox" approach. *International Entrepreneurship and Management Journal, 13* (2), 631–663.

McGrath, R. G. (2013). *The end of competitive advantage: How to keep your strategy moving as fast as your business*. Harvard Business Review Press.

McLeod, A. (2019). Employment and technology in manufacturing. *International Journal of Sociotechnology and Knowledge Development, 11*(3), 14–29.

Merati, M., Salehi, V., & Rafiei, A. (2019). A unique mathematical programming algorithm for performance optimization of organizational indicators in manufacturing sector. *Journal of Industrial and Production Engineering, 36*(7), 461–474.

Moyo, C., & Jeke, L. (2019). Manufacturing sector and economic growth: A panel study of selected African countries. *Journal of Business and Economics Review, 4*(3), 114–128.

Naudé, W., & Nagler, P. (2018). Technological innovation, entrepreneurship and productivity in Germany, 1871-2015. SWPS 2018–02. Available at SSRN: https://doi.org/10.2139/ssrn.3107759, Accessed Dec 20, 2019.

Naudé, W., & Szirmai, A. (2012). The importance of manufacturing in economic development: Past, present and future perspectives. United Nations University-MERIT UNU-MERIT Working Papers, Accessed on Dec 10, 2019.

OWH. (2016). The National Artificial Intelligence Research and Development Strategic Plan. https://www.nitrd.gov/PUBS/national_ai_rd_strategic_plan.pdf, Accessed on April 16, 2016.

Papadakis, V., & Bourantas, D. (1998). The chief executive officer as corporate champion of technological innovation: An empirical investigation. *Technology Analysis and Strategic Management, 10*(1), 89–98.

Porter, M. E. (1985). *Competitive advantage: Creating and sustaining superior performance*. Free Press.

Reinert, E. (2019). *How rich countries got rich ... and why poor countries stay poor*. Public Affairs.

Rostow, W. W. (1960). *The stages of economic growth: A non-communist manifesto*. Cambridge University Press.

Rudinsky, J., & Hvannberg, E. T. (2017). Transferability of voice communication in games to virtual teams training for crisis management. *International Journal of Sociotechnology and Knowledge Development, 9*(1), 1–25.

Scholz, R. (2018). *Industrializing Africa: Understanding the uniqueness of the manufacturing sector*. Nomos Verlag.

Selstad, T. (1990). The rise of the quaternary sector. The regional dimension of wisdom or knowledge-based services in Norway, 1970–1985. *Norsk Geografisk Tidsskrift - Norwegian Journal of Geography, 44*(1), 21–37.

Sivathanu, B., & Pillai, R. (2019). Leveraging technology for talent management: Foresight for organizational performance. *International Journal of Sociotechnology and Knowledge Development, 11*(2), 16–30.

Sobel, R. (1999). *When giants stumble: Classic business blunders and how to avoid them*. Prentice Hall.

Soros, G. (2003). *The alchemy of finance*. Wiley.

State Council of the PRC. (2015). Notice of State Council of the PRC on printing and distributing "Made in China 2025".http://www.gov.cn/zhengce/content/2015-05/19/content_9784.htm, Accessed on May 8, 2015.

Stein, B., & Arnold, D. (Eds.). (2010). *A history of India* (2nd ed.). Wiley-Blackwell.

Szirmai, A., & Verspagen, B. (2015). Manufacturing and economic growth in developing countries, 1950-2005. *Structural Change and Economic Dynamics, 34*, 46–59.

Wen, Y. (2016). *The making of an economic superpower: Unlocking China's secret of rapid industrialization*. World Scientific.

Chapter 18
How Artificial Intelligence Affects Technological Innovations

Abstract This chapter investigates how artificial intelligence (AI) may impact technological innovation. It first defines what technological innovation entails, and then it establishes 10 general propositions. At the end, this chapter empirically validates some of these results by using provincial panel data of China from 2003 to 2015.

Keywords Artificial intelligence · Technological innovation · Regional comparison

The developed analysis of this chapter reveals the working mechanism underlying the effect of AI on technological innovation. In particular, AI stimulates technological innovation through three different channels: (1) it accelerates knowledge creation and creates technology spillover, (2) it advances firms' capacities of learning and absorption, and (3) it boosts investments in R&D and human talents.

The empirical confirmation demonstrates that AI expressively inspires technological innovation when such factors as R&D investment, R&D personnel, the level of openness, and sophistication of infrastructure are controlled. And suggestions and recommendations for managerial decision-making are provided based on the conclusions established in this chapter.

18.1 Introduction

With the current political and geographical developments in the world, patterns of economic development have been changing quickly in unprecedented ways. In order to respond adequately to the fast-changing environment, many leading nations from around the world have made their strategic adjustments. They have started to zoom in on artificial intelligence (AI) as their next strategic direction. For example, the following two important documents, which plainly put out seven chief stratagems for the advancement of AI in the United States of America, were published in 2016 by the Obama White House:

J. Y.-L. Forrest, Y. Liu, *Value in Business*, Contributions to Management Science,
https://doi.org/10.1007/978-3-030-82898-1_18

- Preparing for the Future of Artificial Intelligence (OWH, 2016a) and
- The National Artificial Intelligence Research and Development Strategic Plan (OWH, 2016b).

Again, on May 10, 2018, the White House (WH, 2018) created the Select Committee on Artificial Intelligence for the purpose of joining the strengths and capabilities of different units of the government to improve the efficiency of federal level investment in AI-related areas. By keeping up with such movements started in the United States, in May of 2018, the European Commission (EC, 2018) submitted its corresponding document, entitled "Artificial Intelligence: A European Perspective." It pronounces the EU's place in the international AI competition, while presenting a plan for corresponding actions. That plan involves the goal of increasing investment in AI-related R&D and ideals on how to update the European system of education and training in order to maintain the development strength of European talents. Parallel to the United States and Europe, the Japanese government also assigns great importance to AI and relevant development. In its "Report on the 5th Science and Technology Basic Plan (2016–2020)" (Government of Japan, 2015), the Japanese government recommends to establish at the national level a relatively complete R&D promotion mechanism to center on Internet of things, big data, AI and other technologies, and the materialization of a super intelligent society through extending AI to all aspects of life.

As the fastest growing developing nation in the world, the importance of AI also catches China's attention. The term of AI appears in the 2017 Report of the 19th National Congress of the Communist Party of China (Xi, 2017), reflecting the Chinese ambition of keeping up with the international development and to stimulate the economic restructuring by profoundly fusing internet, big data and AI with the real economy. At the end of 2017, the central government revealed China's "Three-Year Action Plan to Promote the Development of a New Generation of AI Industry 2018–2022" (II, 2017). This document details the jobs and responsibilities related to the development of the next generation of AI in order to promote industrial development and push for "Made in China 2025." It tries to stimulate organic combination of AI technology and the real economy by focusing on the commercialization of new AI technologies through promoting deep fusions of information and manufacturing technologies. In the past few years, China has created great advantages in the world economy in the industrialization of AI with fruitful results. For example, many current Chinese technologies, such as Baidu's face recognition and the speech recognition of iFlytek, are in the international leading positions in their respective market segments. In terms of practical applications, China has also established its great advantages. For instance, China's huge potential market space and large population base have helped its development of advantage in big data supports (Du et al., 2018). Furthermore, AI has been increasingly employed in manufacturing, finance, education, medical care, logistics, and other areas of the economy. It continuously makes people's lives more intelligent than before and the whole society evolves in the direction of intelligence.

Although the importance of AI has been widely recognized and AI technologies have been increasingly employed, scholarly research on how AI impacts technological innovation is still rare. In particular, the intelligentization of manufacturing leads to a common enabling technology for mechanical product innovations, while it also assists with the innovation of new manufacturing technologies (Zhou, 2012). Such intelligentization will develop the current manufacturing industry into an integrated one, where product design, production, and management are more digitized and intelligentized. And the second machine age, where AI and digitization are prevailingly visible, will realize unprecedented technological progresses (Brynjolfsson & McAfee, 2014).

The subject area of AI is composed of two branches. One branch composes of ways to replicate human capabilities; and the other is to increase human capabilities for the purpose of making better decisions through utilizing the power of modern technology (Markoff, 2016). An important factor that affects how AI can help with socioeconomic growth is how AI can positively influence innovation and knowledge creation (Aghion et al., 2017). Because knowledge creation is such a process that either regroups what is known or generates new knowledge, AI can provide its support in both of these fronts; that is why AI promotes significant economic growth (Agrawal et al., 2018).

By glancing through the available literature, it is found that studies on how AI impacts technological innovation are still in the stage of infancy, and their focuses are mainly on how to define relevant concepts. Still open and waiting to be addressed are many questions regarding how AI and technological innovation are related to each other. This is where this chapter comes into being. It contributes to the literature mainly in the following two aspects:

- At the height of theory this chapter explores the inherent mechanism underneath how AI affects technological innovation;
- Based on the panel data of the provincial level from China, this chapter empirically confirms the impact of AI on technological innovation.

The rest of this chapter is organized as follows. Section 18.2 consists of some elementary empirical observations. Section 18.3 investigates the mechanism underneath the impact of AI on technological innovation. Section 18.4 empirically confirms some of the established theoretical results in the previous section. Section 18.5 concludes this chapter with relevant managerial suggestions and recommendations.

18.2 Elementary Empirical Observations

With the implementation of various policies, artificial intelligence (AI) develops rapidly in China with excellent results derived and breakthroughs in key technologies appearing in many areas. However, due to regional differences in scientific and technological strengths, qualities of talents, and infrastructures, the levels of AI development in different regions are also different. From the perspective of input, that is, investment inputs, including those inputs of equipment and talents, this chapter measures aspects of AI. Investment in equipment is the foundation on which AI develops, while providing strong support for firms to introduce relevant technologies and to establish networks. To this end, this chapter employs investment in the fixed assets of telecommunication as a proxy variable. Investment in AI talents is an important guarantee for the development of artificial intelligence. Increasing such investment helps improve the level of AI and the capability of enterprises to innovate. This chapter employs the number of software developers as its proxy variable. By borrowing the thinking logic of Shi et al. (2018), this chapter respectively uses the products of investments in fixed assets of telecommunication and the numbers of software developers to measure the development levels of individual provinces, while these investments and counts of developers are normalized by using the entropy method. The greater the magnitudes of these products, it implies the higher development levels of the relevant AI. Figure 18.1 reflects figuratively the provincial development levels of AI and regional differences of China, which is generated by using geographic information systems software ARCGIS based on the 2015 AI development index (AI) of individual provinces.

As shown in this figure, the provinces with relatively high levels of development in AI are mostly located along China's east coast, including Liaoning, Beijing, Shandong, Jiangsu, Zhejiang, Henan, Sichuan, and Guangzhou. Other than such remote areas as Gansu, Hainan, Ningxia, and Qinghai, the development levels of AI in the rest of the provinces are similar. The main reason that leads to such uneven distribution in AI development is that the eastern provinces located along the coast

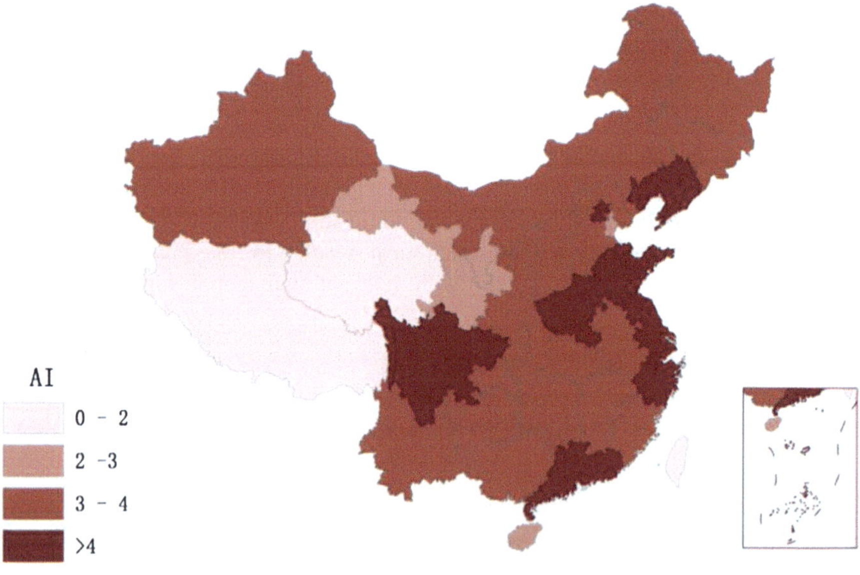

Fig. 18.1 Provincial levels of development of artificial intelligence in 2015

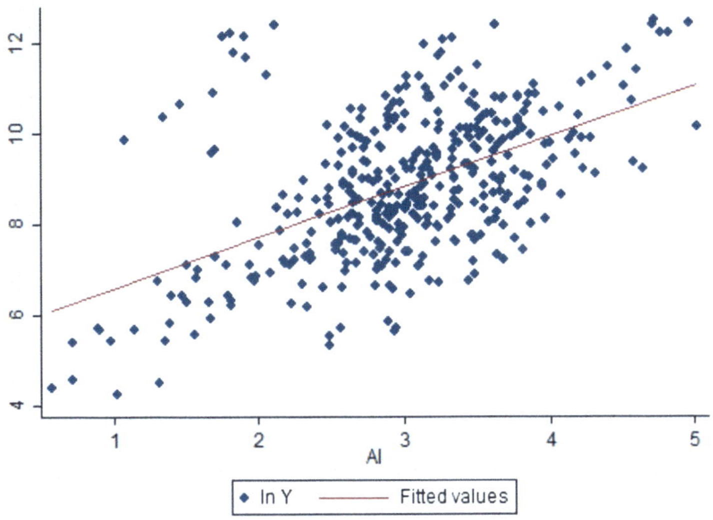

Fig. 18.2 A scatter plot between artificial intelligence and technological innovation

have been enjoying high levels of economic development and rich human capital. These provinces have the necessary infrastructure, technology, and capital to support fast development of AI, leading to their relatively high AI levels. On the other hand, most of the provinces that show relatively low levels of AI development are located in the mid-western remote areas with comparatively backward levels of economic development. Business firms there started relatively later and evolve comparatively slower compared to those located in the eastern provinces, causing the AI technologies in these mid-western provinces also relatively more backward.

To reflect intuitively the relationship between the level of AI development and technological innovation, the scatter plot in Fig. 18.2 is produced by using Stata 13.0. The horizontal axis represents the natural logarithmic values of the development indices of individual provinces during the period of 2003–2015, while the vertical axis the natural logarithmic values of the corresponding numbers of granted patents of the individual provinces. This scatter plot shows an existent positive correlation between AI and technological innovation.

18.3 Influence of Artificial Intelligence on Technological Innovations

As an important driving force of economic development, the core strength of advancement for technology and science, and the essence behind the appearance of smart manufacturing and other emerging industries, AI has been seen as the key to advancing the "fourth" industrial revolution (Cheng & Peng, 2018). The artificially intelligent technology consists of finer areas (Wang & Yang, 2017), such as deep learning, computer vision, speech recognition, virtual personal assistant, intelligent robots, and engine recommendation. In particular, computer vision can identify objects, scenes, and activities out of images and in turn employ these identifications in automated processes of production. It can quickly collect a lot of information, process the information automatically, and complete the monitor of an entire process of production, quality control, product testing, and other tasks. Therefore, it greatly improves the efficiency and automation of production (Acemoglu & Restrepo, 2018). The technology of deep learning can model a specific problem from the real world by utilizing mathematical methods so that similar problems in related fields can be resolved once for all. At the same time of accomplishing realistic applications, the development and application of deep learning expand the applicability range of the entire AI technology, providing technical support for discovering new knowledge (Lecun et al., 2015).

Comparing to ordinary robots, intelligent ones are smarter and can complete complex, high-level works and tasks (Makridakis, 2017), such as quick and convenient modification of the content and method of a job, exploration of extrema, and manufacturing. The wide-range applications and high levels of permeation of these technologies in such areas as computer science, heavy industries, medicine, education, finance, trade, and other areas, have been not only substituting people for completing tedious works, but also identifying new patterns from vast amounts of data for reference (Trajtenberg, 2018). Additionally, applications of these technologies can affect the process of knowledge creation, that of spillovers and the capability of learning and absorption. The consequent effect on technological innovation is produced mainly through the following ways. That is, AI promotes technological innovation by

- Quickening knowledge creation;
- Accelerating the spillover of knowledge and technology;
- Improving the capability of learning and absorption; and
- Increasing investments in R&D and talents.

18.3.1 The Concept of Technological Innovation

To understand the concept of technological innovation, let us first briefly look at how the concept of innovation has evolved in the literature.

In as early as 1776, Adam Smith (1776) acknowledged in his *The Wealth of Nations* the importance of innovation as an activity that fosters wealth. After then, the topic of innovation was not been thoroughly studied until when Joseph Schumpeter (1934) published his book "The Theory of Economic Development." In this book, Schumpeter claims that innovation represents the driving force of economic development. Innovation, be it about products and services, managerial operations, or processes of production, underlies every competitive advantage as a fundamental guiding principle (Forrest et al., 2019). That is both theoretically and practically why business firms have to address the issue of innovation if they desire to ride successfully waves of transient competitive advantages, expand their existing territories, and enter new markets (Brown & Eisenhardt, 1995; OEDC, 1997; Rosenthal, 1992; Stock et al., 2002). Because of its vital importance, innovation has gained a tremendous amount of scholarly attention in recent decades (Becheikh et al., 2006).

In terms of the concept of innovation, Damanpour (1996, p. 694) considers innovation as "a process that includes the generation, development, and implementation of new ideas or behaviors. Further, innovation is conceived as a means of changing an organization, either as a response to changes in the external environment or as a preemptive action to influence the environment. Hence innovation is here broadly defined to encompass a range of types, including new products or services, new process technologies, new organizational structures or administrative systems, or new plans or programs pertaining to organizational members."

Leonard and Swap (1999, p. 7) see innovation as "the embodiment, combination, and/or synthesis of knowledge in novel, relevant, valued new products, processes, or services." From the angle of library and information science, Deiss (2004, p. 18–19) treats innovation as "things that change the way we can do what we want to do; [those that] have added value to our daily lives . . . new, desired, or needed services that add value for university faculty, students, and other scholars ... Innovation is more significantly about what our target audience can do—about the increased capacity of library users to do what they want and need to do in the way that most benefits their productivity, pleasure, and excellence . . . facilitating the work of our primary constituents in ways that are new and useful to them." Baregheh et al. (2009, p. 1334) maintain that "innovation is the multi-stage process whereby organizations transform ideas into new/improved products, service or processes, in order to advance, compete and differentiate themselves successfully in their marketplace."

Rowley (2011, p. 253) defines the concept as follows: "an innovation is a change, in a product, service, process or, more widely, an organization." And, Kaser (2011) sees innovation not "just about doing things that are new or different" but "about doing things that in the eye of the beholder (the user, patron, or customer) meet a need that may not have been appreciated before ... [and] the things that truly alter and improve how we do things; they may even shift our proverbial paradigms." For firms that produce products or provide services, innovation is mainly materialized through introducing original products, services, and processes (Becheikh et al., 2006) that enable firms to enter competitively or create new markets (Smith & Tushman, 2005). Many scholars treat innovation as one major factor of firms' long-term success (Kanter, 2001).

To explore the meaning of innovation from the perspective of front-line managers, starting in 2011, General Electric (2012) commissioned surveys of senior business executives in countries across the world. To single out the key components of different perspectives on innovation, Jeff Dance (2008) lists 30 plus definitions of innovation and provides his summary as follows: Innovation is something fresh (new, original, or improved) that creates value, where "something" can be a process, product, or service that starts as an idea or thought.

To avoid potential distraction by the huge literature on the concept of innovation, let us next look at the concept of technological innovation in order to make this chapter unfold smoothly.

Based on the previously listed attempts of defining the concept of innovation, the definition of technological innovation by Branscomb (2001), and the concept of innovation in manufacturing sector (Forrest et al., 2019), let us define technological innovation as follows:

> Technological innovation is such a set of activities—which could be just one particular activity or several, occurring either simultaneously or temporally—that leads to a fresh technology (new, original or improved) that produces exceptionally added value to a company when compared to other technology-related activities that take place in the same sector as the one the company belongs to.

Because of the emphasis on the comparatively added value, this definition of technological innovation entails many aspects of the concept. For example, it implies

1. An extraordinary level of creativity has to be involved over time;
2. The creativity has to be internally conceived, driven by environmental forces, either current or predicted, and externally adopted, such as market acceptance;
3. Relevant new processes, either productive or managerial or both, have to be implemented to realize the said value;
4. Intended benefits and advantages are materialized;
5. Other than technology-related inventions, this definition also emphasizes on employing the fresh technology to produce marketable fresh products or processes; and
6. Although not explicitly mentioned, the definition includes the roles of relevant processes and outcomes.

Because technological innovations play an increasingly prominent role in the continuous growth of leading economies, governments have been shifting their attention from science and technology policies to a focus on research and innovation policies (Vaughan, 2013). The previous discussion shows that innovation in general and technological innovation in particular is really a socioeconomic process that leads to observable growth in performance of high priority capabilities and growth in values.

18.3.2 *Artificial Intelligence Quickens Knowledge Creation*

The definition of technological innovation, as just introduced above, indicates that such innovation represents either a collection or a process of activities, where the former is the case when the activities take place simultaneously while the latter is the case when the activities take place sequentially. Additionally, technological innovation contains not only purely social activities but also creative ones that involve knowledge recognition and integration on which the freshness of a new technology emerges, as confirmed by (Weitzman, 1998). To this end, knowledge is the core of technological innovation; effective creation of new knowledge is the basis for technological innovation (Chen, 2010). Each technological innovation manifests itself through the emergence of new knowledge and integration and new and existent knowledge.

Proposition 18.1 The fresh technology of technological innovation emerges as a consequence of knowledge creation, such as a new way of understanding existing knowledge, a creative integration of different areas of knowledge, or the production of original knowledge.

Corresponding to knowledge creation, as described above, the development of AI provides data-collection support for revealing brand new ways to look at the existing knowledge, fast speed support for trying different ways to integrate knowledge, and exploration convenience for uncovering new knowledge. That is, we have

Proposition 18.2 The development of artificial intelligence quickens the creation of new knowledge.

In particular, applications of AI technology have clearly helped business firms to increase their capability of data collection, leading to the rapid development of various tools to deal with big data. Interactions between clouds and devices provide unprecedented supplies and carriers for the creation of new knowledge and promote the explicitization of tacit knowledge, while such knowledge can be stored and used in clouds and terminal devices. That in turn accelerates the speed of data acquisition and improves the accuracy and reliability of the consequent understanding and knowledge.

Because of the development of artificial intelligence, introduced are such advanced analytics tools as the technologies of much improved Internet and clouds. For example, the technologies of deep learning and computer vision are able to provide more quickly than ever before a large number of sources of information. That helps create new knowledge and new computational schemes faster than before so that the process of knowledge regrouping accelerates (Agrawal et al., 2018). In other words, what is speeding up is the process that starts from data to information to knowledge creation. For example, when they developed the solid electrolyte new material for lithium ion batteries, Japan's Fujitsu Limited and Chemical RIKEN (or Institute of Physical and Chemical Research) employed related technology of artificial intelligence. This technology helped the scientists control their number of repeated operations and reasonably predict the properties of synthetic compounds. Hence, within a shorter period than expected, the scientists were able to pinpoint out the optimal material combination for high lithium ion conductivity, while simultaneously discovered more efficient high lithium ion conductivity. For details on this case, see (Fujitsu & Riken, 2018).

Additionally, the development of AI promotes the popularization and application of Internet and communication methods of platform economies. It breaks down the boundaries of knowledge within and between companies. It accelerates the dissemination and diffusion of knowledge, and the process of creating new knowledge through knowledge exchange and integration. This end in turn greatly influences the development of AI (Goldfarb & Trefler, 2018). Therefore, we have

Proposition 18.3 Artificial intelligence promotes technological innovation by accelerating the creation of knowledge.

18.3.3 Spillover Effects of Artificial Intelligence

Because of its inevitable spread between various departments of a business firm, the AI technology is promoted simultaneously by all divisions of the firm, leading to the development of interdepartmental technological innovation. With increasingly higher levels of profits, brought forward by AI-based innovations, beneficiary departments will bump up their demands for additional intelligent technologies. After having received the gains from the departments that applied intelligent technologies, the R&D department continues its efforts to improve its firm's technical sophistication and to develop new smart technologies in order to satisfy the new demands. The consequent application of intelligent technologies of a new round once again largely promotes the innovation spillovers for the R&D department. When such a cycle repeats itself, it eventually stimulates a cumulatively cyclic growth of AI and technological innovation.

Proposition 18.4 Artificial intelligence reinforces the following cyclic development: Intelligent technology → growth in interdivisional communication → appearance of interdivisional technological innovation → growth in profits → demand for more intelligent technology → production of new intelligent technology.

At the level of firms, applications of AI technology help reduce the cost of information transfer and sharing between business firms. The formation of knowledge, technology, and information networks makes it more convenient for enterprises to exchange information and sharing know-hows. Hence, the cost of spreading technological innovations across companies is lowered, making inter-firm knowledge and technology spillovers much easier than before. That is subsequently conducive to the diffusion and spread of technological innovations, which further promotes the continuous improvement of technological innovation capabilities of all enterprises involved.

Proposition 18.5 Artificial intelligence makes knowledge sharing and technology transfer easier than ever before while keeping the confidentiality of trade secrets increasingly difficult.

The development of AI accelerates the interactive integration between mobile Internet and Internet of things. That in turn encourages enterprises located at different places to join forces in their processes of production by taking advantage of their respective resources (Lv & Han, 2015). When firms are no longer constrained by their geographical locations, they can apply control remotely and organize production over long distance as long as all equipment work normally. And, in a production process, additional requirements can be added at any time and the predetermined process can be changed adaptively in order to satisfy the varying market conditions. The release from the earlier geographic constraint encourages various talents, firms, and industries to join their efforts and resources on their focal industrial chain. Under the joint effect of the finer division of labor, information sharing, resource integration, etc., firms scattered all over the land will be able to achieve cooperatively their common goals of production. Such change in business operation from the past will gradually break the boundaries of traditional corporate organizations, making firms no longer closed entities, instead a modularized production network. Such networks will greatly facilitate communication and cooperation between enterprises. In short, we have

Proposition 18.6 Artificial intelligence supports the emergence of new philosophies of conducting business, such as remote manufacturing, computer-integrated business system, and platform economy, by removing physical constraints, such as distance, location, uneven distribution of resources, and others. All of these stimulate the development of technological innovation.

For related empirical studies, see (Caputo et al., 2016; Holmstrom et al., 2016; Roos, 2015; Wu et al., 2015).

18.3.4 Artificial Intelligence Improves Learning and Absorption Capabilities

No matter how innovative a firm is, if it cannot appropriately decipher the market information, it is doomed to fail or to be irrelevant. To this end, there are plenty of examples of how corporate giants fell on hard times. One of the notable recent examples is the case of Eastman Kodak. Historically, this firm operated on a culture of innovation and change. However, it did not take necessary actions when the market signaled it to do so. when faced with the challenge of disruptive technology—filmless photos, Kodak did not take decisive actions to combat the inevitable. For more detailed analysis of this firm, see (Kotter, 2012; McGrath, 2013). To this end, we quote Theorem 3.1 from Chap. 3 in order to make the logical reasoning in this chapter self-contained.

Theorem 18.1 For the oligopoly market described below, in the Nash equilibrium, if the consumer surplus β is greater than 0, then at least one new firm would profitably enter the market; and the scale of the entering firm in terms of its operation in this market will be proportional to the magnitude of β. Assume that the market is occupied by m incumbent firms, $m = 1, 2\ldots$, satisfying.

1. They provide the market with mutually substitutable products;
2. Each incumbent enjoys its share of loyal consumers who make purchases from the firm only as long as the price is not more than the reservation price 1;
3. The incumbents compete over the switchers with adjustable prices, where the switchers purchase their products depending on whose price is most competitive; and
4. The management of the incumbents are well aware of the pricing strategies of each other and have established their best responses by playing the Nash equilibrium through pure self-analyses.

This theorem clearly indicates that appropriately understanding market cues is essential in terms of helping a firm to decide the right thing to do and the right product or service to produce and to develop instead of simply investing more heavily in innovation. In particular, the entering firm needs to understand appropriately the switchers in the theorem. Their indecisiveness of which firm's product they would buy could practically mean that they are not happy with the products available in the market; however, due to the necessity of human living, they have to make purchases. That surely opens up a wide door for firms with strong learning and absorption capabilities to explore their appropriate directions to develop their businesses.

Other than knowing what is appropriate to do based on market information, learning and absorption also directly relate to the entry efficiency of external knowledge into an enterprise. That explains why we have the following:

Proposition 18.7 The capabilities of learning and absorption represent a main force behind technological innovation. And the more capable an enterprise is of learning and absorption, the more able the enterprise is to innovate appropriately.

In terms of learning and absorption, the development of AI can help employees to improve these capabilities. Such development constantly brings forward more sophisticated equipment and devices. For example, people have been using continuously improved computers to model and simulate intelligent human activities, which can substitute parts of the brain works widely employed in traditional production systems. These sophisticated equipment and devices have been becoming reliable and efficient analysis tools for employees to make more effective decisions than ever before due to their capabilities of reasoning, learning, association, and problem-solving based on known knowledge (Camerer, 2019). Because of the capabilities of these equipment and devices of constantly improving sophistication, employees' abilities of learning and absorption rise.

Additionally, the development of AI has been practically promoting the learning and absorption capabilities of enterprises. For example, such technologies as deep learning, intelligent image recognition, and applications of big data, can help industrial robots to form independent judgments and to take corresponding actions (Du et al., 2018; Lecun et al., 2015). These robots' independent completions of complex tasks, their capability of transforming acquired information into knowledge, and taking justified subsequent actions play a very positive role in the innovative activities of enterprises. Summarizing these discussions at the level of employees and that of enterprises, we have

Proposition 18.8 By raising the capability of learning and absorption, artificial intelligence helps promote technological innovation.

18.3.5 Artificial Intelligence Increases Investments in R&D and Talents

For a representative firm, its purpose for having an in-house research and development (R&D), if all possible, is to help the firm to achieve one or more of the following:

- Generate, exploit, and convert knowledge, which is newly found either internally or external, into new products and/or processes (Keizer et al., 2002);
- Acquire, integrate, convert, and exploit new technologies (Debackere et al., 1996);
- Entice cooperative partners (Hall & Bagchi-Sen, 2002); and
- Construct new technology settings which are very costly and particularly difficult, even impossible, to acquire from competitors (Lee, 1995).

By employing systems methodology, Forrest et al. (2018) show that when a company grows to a certain scale of strength, it will naturally possess R&D capabilities as described above. So, in such a case, it is strategically significant and important for the firm to officially recognize and organize such a division within the company with sufficient financial and administrative supports. To this end, the literature empirically verifies that in-house R&D has been widely accepted as a crucial determinant of innovation (Hall & Bagchi-Sen, 2002; Parthasarthy & Hammond, 2002). Therefore, we have the following:

Proposition 18.9 Investment in research and development is an effective guarantee for technological innovation to occur.

In fact, investment in R&D plays an important role in every aspect of a R&D process. It attracts talents, covers laboratory costs, helps provide test equipment and devices, etc. All these expenditures potentially lift a firm's innovativeness to elevated levels and lead to the occurrence of technological innovation.

Other than attracting talents, directed investment in scientific and technological talents represents another important component of technological innovation. An effective organization of quality scientific and technological talents with other essential factors can produce a significant impact on the efficiency and capabilities of technological innovation. Within an enterprise, the development of AI potentially leads to additional intelligent industrial robots, high-end intelligent equipment, and a subsequent set of intelligent production processes. Such advances can gradually replace, to a certain degree, human labor, increase productivity, and reduce costs so that enterprises can devote more manpower, material, and financial resources into research and development activities, effectively increasing the investment in R&D and talents.

Additionally, enterprises can create and capture higher levels of value by employing artificially intelligent technologies. With rising levels of profits, enterprises will in turn increase their investment in R&D and talents, based on which the rising levels of profits initially came from. That promotes a new round of technological innovation, forming a new virtuous cycle (Proposition 18.4). Hence, we have the following result.

Proposition 18.10 Artificial intelligence promotes technological innovation through increasing investment in R&D and talents either directly or indirectly.

18.4 An Empirical Case Analysis

This section attempts to confirm some of the established propositions above by an empirical means.

18.4.1 The Model, Variables, and Data

As suggested by the literature, other than AI, other factors, such as the investment intensity in R&D (Patterson & Ambrosini, 2015), foreign direct investment (Feng, 2019), ownership structure (Yi et al., 2017), and technological imitation (Zhang & Li, 2015), also influence technological innovation. For our purpose, these factors are used as control variables in the following measurement model:

$$\begin{aligned} \ln \mathrm{Y}_{i,t} &= \alpha + \beta_1 \ln \mathrm{AI}_{i,t} + \beta_2 \ln \mathrm{K}_{i,t} + \beta_3 \ln \mathrm{FDI}_{i,t} + \beta_4 \ln \mathrm{STATE}_{i,t} \\ &\quad + \beta_5 \ln \mathrm{TETI}_{i,t} + \varepsilon_{i,t}, \end{aligned} \tag{18.1}$$

where the subscripts i and t stand for province i and year t, respectively, AI artificial intelligence, K investment in R&D, FDI foreign direct investment, STATE ownership structure, TETI technical imitation, α is a constant term, β a coefficient of each variable, and ε is a random error.

Specifically, the response variable Y is measured by using the number of patent authorizations as in Carree et al. (2015) in various manufacturing sectors as an indicator of technological innovation. This choice is different from Filippetti et al. (2016) who employ the number of patent applications, Christiansen (1999) who adopts the number of new product developments, and Guan (Guan & Gao, 2009) who uses the sales of new products. The main explanatory variable AI is measured by the number of industrial robots as in Acemoglu and Restrepo (2018) because AI in manufacturing is mainly realized through industrial robots.

As for the control variables, this section measures

1. "investment in R&D" (K), as in Han et al. (2014), by using the radio of internal expenditures of R&D to the total output value;
2. "foreign direct investment" (FDI) in an industry by using the ratio of the total output value of foreign-funded enterprises to the total output value of that industry (Wang et al., 2010);
3. "ownership structure" (STATE) by using the ratio of the output value of state-owned enterprises in an industry to the total output value of the industry (Bai, 2011; Sun et al., 2017); and
4. "technology imitation" (TETI) by using the proportion of enterprises' technical transformation costs to the total industrial output value (Wang & Wang, 2019).

By synthesizing the classification standards of the International Federation of Robotics and the China Industrial Statistics, this section is able to examine a set of panel data for the time period 2008–2017 from International Robot Alliance Website, "China Statistical Yearbook," "China Science and Technology Statistical Yearbook." The data set reflects 14 manufacturing sectors, covering a wide range from textile and clothing to production and supply of electricity, heat, gas, and water.

18.4.2 Results of Empirical Analysis

In terms of the effect of AI on technological innovation, regression analysis is utilized on the available data. The corresponding coefficients are estimated by respectively using Fixed Effects (FE), Random Effects (RE), Feasible Generalized Least Squares method (FGLS), and the System Generalized Method of Moment (System-GMM). By comparing the resultant estimates, Hausman test suggests that FE produces better results than RE. Because FE and FGLS produce estimates of the same signs, while FGLS can possibly eliminate existing heteroscedasticity and serial correlation (Woodridge, 2002), it means that FGLS is a better method than FE. And, because System-GMM can partially resolve the endogenous problem (Roodman, 2009), and because the signs of the coefficients of *ln*AI, as estimated respectively by FGLS and System-GMM, are the same, it becomes clear that this analysis can simply focus on the estimates of System-GMM.

Because the coefficient of AI influence on technological innovation is equal to 0.1945 and passes the test of 5% significance, it means that when all the control variables are fixed, AI positively impacts technological innovation (Propositions 18.4, 18.6, 18.8, and 18.10). In particular, through the application of various related technologies, AI helps increase the flow and sharing of knowledge, boost the capability to learn, absorb and create knowledge. Consequently, AI accelerates the conversion, development, and utilization of knowledge. In short, AI, through various means, assists a firm to significantly increase its productivity and to liberate resources so that unrestricted resources in turn strengthen R&D efforts and activities, leading to much improved capability of technological innovation.

Regarding the control variables, the relatively large coefficient is that of "investment in R&D." It is positive and equal to 0.7586, while passing the 1% significance tests. This fact demonstrates that the amount invested in R&D efforts benefits the innovativeness of a manufacturing firm in the area of technology. In terms of "foreign direct investment," its coefficient is equal to −0.0798, while failing the significance test. Hence, it means that this control variable does not impact technological innovation significantly. As for "ownership structure," its coefficient is equal to −0.4439 and passes the 10% significance test, indicating that "ownership structure" does have a constraining effect on the innovativeness of a manufacturing firm. Specific to the data employed here, this end implies that in the overall Chinese economy, if the segment of state-owned enterprises occupies a large proportion, then the innovativeness of the manufacturing industry will be low. That is because state-owned enterprises tend to affect industrial allocation efficiency and technical efficiency (Sun et al., 2017). These adversely affected efficiencies will surely impact technological innovation negatively. As for the control variable "technical imitation," its coefficient is equal to −0.0644 and passes the significance test of 5%, demonstrating that such imitation has a noteworthy limiting effect on the technology innovation. That is because although enterprises in the 14 manufacturing industries considered in this chapter adopted advanced technologies, they still have not effectively digested and absorbed the true essence of the technologies beyond the stage of imitation. In this sense, quick introduction of advanced technologies generally hinders the development of technological innovation. In short, this empirical study reconfirms the results of Propositions 18.1, 18.7, and 18.9.

18.5 A Few Final Words

This chapter examines how AI impacts technological innovation by such a logical reasoning that is parallel to the one widely used in mathematics and natural science (Kline, 1972) while confirming the established results empirically by using data from China's manufacturing industry. It first looks at the concept of technological innovation by referencing back to numerous definitions of innovation and explores how this concept can be defined generally without the need to make periodic alterations. This new definition stresses simultaneously the originality of a technology in terms of its newness, or freshness, or enhancement, and the following creation of value. Starting from this general definition of technological innovation and the specific logical reasoning, 10 generally true propositions are developed. Specifically, this chapter is able to discover the root on which an original technology can potentially appear out of the R&D effort of technological innovation (Proposition 18.1), how AI stimulates technological innovation (Propositions 18.3, 18.6, 18.8, and 18.10).

In terms of empirical analysis, by controlling "investment in R&D," "foreign direct investment," "ownership structure," and "technology imitation," it is found that advances in AI assist the development of technological innovation. At the same time, the control variable "investment in R&D" indeed benefits the technological innovativeness of a manufacturing firm, while "foreign direct investment" does not impact technological innovation significantly, and other two control variables have constraining effects on the innovativeness of a manufacturing firm. These empirical results reconfirm the results of Propositions 18.1, 18.7, and 18.9.

The theoretical conclusions established in this chapter naturally lead to the following policy suggestions and recommendations.

1. Raise the level of sophistication of AI for the purpose of promoting the technological innovativeness of manufacturing enterprises.

   First, invest increasingly in the AI effort and activities that result in new service formats, such as cloud computation, management of big data, Internet of things, and intelligent manufacturing. In its penetration into various industrial sectors, employ AI technology to upgrade and transform traditional industries, while encouraging the creation of new knowledge and the development of novel technologies.

   Second, organize by using AI-based technologies interorganizational networks and strategic blocks, consisting of government agencies, business firms, research institutes and universities, as platforms for coordinated innovations to occur. Within such networks and blocks, the government can help coordinate joint efforts of industry, academia, and R&D by associating, exchanging, and sharing information and knowledge among innovation entities. Such governmental guidance can either directly or indirectly promote rapid turnover and spread of AI technologies between and among individual business firms and different industries. The consequently shortened R&D periods in turn stimulate the creation and spillover of new knowledge and new technology, which helps raise the capability of technological innovativeness.

   Third, considering the present challenges facing Chinese manufacturing firms, it becomes more necessary than ever before that communications between the government, industry, and enterprises need to be promoted. By doing so, China will be enabled to

   - Uninterruptedly advance its technological innovativeness,
   - Develop needed mechanisms for producing and discovering talents, and
   - Effectively coordinate the development of AI industry.

     With such capabilities well developed, China will lead the global development of the AI industry and will acquire its dominance within the increasingly severe international competition.

2. Create a mechanical way to encourage the development of AI technology and the AI industry.

Adopt well-designed measures to promote the development of AI technology and the AI industry so that relevant new knowledge can be formed and novel technologies can be created. Well-designed measures need to address the problem of how to smoothly and effectively facilitate exchanges of ideas and information, and that of how to coordinate the demands of local, regional, national, and international markets. Addressing these issues holistically will assist business enterprises to engage in innovative activities in order to provide creative offers to different markets of various levels (Theorem 18.1). Successfully doing so fortifies cross-regional and cross-national exchanges and cooperation, which in turn support the introduction of advanced knowledge and technology, and promote the healthy development of an AI industry.Design and implement necessary coping policies, speed up the systematic advancement of the underlying industrial chain of AI and continuously improve the quality and scale of technological innovation, while promoting the sustainability of the economy.As stated in other parts of this book, the validity of the conclusions established in this chapter is also constrained. For example, AI's impact on technological innovation is analyzed in four different angles: the creation of knowledge, the overflow of knowledge, capabilities of learning and absorption, and investment in R&D efforts. However, if one likes to directly confirm these established conclusions empirically, he/she will find that it is difficult, if not impossible, to design appropriate proxies and to collect adequate data for such variables as "creation of knowledge," "overflow of knowledge," and "capabilities of learning and absorption." This fact opens a wide door for future empirical research.

On the other hand, the section of empirical confirmation only considers the panel data of China's manufacturing sector during a particular time period. So, the developed confirmations of this chapter might not hold true if data are from other countries or regions in the world. Hence, similar empirical confirmations can be carried out with data from outside China.

References

Acemoglu, D., & Restrepo, P. (2018). Artificial intelligence, automation and work. NBER working paper no. 24196.

Aghion, P., Jones, B. F., & Jones, C. I. (2017). Artificial intelligence and economic growth. NBER working paper no. 23928.

Agrawal, A. K., Gans, J. S., & Goldfarb, A. (2018). Prediction, judgment and complexity: A theory of decision making and artificial intelligence. NBER working paper no. 24243.

Bai, J. H. (2011). Are government R&D subsidies efficient in China! Evidence from large and medium enterprises. *China Economic Quarterly, 10*(04), 1375–1400. https://doi.org/10.4236/ajibm.2014.49056

Baregheh, A., Rowley, J., & Sambrook, S. (2009). Towards a multidisciplinary definition of innovation. *Management Decision, 47*(8), 1323–1339.

Becheikh, N., Landry, R., & Amara, N. (2006). Lessons from innovation empirical studies in the manufacturing sector: A systematic review of the literature from 1993 to 2003. *Technovation, 26*(5), 644–664.

Branscomb, L. M. (2001). Technological innovation. In *International Encyclopedia of the social & behavioral sciences* (pp. 15498–15502). Elsevier.

Brown, S. L., & Eisenhardt, K. M. (1995). Product development: Past research, present findings, and future directions. *Academy of Management Review, 20*(2), 343–378.

Brynjolfsson, E., & Mcafee, A. (2014). *The second machine age*. Norton.

Camerer, C. F. (2019). Artificial intelligence and behavioral economics. In A. K. Agrawal, J. Gans, & A. Goldfarb (Eds.), *The economics of artificial intelligence: An agenda*. University of Chicago Press.

Caputo, A., Marzi, G., & Pellegrini, M. M. (2016). The internet of things in manufacturing innovation processes: Development and application of a conceptual framework. *Business Process Management Journal, 22*(2), 383–402.

Carree, M., Piergiovanni, R., Santarelli, E., et al. (2015). Factors favoring innovation from a regional perspective: A comparison of patents and trademarks. *International Entrepreneurship and Management Journal, 11*(4), 793–810.

Chen, C. J. (2010). The effects of knowledge attribute, alliance characteristics, and absorptive capacity on knowledge transfer performance. *R & D Management, 34*(3), 311–321.

Cheng, C. P., & Peng, H. (2018). The mechanism of artificial intelligence affecting employment and China's countermeasures. *China Soft Science Magazine, 2018*(10), 62–70.

Christiansen, J. A. (1999). *Competitive innovation management: Techniques to improve innovation performance*. Palgrave Macmillan.

Damanpour, F. (1996). Organizational complexity and innovation: Developing and testing multiple contingency models. *Management Science, 42*(5), 693–716.

Dance, J. (2008). What is innovation? 30+ definitions lead to one fresh summary. *Fresh Consulting* (blog), May 22, 2008., http://freshconsulting.com/what-is-innovation, Accessed on May 2, 2019.

Debackere, K., Clarysse, B., & Rappa, M. A. (1996). Dismantling the ivory tower: The influence of networks on innovative output in emerging technologies. *Technological Forecasting and Social Change, 53*, 139–154.

Deiss, K. J. (2004). Innovation and strategy: Risk and choice in shaping user-centered libraries. *Library Trends, 53*(1), 17–32.

Du, C. J., Hu, J., & Chen, W. X. (2018). Development model and countermeasures of China's new generation of artificial intelligence industry. *Economic Review Journal*, (04), 41–47+2.

EC. (2018). http://publications.jrc.ec.europa.eu/repository/bitstream/JRC113826/ai-flagship-report-online.pdf, Accessed on April 16, 2019.

Feng, Y. C. (2019). Effects of environmental regulation and FDI on urban innovation in China: A spatial Durbin econometric analysis. *Journal of Cleaner Production, 235*. https://doi.org/10.1016/j.jclepro.2019.06.184

Filippetti, A., Frenz, M., & Ietto-Gillies, G. (2016). The impact of internationalization on innovation at countries' level: The role of absorptive capacity. *Cambridge Journal of Economics, 41* (2), 413–439.

Forrest, J. Y. L., Lin, C. C., Mondal, S., & Tucker, R. (2019). Environmental forces underneath the innovativeness of manufacturing firms. *Theoretical Economics Letters, 9*, 1353–1382. https://doi.org/10.4236/tel.2019.95088

Forrest, J. Y. L., Mondal, S., Tucker, R., & Lin, C. C. (2018). Effects of manufacturing firms' strategies on innovation: A holistic view. *Proceedings of the 2018 annual conference of NABET*, November 1–2, 2018, State College, PA (pp. 74–94).

Fujitsu Limited & Riken. (2018). Fujitsu and RIKEN Demonstrate AI's Utility in Material Design. https://www.fujitsu.com/global/about/resources/news/press-releases/2018/0316-01.html, Accessed on April 29, 2019.

General Electric. (2012). GE Global Innovation Barometer: Global Research Report, 30, http://files.gecompany.com/gecom/innovationbarometer/GE_Global_Innovation_Barometer_Report_January_2012.pdf.

Goldfarb, A., & Trefler, D. (2018). AI and international trade. NBER working paper no. 24254.

Government of Japan. (2015). https://www8.cao.go.jp/cstp/kihonkeikaku/5basicplan_en.pdf, Accessed on April 16, 2019.

Guan, J. C., & Gao, X. (2009). *Exploring the h-index at patent level*. John Wiley & Sons.

Hall, L. A., & Bagchi-Sen, S. (2002). A study of R&D, innovation, and business performance in the Canadian biotechnology industry. *Technovation, 22*, 231–244.

Han, X. F., Hui, N., & Song, W. F. (2014). Can informantization improve the technology innovation efficiency of Chinese industrial sectors. *China Industrial Economics, 2014*(12), 70–82.

Holmstrom, J., Holweg, M., Khajavi, S. H., & Partanen, J. (2016). The direct digital manufacturing (r) evolution: Definition of a research agenda. *Operations Management Research, 9*, 1–10.

II - Industrialization and Informationalization. (2017). Three-Year Action Plan to Promote the Development of a New Generation of AI Industry 2018–2022. http://www.miit.gov.cn/n1146285/n1146352/n3054355/n3057497/n3057507/c5979554/content.html, Accessed on April 22, 2019.

Kanter, R. M. (2001). *Evolve! Succeeding in the digital culture of tomorrow*. Harvard Business School Press.

Kaser, D. (2011). Editor's notes: Innovation can be fun. *Computers in Libraries, 31*(5), 4.

Keizer, J. A., Dijkstra, L., & Halman, J. I. M. (2002). Explaining innovative efforts of SMEs. An exploratory survey among SMEs in the mechanical and electrical engineering sector in the Netherlands. *Technovation, 22*, 1–13.

Kline, M. (1972). *Mathematical thought from ancient to modern times*. Oxford University Press.

Kotter, J. (2012). Barriers to change: The real reason behind the Kodak downfall. *Forbes*, May 2, 2012, www.forbes.com/sites/johnkotter/2012/05/02/barriers-to-change-the-real-reason-behind-the-kodak-downfall

Lecun, Y., Bengio, Y., & Hinton, G. (2015). Deep learning. *Nature, 521*(7553), 436.

Lee, J. (1995). Small firms' innovation in two technological settings. *Research Policy, 24*, 391–401.

Leonard, D. A., & Swap, W. C. (1999). *When Sparks fly: Igniting creativity in groups*. Harvard Business School Press.

Lv, T., & Han, N. (2015). Intelligent manufacturing: The global trend and Chinese strategy. *Frontiers, 2015*(11), 6–17.

Makridakis, S. (2017). The forthcoming artificial intelligence (AI) revolution: Its impact on society and firms. *Futures, 90*(June), 46–60.

Markoff, J. (2016). *Machines of Loving Grace: The quest for common ground between humans and robots*. HarperCollins.

McGrath, R. G. (2013). *The end of competitive advantage: How to keep your strategy moving as fast as your business*. Harvard Business Review Press.

OECD. (1997). *The measurement of scientific and technological activities, proposed guidelines for collecting and interpreting technological innovation data*. Organization for Economic Cooperation and Development.

OWH. (2016a). *Preparing for the Future of Artificial Intelligence*. https://obamawhitehouse.archives.gov/sites/default/files/whitehouse_files/microsites/ostp/NSTC/preparing_for_the_future_of_ai.pdf, Accessed on April 20, 2019.

OWH. (2016b). *The National Artificial Intelligence Research and Development Strategic Plan*. https://www.nitrd.gov/PUBS/national_ai_rd_strategic_plan.pdf, Accessed on April 16, 2016.

Parthasarthy, R., & Hammond, J. (2002). Product innovation input and outcome: Moderating effects of the innovation process. *Journal of Engineering and Technology Management, 19*, 75–91.

Patterson, W., & Ambrosini, V. (2015). Configuring absorptive capacity as a key process for research intensive firms. *Technovation, 36-37*, 77–89.

Roodman, D. (2009). How to do Xtabond2: An introduction to difference and system GMM. *Stata Journal, 9*(1), 86–136.

Roos, G. (2015). Servitization as innovation in manufacturing: A review of the literature. In *The handbook of service innovation* (pp. 403–435). London. https://doi.org/10.1007/978-1-4471-6590-3_19

Rosenthal, S. R. (1992). *Effective product design and development*. Irwin.

Rowley, J. (2011). Should your library have an innovation strategy? *Library Management, 32*(4/5), 251–265.

Schumpeter, J. A. (1934). *The theory of economic development*. Harvard University Press.

Shi, X. A., Li, L. S., Cheng, Z. H., & Liu, J. (2018). Impact analysis of "internet+" on the value-chain improvement of China's manufacturing. *Studies in Science of Science, 36*(08), 1384–1394.

Smith, A. (1776). *The wealth of nations, books I-III, (1986 printing)*. Penguin Books.

Smith, W. K., & Tushman, M. L. (2005). Managing strategic contradictions: A top management model for managing innovation streams. *Organization Science, 16*(5), 522–536.

Stock, G. N., Greis, N. P., & Fischer, W. A. (2002). Firm size and dynamic technological innovation. *Technovation, 22*, 537–549.

Sun, Z., Xiao, L. P., & Liu, L. H. (2017). The changes of industrial ownership structure and innovation: Is the state-owned enterprise dominant in favor of innovation? *Nankai Economic Studies, 2017*(06), 3–19.

Trajtenberg, M. (2018). AI as the next GPT: A political-economy perspective. NBER working paper no. 24245.

Vaughan, J. (2013). Technological innovation: Perceptions and definitions. ALA TechSource, a publishing unit of the American Library Association.

Wang, L. J., & Wang, Q. L. (2019). Study on the relationship of input and output of re-innovation after digesting the introduced technology in high-tech industry: An empirical study based on different industry data. *China Soft Science, 01*, 184–192.

Wang, R., Yan, B., & Deng, W. G. (2010). The impact of FDI on the independent innovation capability of Chinese indigenous industries: From the perspective of industrial linkages. *China Industrial Economics, 2010*(11), 16–25.

Wang, Z. H., & Yang, Z. (2017). Thoughts on artificial intelligence technology research and future intelligent information service system. *Telecommunications Science, 33*(05), 1–11.

Weitzman, M. L. (1998). Recombinant growth. *Quarterly Journal of Economics, 113*(2), 331–360.

WH. (2018). https://www.fedscoop.com/white-house-artifical-intelligence-committee-kratsios/, Accessed on April 16, 2019.

Woodridge, J. M. (2002). *Econometric analysis of cross section and panel data*. The MIT Press.

Wu, D., Rosen, D. W., Wang, L., & Schaefer, D. (2015). Cloud-based design and manufacturing: A new paradigm in digital manufacturing and design innovation. *Computer-Aided Design, 59*, 1–14.

Xi, J. P. (2017). http://www.gov.cn/zhuanti/2017-10/27/content_5234876.htm, Accessed on April 16, 2019.

Yi, J., Hong, J., & Hsu, W. C. (2017). The role of state ownership and institutions in the innovation performance of emerging market enterprises: Evidence from China. *Technovation, 2017*, S016649721730247X.

Zhang, Y. C., & Li, X. T. (2015). Measurement of technological innovation efficiency and its influencing factors in China's high-tech transformation of traditional industries: An empirical analysis based on the transcendental logarithmic stochastic frontier model. *Technology Economics, 34*(03), 18–26+111.

Zhou, J. (2012). Digitization and intelligentization of manufacturing. *China Mechanical Engineering, 23*(20), 2398–2400.

Part VI
Government Policies and Supports

Chapter 19
How the Government Can Help Stimulate Economic Growth

Abstract This chapter, which is mainly based on Forrest et al. (to appear, The role government policy and supports play to stimulate economic growth), studies the following three issues: (1) why policy tools of the government might work in real life; (2) when they do work, what the underlying mechanism is through which implemented policies play their roles; and (3) why it is necessary for the government to design and implement policies and provide supports for the purpose of stimulating economic growth.

Keywords Economic agent · Feedback · Market · Momentum · Policy tool · System

Based on the logic and methods of systems science and the logical reasoning that is parallel to that commonly used in mathematics and natural science, this chapter establishes six formal and generally true propositions to address these listed and some other related issues. Then, this chapter concludes with a list of recommendations for policymakers and government officers. In particular, these recommendations outline at what time and in what ways the government can expect their implemented policies to produce their desired outcomes. At the very end, some possible directions and open problems for future research are given.

19.1 Introduction

At the same time when the development of communication technology makes daily living convenient, it also changes how enterprises conduct their business. Such changes in the business world challenge several ambitious nations, either developed or developing, to think about how they can employ policy tools to transform their states of industries. For these nations to maintain and/or to acquire their leadership in the world economy and international politics, they desire to help their industries to rise from automated manufacturing, the current state, to intelligent manufacturing, an expected future state, such as Industry 4.0 (Hermann et al., 2016). Along with this

J. Y.-L. Forrest, Y. Liu, *Value in Business*, Contributions to Management Science,
https://doi.org/10.1007/978-3-030-82898-1_19

trend of development from around the world, the following questions arise unsurprisingly:

1. Although governments have been introducing and implementing various policies for different purposes, what is the reason, if any, for why government policies are anticipated to play their roles in an economic system? and
2. Are policy tools and governmental supports fundamentally necessary to practically stimulate economic growth?

Evidently, these are important questions that face all nations from around the world. Specifically, due to the momentum of historical development, developed nations like to maintain their leading and/or dominating positions in the world economy and international polities, especially within the current globalization. Some of these developed nations include Great Britain (Hall & Pesenti, 2017), the USA (OWH, 2016), Germany (Industry 4.0), Japan (Government of Japan, 2015), and France (Marc et al., 2018). With their unhappy fiascos of modernization in the past 100 years (Wen, 2016), all other nations, either developing or undeveloped, face an increasing possibility of plummeting economically, socially, and politically further and further behind the developed nations. To potentially avoid such likelihood, some developing nations, such as China (State Council of the PRC, 2015), also actively assist their domestic economies to make necessary leaps.

Because of their theoretical and practical importance, this chapter attempts to investigate these and related questions. At the same time, it fills a gap that exists in the existent literature regarding the reason why policies and supports from the government are needed in stimulating economic development. The rest of this chapter is organized as follows. Section 19.2 reviews the literature and outlines the contribution this chapter makes. Section 19.3 studies the mechanism through which policy tools potentially work. Section 19.4 investigates why governmental policies and supports are practically needed in economic development. Section 19.5 concludes this chapter and provides related practical recommendations and important questions for future considerations.

19.2 Literature Review

This chapter touches on two areas of the literature. One involves studies on how governmental policies actually work; and the other explorations on why governmental policies and supports are necessary for economic development.

For the first area of the literature—how governmental policies actually work, it is found that policy instruments can help advance the functionality of an innovation system (Wieczorek & Hekkert, 2012). As for monetary policy, the following three channels are found (Auclert, 2019) to amplify the effects of the policy: the heterogeneity channel of earnings from unequal income gains, the Fisher channel from unexpected inflation, and the channel of interest rate exposure from changes in real interest rate. By considering credit support policies of the Euro-system, and by examining the effectiveness of the system and the underlying mechanism of transmission, policies in stimulating credit flows from banks to the private sector are found to be successful (Boeckx et al., 2016). By checking how government spending might lead to the crowding-in effect on private consumption, Rebei (2017) empirically confirms that the appearance of such effects in early times; however, such past relationship for some reason degenerated in recent times. An investigation on the responses and consequences of a variable credit spread optimal policy shows (Cúrdia & Woodford, 2016) that an optimal policy can be well approximated by using a simple target criterion. By examining the formulation of labor clauses in various stages of the policy process, Jaehrling et al. (2018) show how local politicians, unions, and employers need to form alliances.

Contributing to and enriching this area of the literature, this chapter reveals the reason why it is very important for the government to design and adopt appropriate policies, under what conditions a government can bring into line the objectives of a hefty percentage of individual business enterprises, and what kinds of policy tools can assist the government to accomplish such a goal.

For the second area of the literature—why governmental policies and supports are necessary for economic development, existent studies can be naturally classified into the following two contradictory classes:

1. Yes, governmental policies and supports are necessary;
2. No, governmental policies and supports do not work generally.

In particular, in support of (1), a framework is developed by Andreoni and Chang (2019) for the purpose of strategically coordinating interactive industrial policy measures; and it is argued (Howell et al., 2019) that for the United States to maintain its future competitiveness internationally, it needs to introduce realistic federal, industry, and company policies regarding microelectronics. Conversely, in support of (2), there are eight Asian economies, including Japan, South Korea, Taiwan, Singapore, Hong Kong, Malaysia, Thailand, and Indonesia, that achieved rapid economic growth between 1965 and 1990, a statistically unlikely event. In all these cases, government interventions clearly played their respective roles. However, Jomo (2019) notes that it is really ambiguous as to whether or not adopted industrial policies actually led to the miracle gains. As for the concerted effort of the US government to coordinate the Advanced Manufacturing Initiative for America's Future, Hemphill (2014) finds that such effort would not blend well with the Obama administration.

Contributing this area of the literature, this chapter establishes several theoretical results that are generally true. In particular, what are established include: (1) the need for a nation to provide policy supports in order to sustain the momentum of economic growth, if such momentum already exists; (2) under what conditions an adopted and implemented policy can directly help manufacturing enterprises advance their economic performance, and (3) why policy tools need to be flexible, reflecting the differences of geographical regions, if they are expected to positively promote economic growth in these regions.

To make this chapter self-contained and the logical reasoning flow smoothly, each business firm is assumed to exist for the purpose of satisfying a market demand; and its operation is financially maintained by cash flows from the market exchanges of its products and services. By consumer, it stands for an end user of a market offer. By customer, it represents a firm that uses its supplies to produce its outputs that might be supplies for another customer or a consumer. If a product can be simultaneously used by consumers and customers, then these consumers and customers will be jointly referred to as customers.

19.3 The Mechanism through Which Policy Tools Potentially Work

The focus of this section is on the question of why policy tools are expected to practically work in the focal economic system. This section consists of three subsections with the first one on the systemic modeling of the economy, the second one explaining why it is important to have a functional government, and the third one on conditions under which policy tools actually work.

19.3.1 The Economy: Seen Systemically

The concept of numbers is abstracted out of the counting of things by ignoring what the things are and what internal structures the things possess. However, when the internal structure of things can no longer be ignored, the concept of systems come into being. In particular, by system, it stands for such a notion that models an organization or a structure as a collection of some components or objects and a collection of relations that associate the objects in certain specified ways.

Within this concept of systems, natural science has treated the components of the system under consideration as isolated objects, believing, falsefully, that if each of the isolated components is understood, the system then as a whole is also comprehended. That is, what matters in terms of understanding the system is the set of relations or associations that help organize the components into an invisible whole (or system) (Lin, 1999). Therefore, system can be found everywhere in life; and the concept of systems can be applied in studies of business entities, events, and processes. For instance,

- Each household is a system, consisting of people and all things acquired to make the living of the people possible. Only when the people and the things are somehow associated with each other, the household becomes a family.
- Each business firm is a system, consisting of employees and various necessary resources and hardware. It is these people, resources, and hardware that a recognizable firm appears.

Through the lens of systems, one can readily see that the world is mainly composed of systems of various kinds and interactions of systems. That is, a key essence of the world is the systemness of various kinds of structures (or organizations), when looking at physical, nonliving things, and various types of organizations, when glaring at living organisms. Such intellectual recognition elucidates the reason why the concept of systems can be appropriately and naturally employed in researches of business evolutions, processes, events, and interactions of business entities. For relevant details and stories of application successes, see, for example, Forrest (2018), Porter (1985), and Rostow (1960).

As is well-known, for a concept to be useful either theoretically or practically, the concept has to be formalized in symbolic terms. For this reason, a system is identified as an ordered pair $S = (M, R)$ of two sets, where set M contains all the isolated objects or elements of system S and the set R demonstrates all the relations, particular to the system, that connect the elements in M into a whole (Lin, 1999). To help understand this symbolic expression of an abstract system, let us decompose in theoretical terms a business organization into a set of components, such as employees, real-estate properties, equipment, and others, and a set of relations that associate the components into the recognizable organization. This decomposition of the organization reveals that it is these particular relations that the resultant whole can be realistically acknowledged as a business enterprise.

By continuing the thinking logic above, one can naturally see that each object in M is a component of the system and can also be a system itself. That is exactly as in the case that a family consists of people (along with other things) as its components and objects, while each person is a biological system of its own right. Such leveled structure continues downward. For example, as a biological system, each person is made up of organs. And each organ is a system, made up of even smaller components. This hierarchical structure continues, depending on when a research likes to stop from going further downward.

As for the set R of relations, it does not have to contain only numerical relations or symbolic expression written in conventionally recognizable mathematical or statistical forms. For example, the relationship of siblings should not be expressed mathematically or statistically unless one is willing to sacrifice most of the characteristics of the sibling relationship.

Let $S = (M, R)$ and $S_1 = (M_1, R_1)$ be two given systems. System S_1 is said to be a partial system of S, if the following condition holds true: M_1 is a subset of M and for any relation $r_1 \in R_1$ there is a relation $r \in R$, satisfying that relation r_1 of S_1 is the restriction of a relation r in S. Symbolically, this end is written as follows:

$$r_1 = r|M_1.$$

Given a system $S = (M, R)$, it is said to be multileveled, if the system S satisfies the following condition:

- There is an object $S_1 \in M$ such that $S_1 = (M_1, R_1)$ is a system. In this case, S_1 is known as a first-level object system of S.

- There is an object $S_2 \in M_1$ such that $S_2 = (M_2, R_2)$ is a system. In this case, S_2 is referred to as a second-level object system of S.
- Such layered structure continues until reaching an nth-level object system of S so that the objects in S are no longer systems, where n is a natural number.

Given a system $S = (M, R)$, it is said to be centralized if the following conditions hold true: (1) every object of the system S is once again a system; and (2) there exists another system $C = (M_C, R_C)$, satisfying that $M_C \neq \varnothing$ and for any different objects $x = (M_x, R_x)$, $y = (M_y, R_y) \in M$,

$$M_C = M_x \cap M_y \text{ and } R_C \subseteq R_x|M_C \bigcap R_y|M_C,$$

where $R_x|M_C = \{r| M_C : r \in R_x\}$ and $R_y|M_C = \{r| M_C : r \in R_y\}$ represent respectively the restrictions of the relations sets R_x and R_y on M_C.

This system C is referred to as a center of the centralized system S. An intuitive understanding of such a centralized system S is that S is centralized, if, and only if, another system C exists such that C is a partial system of each and every object of S.

In order to answer the questions stated above, let us see how the focal nation's economy can be theoretically modeled as an abstract system composed of all economic agents as component parts and associations that join the component parts in ways that are specific to the focal nation into a functional economy. The component parts of this systemic modeling include, but are not limited to, consumers, families, business firms, levels of the government, various kinds of industries, economic sectors, and markets. Because each nation in real life possesses its individually specific economy, it can be imagined that other than component parts are different from one nation to another, for example, people in one nation are different from those of another nation, how their component parts are associated to each other are also different from one nation to another.

In this highly abstract systemic model of the focal nation's economy, the associations describe how economic factors, such as money, information, knowledge, and talent, flow within the economy. They demonstrate how market demands and relevant supplies change constantly in terms of their magnitudes and types. They portrait how preferences and tastes of consumers evolve relentlessly.

To help with the understanding of this systemic model, one can naturally see that

- Every family is a holistic system, consisting of family members and all necessities of living acquired over time;
- Every business firm is a holistic system, made up of members of various families, organized towards the realization of certain economic objectives;
- Every economic sector is a holistic system, composed of many firms that trade with each other in one way or another;
- Every market is a holistic system, which gathers consumers and business firms through exchanges of products, services, information, money, etc.; and
- Every level of the government is a holistic system, which helps provide stability and security for the entire economy.

That is, the economy of the focal nation is really or can be well modeled as a system of multiple levels. The functional, hierarchical structure of this economy, as seen as a multileveled system, is guaranteed by the government through employing various tools, such as policies, regulations, judicial procedures, law reinforcement, and the military might. It is the stable and smooth functions of the government that provides a formal structure for the economic system to exist and to operate.

In terms of the concept of centralized systems, the topmost level of the government functions as the center of the economy. Any slight change at this level will affect, either positively or negatively, most areas of the economic system. The formal structure of the economy, in abstract terms, can be envisioned as a three-dimensional hierarchy. The bottom of the hierarchy consists of individual consumers, whose consumptions, either for the purpose of basic living or that of enjoying a luxury lifestyle, mobilize all other layers of the economy. The middle of the hierarchy is occupied by business firms that respectively serve their particular segments of consumers. The top of the hierarchy appears in the form of various markets, through which exchanges of information, knowledge, money, products, services, etc., happen.

19.3.2 Why Is Government Important?

By continuing the discussion of the previous section, the economy of a region or a nation can be studied as a multileveled, centralized, and large-scale complex system. Hence, each economy surely satisfies Theorem 2.5. For the sake of convenience for the reader, that theorem is rephrased as follows:

Theorem 19.1 (Lin, 1988)

Assume ZFC. Let $S = (M, R)$ be a system that satisfies the following three conditions:

- $|M|$, the cardinality of M, $\geq c$, where c is the cardinality of the set of all real numbers,
- Each object in S is also a system with a finite set of objects, and
- Such an element G exists that belongs to at least c many objects of S,

then a partial system B of S exists with the following properties:

- The cardinality of the object set of B is equal to or greater than c,
- Element G belongs to every object of B, and
- System B is centralized.

In the language of the current discussion, Theorem 19.1 indicates that although there might be areas within a nation's economy over which the government has either no or little influence, the government can still exert its influence over at least one portion of the economy that is roughly the same magnitude as the entire economy. To facilitate a better understanding of this sentence, let us employ the

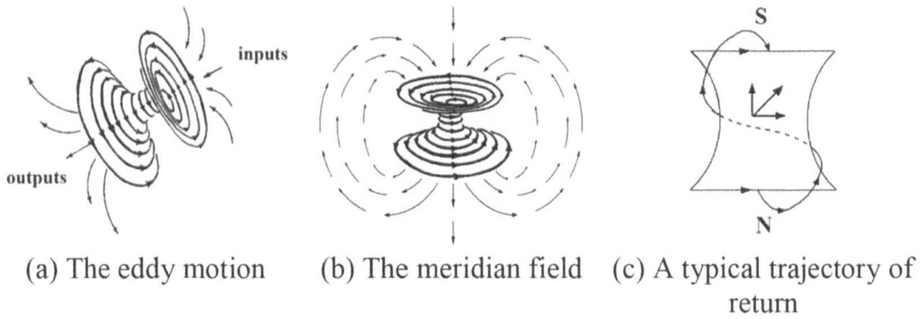

Fig. 19.1 The yoyo model of a general system. (**a**) The eddy motion. (**b**) The meridian field. (**c**) A typical trajectory of return

yoyo model to intuitively see how systems generally evolve and potentially intermingle with each other. For all related details on this model, please consult with Chap. 2. And, to make this chapter self-contained, let us briefly outline what is needed for our discussion here.

This systemic model says that each and every system can be intuitively seen or imagined as an abstract yoyo, as shown in Fig. 19.1 (Lin, 2009), which is Fig. 2.2, provided here for convenience. In other words, every system can be modeled as a multidimensional entity that gyrates around its axis. If such a multidimensional entity is fathomed in our three-dimensional space, its special structure as shown in Fig. 19.1a, appears. The input side describes how the entity pulls in "*things*," such as information, knowledge, investment, supplies, and human talents. After these "*things*" are pushed through the "neck," which models the production process of the entity that converts various inputs into market offers, things are given out in the form of outputs, such as products and services. Some of the outputs do not return back into the entity from the input side, while some do, Fig. 19.1b. To help fully understand this model of general systems, Fig. 19.1c shows along what a trajectory some outputs return back into the input side. Because of the general shape of this systemic model, it is known as a *yoyo*.

In this systemic yoyo model, the government is modeled as the center arrow in the yoyo body. This arrow not only provides the orientation for the system, but also affects all other component elements, such as industries, companies, and markets, of economy. Hence, the non-negligible role played by the government in the nation's economy is clearly shown. In other words, when a nation's government desires to promote the advancement of the economy, it can conveniently employ policy tools to increase the degree of marketization, deepen political reform, and encourage dramatic rise of private enterprises. But only so, such effort of the government is most likely successful due to its particular position as the center of the economy. According to Theorem 19.1, this approach of employing policy tools indeed practically works in life, because the government can effectually levy its will on a large segment of the economy although not the entire economy. Therefore, the conclusion below follows:

Proposition 19.1 Each commitment of the government represents a course of social influence and a relevant sequence of supports. Within such a process, government officials are able to solicit a large number of economic agents to offer their assistance so that a predetermined national objective can be accomplished.

Joint with Theorem 19.1, this proposition implies that a governmental commitment can eventually lead to some sort of collaboration and coordination of individual economic agents in its effort to materialize the desired outcome. In terms of the literature, an analogous, but more specifically stated, conclusion than this proposition is empirically derived by Chemers (2001) in terms of leadership and by Kouzes and Posner (2007).

A system is known as goal-orientated, if the system has a focus on reaching a determined collection of objectives and consequences. Examples of goal-oriented systems include regulation, control, self-organization, learning, autopoiesis, self-reproduction, self-correction, adaptation, evolution, etc. And, the literature on goal-oriented systems is capacious Klir (2001).

By combining the concept of goal-oriented systems and the systemic study of leadership (Lin & Forrest, 2011), the conclusion below follows:

Proposition 19.2 The organizational structure of a nation can be modeled as a goal-oriented system, from which the importance of the government is shown in its capability to adjust the system's orientation. To align with such capability of the government, individual economic agents alter and fine-tune their systemic orientations and operations.

Speaking both realistically and systemically, in any nation there are natural unevenness and imbalance within the nation's economic development due to uneven distributions of natural resources and human talents. These unevenness and imbalance can be vividly seen in the flow-pattern evolution of the dishpan experiment (Fultz et al., 1959). In particular, the well-known dishpan experiment is composed of a spinning dishpan that is filled with fluid. Corresponding to our discussion here, this dish can be used to model a nation's economy, where the "fluid" models the movement of such economic factors as money, information, knowledge, goods, services, and others within the economy. As the dish spins, the flow pattern within the fluid, which can be imagined to be seen from either the input or output side of the yoyo body of the economy, alternates between a uniform one, Fig. 19.1a, and a chaotic one, Fig. 19.2. Such flow-pattern alterations demonstrate the fact that imbalances in economic development of geographical regions are commonly existing phenomena.

Now, the systemic yoyo model of the economy, as shown in Fig. 19.2, illustrates the reason why Proposition 19.2 holds true. In particular, in this figurative illustration, the entire national economy is modeled as a spinning yoyo with eddy and meridian fields around. The orientation of the overall system, as desired by the government, is shown as the large dark-colored arrows that pass through the middle of the yoyo body. It can be seen from Theorem 19.1 that when a local spin field, say A_1, does not spin and is not oriented in a way that agrees with that of the overall yoyo

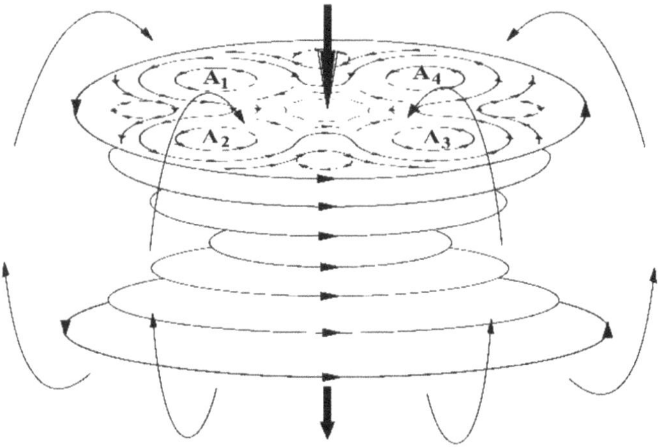

Fig. 19.2 How a goal-oriented system looks like

body, then this local spin field will be either thawed under the joint forces of its adjacent local fields or sidelined by being forced to the periphery of the overall yoyo field. In such a case, local field A_1 will not be a part of the centralized partial system B of the original economic system S.

19.3.3 When Will Policy Tools Actually Work?

As the title suggests, this subsection studies conditions under which policy tools will actually work.

Proposition 19.3 Assume that a nation is able to aim at accomplishing an objective that will potentially benefit a large number of economic agents. If the nation's government is able to obtain public support by using policy tools to exert its organizational influence, then a joint ambition, which will involve a major portion of the economy, will appear.

This result follows directly from Theorem 19.1, where what are assumed matches the conditions satisfied by the system $S = (M, R)$. Hence, the conclusion of Proposition 19.3 follows from the presence of the centralized partial system B in that theorem. In terms of the literature, Chemers (2001) presents a relevant but different study.

Proposition 19.4 If a nation's government is capable of executing its policies effectively, then the government is capable of forming a united front of effort for a large proportion of individual economic agents. In other words, the government is able to line up the objectives of a large segment of the economy, even though the involved enterprises might aim at achieving their respectively different and often conflicting business goals.

This result is a corollary of Theorem 19.1 on the basis of the intuition in Fig. 19.2. In particular, the regional fields located in between A_1, A_2, A_3, and A_4 swirl in opposite directions when compared to the spin directions of A_1, A_2, A_3, and A_4. Such differences in spin directions intuitively demonstrate the possibility that business firms that aim at achieving different or even conflicting goals can peacefully coexist and can help strengthen each other economically.

Proposition 19.5 If a nation's government aims at advancing the national economy, then it can accomplish this initial ambition by adopting and implementing appropriate policy tools.

Jointly, Propositions 19.2–19.4 directly lead to this conclusion.

The essence of the propositions established above is that for an adopted and implemented policy to produce the desired outcome, it has to aim at potentially producing the ideal outcome of benefiting a large segment of the national economy. With such a consistent aim, although conceptual and practical specifications might be different from the implementation of one policy to another, each policy, implemented, provides a common goal for a large percentage of economic agents to target at. Over time, the consistent aim of the government becomes a central highlight for the nation's citizens to feel enthusiastic about.

The following theorem, which is Theorem 3.1 rewritten here for convenience, explains how firms compete against each other.

Theorem 19.2 Assume that a market satisfies the following conditions:

- There are m firms, $m = 1, 2, \ldots$, that serve the market with their substitutable products;
- Every incumbent firm has a base of loyal customers who purchase only from the firm if the price is no more than their reservation value;
- There are such customers, collectively known as the consumer surplus, that switch from the product of one firm to that of another depending on whose price is more competitive;
- The production cost of each firm is constant; and
- These firms know others' strategies of pricing and respond by using the Nash equilibrium through pure self-analyses in order to compete over the consumer surplus.

Then, at least one new firm can profitably and competitively enter the market with its version of substitute product, while collectively expecting to earn a profit that is directly proportional to the magnitude of the consumer surplus.

In real-life terms, the existence of consumer surplus is guaranteed by the forever-evolving customer preferences and tastes, which generally turn some of the once loyal customers into switchers. All entering firms initially target these switchers, while hoping to attract some of the loyal customers from the incumbent firms. Additionally, the listed assumptions about the market generally represent that the procedure, technology, and managerial routine used in producing the said products have been more or less standardized. Therefore, if a firm wants to enter the said market profitably and competitively, it has to mean that this firm has successfully.

- Designed a different procedure of production,
- Introduced a novel technology, and/or
- Come up with a new managerial routine that can expressively lower its business expenditure in making its product of improved complexity and functionality, as empirically confirmed by Christensen et al. (1998).

19.4 Why Are Governmental Policies and Supports Practically Needed?

This subsection attempts to address the question given in the title by showing how members of a supply-chain ecosystem can be systemically threaded together, and how an existing momentum of economic growth can be practically sustained.

19.4.1 Systemic Threads within a Supply-Chain Ecosystem

To illustrate why policies and government supports are essentially necessary in terms of promoting economic growth, it can be seen from Theorem 19.2 that if a firm can innovatively decipher a market signal, then it is in a competitively advantaged position over its rivals. However, can the firm practically enhance its consequent performance? The answer to this question really depends on how other players in the firm's supply-chain ecosystem are doing (Adner, 2006). Specifically, for the focal firm to produce its imagined product based on its innovative understanding of the market signal, it has to obtain needed supplies from upstream suppliers. That in real life can represent challenges to upstream suppliers within the supply-chain ecosystem. Even after overcoming all relevant technical difficulties jointly with suppliers and created values, the firm can still be hindered with its effort to capture value. It is because for the produced offer to reach the ultimate consumers in the marketplace, it is very likely that some infrastructures, which do not exist presently, are needed from downstream complementors (Adner, 2006).

For any firm, its supply-chain ecosystem is composed of upstream components, such as suppliers, and/or downstream complements, such as customers, supporters, and assistants, known as complementors, who help make the firm's product usable by consumers (Adner, 2006). Although they are not any direct part of the supply chain, complementors need to invest in the construction of the necessary infrastructures in order to help the firm's offer to reach the targeted consumers. For example, although governments, regulators, etc., are not part of the supply chain of any firm,

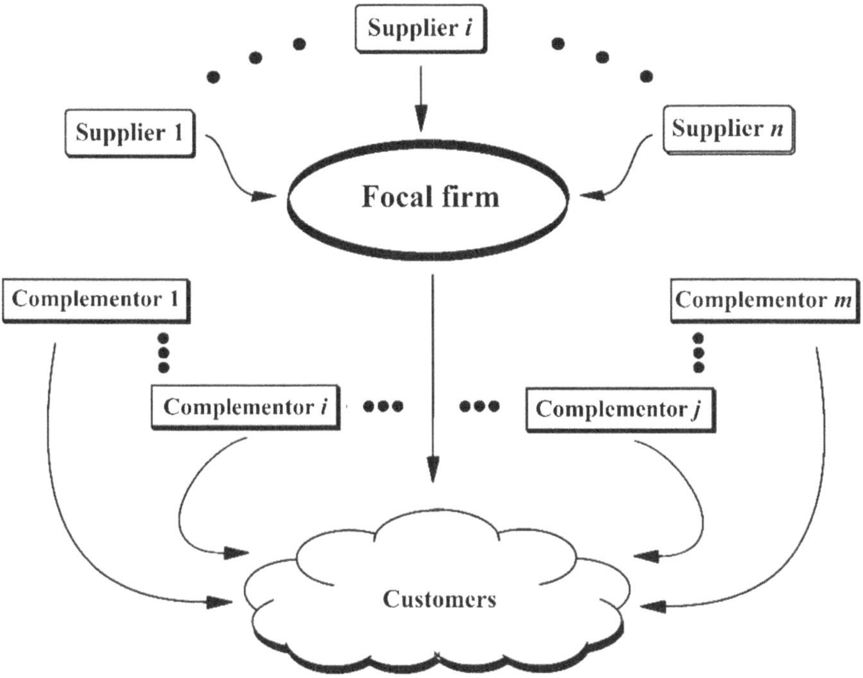

Fig. 19.3 The focal firm's supply-chain ecosystem

they are typical complementors that invest in building and maintaining, say, roads necessary for transportation or specifications of new safety procedures.

This supply-chain ecosystem is portrayed systemically in Fig. 19.3. In particular, the focal firm uses supplies (or components) from n supplies to produce its offer (s) and distributes its offers to targeted customers with the support and assistance of m complementors, $m, n = 1, 2, 3, \ldots$, which provide the necessary means to connect the focal firm and the customers. Although Fig. 19.3 only shows the first-tier upstream components and downstream complements, the hierarchical structure of the ecosystem realistically extends upward and downward along the value-creation chain to include other upstream tiers of components and downstream tiers of complements. These components and complements include, for example, suppliers' suppliers, customers' customers, complementors' complementors, etc.

By thinking about, at least mentally, the totality of all the different tiers, the systemic nature of such a supply-chain ecosystem can be richly and vibrantly depicted. At the same time, internally the focal firm also possesses its systemic structure, within which different units work holistically together, while the provided managerial supports encourage employees to innovatively integrate components into the market offer(s) that have been imagined based on the firm's creative comprehension of the market signal. For similar, while drastically different, systemic constructs, see, for example, Iansiti and Levien (2004) and Adner (2006).

Let us denote the supply-chain ecosystem of the focal firm of concern as E, and model firm i of the ecosystem as system $S_i = (M_i, R_i)$. If T represents the index set of the firms in the supply chain, then this index set T satisfies:

- Index set T and the set of all firms in the supply-chain ecosystem are 1–1 corresponding to each other. In other words,
- For any $t \in T$, there corresponds to a unique firm $S_t = (M_t, R_t)$ in the ecosystem; and for any firm $S_s = (M_s, R_s)$ in the ecosystem, there is a unique index $s \in T$ that labels the system S_s.

That is, E can be modeled as a hierarchy $E = \{S_k : k \in T\}$ of firm systems S_k, for $k \in T$.

Given two systems $S_i = (M_i, R_i)$, $i = 1, 2$, and a mapping h: $M_1 \to M_2$, each relation $r \in R_1$, let us define $h(r) = \{h(x) : x \in r\}$. Without causing confusion, let us write h: $S_1 \to S_2$. A mapping h: $S_1 \to S_2$ is said to be partial, provided that for some object $x \in M_1$, $h(x)$ is not defined. The so-called identity mapping $id : S_1 \to S_1$ is defined by $id(x) = x$, for $x \in M_1$.

Assume that the hierarchical expression $E = \{S_k : k \in T\}$ of the ecosystem, as described above, satisfies the following condition: It does not include such systems S_{k_1} and S_{k_2}, for $k_1, k_2 \in T$ and $k_1 \neq k_2$, that

- Some products of S_{k_1} are used as supplies of S_{k_2}, and
- Some products of S_{k_2} are employed by S_{k_1} as supplies.

Then, the result below follows:

Theorem 19.3 The supply-chain ecosystem $E = \{S_k : k \in T\}$ of the focal firm can be theoretically made into a partially linked hierarchy $\{E, \ell_{ts}, T\}$ of systems with the family $\{\ell_{ts} : S_s \to S_{it} : t, s \leq T, s \geq t\}$ of partial linkage mappings.

To show this theorem, it suffices to specifically define each of the partial linkage mapping $\ell_{ts} : S_s \to S_{it}$, for any t, $s \leq T$, satisfying $s \geq t$. To this end, the input-output relationship among the firms in hierarchy E of systems implies that T can be partially ordered as follows naturally: For $s, t \in T$, $s \geq t$ if, and only if, some products of firm $S_s = (M_s, R_s)$ are used as supplies of firm $S_t = (M_t, R_t)$.

For $s, t \in T$, satisfying $s \geq t$, let us define the partial linkage mapping ℓ_{ts} from firm $S_s = (M_s, R_s)$ into customer $S_t = (M_t, R_t)$ as follows

$$\ell_{ts}(m_s) = m_t \in M_t,$$

for any $m_s \in M_s$ such that system S_s's product m_s is applied as a supply in S_t's product m_t.

It can be checked readily that for any $r, s, t \in T$, satisfying $s \geq r \geq t$, the following holds true:

$$\ell_{ts} = \ell_{tr} \circ \ell_{rs}, \text{and } \ell_{tt} = id_{S_t}.$$

This systemic modeling of the supply-chain ecosystem of the focal firm implies that ecosystem E of the firm can be theoretically treated as a partially linked hierarchy of firm systems $\{E, \ell_{ts}, T\}$ with E being the collection of all players in the ecosystem. For relevant terminologies, details, and symbolic expressions, see (Lin, 1999).

19.4.2 Momentum of Economic Growth: How to Sustain It

A good tool from system science that is useful for the investigation of how to maintain an existing momentum of economic growth is the well-known concept and analysis of feedback systems (Lin, 1999). Before possibly employing this tool in the current study, let us briefly examine the concept.

Let X and Y be two linear spaces and $S : X \to Y$ and $S_f : Y \to X$ and linear functions from X to Y and from Y to X, respectively. A system $S' \subseteq X \times Y$ is said to be the feedback system of S by S_f, provided that S' is an input-output system, satisfying that

$$\forall (x, y) \in X \times Y\big((x, y) \in S' \leftrightarrow (\exists z \in X)\big((x + z, y) \in S \text{ and } (y, z) \in S_f\big)\big).$$

System S_f is known as the feedback component of S'. For more details, please consult with Chap. 2. For the sake of convenience, the intuition of the concept of feedback systems is given in Fig. 2.1, which is once again shown here as Fig. 19.4.

In the language of our current discussion, what Theorem 19.3 means is that the existing consumer surplus, as assumed in Theorem 19.2, reflects the reality that the firms that currently serve the market no longer offer adequate supplies due to the forever evolution of consumer preference and tastes. This end of course can be understood very differently from one entrepreneur to another. When the focal firm has introduced its original product based on its innovative comprehension of the market call, it means that each tree-like subset of the partially ordered set T has minimal elements, consisting of various market demands. Here, a subset D of T is said to be tree-like, if it satisfies the following condition:

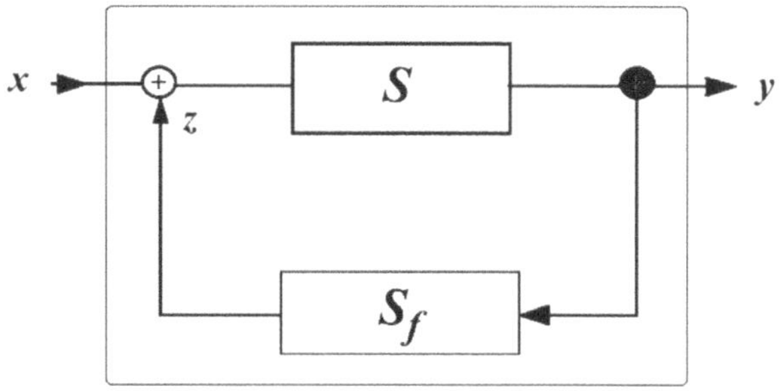

Fig. 19.4 The general diagram of a feedback system

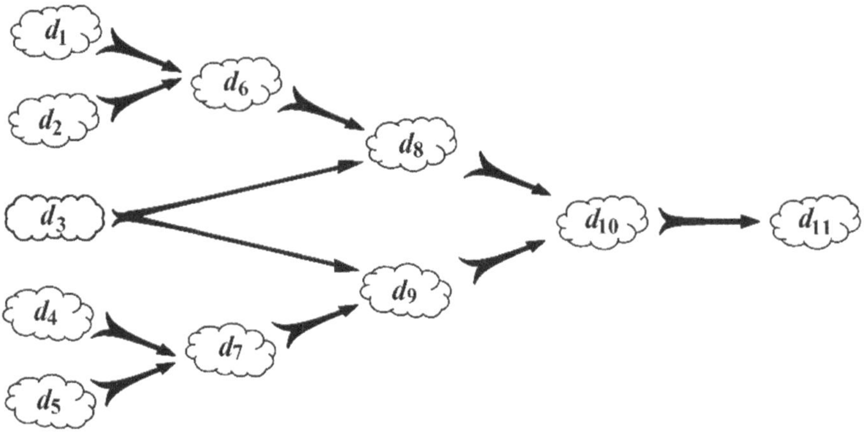

Fig. 19.5 How tree-like subset looks like

$$\forall d_1, d_2 \in D \exists d_s \in D \ni (d_1 \geq d_s \text{ and } d_2 \geq d_s),$$

where $\geq$ is the partial ordering in T, as defined above. In nontechnical terms, this expression says that for any chosen elements d_1 and d_2 from D, there is at least one d_s from D such that d_s is a downstream customer of both d_1 and d_2. Graphically, Fig. 19.5 described how such a tree-like subset D looks like. At the same time, the appearance of new entrant(s) into the market also implies that suppliers of the focal firm in the supply-chain ecosystem are confronted to produce their innovative outputs to jointly answer the market call with the focal firm. In real life, it is very possible that a single firm's offer may not adequately answer the market call. However, the collective answers offered by various entrepreneurial firms tend to satisfy the market demand.

This discussion, on the basis of the concept of feedback systems, leads to the following conclusion regarding the importance of government's policies.

Proposition 19.6 For an extant momentum of economic growth to sustain, the prevailing feedback system in the economic system, which underpins the horse race between market demands and manufacturing production, needs to be constantly supported by policies in order to continuously intensify the race.

To see why this proposition holds true, let us ride on the previous discussion, which indicates that market exchange helps reveal consumer demands that in turn stimulates and encourages manufacturing production; and on the other hand, the production of competitively evolving features and qualities of products boosts consumers to further develop their preferences and tastes. Such cyclical, mutually reinforcing market evolution endures, as long as the normal operation of the market is secured so that market demands and manufacturing production enter into a horse race against each other. In other words, a feedback system either implicitly or explicitly appears in the economy. In particular, through forever changing market demands this feedback system forces manufacturers to produce more of their products and make their products possess additional functionalities. And, through the growing magnitude and improving quality of manufacturing production the feedback system makes firms hire more employees with rising salaries that helps deepen the purchasing power of consumers. Such mutual reinforcements between market demands and manufacturing production lift the demands and production continuously onto a higher level.

The strength of the prevailing feedback system is maintained and powered by as many entrepreneurial firms as possible to participate in providing their original products. That is where the government, representing the whole of the economy, needs to play its role by implementing appropriate policies and encourages either directly or indirectly the construction and maintenance of adequate infrastructures in order to help these original products reach their end users. In other words, the magnitude of the prevailing feedback system is too colossal to be effectively affected by individual business enterprises other than the government. In comparison, the government is in such a unique position that makes it possible for the government to reach and to effectively mobilize a large proportion of the economic system (Theorem 19.1). For example, the government can employ various policy tools and supports to encourage coordination of many individual firms in very large magnitudes to form joint efforts, shared resources, exchanges of information and knowledge, etc. In terms of the literature, studies demonstrate that the government can employ policy tools to stimulate economic growth by

- Endorsing the creed of commercialization at all societal levels, especially in areas of dense populations (Mendels, 1972; Wen, 2016);
- Advancing entrepreneurship and the spirit of risk-taking (Petrovslaya et al., 2017); and
- Offering the freedom for people to make choices, such as where to live, which profession to go into, and which career path to take (Huntington, 1996).

And, the government can readily offer various kinds of supports (Fleischman & Parker, 2017; Forrest et al., 2018; Rostow, 1960; Wen, 2016), such as

- development and maintenance of appropriate domestic and international markets;
- continuous growth of these markets' depth and sophistication of related distribution networks;
- orderly operations of these markets and related supportive networks;
- regional and national stability, be they societal, or financial or political;
- construction and maintenance of appropriate infrastructures, such as roads, irrigation systems, and energy supplies;
- effective exchange of knowledge and information that promote coordination, innovation, and division of labor;
- development of unconventional educational programs necessary for training workers;
- availability of resources and necessary synchronization of resources.

Furthermore, what has been discussed above in this chapter collectively illuminates the reason why to further grow the economy, the focal nation has to simultaneously increase the scale of marketization, deepen its political reform, and raise dramatically the size of the economic sector of private enterprises. In particular, increasing the scale of marketization can adequately signal market demands through unprecedented market phenomena and forces. Deepening political reform can bring the government closer to the current state of the intensifying horse race between the market demands and manufacturing productions. And, dramatically raising the size of the economic sector of private enterprises can help more entrepreneurs to respectively comprehend market signals and take their correspondingly appropriate actions. These entrepreneurial actions generally lead to the design and production of varied original products and services to the marketplace.

For the overall organizational structure of the focal nation to stay as a stable, smoothly functional system, the focal nation needs to synchronize its political system with the prevalent working state of the feedback system of the economy. It is because the political system underlies the composition and operation of the government, while the feedback system reinforces the horse race between market exchanges and manufacturing productions through raising the purchasing power of the population and forever-evolving consumption preferences and tastes.

19.5 A Few Final Words

By employing systems science, systems methodology and the logical reasoning that parallels the one commonly used in mathematics and natural science, this chapter is able to creatively study the following two questions, which are both academically and practically important: (1) what is the reason, if any, for why government policies are anticipated to play their roles in an economic system? And (2) why are policy tools and governmental supports fundamentally necessary to practically stimulate economic growth? As a consequence, the following results, among others, are developed:

- A government's ability to design and implement appropriate policy tools can lead to further advancing the nation's economy because the government in this case can adjust the organizational structure that underlies the economy as a goal-oriented system. Within such a goal-oriented system, a determinant number of economic agents can alter their orientations and operations (Propositions 19.2–19.5).
- The existing momentum of economic growth can be maintained by the government, if the government constantly uses policies and supports to strengthen the prevailing horse race between market demands and manufacturing production. In such a case, this horse race will continue to intensify the competition between demands and production, while further lubricating the smooth function of the economy (Proposition 19.6).

Because of the novelty in the methodology employed in this chapter that is drastically different from those employed in the literature, the conclusions theoretically established above can provide policymakers with dependable bases to make their policy decisions. For example, the following general recommendations for a nation's government officials can be derived naturally:

1. The nation needs to regularly adjust itself to stay in sync with the underlying organizational structure of its economy. Only with such flexibility, the government is able to help orient the economy with its future evolution of (Proposition 19.2).
2. The nation needs to frequently adjust its economic objectives in order to potentially benefit most economic agents. Doing so successfully will enable policymakers of the nation to derive a national ambition supported by a majority of the nation (Proposition 19.3).
3. The nation needs to be in constant look out for effective ways to implement national policies, if it desires to achieve the goal of further advancing the nation's economy (Propositions 19.4 and 19.5).
4. The nation needs to purposefully create and maintain such a feedback mechanism that strengthens the race between consumer demands and production (Proposition 19.6). In particular, initially creating such a feedback mechanism can potentially develop an originally impoverished nation into one of proto-industrialization (Coleman, 1983; Forrest et al., 2018). Hence, other than satisfying local markets, developed handicraft productions and commercialized agriculture can also meet the demands of external markets. And, by maintaining such a feedback mechanism that is already in existence, the proto-industrial economy will evolve into much higher levels of economic development (Wen, 2016).

As with any scholarly work, this chapter suffers from its limitations and leaves many important issues unaddressed. For example, this chapter develops several insightful conclusions based on one methodology of systems science. However, there are many other methods and approaches available in systems science developed for investigating organizations, their evolutions, and interactions (Forrest et al., 2013). Hence, it can be expected that when other available tools of systems science are utilized one by one to study business-related issues and problems, additional finer and reliably applicable conclusions can be established.

Underneath all the reasoning and discussions, one implicit assumption is that on why a business firm exists—it exists to satisfy a specific market niche by positive cash flows from the product marketplace. Indeed, this assumption used to be true in the business world (Sobel, 1999). However, its validity no longer holds true in the present business landscape. For example, in recent years, there have appeared such e-business operations that stay financially viable by pumping up their future promises and potentials. What follows these future promises and potentials is large cash inflows in the form of venture capitals (Li & Ma, 2015). Hence, when practically applying conclusions developed in this chapter, failures may very well appear due to a violation of this assumption—that is, the firms considered exist for purposes other than what is assumed above.

Such limitations, either listed here or not, of this chapter point to potential topics and directions for future research.

References

Adner, R. (2006). Match your innovation strategy to your innovation ecosystem. *Harvard Business Review, 84*(4), 98–107.

Andreoni, A., & Chang, H. J. (2019). The political economy of industrial policy. *Structural Change and Economic Dynamics, 48*(March), 136–150.

Auclert, A. (2019). Monetary policy and the redistribution channel. *American Economic Review, 109*(6), 2333–2367.

Boeckx, J., de Sola Perea, M., & Peersman, G. (2016). The transmission mechanism of credit support policies in the Euro Area. *National Bank of Belgium Working Paper, No. 302.*

Chemers, M. M. (2001). Cognitive, social, and emotional intelligence of transformational leadership: Efficacy and effectiveness. In R. E. Riggio, S. E. Murphy, & F. J. Pirozzolo (Eds.), *Multiple intelligences and leadership* (pp. 139–160). Lawrence Erlbaum.

Christensen, C. M., Suárez, F. F., & Utterback, J. M. (1998). Strategies for survival in fast-changing industries. *Management Science, 44*(12-part-2), S207–S220.

Coleman, D. C. (1983). Proto-industrialization: A concept too many. *The Economic History Review (New Series), 36*(3), 435–448.

Cúrdia, V., & Woodford, M. (2016). Credit frictions and optimal monetary policy. *Journal of Monetary Economics, 84*(December), 30–65.

Fleischman, R. K., & Parker, L. D. (2017). *What is past is prologue: Costing accounting in the British industrial revolution, 1760-1850.* Routledge.

Forrest, J. Y. L. (2018). *General systems theory: Foundation, intuition and applications in business decision making.* Springer.

Forrest, J. Y. L., Duan, X. J., Zhao, C. L., & Xu, L. D. (2013). *Systems science: Methodological approaches.* CRC Press, an imprint of Taylor and Francis.

Forrest, J. Y.-L., Jallow, A. K., Wen, C., Shi, J. H., & Guo, H. (n.d.). (to appear). The role government policy and supports play to stimulate economic growth. Submitted for publication.

Forrest, J. Y. L., Zhao, H. C., & Shao, L. (2018). Engineering rapid industrial revolutions for impoverished agrarian nations. *Theoretical Economics Letters, 8*, 2594–2630.

Fultz, D., Long, R. R., Owens, G. V., Bohan, W., Kaylor, R., and Weil, J. (1959). *Studies of thermal convection in a rotating cylinder with some implications for large-scale atmospheric motion.* Meteorol. Monographs (American Meteorological Society), *21*(4).

Government of Japan. (2015). https://www8.cao.go.jp/cstp/kihonkeikaku/5basicplan_en.pdf, Accessed on April 16, 2019.

Hall, W., & Pesenti, J. (2017). Growing the artificial intelligence industry in the UK. https://assets.publishing.service.gov.uk/government/uploads/system/uploads/attachment_data/file/652097/Growing_the_artificial_intelligence_industry_in_the_UK.pdf, Accessed on October 15, 2017.

Hemphill, T. A. (2014). The US advanced manufacturing initiative: Will it be implemented as an innovation – Or industrial – Policy? *Innovation: Organization & Management, 16*(1), 67–70.

Hermann, M., Pentek, T., & Otto, B. (2016). Design principles for industry 4.0 scenarios. *The 49th Hawaii international conference on system sciences (HICSS)*, Koloa, HI (pp. 3928–3937).

Howell, T. R., Noellert, W. A., MacLaughlin, J. H., & Wolff, A. W. (2019). *The microelectronics race: The impact of government policy on international competition.* Routledge.

Huntington, S. P. (1996). *The clash of civilizations and the remaking of world order.* Simon & Schuster.

Iansiti, M., & Levien, R. (2004). *The keystone advantage.* Harvard Business School Press.

Jaehrling, K., Johnson, M., Larsen, T. P., Refslund, B., & Grimshaw, D. (2018). Tackling precarious work in public supply chains. *Work, Employment and Society, 32*(3), 546–563.

Jomo, K. S. (2019). *Southeast Asia's misunderstood miracle: Industrial policy and economic development in Thailand, Malaysia and Indonesia*. Routledge.

Klir, G. (2001). *Facets of systems science*. Springer.

Kouzes, J., & Posner, B. (2007). *The leadership challenge*. Jossey Bass.

Li, T., & Ma, J. H. (2015). Complexity analysis of dual-channel game model with different managers' business objectives. *Communications in Nonlinear Science and Numerical Simulation, 20*, 199–208.

Lin, Y. (1988). An application of systems analysis in sociology. *Cybernetics and Systems: An International Journal, 19*, 267–278.

Lin, Y. (1999). *General systems theory: A mathematical approach*. Kluwer Academic/Plenum Publishers.

Lin, Y. (2009). *Systemic yoyos: Some impacts of the second dimension*. Auerbach Publication, an imprint of Taylor and Francis.

Lin, Y., & Forrest, B. (2011). *Systemic structure behind human organizations: From civilizations to individuals*. Springer.

Marc, S., Yann, B., Charly, B, et al. (2018). Mission Villani: for a meaningful artificial intelligence towards a French and European strategy. https://www.aiforhumanity.fr/pdfs/MissionVillani_Report_ENG-VF.pdf, Accessed on March 8, 2018.

Mendels, F. F. (1972). Proto-industrialization: The first phase of the industrialization process. *The Journal of Economic History, 32*, 241–261. https://doi.org/10.1017/S0022050700075495

OWH. (2016). *Preparing for the Future of Artificial Intelligence*. https://obamawhitehouse.archives.gov/sites/default/files/whitehouse_files/microsites/ostp/NSTC/preparing_for_the_future_of_ai.pdf, Accessed on April 20, 2019.

Petrovslaya, I., Zaverskiy, S., & Kiseleva, E. (2017). Attribute to entrepreneurship in Russia. *Advances in Systems Science and Applications, 17*(2), 29–43.

Porter, M. E. (1985). *Competitive advantage: Creating and sustaining superior performance*. Free Press.

Rebei, N. (2017). Evaluating changes in the transmission mechanism of government spending shocks. *International Monetary Fund*, WP/17/49.

Rostow, W. W. (1960). *The stages of economic growth: A non-communist manifesto*. Cambridge University Press.

Sobel, R. (1999). *When giants stumble: Classic business blunders and how to avoid them*. Prentice Hall.

State Council of the PRC. (2015). Notice of State Council of the PRC on printing and distributing "Made in China 2025".http://www.gov.cn/zhengce/content/2015-05/19/content_9784.htm, Accessed on May 8, 2015.

Wen, Y. (2016). *The making of an economic superpower: Unlocking China's secret of rapid industrialization*. World Scientific.

Wieczorek, A. J., & Hekkert, M. P. (2012). Systemic instruments for systemic innovation problems. *Science and Public Policy, 39*(1), 74–87.

Chapter 20
When Government Policies Improve Firm Performance

Abstract Based on the theoretical studies in Chap. 3, this chapter, which is mainly based on (Forrest et al., to appear, Market competition and economic effects of government policies), investigates conditions under which government policies can be expected to improve the economic performance of firms through their market offers that can be either physical products, services, or informational goods.

Keywords Manufacturing · Market · Nash equilibrium · Network · Performance · Supply-chain ecosystem

Other than establishing generally true conclusions, a numerical scenario is analyzed in order to confirm some of the six formally developed results in this chapter. Then, recommendations are formulated based on the established results for policymakers and government officers regarding when and how their adopted and implemented policies can potentially lead to their predetermined, desired outcomes. This chapter concludes by pointing to potential directions and open problems for future research.

20.1 Introduction

Presently, the governments of several leading countries from around the world are proactively involved in the transformation of their manufacturing industries from the current state of automated manufacturing to a future state of intelligent one. Some scholars refer to the future state of manufacturing as the fourth industrial revolution (Industry 4.0). Manufacturing companies in the imagined future state would employ intelligent production technologies and processes. The aspect of intelligence would be reflected in various cyber-physical systems (CPS), the Internet of things (IoT), cloud computing, cognitive computing, and artificial intelligence (AI). As what the past rounds of industrial revolutions revealed (Rostow, 1960; Wen, 2016), the governments of these leading nations have been, since the turn of this new century, proactively promoting and stimulating their desired, while imagined,

J. Y.-L. Forrest, Y. Liu, *Value in Business*, Contributions to Management Science,
https://doi.org/10.1007/978-3-030-82898-1_20

transformations of the manufacturing industries by designing and implementing appropriate policies and by offering governmental supports. The following list showcases some of those relevant efforts:

- Starting in 2011, the United States has introduced various strategic plans and acts to facilitate the transformation of manufacturing industries. For example, the "Advanced Manufacturing Partnership Program" (PCAST, 2012) was published in 2012; and in 2014 and 2016, the following were respectively published: "Industrial Internet Alliance," "Revitalizing the US Manufacturing and Innovation Act 2014," and the "National Manufacturing Innovation Network Strategic Plan" (OWH, 2016).
- Jointly with industry and academia, Germany in 2013 developed an "Industry 4.0" platform.
- To ensure its dominance in manufacturing, Japan in 2015 introduced its ultra-smart "Society 5.0" by joining its technology and various systemic elements (Government of Japan, 2015).
- The Chinese plan of "Made in China 2025" was announced in 2015 (State Council of the PRC, 2015). After that in 2016, China presented the following two documents: (1) "Guiding Opinions on Deepening the Integral Development of Manufacturing Industry and Internet" (State Council of the PRC, 2016), and (2) "Development Plan of Intelligent Manufacturing (2016-2020)" (MIIT, 2016a). and
- The government of the United Kingdom in 2017 developed its plan of "Growing the Artificial Intelligence Industry in the UK" (Hall & Pesenti, 2017).

Because these leading nations all had prior experiences of successes with the industrial revolution, while now they are taking similar actions, the following question appears naturally:

> How is a government affecting the economic performance of the manufacturing sector through policies?

The answer, if there is one such answer, to this question is both theoretically and practically important to all countries. In particular, for developed nations, large percentages of their economies have been dominantly composed of services, while the amount of employment created by the manufacturing sector has been dwindling (Szirmai & Verspagen, 2015). At the same time, after having enjoyed the benefit of being on the top, these nations surely desire to maintain their leading positions in the world economy. Simultaneously, for developing and undeveloped nations, the answer to this question is also important. While the recent bloody history of being helplessly exploited by developed nations is still flesh (Wen, 2016), these nations do not really want to face the strong possibility of economically, socially, and politically falling further behind the developed nations. The objective of this chapter is to address this question by providing relevant answers based on systemic logic of reasoning, an unconventional approach in terms of the relevant literature of concern.

Before charging forward towards answering the previous question, let us first pay some attention to the methodology used here, because of its clear contrast with all other approaches employed by various scholars in the past in their attempts to study-related issues. In particular, the specific logic of reasoning, employed in this chapter, is parallel to that used in mathematics (Kline, 1972), where theoretical conclusions are logically derived from basic axioms without being constrained by specific sets of data, anecdotes, and relevant methodological limitations (Kuhn, 1962). Different from conclusions derived by using such an approach, those developed empirically are closely tied to the specific data employed in related studies. When such data-specific conclusions are utilized in situations beyond where the data were from, derived suggestions do not have much potential of reliability. Hence, one other objective of this chapter is to make sure the propositions established here allow for the development of reliable policy recommendations. If one looks at the entire literature of business and economics studies, he/she can see clearly that an increasing number of scholars have employed the approach of systemic reasoning (e.g., Forrest, 2018; Mazzei et al., 2017; Porter, 1985).

The rest of this chapter is organized as follows. Section 20.2 reviews relevant literature, while showing how this chapter contributes to the literature. Section 20.3 derives six generally true theoretical propositions about how government policies bring forward economic effects. Section 20.4 empirically confirms some of the established general conclusions. And Sect. 20.5 concludes this chapter with practical recommendations and open questions for future studies.

20.2 Literature Review

Innovation policies are found (Borrás & Edquist, 2013) to be effective if adopted in mixes to resolve problems of an innovation system. Policy tools, when consistently utilized across different communities, could help build capacities, as discovered by Stidham et al. (2014) when they examine how six communities in Idaho, Oregon, and Utah mitigated wildfire risks at two points in time. It seems that the bank lending channel played a nontrivial role in the transmission mechanism of monetary policy in the United States until the early 1980s (Endut et al., 2018) in terms of interest rate, exchange rate, and bank lending channel. To better matching policy-tool mixes and target behaviors, Howlett (2018) suggests a new research and practice agenda without the presumption that policy targets are rational utility maximizers.

In light of how various shocks affect the economy and how policies could help optimize responses to shocks, simple target criteria seem to provide an appropriate estimate of optimal policies (Cúrdia & Woodford, 2016). An investigation of how unconventional monetary policies impact the financial market in Europe suggests (Fratzscher et al., 2016) that such policies advanced equity prices, reduced the fragmentation of bond market, and positively affected equity markets and market confidence. Interacting fiscal policy, sovereign debt, and financial fragility appears to reveal (Kirchner & van Wijnbergen, 2016) that the effectiveness of fiscal stimuli is lost when banks are deeply in sovereign debts and that applicable gains fall expressively when interest rates are close to zero. Alliances of local politicians, unions, and employers are demonstratedby analyzing the association between labor clauses and issues in public supply chains, to be important by Jaehrling et al. (2018) in terms of ensuring socially responsible procurement.

By comparing with this literature, the contribution this chapter makes is the theoretical development of conditions under which the economic performance of manufacturing firms improves. Methodologically, the systemic thinking employed here helps prevent this chapter from falling into the traps witnessed by rounds of debates on whether it is necessary to apply government policies to boost economic development since the time before Adam Smith (Andreoni & Chang, 2019). Specifically, these debates replied on anecdotes and statistics of limited validity to produce their respectively inconclusive and noncompromising conclusions either for or against the use of policies. Squarely different from these inconclusive and noncompromising conclusions, results theoretically developed in this chapter are scientifically definite.

20.3 How Government Policies Affect Market Competition

This section looks at the economic effects of government policies from the angle of market demands for new competition and innovation, where an innovation is identified with such an activity (or a set of activities) that leads to the design and production of a product of exceptionally added value for a firm when compared to the case without the activity (Forrest et al., 2018). That is, an innovation stands for all such actions that lead to the offering of an original or much improved product. Although since the time of Adam Smith (1776) innovation has been well recognized as a very important factor for wealth creation, Theorem 3.2 represents the first theoretical reason for why that is the case. For the sake of convenience, let us recite this theorem.

Theorem 20.1 In Nash equilibrium, if the proportion of switchers in the closed market of free competition defined below is greater than or equal to the percentage magnitude of the loyal-customer base of one incumbent firm, then the market will attract at least another firm to profitably enter with their versions of substitute offers so that the expected total profit of all entrants can be potentially larger than the minimum profit of the incumbent firms.

The market in Theorem 20.1 is assumed to satisfy five conditions: (1) It is occupied by m incumbent firms, $m = 1, 2, \ldots$, which offer their mutually substitutable offers. (2) Each incumbent's total cost of production and marketing is constant. (3) Each incumbent has a base of loyal customers who make purchases only from their favorite firm if the price is no more than their reservation value. (4) These firms compete over switchers who purchase from the firm whose price is the lowest. And (5) these incumbents know others' pricing strategies and play the Nash equilibrium through pure self-analysis.

With this theorem serving as the starting basis, let us look at the economic effects of government policies.

20.3.1 Potentials for Improving Firm Performance

The term "product" in economics is defined to be any market offer, be it tangible or intangible. Examples can include any physical thing, such as a table or chair, or a nonphysical thing, such as a promise, say, a life insurance policy, or a service, such as a coin-operated laundry facility, or an informational good, such as managerial advice. Such convention is evidently not true in areas of management and marketing, where "product" seems to represent simply a physical thing that customers demand (Porter, 1985; Rayport & Sviokla, 1995; Stabell & Fjeldstad, 1998).

For the need of this chapter, the following understanding is adopted: a product stands for a thing, be it tangible or intangible, a business entity produces and consumers demand for (Forrest, Liu, et al., 2020). In terms of the current context, by manufacturing firm, it means an enterprise that offers a product to the market by designing and/or producing the product. Therefore, a firm, which builds physical things for the product market, is a manufacturing firm. At the same time, an insurance company is considered also as a manufacturing firm, because it proposes insurance policies to customers. In this case, each policy is an intangible product that is made up of promises that a customer can depend on when the predetermined condition occurs. Additionally, if a firm provides a service, such as a self-operated laundry facility, indoor tanning, then the firm is also considered as a manufacturing firm. In this case, customers can "receive" the service by enjoyment and experience. The same reasoning applies to financial firms because they offer investment recommendations, produced by gathering, organizing, and analyzing data.

To make the logical reasoning in this chapter plausible, assume that each manufacturing firm exists for the objective of filling a market need through positive cash flows in the marketplace from selling its product(s) to consumers (or customers). Although most manufacturing firms in the past did conform to this assumption (Sobel, 1999), this norm has been violated in recent years (Li & Ma, 2015). For example, in the current world of business, some of the very well-known companies stay financially afloat mainly through continued inputs of venture capitals.

Proposition 20.1 Increasing efficiencies of management and resources can potentially help improve the economic performance of a manufacturing firm.

This result is a direct corollary of Theorem 20.1. Specifically, when the external market environment is fixed, a firm can potentially improve its economic performance only by increasing the efficiency of its operation and/or production. Otherwise, the firm's level of expected profits would be either stagnant or falling according to Theorem 3.2 in Forrest, Nicholls et al. (2020). At the same time, the market segment of switchers is realistically composed of dissatisfied customers. Hence, to potentially attract as many switchers as possible from this market segment, while preventing new competitors from entering the market, Theorem 20.1 indicates that although the incumbent firms might be locked in their state of mutual forbearance (Bernheim & Whinston, 1990), these firms have to individually improve the efficiency of their operational routines, processes and available resources. Only by accomplishing this end, each incumbent has the potential to lower its business cost and raise its expected maximum selling price so that its boundary conditions will be pushed further outward. That is, in order to strengthen its market share, each incumbent firm needs to go beyond pricing strategies and discover novel ways to make its management and resources more efficient than before.

When a manufacturing firm challenges to strengthen its existing market territory by making its management and resources more efficient, it also naturally acquires the ability to discover new segments of the product market that bring forward different potentials for the firm. As what is implied by Theorem 20.1, to effectively find potentials in a new market segment, a firm needs to identify such a segment in the marketplace where the number of switchers increases rapidly. Such a rapidly increasing number of switchers can be the consequence of two possible scenarios: (1) the segment of the product market expands quickly; or (2) the segment calls for much improved products or brand-new ones. Due to the characteristics of the present world of business—once sustainable competitive advantages become transient (McGrath, 2013), the ability to go into new markets or new segments of the product market both competitively and profitably, as described in Theorem 20.1, provides a manufacturing firm with potentials to improve its economic performance (Wagner, 2012).

Proposition 20.2 The ability to engage in the sharing of information and knowledge can lead a firm to the potential of improving its economic performance.

From the definition of innovation, it follows that novelties and creativities tend to bring forward new and exceptional value, while they can be introduced in any aspect of a firm's operation. For example, both novelties and creativities can appear in the introduction of new production methods of existing products, in the design of new and original products, in the adoption of procedures of much improved efficiency, in the creation of new markets, and in the introduction of new resources (Schumpeter, 1934). The practical difficulty of having novel and creative ideas, as the recorded history suggests (Ridley, 2016; Ye et al., 2012), is that they tend to appear when certain intelligent people unexpectedly and fitfully find themselves in the right place at the right time. Contrary to this historical record of how innovations passively appear, recent studies (Yang, 2018) show theoretically that novelties and creativities can be logically and systemically generated.

Hence, Theorem 20.1 implies that when a segment of the product market signals for new innovation and additional competition by showing a significantly increasing number of switchers, many entrepreneurs will take advantage of the opportunity to offer their solutions to satisfy the market demand. However, differences in their knowledge structures and backgrounds generally direct these entrepreneurs to offer their individually different solutions to possibly satisfy different aspects of the signaled market demand. Hence, on the basis of such a background, the ability to engage in the sharing of information and knowledge will assist a manufacturing firm to investigate the varied products offered by these entrepreneurial attempts and to introduce its own holistically more encompassing product that closely fits customers' real need. That is the underlying reason why the ability to engage in the sharing of information and knowledge can lead a firm to the potential of improving its economic performance.

In early times, the business world used to be composed of firms that autonomously competed against each other in the marketplace in order to acquire advantages in general and to make profits in particular. However, that scene of autonomous competition has been largely gone from the world of business. Instead, firms in the presently globalizing economy tend to be parts of interorganizational networks and strategic blocks. They find different opportunities to collaborate with each other both socially and professionally in order to achieve a state of mutual forbearance for the purpose of maintaining their market positions and profit pictures (Galaskiewicz & Zaheer, 1999).

By a strategic network, it stands for a network of firms, strategically associated with each other to establish certain advantages for all partners (Gulati et al., 2000). For example, each firm is an input-output system. So, some firms are naturally threaded together by their input-output chains, leading to the appearance of a supply-chain ecosystem (or network), consisting of upstream components and downstream complements (Adner & Kapoor, 2010). In such a supply-chain ecosystem, suppliers represent some of the upstream components; and customers, supporters, and assistants, who help make a market offer available to consumers, are some of the downstream complements. Indeed, supporters and assistants are literally, according to how a supply chain is defined, not any part of the supply chain. However, without these players, many products cannot practically reach the ultimate consumer. These supporters and assistants have to invest in and develop the necessary infrastructure to make the supply-chain ecosystem practically functional.

Proposition 20.3 The membership in a network of organizations that involves different supply-chain ecosystems provides a firm with a great potential to improve its economic performance.

If a firm wants to strategically join such a network as described in the proposition, then it has to meet certain criteria by achieving certain well-formulated objectives. Theorem 20.1 indicates in the current context that members of a supply-chain ecosystem collectively work together to satisfy a targeted market demand in a specific way. Therefore, the membership in such a network increases the chance for a firm to learn (Shapiro & Varian, 1999) information (such as varied understandings of a market demand, how different solutions are developed), and to gain access to different markets or market segments, and knowledge of relevant technologies (Gulati et al., 2000). Moreover, each member firm is provided with opportunities for risk sharing, economies of scale, and scope (Shapiro & Varian, 1999). Members' reciprocal activities help them to reap the resultant benefits (Blankenburg et al., 1999). That is, by applying Proposition 20.2 one can conclude that strategically joining a network, as described, provides all member firms with a great potential to elevate their economic performance.

20.3.2 Potentials of Government Policies

As the title suggests, this subsection studies conditions under which government policies lead to economic potentials for manufacturing enterprises.

Proposition 20.4 By aiming at encouraging manufacturing enterprises to individually improve the efficiencies of their management and resources, the adopted government policy will positively influence these enterprises' economic performance.

This result is a direct corollary of Proposition 20.1. Practically, an improvement in efficiencies of management and resources can be achieved by sharing information and knowledge, especially those information and knowledge about how to innovatively understand market signals. This end can be effectively accomplished by engaging in useful networks of collaborations, be they internally or externally, and by introducing advanced technologies to improve productivity. Jointly, Propositions 20.2 and 20.3 together imply that enhanced management and resource efficiency, productive collaborations, and advanced technologies all produce positive effects on the performance of manufacturing enterprises.

In reality, the performance of a firm is also a function of how well other players in its supply-chain ecosystem do (Adner, 2006). For instance, if a firm plans to offer its imagined product to satisfy the market demand, as how it innovatively recognizes, it has to consider whether its envisioned product will technologically make it impossible for suppliers to produce the needed supplies. Even if there is no problem with supplies, the firm still has to consider whether its intended product can successfully reach the end users in the product market. That really depends on whether or not downstream complementors can materialistically provide the needed infrastructures (Adner, 2006).

Proposition 20.5 By aiming at assisting the formation of a manufacturing agglomeration, a government policy will positively influence manufacturing enterprises' economic performance more than any other policy that does not promote the development of industrial agglomerations.

From the description of the supply-chain ecosystem of any firm, as given above, Proposition 20.3 and the herd effect of organizational behaviors (Banerjee, 1992; Burtch, 2011), this proposition follows naturally. Simultaneously, this proposition supports at the theoretical height the empirical findings that industrial agglomeration could possibly promote economic growth (Ciccone & Hall, 1996; Helsley, 2004; Martin, 2001), potentially produce knowledge and technology spillover effects, and conceivably promote labor productivity (Brülhart & Mathys, 2008).

As for a nation, the unevenness existing in its economic development naturally exists between different regions. The confirmation for this claim can be provided by the following systemic model: First, let us respectively use the dish and fluid in the famous dishpan experiment (Fultz et al., 1959) to model the nation and its economy. Then, the development of the national economy is vividly depicted by how the fluid

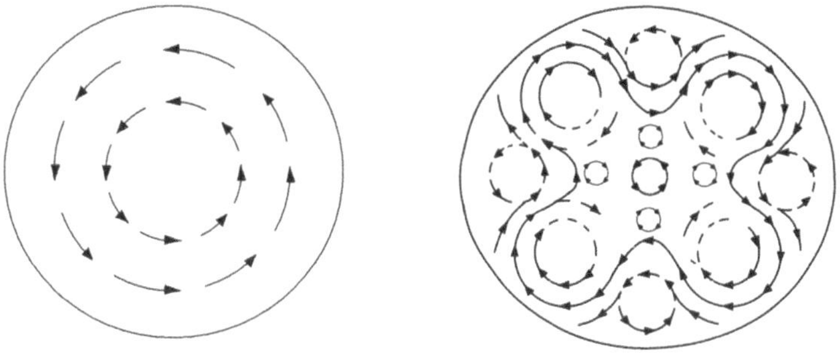

(a) The fluid flows uniformly (b) The fluid flows chaotically

Fig. 20.1 The alternation of flow patterns in dishpan experiment. (**a**) The fluid flows uniformly. (**b**) The fluid flows chaotically

evolves within the spinning dish, where the flow pattern of the fluid alternates between the uniform and chaotic pattern shown in Fig. 20.1a and b. That is, unevenness in flow directions and that in the scales of regional pools appear.

In this systemic modeling, the term "fluid" models all such economic factors as money, information, knowledge, goods, etc., and flow pattern of the fluid models how each of these factors evolves within the economy. The flow pattern alternation reveals the fact that unevenness and imbalance in economic development in different regions are some of the common phenomena in economic development. This end is anecdotally confirmed by Lyson and Falk (1993) in their study on how rural America experiences unevenness in its economic development, and by Pike and Tomaney (2009) who find uneven economic development in the post-devolution UK.

Both the logical reasoning based on the systemic modeling and anecdotal confirmations support the general conclusion that different geographical regions generally experience varied and imbalanced rates of economic development. To this end, it is found that imbalances in multilateral merchandise trades could lead to uneven economic development (Frank, 1978).

By expanding this reasoning, it reveals that every time when a nation adopts an economic policy, the policy will lead to varied incentive effects for different regions within the nation, although the policy is initially implemented uniformly across all these regions. Therefore, for manufacturing enterprises located in relatively undeveloped geographic areas, they need preferential policy treatments and supports in order to improve their managerial and production efficiencies and to enhance their levels of economic performance. The marginal effect of adopted policies on these enterprises will be higher than that on enterprises located in developed regions. Because managerial and production efficiencies of these latter enterprises have already developed to a beneficial level, influences of a newly adopted policy on these enterprises will not be as significant as those on the former enterprises located in relatively undeveloped geographic areas. That is, the marginal effect an adopted economic policy produces on enterprises located in advanced economic areas will be relatively low. This logical reasoning naturally leads to the following result.

Proposition 20.6 Geographical and regional differences always exist in how much an adopted economic policy affects the economic performance of manufacturing enterprises. Specifically, the positive effect of an economic policy tends to be more noticeable for manufacturing enterprises located in less economically developed regions than those in developed areas.

20.4 An Empirical Confirmation

In this section, real-life data from China's manufacturing sector, as defined conventionally, are employed to empirically double check the validity of some of the established results in the previous section of this chapter, especially Propositions 20.4–20.6.

20.4.1 Source of Data and Econometric Model

To assist the implementation of a national strategy, China's Ministry of Industry and Information Technology (MIIT) has been conducting "The China's pilot projects for intelligent manufacturing (IM)" since 2015. The objective of these projects is to promote the development of intelligent manufacturing in China, while they help optimize the system of IM standards, and provide a necessary foundation for the future adoption of a nationwide intelligent policy.

The difference-in-difference (DID) method (Card & Krueger, 1994) is chosen here for evaluating the effect of this pilot project in terms of the improvement in the performance of manufacturing companies. The subject companies used in this study are from the MIIT's publicity list of the pilot enterprises published in 2015–2017, paired with their annual report data approved by China's Securities Regulatory Commission for the period 2011–2017 from Shanghai Stock Exchange (http://www.sse.com.cn/), Shenzhen Stock Exchange (http://www.szse.cn/), and the MIIT (http://www.miit.gov.cn/). The experimental grouping variable (intel) is divided into "treatment group" and "control group." The former contains 89 listed pilot companies with an assigned value of 1, while the latter includes a random sample of 100 manufacturing companies, selected from Shanghai Stock Exchange, with an assigned value of 0. After removing duplicates in the "treatment group," the "control group" contains 92 companies.

The variable (time) of test period takes values either before or after the implementation, depending on when the IM policy is implemented with individual pilot enterprises selected. In particular, for a pilot enterprise, time = 1 for the year and afterward when the enterprise is selected; and time = 0 for the years before the selection. So, the total sample is divided into four quadrants: (1) (intel = 1, time = 0) = "treatment group" and before the policy is implemented; (2) (intel = 1, time = 1) = "treatment group" and after the policy is implemented; (3) (intel = 0, time = 0) = "control group" and before the policy is implemented; and (4) (intel = 0, time = 1) = "control group" and after the policy is implemented.

Let us now employ the following fixed-effect model to test the net impact of the IM policy on the economic performance of the chosen manufacturing companies:

$$\mathrm{roa}_{it} = \alpha + \beta_1 \mathrm{did}_{it} + \beta_2 \mathrm{lcv}_{it} + \beta_3 \mathrm{manage}_{it} + \beta_4 \mathrm{ocf}_{it} + \beta_5 \mathrm{top1}_{it} + \beta_6 \mathrm{age}_{it} + \beta_7 \tan_{it} + \delta_{it},$$

where $\mathrm{roa}_{\mathrm{it}}$ is the return on assets of the enterprise in region i and year t, $\mathrm{did}_{it}(= \mathrm{intel} \times \mathrm{time})$ $\mathrm{did}_{\mathrm{it}}$ the variable of the pilot project IM policy, which represents the net effect of implementing the intelligent policy. And $\mathrm{lcv}_{\mathrm{it}}$, manage_{it}, ocf_{it}, $\mathrm{top1}_{\mathrm{it}}$, $\mathrm{age}_{\mathrm{it}}$, $\mathrm{tang}_{\mathrm{it}}$ are the control variables, representing the asset-liability ratio, asset management capacity, operating cash flow ratio, equity concentration, company age, and fixed assets, respectively; and $\propto\alpha$, β, and δ respectively the constant term, the coefficient of a variable, and the residual.

The output roa_{it} is the ratio of the net profit in the financial statements of a listed company over the total assets. The value of the explanatory variable *did* is assigned based on the "Public Manufacturing Pilot Demonstration Project List," published by MIIT from 2015 to 2017 (MIIT, 2015; MIIT, 2016b; MIIT, 2017), and the financial data, according to when a company is selected as a pilot enterprise of IM.

Asset-liability ratio "*lcv*" can be employed to investigate the impact of risk management policies on corporate values (Hoyt & Liebenberg, 2011). This chapter assumes that this ratio has a negative impact on the performance of any company considered here. The capability "*manage*" of managing asset, defined to be the ratio of operating income to total assets (Higgins, 1981), indicates a firm's ability to generate profits by using its assets. It is expected to positively impact the performance of a manufacturing company (Forrest & Nightingale, 2017). The cash flow ratio "*ocf*" of operation, which is the ratio of cash flow from operational activities over total assets, reflects the ability of a company to create cash from its invested capital.

The equity concentration "*top*1" is defined to be the proportion of the largest shareholder's equity of a company, which is significantly correlated to firm performance in developing countries (Lins, 2003; Makhija & Spiro, 2000). From Forrest et al. (2019), it follows that this variable is negatively correlated to the performance of a listed company; and the enterprise age "age" does not have much visible influence on the performance of a company (Forrest et al., 2019). The ratio "*tan*" of fixed assets is defined to be the ratio of fixed assets over total assets; and it is expected to negatively influence the performance of a company.

20.4.2 A Statistics-Based Analysis

The regression results of a fixed-effect model show the IM policy positively impacted the economic performance of listed manufacturing companies (with 5% significance). That confirms the conclusion of Proposition 20.4. Beyond this confirmation, other results include that

- Both asset-liability ratio and company age have significant negative impacts on a manufacturing company's performance.
- All other variables, including debt-to-equity ratio, capacity of asset management, cash flow ratio of operation, equity concentration, and the ratio of fixed assets, do not significantly affect a manufacturing enterprise's performance.

To reconfirm the impact of the IM policy variable *did*, it is found that when the pilot project of the IM policy was not implemented, the trend of the control group was the same as that of the treatment group. In particular, the reconfirmation indicates that before the implementation of the pilot project of the IM policy, the variable "*did*" did not affect the performance of the subject companies; and after the implementation, the variable significantly affected the economic performance of these companies. In other words, the impact of the IM policy, as revealed above, is credible.

To reconfirm distributional heterogeneity, the sample of this study is separated into two subsamples. One subsample, labeled as "without agglomeration," contains such pilot enterprises that each of them is the sole enterprise selected to the pilot project within its entire province. And the other subsample, labeled as "with agglomeration," includes all the other pilot enterprises of the original sample. Then, the test outcomes demonstrate that for the case of "without agglomeration," the IM policy does not show a significant impact on the economic performance of the pilot enterprises, while the opposite is true for the case of "with agglomeration." In other words, the IM policy indeed shows a significant positive impact. That is, this test confirms Proposition 20.5—the spillover effect of knowledge and technology makes a significant agglomeration effect on enterprises.

To empirically confirm Proposition 20.6—the regional heterogeneity, 31 provinces, autonomous regions, and municipalities that are under the direct administration of the central government, excluding Hong Kong, Macao, and Taiwan, are grouped into two regions: the central/western region and eastern region. The former represents less developed areas of China, including Shanxi, Jilin, Heilongjiang, Anhui, Jiangxi, Henan, Hubei, Hunan, Sichuan, Chongqing, Guizhou, Yunnan, Shaanxi, Gansu, Qinghai, Ningxia, Tibet, Xinjiang, Guangxi, and Inner Mongolia. And the latter region represents more developed regions of China, including Beijing, Tianjin, Hebei, Liaoning, Shanghai, Jiangsu, Zhejiang, Fujian, Shandong, Guangdong, and Hainan.

Based on this geographical division of China, the regional heterogeneity of the pilot IM projects is tested. The regression results demonstrate that for the central/western region, the IM policy has a positive impact on enterprises' performance, which passes the 10% significance; and for the eastern region, the effect of the IM policy is not significant. That is, Proposition 20.6 is empirically confirmed.

20.5 A Few Final Words

This chapter employs systems science and such a logical reasoning that is parallel to the one commonly used in mathematics and natural science to investigate when government policies positively influence the economic performance of business firms that produce market offer(s), be they physical products, services, or informational goods. Along with other results, established are the following main conclusions:

- A business firm that produces market offer(s) can possibly improve its economic performance, as concluded in Propositions 20.1–20.3, if it is able to:
  - Elevate the efficiency of its management and resources;
  - Purposefully exchange information and knowledge with other firms; or
  - Strategically join a network of different supply-chain ecosystems.
- By either aiming at improving manufacturing enterprises' efficiency of management and resources or assisting the formation of manufacturing agglomerations, an adopted policy will help improve manufacturing enterprises' economic performance (Propositions 20.4 and 20.5).

Because conclusions established in this chapter are not subject to the constraints of data- or anecdote-based methods, they are expectedly able to provide dependable bases on which government officers and policymakers make their decisions. Especially, at any time when the government of a nation is adopting and implementing an economic policy, it is highly recommended that

1. No matter what the policy is for, it needs to aim at improving the efficiency of manufacturing enterprises' management and resources (Proposition 20.1), stimulating these enterprises to share information and knowledge (proposition 20.2), and developing manufacturing agglomerations (Proposition 20.3); and
2. Different policies and different levels of supports need to be implemented and provided, respectively, to regions of varied levels and magnitudes of economic development. If identical policies and supports are uniformly applied across the board, their effects tend to be more pronounced for less developed regions than advanced areas (Proposition 20.6).

As for ideas for possible future research, the following observations are worth particular attention. First, the methodology underneath this chapter stands for a very small part of systems science. There are many other tools available for use in systems science. They are developed for different purposes of investigation, such as understanding how functional organizations work, how organizations evolve, and how organizations interact with each other (Forrest et al., 2013). So, it is a worthy endeavor to employ other well-developed systems-scientific tools for the study of policies and their effects. Second, all conclusions established in this chapter are logically backed on the assumption of why a business firm exists. Although this assumption used to be true (Sobel, 1999), it is no longer generally the case in the present business world (Li & Ma, 2015). Hence, future research may examine whether similar conclusions as the ones developed in this chapter apply to business enterprises established for other purposes than the one assumed here.

References

Adner, R. (2006). Match your innovation strategy to your innovation ecosystem. *Harvard Business Review, 84*(4), 98–107.

Adner, R., & Kapoor, R. (2010). Value creation in innovation ecosystems: How the structure of technological interdependence affects firm performance in new technological generations. *Strategic Management Journal, 31*(3), 306–333.

Andreoni, A., & Chang, H. J. (2019). The political economy of industrial policy. *Structural Change and Economic Dynamics, 48*(March), 136–150.

Banerjee, A. V. (1992). A simple model of herd behavior. *The Quarterly Journal of Economics, 107* (3), 797–817.

Bernheim, B. D., & Whinston, M. D. (1990). Multimarket contact and collusive behavior. *RAND Journal of Economics, 21*(1), 1–26.

Blankenburg, H. D., Eriksson, K., & Johanson, J. (1999). Creating value through mutual commitment to business network relationships. *Strategic Management Journal, 20*(5), 467–486.

Borrás, S., & Edquist, C. (2013). The choice of innovation policy instruments. *Technological Forecasting and Social Change, 80*(8), 1513–1522.

Brülhart, M., & Mathys, N. A. (2008). Sectoral agglomeration economies in a panel of European regions. *Regional Science & Urban Economics, 38*(4), 348–362.

Burtch, G. (2011). Herding behavior as a network externality. *Proceedings of the international conference on information systems*, ICIS 2011, Shanghai, China, December 4–7, 2011.

Card, D., & Krueger, A. B. (1994). Minimum wages and employment: A case study of the fast-food industry in New Jersey and Pennsylvania. *American Economic Review, 84*(4), 772–793.

Ciccone, A., & Hall, R. E. (1996). Productivity and the density of economic activity. *American Economic Review, 86*, 54–70.

Cúrdia, V., & Woodford, M. (2016). Credit frictions and optimal monetary policy. *Journal of Monetary Economics, 84*(December), 30–65.

Endut, N., Morley, J., & Tien, P. L. (2018). The changing transmission mechanism of US monetary policy. *Empirical Economics, 54*, 959–987.

Forrest, J. Y. L. (2018). *General systems theory: Foundation, intuition and applications in business decision making*. Springer.

Forrest, J. Y. L., Duan, X. J., Zhao, C. L., & Xu, L. D. (2013). *Systems science: Methodological approaches*. CRC Press, an imprint of Taylor and Francis.

Forrest, J. Y.-L., Liu, J., Guo, H., Liu, Y., Phipps, T. M., Yang, Y. J., & Cao, Y. R.. (n.d.). (to appear). Market competition and economic effects of government policies. Submitted for publication.

Forrest, J. Y.-L., Liu, Y., Golden, J., Sissoko, Y., & Qiaoxing Li, Q. X. (2020). A reformulated value chain framework useful for different kinds of markets. *Journal of Business and Economics Studies, 24*(1), 33–53.

Forrest, J. Y.-L., Nicholls, N., Schimmel, K., & Liu, S. F. (2020). *Managerial decision making: A holistic approach*. Springer.

Forrest J. Y. L, & Nightingale, J. (2017). Successfully transition into the era of transient competitive advantages. *Proceedings of 2017 NABET Annual Conference*, October 26–27, 2017, State College, PA (pp. 57–79).

Forrest, J. Y.-L., Tucker, R., Lin, C.C., & Mondal, S. (2018). Impacts of manufacturing firms' culture, structure and leadership on innovation: A systemic approach. Proceedings of the 41st annual meeting of northeastern association of business, Economics and Technology, State College, PA.

Forrest, J. Y. L., Tucker, R., Lin, C. C., & Mondal, S. (2019). The innovativeness of manufacturing firms from the angles of organizational culture, structure and leadership. *Journal of Business, Economics and Technology, 22*(1), 41–66.

Frank, A. G. (1978). Multilateral merchandise trade imbalances and uneven economic development. In *Dependent accumulation and underdevelopment*. Palgrave Macmillan.

Fratzscher, M., Lo Duca, M., & Straub, R. (2016). ECB unconventional monetary policy: Market impact and international spillovers. *IMF Economic Review, 64*, 36–74.

Fultz, D., Long, R. R., Owens, G. V., Bohan, W., Kaylor, R., Weil, J. (1959). *Studies of Thermal Convection in a Rotating Cylinder with Some Implications for Large-Scale Atmospheric Motion.* Meteorol. Monographs (American Meteorological Society), *21*(4).

Galaskiewicz, J., & Zaheer, A. (1999). Networks of competitive advantage. In S. Andrews & D. Knoke (Eds.), *Research in the sociology of organizations* (pp. 237–261). JAI Press.

Government of Japan. (2015). https://www8.cao.go.jp/cstp/kihonkeikaku/5basicplan_en.pdf. Accessed on April 16, 2019.

Gulati, R., Nohria, N., & Zaheer, A. (2000). Strategic networks. *Strategic Management Journal, 21* (3), 203–215.

Hall, W., & Pesenti, J. (2017). Growing the artificial intelligence industry in the UK. https://assets.publishing.service.gov.uk/government/uploads/system/uploads/attachment_data/file/652097/Growing_the_artificial_intelligence_industry_in_the_UK.pdf, Accessed on October 15, 2017.

Helsley, R. W. (2004). Economics of agglomeration: Cities, industrial location and regional growth. *Journal of Economic Geography, 4*(3), 345–349.

Higgins, R. C. (1981). Sustainable growth under inflation. *Financial Management, 10*(4), 36–40.

Howlett, M. (2018). Matching policy tools and their targets: Beyond nudges and utility maximisation in policy design. *Policy & Politics, 46*(1), 101–124.

Hoyt, R. E., & Liebenberg, A. P. (2011). The value of enterprise risk management. *The Journal of Risk and Insurance, 78*(4), 795–822.

Jaehrling, K., Johnson, M., Larsen, T. P., Refslund, B., & Grimshaw, D. (2018). Tackling precarious work in public supply chains: A comparison of local government procurement policies in Denmark, Germany and the UK. *Work, Employment and Society, 32*(3), 546–563.

Kirchner, M., & van Wijnbergen, S. (2016). Fiscal deficits, financial fragility, and the effectiveness of government policies. *Journal of Monetary Economics, 80*(June), 51–68.

Kline, M. (1972). *Mathematical thought from ancient to modern times*. Oxford University Press.

Kuhn, T. (1962). *The structure of scientific revolutions*. University of Chicago Press.

Li, T., & Ma, J. H. (2015). Complexity analysis of dual-channel game model with different managers' business objectives. *Communications in Nonlinear Science and Numerical Simulation, 20*, 199–208.

Lins, K. V. (2003). Equity ownership and firm value in emerging markets. *The Journal of Financial and Quantitative Analysis, 38*(1), 159–184.

Lyson, T. A., & Falk, W. W. (Eds.). (1993). *Forgotten places: Uneven development in rural America. Rural America Series*. University Press of Kansas.

Makhija, A. K., & Spiro, M. (2000). Ownership structure as a determinant of firm value: Evidence from newly privatized czech firms. *Financial Review, 35*(3), 1–31.

Martin, O. (2001). Growth and agglomeration. *International Economic Review, 42*(4), 947–968.

Mazzei, M. J., Ketchen, D. J., Jr., & Shook, C. L. (2017). Understanding strategic entrepreneurship: A "theoretical toolbox" approach. *International Entrepreneurship and Management Journal, 13* (2), 631–663.

McGrath, R. G. (2013). *The end of competitive advantage: How to keep your strategy moving as fast as your business*. Harvard Business Review Press.
MIIT. (2015). Circular of Ministry of Industry and Information Technology of the People's Republic of China on Publishing the List of Pilot Demonstration Projects of Intelligent Manufacturing in 2015. http://www.miit.gov.cn/n1146285/n1146352/n3054355/n3057585/n3057589/c3590738/content.html. Accessed on July 21, 2015.
MIIT. (2016a). Circular of MIIT and MOF on the Issuance of Development Plan of Intelligent Manufacturing (2016–2020). http://www.miit.gov.cn/n1146285/n1146352/n3054355/n3057585/n3057590/c5406038/content.html. Accessed on December 8, 2016.
MIIT. (2016b). Circular of Ministry of Industry and Information Technology of the People's Republic of China on Publishing the List of Pilot Demonstration Projects of Intelligent Manufacturing in 2016. http://www.miit.gov.cn/n1146285/n1146352/n3054355/n3057585/n3057590/c5032698/content.html. Accessed on July 4, 2016.
MIIT. (2017). Circular of Ministry of Industry and Information Technology of the People's Republic of China on Publishing the List of Pilot Demonstration Projects of Intelligent Manufacturing in 2017. http://www.miit.gov.cn/n1146285/n1146352/n3054355/n3057585/n3057590/c5861997/content.html. Accessed on October 16, 2017.
OWH. (2016). The National Artificial Intelligence Research and Development Strategic Plan. https://www.nitrd.gov/PUBS/national_ai_rd_strategic_plan.pdf. Accessed on April 16, 2016.
PCAST. (2012). Report to the President on capturing domestic competitive advantage in advanced manufacturing. https://permanent.access.gpo.gov/gpo34116/pcast_amp_steering_committee_report_final_july_17_2012.pdf. accessed on July, 2012.
Pike, A., & Tomaney, J. (2009). The state and uneven development: The governance of economic development in England in the post-devolution UK. *Cambridge Journal of Regions, Economy and Society, 2*(1), 13–34.
Porter, M. E. (1985). *Competitive advantage: Creating and sustaining superior performance*. Free Press.
Rayport, J. F., & Sviokla, J. J. (1995). Exploiting the virtual value chain. *Harvard Business Review, 73*(6), 75–85.
Ridley, M. (2016). *The evolution of everything: How new ideas emerge*. Harper Perennial.
Rostow, W. W. (1960). *The stages of economic growth: A non-communist manifesto*. Cambridge University Press.
Schumpeter, J. A. (1934). *The theory of economic development: An inquiry into profits, capital, credit, interest, and the business cycle*. Harvard University Press.
Shapiro, C., & Varian, H. R. (1999). *Information rules: A strategic guide to the network economy*. Harvard Business School Press.
Smith, A. (1776). *The wealth of nations, books I-III, (1986 printing)*. Penguin Books.
Sobel, R. (1999). *When giants stumble: Classic business blunders and how to avoid them*. Prentice Hall.
Stabell, C. B., & Fjeldstad, O. D. (1998). Configuring value for competitive advantage: On chains, shops, and networks. *Strategic Management Journal, 19*(5), 413–437.
State Council of the PRC. (2015). Notice of State Council of the PRC on printing and distributing "Made in China 2025".http://www.gov.cn/zhengce/content/2015-05/19/content_9784.htm. Accessed on May 8, 2015.
State Council of the PRC. (2016). Guidance of State Council of the PRC on Deepening the Integration of Manufacturing Industry and Internet. http://www.gov.cn/zhengce/content/2016-05/20/content_5075099.htm. Accessed on May 20, 2016.
Stidham, M., McCaffrey, S., Toman, E., & Shindler, B. (2014). Policy tools to encourage community-level defensible space in the United States: A tale of six communities. *Journal of Rural Studies, 35*(July), 59–69.
Szirmai, A., & Verspagen, B. (2015). Manufacturing and economic growth in developing countries, 1950-2005. *Structural Change and Economic Dynamics, 34*, 46–59.

Wagner, J. (2012). International trade and firm performance: A survey of empirical studies since 2006. *Review of World Economics, 148*, 235–267.

Wen, Y. (2016). *The making of an economic superpower: Unlocking China's secret of rapid industrialization*. World Scientific.

Yang, C. Y. (2018). *Extension innovation method*. CRC Press.

Ye, G. L., Priem, R. L., & Alshwer, A. A. (2012). Achieving demand-side synergy from strategic diversification: How combining mundane assets can leverage consumer utilities. *Organization Science, 23*(1), 207–224.

Afterword: Need for a Multi-Approach Methodology and Economic Induction

To wrap up the presentation of this volume, this afterword attempts to close up several holes existing in the methodology employed in the previous chapters. In particular, what has been methodologically emphasized throughout this book is that for studies of value creation and capture, there is a need to introduce the intuition and methods of systems science and to develop the necessary logical reasoning on a rigorously developed foundation so that the logical reasoning can flow smoothly in a parallel fashion as that commonly used in mathematics and natural science.

However, with such emphasis, scholars who are either more methodologically oriented or conscious about which method they use would naturally ask such relevant questions as those below:

- Does intuition create incorrect imaginations?
- Can intuition-based logical reasoning lead to incorrect conclusions?
- How can analytical approaches produce incorrect answers?
- Compared to mathematics, what method is missing that can bridge individual cases to general conclusions?
- What will the possible next steps be following the theory established in this book in terms of research and applications?

To address these questions one by one, the rest of this afterword is organized as follows. Section A.1 looks at how intuition and intuition-based logical reasoning can lead to incorrect conclusions, while an analytical approach comes for rescue. Section A.2 examines how analytical approaches can produce inconsistent conclusions and analyzes what really happens that lead to inconsistent conclusions. Section A.3 demonstrates the scholarly need to introduce a principle of economic induction that can potentially broaden empirical conclusions into generally true statements. Section A.4 details possible next steps of research and application following what are established in this book. This afterword and this book conclude in Section A.5.

J. Y.-L. Forrest, Y. Liu, *Value in Business*, Contributions to Management Science,
https://doi.org/10.1007/978-3-030-82898-1

How Intuition-Based Logical Reasoning Can Lead to Incorrect Conclusions

This section first looks at a scenario that show how wrong human sensing capability can be. And then it uses theoretical examples from plane geometry to show that although with the help of intuition, logical reasoning can lead to generally true conclusions, if the intuition used is incorrect, then the same logical reasoning will likely lead to false conclusions. What this section demonstrates indicates that to possibly produce reliable conclusions, different approaches need to be applied simultaneously.

When the term "intuition" is concerned with, it basically means that we use one of our sense organs to detect what is happening in the environment and what it implies to us. For example, by looking at the line drawing in Fig. A1 (Lin & Fan, 1997), one might ask out of his/her curiosity the following question: which line segment is longer—the vertical one or the horizontal one?

It seems to most people that the vertical segment is longer. Not only so, but it is also obviously longer than the horizontal one by quite a bit. However, actual measurements of the segments reveal that these two segments are of the same length. Hence, this simple drawing indicates that human intuition can produce incorrect information and beliefs.

The next case analysis demonstrates that the widely employed intuition in plane geometry can also readily lead to incorrect consequences.

Theorem A1 (Kay, 1994, p. 67) For any obtuse angle $\angle ABC$, $m\angle ABC = 90^{o}$.

Proof Suppose that an obtuse angle $\angle ABC$ is given as in Fig. A2a. Construct the rectangle ABDE in Fig. A2b so that $mAE = mBD = mBC$, and complete the quadrilateral ABCE.

Next, draw the perpendicular bisector LO of segment EC and that MO of the segment AB, Fig. A2c. Because the lines AB and EC are not parallel to each other, the bisectors LO and MO must intersect at a point, labeled O. Next, draw segments AO, EO, CO, and BO, Fig. A2d. Because $mAO = mBO$ (since O is equidistant A and B), $mEO = mCO$ (since O is equidistant E and C), and $mAE = mBC$

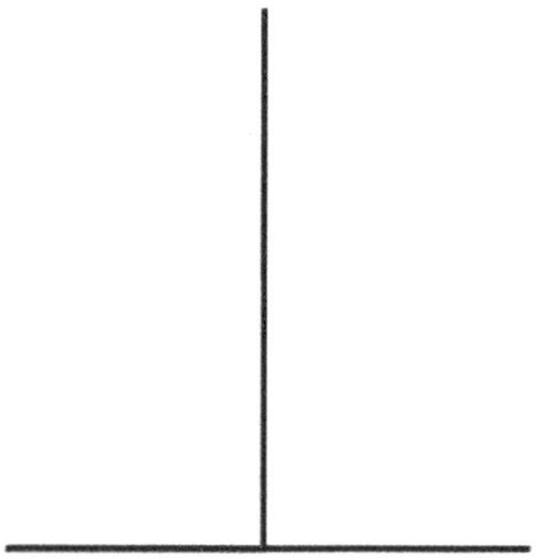

Fig. A1 Information might be possibly extracted incorrectly

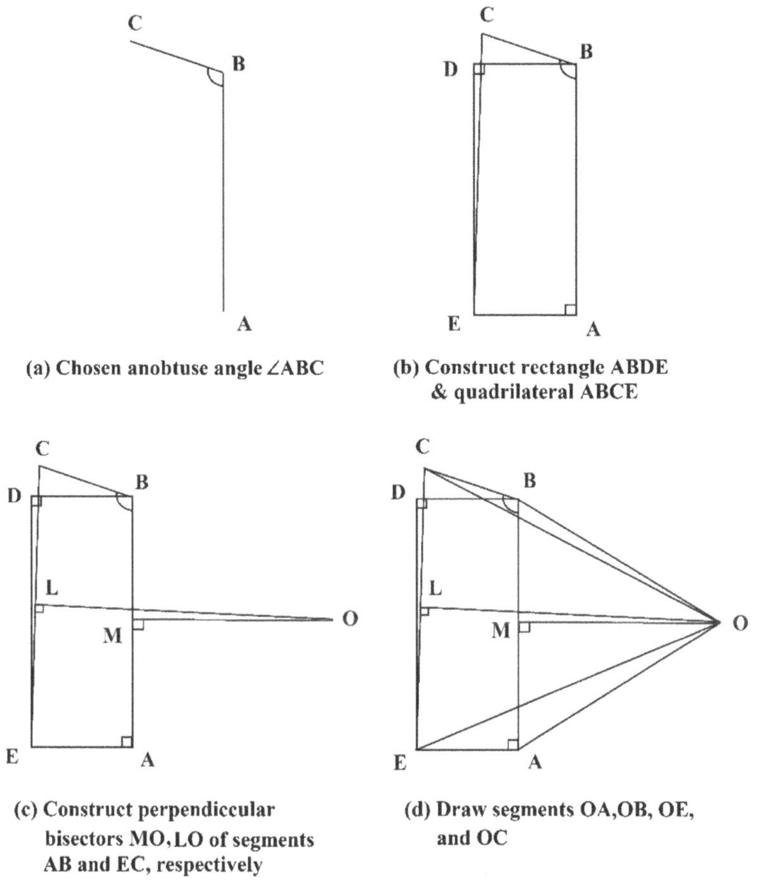

Fig. A2 In support of Theorem A1. (**a**) Chosen an obtuse angle ∠ABC. (**b**) Construct rectangle ABDE and quadrilateral ABCE. (**c**) Construct perpendicular bisector MO, LO of segments AB and EC, respectively. (**d**) Draw segment OA, OB, OE, and OC

(as guaranteed by the construction of the rectangle ABDE), the rule of side-side-side of plane geometry imply that ΔOBC is congruent to ΔOAE. Therefore, as corresponding angles, the following holds true:

$$m\angle \text{OBC} = m\angle \text{OAE}. \tag{A1}$$

Because $m\angle\text{OBC} = m\angle\text{ABC} + m\angle\text{OBA}$ and $m\angle\text{OAE} = m\angle\text{OAB} + m\angle\text{BAE} = m\angle\text{OAB} + 90^{\circ}$, Eq. (A1) can be rewritten as follows:

$$m\angle \text{ABC} + m\angle \text{OBA} = m\angle \text{OAB} + m\angle \text{BAE} = m\angle \text{OAB} + 90^{\circ}. \tag{A2}$$

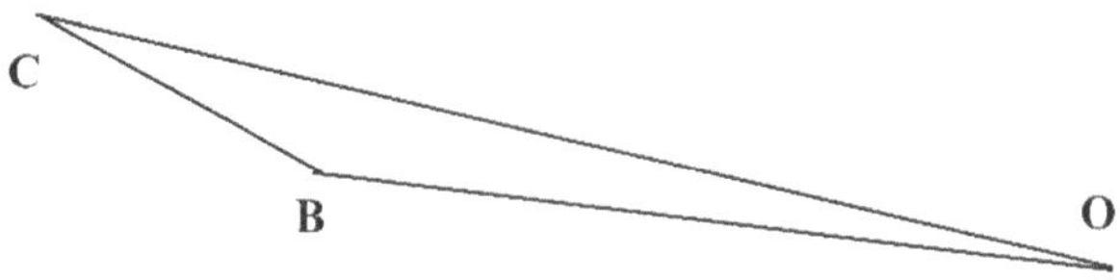

Fig. A3 The true appearance of ΔOBC

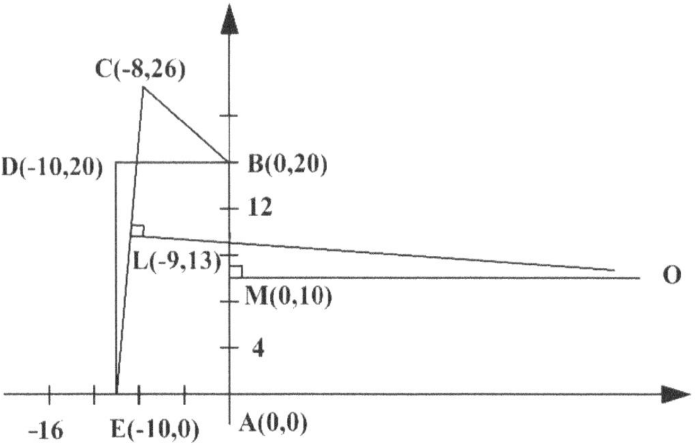

Fig. A4 Analytical approach to the proof of Theorem A1

Because ΔABO is isosceles, $m\angle \mathrm{OBA} = m\angle \mathrm{OAB}$. That simplifies Eq. (A2) as follows:

$$\mathrm{m}\angle \mathrm{ABC} = \mathrm{m}\angle \mathrm{BAE} = 90^{\circ}. \tag{A3}$$

In other words, what is shown here is that any obtuse angle $m\angle \mathrm{ABC}$ is a right angle. End of proof.

Evidently, the proof of Theorem A1 is flawed, because the conclusion is incorrect. In particular, the interaction point O of the perpendicular bisectors LO and MO can be located far from the rectangle ABDE so that instead of the appearance in Fig. A2, ΔOBC really looks like that in Fig. A3. Therefore, $m\angle \mathrm{OBC} \neq m\angle \mathrm{ABC} + m\angle \mathrm{OBA}$ and Eq. (A2) no longer holds true. The reason why the interaction point O in Fig. A2 is not that far from the rectangle ABDE is because when the perpendicular bisectors LO and MO are drawn, these segments naturally have a thickness. However, when they are drawn theoretically, they do not have such thickness.

What has been discussed here points to the following two observations.

Observation A1 Intuition-based arguments can lead to incorrect conclusions.

Observation A2 As soon as a mental intuition is drawn on paper, some aspects of the original intuition are lost, which potentially leads to incorrect conclusions.

What is described above about the flaw of the proof of Theorem A1 can be analytically confirmed by using the example in Fig. A4. All details are omitted.

Observation A3 Analytical methods can help clear incorrect conclusions drawn on intuition-based logical reasoning.

Following this observation, the following question naturally arises: instead of geometrical intuition, can one simply employ analytical means to establish conclusions? The next section will address this question and produces the following answer: No, analytical methods can still lead to inconsistent conclusions.

How Analytical Analysis Can Lead to Inconsistent Conclusions

This section looks at two different analytical arguments with one for and the other against the existence of the delta (or δ) function. Doing so enables us to see that even analytical methods do not always produce consistent conclusions.

Existence of the δ Function

The well-known delta (or δ) function is initially introduced by physicist Paul Dirac (1958) to model the density of an idealized point mass or point charge. It is a function $\delta(x)$ defined for every real number x such that it satisfies

$$\delta(x) = 0, x \neq 0; \text{and} \int_{-\infty}^{+\infty} \delta(x)dx = 1. \tag{A4}$$

The existence of this delta function can be shown in many different ways. For example, Boyce and DiPrima (2001) provide the following argument. Let γ be a positive number and define

$$d_\gamma(x) = \begin{cases} \frac{1}{2\gamma}, & \text{if } -\gamma < x < \gamma \\ 0, & \text{otherwise.} \end{cases} \tag{A5}$$

Then this function satisfies the following for any positive value γ

$$\int_{-\infty}^{+\infty} d_\gamma(x)dx = 1. \tag{A6}$$

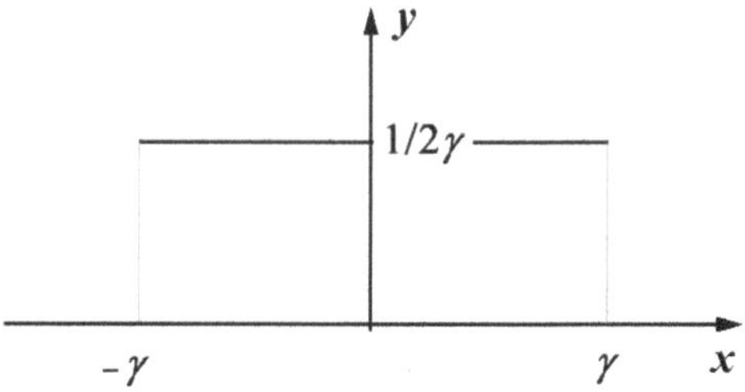

Fig. A5 The Graph of $y = d_\gamma(x)$

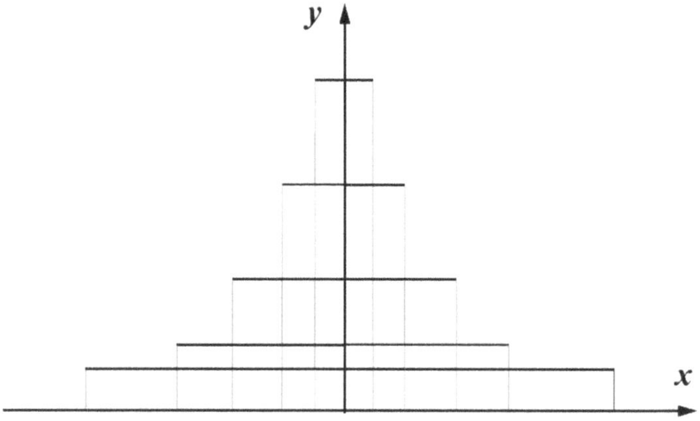

Fig. A6 The graph of $y = d_\gamma(x)$ as $\gamma \to 0$

Because $\lim_{\gamma \to 0} d_\gamma(x) = 0$, if $\gamma \neq 0$, and because Eq. (A6) has nothing to do with γ, that is

$$\lim_{\gamma \to 0} \int_{-\infty}^{+\infty} d_\gamma(x)dx = 1, \tag{A7}$$

the δ function in Eq. (A4) can be defined as follows: $\delta(x) = \lim_{\gamma \to 0} d_\gamma(x)$.

In this entire argument for the existence of the δ function, the intuitions of $d_\gamma(x)$ and Eq. (A7) are given in Figs. A5 and A6.

Nonexistence of the δ Function from the Angle of Riemann Integrals

To show the nonexistence of the δ function from the angle of Riemann integrals, it suffices to demonstrate that the second part in Eq. (A4) cannot be satisfied if a

Fig. A7 Polygonal estimate of the area under $f(x)$

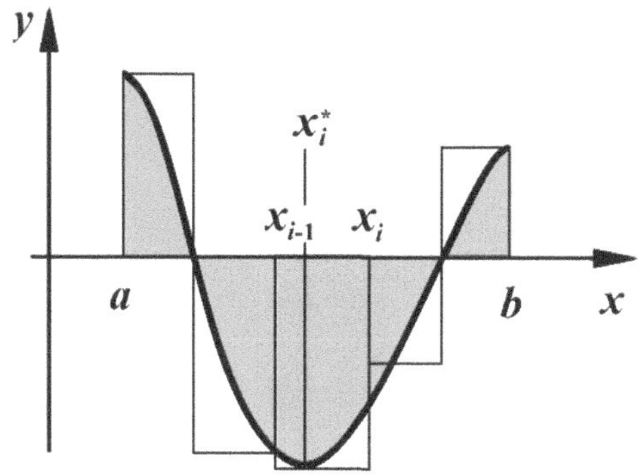

function $\delta(x)$ satisfies the first part. To this end, let us briefly look at how the concept of Riemann integrals is defined, for more details, see Steward (1987).

Let $f(x)$ be a function defined on a closed interval $[a, b]$ and P a partition of $[a, b]$ with $a = x_0 < x_1 < \ldots < x_{n-1} < x_n = b$. Let $x_i^* \in [x_{i-1}, x_i]$ be arbitrary, $\Delta x_i = x_i - x_{i-1}$ and $\|P\| = \max_{1 \leq i \leq n} \{\Delta x_i\}$. Then the (definite) integral of $f(x)$ from a to b is the following limit, if it exists:

$$\int_a^b f(x)dx = \lim_{\|P\| \to 0} \sum\nolimits_{i=1}^{n} f(x_i^*)\Delta x_i. \tag{A8}$$

In this case, $f(x)$ is said to be integrable on $[a, b]$. The right-hand side of Eq. (A8) is the limit of the polygonal estimate of the area of the (shaded) region under the curve of the function $f(x)$, Fig. A7.

If $\int_a^u f(x)dx$ exists for every $u \geq a$, then the integral of $f(x)$ from a to $+\infty$ is defined as follows, if the limit on the right-hand side exists:

$$\int_a^{+\infty} f(x)dx = \lim_{u \to +\infty} \int_a^u f(x)dx. \tag{A9}$$

Similarly, if $\int_u^b f(x)dx$ exists for every $u \leq b$, then the integral of $f(x)$ from $-\infty$ to b is defined as follows, if the limit involved exists:

$$\int_{-\infty}^b f(x)dx = \lim_{u \to -\infty} \int_u^b f(x)dx. \tag{A10}$$

And, if both $\int_a^{+\infty} f(x)dx$ and $\int_{-\infty}^a f(x)dx$ exist for a real number a, define

$$\int_{-\infty}^{+\infty} f(x)dx = \int_{-\infty}^{a} f(x)dx + \int_{a}^{+\infty} f(x)dx. \tag{A11}$$

So, if a function δ satisfies $\delta(x) = 0$, $x \neq 0$, as stated in Eq. (A4), then

$$\begin{aligned}\int_{-\infty}^{+\infty} \delta(x)dx &= \int_{-\infty}^{-1} \delta(x)dx + \int_{-1}^{+1} \delta(x)dx + \int_{+1}^{+\infty} \delta(x)dx \\ &= \lim_{u\to 0^-} \int_{-1}^{u} \delta(x)dx + \lim_{u\to 0+} \int_{u}^{+1} \delta(x)dx = 0.\end{aligned} \tag{A12}$$

That is, in terms of Riemann integrals, the two parts in Eq. (A4) cannot be satisfied simultaneously. Or, the δ function, as given in Eq. (A4), cannot exist.

Observation A4 Analytical approaches, as precise as calculus, can also produce inconsistent conclusions.

This discussion leads to the following natural question: Can the concept of Riemann integrals be generalized so that the δ function will become valid? To help address this question, the following subsection looks at the concept of Lebesgue integrals.

Nonexistence of the δ Function from the Angle of Lebesgue Integrals

Contrary to the previous argument for the existence of the δ function, the following argument from the perspective of Lebesgue integrals (Jiang & Wu, 1979) shows that such a particular δ function cannot actually exist.

In the n-dimensional Euclidean space $\mathbf{R}^n$,where $\mathbf{R}$ is the real-number line, the following set of n-tuples of real numbers is known as an open interval:

$$D = \{\mathbf{x} = (x_1, x_2, \ldots, x_n) \in \mathbf{R}^n : a_i < x_i < b_i, i = 1, 2, \ldots, n\}, \tag{A13}$$

for some $\boldsymbol{a} = (a_1, a_2, \ldots, a_n)$, $\boldsymbol{b} = (b, b_2, \ldots, b_n) \in \mathbf{R}^n$. If some of the inequality signs in Eq. (A13) include equal sign, then D is simply called an interval. No matter whether D is an open interval or not,

$$|D| = \prod_{i=1}^{n} (b_i - a_i) \tag{A14}$$

is referred to as the "volume" of the interval D. For any set $E \subseteq \mathbf{R}^n$, $\boldsymbol{D} = \{D_i \subseteq \mathbf{R}^n : i \in \boldsymbol{I_D}\}$ is an interval cover of E, provided that each D_i is an interval and

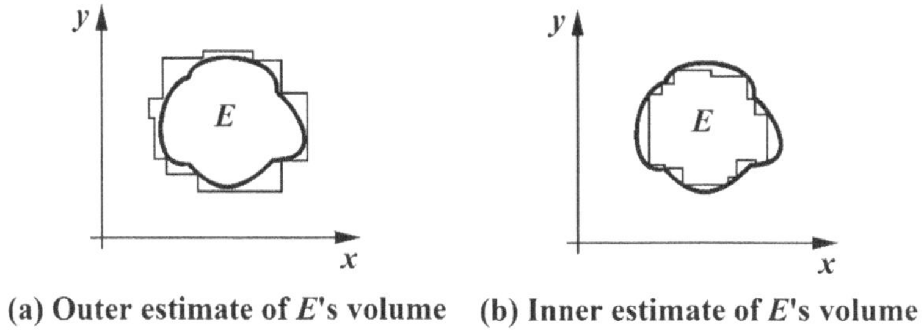

Fig. A8 Geometry of Eqs. (A16) and (A17). (**a**) Outer estimate of E′s volume. (**b**) Inner estimate of E′s volume

$$E \subseteq \bigcup_{i \in \boldsymbol{I}} D_i, \tag{A15}$$

where $\boldsymbol{I_D}$ is an index set that defines the collection $\boldsymbol{D}$. The Lebesgue outer measure of E, denoted as m^*E, is defined as follows:

$$m^*E = inf\left\{\sum\nolimits_{i \in \boldsymbol{I}} |D_i| : \boldsymbol{D} = \{D_i \subseteq \mathbf{R}^n : i \in \boldsymbol{I_D}\} \text{ is an interval cover of } E\right\}. \tag{A16}$$

If $E \subseteq \mathbf{R}^n$ is bounded, then there is an interval $D \subseteq \mathbf{R}^n$ such that $E \subseteq D$. The Lebesgue inner measure of E, denoted as m_*E, is defined as follows:

$$m_*E = sup\left\{|D| - \sum\nolimits_{i \in \boldsymbol{I}} |D_i| : \boldsymbol{D} = \{D_i \subseteq \mathbf{R}^n : i \in \boldsymbol{I_D}\} \text{ is an interval cover of } D - E\right\}. \tag{A17}$$

The idea behind Eqs. (A16) and (A17) is given in Fig. A8, where for convenience the arbitrary set E is imagined to be in the second dimensional Euclidean space $\mathbf{R}^2$, which is the well-known xy-plane. In particular, the outer measure of E is equal to the infimum of all the approximations of polygonal areas that totally cover E, while the inner measure is the supremum of all those that are inside E.

If a bounded set $E \subseteq \mathbf{R}^n$ satisfies $m_*E = m^*E$, then the set E is said to be measurable with its measure being equal to $mE = m^*E = m_*E$. If $E \subseteq \mathbf{R}^n$ is an unbounded set, E is measurable, provided that for any interval $D \subseteq \mathbf{R}^n$, $E \cap D$ is measurable.

Theorem A2 The set $D = \{0\}$ is a measurable set in $\mathbf{R}^1 = \mathbf{R}$ with Lebesgue measure 0.

Proof Evidently, $D_i = (-1/i, 1/i)$, for natural number i, is an interval that covers the given set D. It is readily seen that

$$m^*D = inf\{|D_i| : i = \text{a natural number}\} = 0. \tag{A18}$$

Let us take the unit interval $U = (-1, 1)$. Then $E_i = (-1 - 1/i, 1 + 1/i)$ is an interval cover of $U - D$, i is a natural number, satisfying that

$$\begin{aligned} m_*D &= sup\{|U| - |E_i| : i = \text{a natural number}\} \\ &= sup\left\{1 - \frac{1}{2i} : i = \text{a natural number}\right\} = 0. \end{aligned} \tag{A19}$$

By joining Eqs. (A18) and (A19), Theorem A2 follows. End of proof.

Let $E \subseteq \mathbf{R}^n$ be a measurable set with $mE < \infty$, $f(x)$ a real-valued function (so assumed for the rest of this subsection) defined on E with lower bound b and upper bound B, and $P = \{E_i\}_{i=1}^n$ a partition of E, satisfying $E_i \bigcap E_j = \emptyset$ $(i \neq j)$ and $E = \bigcup_{i=1}^n E_i$. Then, the Lebesgue integral of $f(x)$ on E is defined as follows: First, similar to the summation in Eq. (A8), define

$$s = \sum_{i=1}^n b_i m E_i \text{ and } S = \sum_{i=1}^n B_i m E_i, \tag{A20}$$

where $b_i = inf\{f(x) : x \in E_i\}$ and $B_i = \sup\{f(x) : x \in E_i\}$. Because $bmE \leq s \leq S \leq BmE$, for any partition P of E, the following supremum and infimum always exist:

$$\int_{-E} f(x)dx = \sup_P\{s\} \text{ and } \bar{\int_E} f(x)dx = \inf_P\{s\}. \tag{A21}$$

If the two expressions in Eq. (A21) are equal, then $f(x)$ is said to be integrable on E, written as

$$\int_E f(x)dx = \int_{-E} f(x)dx = \bar{\int_E} f(x)dx. \tag{A22}$$

In this discussion, if $f(x)$ is any function defined on E, then define

$$\{f(x)\}_n = \begin{cases} f(x), & \text{if } |f(x)| \leq n \\ n, & \text{if } f(x) \geq n \\ -n, & \text{if } f(x) \leq -n \end{cases}. \tag{A23}$$

If for any natural number n, $\{f(x)\}_n$ is integrable on E, then $f(x)$ is said to be integrable on E, satisfying the following if the right-hand side is a finite number:

$$\int_E f(x)dx = \lim_{n\to\infty} \int_E \{ f(x)\}_n dx. \tag{A24}$$

Next, let us remove the condition $mE < \infty$ from the previous discussion. If $f(x)$ is a function defined on a measurable set $E \subseteq \mathbf{R}^n$, define

$$F(x) = \begin{cases} f(x), & \text{if } x \in E \\ 0, & \text{otherwise} \end{cases}, \tag{A25}$$

then $F(x)$ becomes a function defined on the entire $\mathbf{R}^n$. Without loss of generality, the rest of this subsection considers Lebesgue integral on the entire space $\mathbf{R}^n$. To this end, let $f(x)$ be a function defined on $\mathbf{R}^n$, and $\{R_i\}_{i=1}^{\infty}$ a sequence of bounded intervals in $\mathbf{R}^n$ such that

$$R_i \subseteq R_j, i \leq j, \text{and} \bigcup_{i=1}^{\infty} R_i = \mathbf{R}^n. \tag{A26}$$

If $f(x)$ is integrable on each R_i, then $f(x)$ is said to be integrable on $\mathbf{R}^n$ with integral given by Eq. (A27), if the limit on the right-hand side exists.

$$\int_{\mathbf{R}^n} f(x)dx = \lim_{i\to\infty} \int_{R_i} f(x)dx. \tag{A27}$$

With the concept of Lebesgue integrals introduced, the following can be shown (all details are omitted):

Theorem A3 If a function defined on $\mathbf{R}^n$ is Riemann integrable, then it is also Lebesgue integrable and both integrals are the same.

To demonstrate that the concept of Lebesgue integrals is more general than that of Riemann integrals, let us consider the following Dirichlet function defined on the unit interval [0,1]:

$$f(x) = \begin{cases} 1, & \text{if } x \text{ is a rational number} \\ 0, & \text{if } x \text{ is an irrational number} \end{cases}$$

It can be shown that this function is not integrable in the Riemann sense, while it is integrable in the Lebesgue sense. Now, the following theorem is the key for us to show that the δ function does not exist. Once again, the detailed proof is omitted.

Theorem A4 Assume that $f(x)$ and $g(x)$ are two functions defined on $\mathbf{R}^n$. If $f(x) = g(x)$ holds true except for $x \in N \subseteq \mathbf{R}^n$, where $mN = 0$, then $\int_{\mathbf{R}^n} f(x)dx = \int_{\mathbf{R}^n} g(x)dx$.

Now, Theorems A2 and A4 jointly imply that no function $\delta(x)$ can satisfy the first part of Eq. (A4) and also the second part simultaneously no matter whether it is in the sense of Riemann integrals or that of Lebesgue one. To address this problem, Schwartz (2008) treats this δ function as a distribution from which the fore-mentioned mathematical difficulty is successfully revolved. This discussion naturally leads to

Observation A5 When known analytical methods produce inconsistent outcomes, a new methodological approach, analytical or not, needs to be introduced.

In the context of this book, there are countless many instances in the literature where employed methodology, which tends to be statistics-based with data collected from specific environments, does not produce consistent conclusions. Regarding this phenomenon, what Observation A5 implies is that every time when the same methodology leads to inconsistent conclusions, it is a good idea to introduce a brand-new methodology. In other words, when an identical statistics-based approach is employed to analyze a business situation, if slight change in the data used leads to drastically different outcomes, it means a nonstatistical method needs to be employed.

What Has Gone Wrong?

It is a well-known fact in the business-related literature that an identical statistics-based approach can readily produce drastically different outcomes. One of the possible reasons for such a phenomenon to occur is that changes in the data used, although they might be only minor and slight, generally lead to different inferences. This scenario is very different from those methods developed out of the branch of deterministic mathematics, including calculus and related methods and theories. In this traditional branch of mathematics, the conventional belief or ideal situation is that the entire system of mathematics, meaning the deterministic part, is consistent does not suffer from any inconsistent and contradictory results (Kline, 1972; Kunen, 1980). Evidently, Sects. A.2.1 and A.2.2 demonstrate a completely different situation from this convention and ideal, where the theory of calculus indeed contains completely opposite conclusions—existence and nonexistence of the δ function. So, a natural question arises here: what might have potentially gone wrong within the reasonings?

According to Forrest (2013), what has gone wrong is closed related to the concept of infinity, which is implicitly employed in the operation of taking limits and definite integrals. To fully understand what is involved, let us first look at the following vase puzzle.

Vase Puzzle (Lin, 1999, p. 178): Suppose that a vase and an infinite number of pieces of paper are available. The pieces of paper are labeled by natural numbers 1, 2, 3, … so that each piece has at most one label on it. The following recursive procedure is performed:

Step 1: Put the pieces of paper, labeled from 1 to 10 into the vase; then remove the piece labeled 1.
Step n: Put the pieces of paper, labeled from $10n - 9$ to $10n$ into the vase; then remove the piece labeled n, where n is any natural number 1, 2, 3, ...

One question that may be wondered is: after the recursive procedure is completed, how many pieces of paper are left in the vase? The existence of such a vase and the pieces of paper is readily and materially possible, see Lin (1999, p. 178–179) for details.

To answer this question, let us approach it in two different ways: an elementary modeling and a set-theoretic modeling. For the first way, upon the completion of step n, the number of pieces of paper left in the vase is

$$f(n) = 9n, \text{for } n = 1, 2, 3, \ldots \quad \text{(A28)}$$

To complete the recursive procedure, the value of n has to finish going through each and every natural numbers. So, the answer to the question is: $\lim_{n\to\infty} f(n) = +\infty$.

For the second set-theoretic modeling, the answer to the question is zero (0). In fact, if there are pieces of paper left in the vase upon the completion of the recursive procedure, then let us take any one piece of those left in the vase. According to how pieces of paper are put into and removed from the vase, this piece of paper has a natural number label. Say, the label is k. However, this particular piece of paper has been already taken out of the vase at step k. That is a contradiction, indicating that the vase is empty.

By carefully checking the entire situation involved in this vase puzzle, what really happens is the assumption that the recursive procedure can be completed. In particular, the first approach of answering the question implicitly assumed that the completion of the procedure is accomplished by numerating through the set of all natural numbers without actually reaching the end of the procedure, while the second approach assumed that the procedure is completed and then the consequence is looked at with 20–20 hindsight.

This recognition leads to the separation of the concept of conventional infinity into those of actual infinity and potential infinity (Forrest, 2013). The former stands for the case that an infinite process can actually been finished, while the latter the case that an infinite procedure can only approach its target as closely as imaginably possible without being actually arrived. The vase puzzle above demonstrates that actual infinity and potential infinity represent two different concepts.

Corresponding to the separation of the conventional infinity into actual and potential infinities, the concept of limits is also separated into that of realizable limits and that of not realizable limits. Specifically, for a given function $f(x)$ and a real number x_0, if in the expression $\lim_{x\to x_0} f(x) = A$ the notation of limit satisfies the following conditions:

- The function $f(x)$ is defined at $x = x_0$;
- Both x and $f(x)$ approach their respective limits in the fashion of actual infinities; and
- $f(x) = A$ if and only if $x = x_0$,

then the expression $\lim_{x \to x_0} f(x) = A$ is said to be realizable. Otherwise, this limit expression is said to be not realizable.

Next, let us examine the analysis in Sect. A.2. in order to find where things went wrong. First, Eq. (A4) involves the following limits:

$$\int_{-\infty}^{+\infty} \delta(x)dx = \lim_{\substack{u_L \to -\infty \\ \gamma \to 0 \\ u_R \to +\infty}} \left\{ \int_{u_L}^{-\gamma} \delta(x)dx + \int_{-\gamma}^{\gamma} \delta(x)dx + \int_{\gamma}^{u_R} \delta(x)dx \right\} = \lim_{\gamma \to 0} \int_{-\gamma}^{\gamma} \delta(x)dx. \tag{A29}$$

And, similarly, Eq. (A6) consists of the limits below:

$$\int_{-\infty}^{+\infty} \lim_{\gamma \to 0} d_\gamma(x)dx = \lim_{\substack{u_L \to -\infty \\ \gamma \to 0 \\ u_R \to +\infty}} \left\{ \int_{u_L}^{-\gamma} d_\gamma(x)dx + \int_{-\gamma}^{\gamma} d_\gamma(x)dx + \int_{\gamma}^{u_R} d_\gamma(x)dx \right\} = \lim_{\gamma \to 0} \int_{-\gamma}^{\gamma} d_\gamma(x)dx, \tag{A30}$$

for any real number $\gamma > 0$. Because Eq. (A29) represents a not realizable limit, while Eq. (A30) a realizable limit, it means that the δ function cannot be modeled by $\lim_{\gamma \to 0} d_\gamma(x)$, because $\delta(x) \neq \lim_{\gamma \to 0} d_\gamma(x)$.

Observation A6 Assume that different analytical approaches are employed to explain a given phenomenon. If inconsistent outcomes are produced, then some of the approaches are either applied inappropriately or some relevant concepts need to be refined.

On the surface, this subsection seems to be entirely mathematical in nature. However, it is really not so. The concepts of realizable and not realizable limits are in fact widely used in macroeconomics indiscriminately and implicitly. In particular, to help materialize leaps from individual components (such as firm-level phenomena) to an organic whole (such as the economy), a standard reasoning generally discards the possibility for idiosyncratic shocks at microlevel of firms to cause aggregate fluctuations at the macrolevel of the economy (Lucas, 1977). The argument is that firm-level micro-shocks, when seen at the height of the economy, are averaged out so that they do not cause noticeable aggregate fluctuations in the economy. However, it is found that such leaps, seen as a limiting stage of micro-shocks in the whole of the economy, are not always realizable. For example, Gabaix (2011) empirically confirms that idiosyncratic firm-level shocks can indeed explain an import part of aggregate economic movements and provide a micro-foundation for aggregate shocks. Specifically, Gabaix finds that the idiosyncratic movements of the largest 100 US firms do seem to explain around one-third of variations in the output growth of the US economy.

Moment to Introduce a Principle of Economic Induction

A lot of studies in business-related disciplines and economics are data-based, while many of the authors like to state their conclusions in as general terms as possible. This phenomenon naturally leads one to think about mathematical induction, which lays out a reliable way for a scholar to draw general conclusions based on a few individual case studies. Hence, a natural question arises here as follows.

> Can a similar principle of induction be established for business and economics disciplines?

In this section, let us dive into some of the details that are underneath this important theoretical question. Evidently, if such a principle of induction can be established for business and economics disciplines, many data-based discoveries can be readily converted into generally true statements. Consequently, relatively more reliable recommendations and predictions, instead of suggestions, can be provided.

Symbolically, mathematical induction is given in one of the following two versions (Forrest, 2013, p. 192):

Mathematical Induction (version 1): Assume that $\Phi(n)$ is a proposition regarding natural number n. If the following two steps can be shown,

Initial step: $\Phi(k_0)$ holds true for a certain natural number k_0,

Inductive step: If $\Phi(n)$ holds true for $n = k \geq k_0$, then $\Phi(k + 1)$ holds true,

then the proposition $\Phi(n)$ holds true for all natural numbers $n \geq k_0$.

Mathematical Induction (version 2): Assume that $\Phi(n)$ is a proposition regarding natural number n. If the following two steps can be shown,

Initial step: $\Phi(k_0)$ holds true for a certain natural number k_0,

Inductive step: If $\Phi(n)$ holds true for all natural number n satisfying $k_0 \leq n \leq k$, then $\Phi(k + 1)$ holds true,

then the proposition $\Phi(n)$ holds true for all natural numbers $n \geq k_0$.

This mathematical induction can be intuitive explained as follows: Treat the proposition $\Phi(n)$ as a ladder of infinite length so that natural number n represents a step so that the greater such a number is, the higher the step one is on the ladder. Hence, what the mathematical induction says is that one can get on the ladder, say, at step k_0. Assume that the person can climb the ladder to step k ($\geq k_0$). If now it can be shown that the person can climb to step $k + 1$, then the person can finish climbing the infinitely long ladder $\Phi(n)$.

To describe what is meant by a similar principle of induction established for business and economics disciplines, let us lose up the understanding of $\Phi(n)$ and n, as given in the previously stated mathematical induction, as follows. Instead of a natural number, let us treat n as, say, a moment of time or a particular economic state. And instead of a proposition regarding a natural number variable n, let us see $\Phi(n)$ as the conclusion produced out of an empirical study at moment n or economic state n. By using such a parallelism, the previously stated mathematical induction can be readily reformulated for business and economic disciplines.

Of course, for this end to play out smoothly, all possible states of the variable n need to be defined with an appropriate linear ordering $\leq$ introduced. And if the variable n takes values from a continuum, instead of a discrete set, then we can reformulate the following continuous mathematical induction for business and economics disciplines:

Mathematical Induction over Real Numbers (Forrest, 2013, p. 289): Assume that $\Phi(x)$ is a proposition in terms of real number x. If the following steps hold true:

The initial step: There is a certain real number x_0 such that $\Phi(x)$ holds true for any $x < x_0$;

The inductive step: If $\Phi(x)$ holds true for any $x < y$, then there is $z > y$ such that $\Phi(x)$ holds true for any $x < z$,

then $\Phi(x)$ holds true for any real number x.

To conclude this section, let us look at two simple examples. First, it is well-known that some economic phenomena are time-dependent. If we like to establish a generally true conclusion $\Phi(x)$ regarding one of such economic phenomena, then the variable x will most likely be time. In this case, to show the truthfulness of the conjectured proposition $\Phi(x)$, we will need to employ a reformulated induction from that over real numbers.

Second, it is also well-known that some economic phenomena are dependent on people's beliefs. If we like to establish a generally valid statement $\Phi(n)$ about one of such belief-specific phenomena, then the variable n will take different systems of beliefs as its potential values. In this case, to show the general validity of the conjectured statement $\Phi(n)$, we will need to first construct the entire spectrum of all possible belief systems, from which the variable n will take its values, and then introduce a linear order relation $\leq$ between any two belief systems. If this spectrum of different beliefs is discrete, then we will need to employ a reformulated induction from those over natural numbers.

Next Steps for the Foreseeable Future

Evidently, this book has pointed to a brand-new direction and introduces a novel methodology for research and creative thinking for business and economics disciplines in general and front-line managers and marketers in particular. Hence, a natural question at this junction arises as follows:

> What will the next steps be from here for the foreseeable future for theorists and practitioners? More specifically, how can future scholarly research and managerial decision-making use and test the theory developed in this book?

The answer to this question is that there are literally infinite many different steps academicians and decision-making managers and entrepreneurs can take in their works, while they can also expect to produce important theoretical discoveries and/or important economic benefits. First, in terms of applying conclusions established in this book in real life to produce tangible economic benefits, relevant details can be found with the recommendations given at the end of each chapter. So, all relevant details are omitted here. Second, let us look at the previous question in two different angles: theoretical possibilities and possibilities in relation to real-life scenarios.

Theoretical Possibilities

In terms of theoretical development, this book represents a first step and initial effort towards the development of a generally applicable theory of value creation and capture. Indeed, it has touched on a wide range of related topics, such as demand- and supply-side strategies, value chains, customer values, roles of manufacturing and artificial intelligence, and government policies and supports. However, the presented discussion on each topic merely signifies a potentially rich and worthwhile endeavor, on which much more works need to be done.

The overwhelming majority of the literature examines specific cases with involved particulars different from one scenario to another by using anecdotes and/or data of limited validity. Opposite to such literature, this book attempts to develop a generally applicable theory by focusing on those elementary properties that are underlying many specific cases no matter how different they appear to be from each other. This objective might seem impossible on the surface. However, it has been successfully achieved in disciplines of natural science. For example, mathematics can be seen as a collection of various methods and theories. Although it covers a wide range of seemingly unrelated topics, people still find that these topics possess a common foundation—the set theory. This theory, as the foundation of the entire mathematics, is developed on the following ten basic axioms (Lin, 1999, p. 63). In other words, the entire edifice of mathematics and applications in various areas of life can be constructed theoretically on these basic building blocks.

- Axiom of existence: There exists a set.
- Axiom of extensionality: Two sets are equal if they have the same elements.
- Axiom of regularity: Every non-empty set x contains an element y such that x and y are disjoint.
- Axiom of comprehension scheme: For each formula φ and each set z, there is a set y such that $y = \{x \in z : \varphi(x)\}$.
- Axiom of pairing: For any sets x and y, there is a set z such that $x \in z$ and $y \in z$.
- Axiom of union: For set given set S of sets there exists a set A which contains all elements belonging to at least one element set in S.
- Axiom of replacement scheme: For each formula φ and set A, if to each element $x \in A$ there corresponds exactly one y satisfying $\varphi(x, y)$, then there exists a set Y consisting of all y satisfying $\varphi(x, y)$ for some $x \in A$.
- Axiom of infinity: There exists a set x satisfying (1) $\varnothing \in x$, and (2) for each $y \in x$ there exists a set $S(y) \in x$ which contains all elements of y and y itself.
- Axiom of power set: For any set x, there exists a set y, called the power set of x and denoted as $p(x)$, such that all subsets of x are contained in y.
- Axiom of choice: Any set can be well-ordered.

Similarly, physics and chemistry are developed so that although mechanical devices, designed for varied purposes, can be very different from each other, their underlying principles of design are the same; and although chemicals, produced for diverse real-life applications, are very different from one another, the fundamental principles of how the basic chemical elements interact with each other stay the same. Hence, what is labored in this book is to demonstrate such an effort that is parallel to those in mathematics, physics, chemistry, and other disciplines in natural science so that the produced theory can have a wide range of practical applications.

In particular, although firms have their individually very specific organizations and cultures, fundamental characteristics of these drastically different organizations and cultures can be theoretically modeled by using the concept of systems (Chap. 2). Hence, the very fundamental organizational features of firms can be investigated. Similarly, although market competitions take different forms and each competition is different from others, how competing firms interact with each other can still be studied in a unified fashion by using systemic modeling and game theory (Chap. 3). It is evident that each firm possesses its specific set of resources and deploys these resources in its particular ways. By employing the planned axiomatic approach, the entire resource-based theory of the firm can be more rigorously developed on three basic and intuitive axioms (Chaps. 4 and 5). And all the chapters in the entire book attempt to materialize such a pattern for various chosen topics of value creations and capture. By comparing what is achieved in this book to those in mathematics, physics, chemistry, etc., as described above, a major area for theoretical research appears: further elevate the height of axiomatization from what has been accomplished in this volume.

Contrary to this lofty next step of research, one can also be more specific. For example, in Chap. 3, the following theorem is shown:

Theorem 3.2. If the market described below is in Nash equilibrium, then the following hold true:

1) If the market's consumer surplus is larger than the loyal-customer base of one incumbent firm, then the market calls for new innovation and additional competition.
2) If some firms answer the call by entering the market with their versions of substitute offer, then the expected profit of all these firms can be potentially larger than that of at least one incumbent firm.

Some very important research questions include:

- How can an entrepreneur effectively receive a market invitation with sufficient market depth and width? The key here is how to find the relevant market depth and width and how to judge if they are sufficient.
- How do entrepreneurs formulate their different customer value propositions after they successfully answered the previous question? The essence here is that formulated customer value propositions are mostly dictated by entrepreneurs' individually varied experiences and background knowledge.
- How will these individually different customer value propositions compete against each other for market success?
- How will a firm organize its resources in innovative ways in order to deliver what is promised in its customer value proposition?

Of course, attempts to address these and other related questions need to be carried out in the fashion of achieving generality without using anecdotes and data of limited validity. And then as practical applications, developed conclusions need to be examined against various real-life scenarios in order to find how well they can lead to tangible economic benefits or reliable managerial recommendations.

Such critical look at Theorem 3.2 needs also to be done to each and every proposition developed in this book in order to advance the imagined general theory with new sets of generally true propositions. Recursively, such critical look can be repeatedly done to each and every of the newly established propositions. As a matter of fact, this process of theoretical development is what is practiced in the area of mathematics and other disciplines of natural science.

Possibilities in Relation to Real-Life Scenarios

For possibilities that arise from works based on this book that are more or less directly related to real life, let us focus on either empirically testing established theoretical propositions or discovering when some of these propositions might go wrong when applied.

For the question of when an established proposition in this book might go wrong when applied to revolve or to explain an issue in real life, there are several possible scenarios. First, there are some assumptions or conditions that are unconsciously

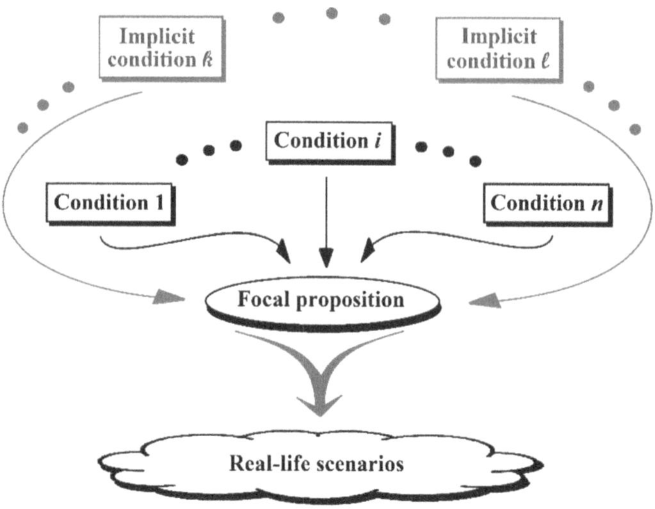

Fig. A9 There exist implicit conditions

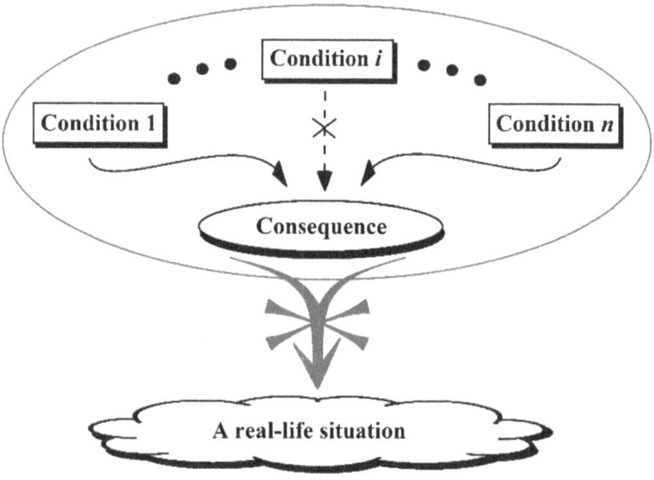

Fig. A10 Some necessary condition(s) are violated

employed in the logical reasoning in the supporting argument of the proposition, Fig. A9. In this graphical presentation, implicit conditions k and l are not considered in the formulation of the focal proposition, although conditions 1, 2, . . ., n are indeed examined. A good example here is what is analyzed in Subsection A.2.4, where two different concepts were mistakenly assumed as the same. In such a case, correcting the problem can easily lead to a more advanced theory with refiner propositions.

Second, some of the conditions listed in the focal proposition are violated so that the stated conclusion does not follow (Fig. A10), making an application of the proposition void. In Fig. A10, the entire oval area represents the focal proposition; and Condition i is violated so that the consequence stated in the focal proposition is no longer assuredly true. Such scenarios often appear when a theoretical proposition is applied in real life due to the reason that some easily stated conditions involve one or several variables whose values cannot be readily collected. For example, although the concept of innovation can be theoretically defined, but in real life, how to measure the innovativeness of a firm represents a real challenge. For relevant references, see Chaps. 8 and 13.

Third, some variables that appear in the focal proposition cannot be numerically measured exactly in the real-life scenario of concern. So, whichever proxy measures are employed, they only partially reflect what the proposition really says. And fourth, after relevant data are collected, limitations of the tool employed to analyze the data also distort what the proposition actually implies.

Next, let us look at potential issues related to empirically testing established theoretical propositions. Generally, such tests can be carried out for at least two purposes:

1. Empirically confirm what is theoretically claimed, such as those in this book; and
2. Discover real-life or real-life-like situations where a theoretical claim does not hold true.

Works for Purpose (1) are necessary for illustrating propositions that are logically established on solid ground. They are theoretically needed, if the claims of concern are established otherwise. The former case stands for how homework problems in a textbook are developed, even though it is not needed for the purpose of further developing the underlying theory. For example, after having rigorously established the formula $n + n = 2n$, for each natural number n, there is no need to empirically confirm the formula's correctness by plugging in specific natural numbers one by one for n in the formula. However, if the formula appears in a textbook, it may be needed for one to replace n by several particular natural numbers in order to strengthen the understanding of the concept of variables.

On the other hand, if attempts to apply the formula $n + n = 2n$ in real life reveals scenarios where this formula does not hold true, say, $1 + 1 \neq 2$, then such a particular scenario represents a "gold mine" in the sense that one has a chance to enhance the established theory by checking where the existing theory experiences a gap in the underlying reasoning. This end is different from rounds of debates surrounding a particular social phenomenon that often appears in the literature of economics and business studies. For example, corresponding to the current governmental pushes for industrial transformation from the current state of automated manufacturing to that of intelligent one (Chap. 19), the question of whether or not government policies are needed in such efforts is back as a focal point of mainstream debates after many rounds of previous debates without producing any consistent and consecutive enriched conclusions. Based on what is presented in Andreoni and Chang (2019) and Rostow (1960), the word "debate" used in this context implicitly means that the starting points of discussions are not rigorously or scientifically established so that consequent arguments become debates instead of attempts to make an established theory more solid and more scientifically sound.

As a matter of fact, based on the formula $n + n = 2n$, for any natural number n, all the frequently employed calculus- and statistics-based methods and approaches in the literature of economics and business studies are established. From $1 + 1 > 2$ that, in words, means that the whole is greater than the sum of parts, appears the well-known systems science (Lin, 1999), on which this book is based.

What is discussed above represents that each time when an empirical testing or real-life application of an established proposition in this book demonstrates that the proposition is invalid, one needs to go back to the supporting argument of the proposition to find out where the logical reasoning went wrong and how the problem can be fixed. This process of repeatedly correcting and improving a scientifically sound theory is different from those found in many business and economics studies where either

- Different empirical studies are repetitively employed to confirm and reconfirm some known inconsistent or even contradictory conclusions, or
- New theories, mostly data or anecdote-based, are used to refresh previous rounds of debates without producing anything definite other than additional conjectures.

A Few Final Words

To conclude this volume, we like to point out the history that every time when a new methodology is introduced into the study of business and economics, a whole new era of research appears. So, if you, the reader, have reached this far in this book, we like to emphasize that you have found a "gold mine" in your research career, where you can potentially establish generally true theoretical conclusions based on those empirically established either by you or by others, especially those that are thought-provoking, by using the methodology and way of thinking adopted in this book.

At the very end of this book, we like to use this opportunity to thank you for being with us all the way through this exciting journey of knowledge exploration. If you have any comments, questions, or suggestions, please contact us through email. Our addresses can be found at the end of the Preface of this book. We love to hear from you.

Bibliography

Aas, T. H., Breunig, K. J., Hydle, K. M., & Pedersen, P. E. (2015). Innovation management practices in production-intensive service firms. *International Journal of Innovation Management, 19*(05). https://doi.org/10.1142/S1363919615500553

Abbate, C. S., Boca, S., & Gendolla, G. H. E. (2016). Self-awareness, perspective-taking, and egocentrism. *Self and Identity, 15*(4), 371–380.

Acemoglu, D., & Restrepo, P. (2016). The race between man and machine: Implications of technology for growth, factor shares, and employment. NBER Working Papers No. 22252.

Acemoglu, D., & Restrepo, P. (2018). Artificial intelligence, automation and work. NBER working paper no. 24196.

Acemoglu, D., & Robinson, J. (2012). *Why nations fail – The origins of power, prosperity, and poverty*. Crown Business.

Adner, R. (2006). Match your innovation strategy to your innovation ecosystem. *Harvard Business Review, 84*(4), 98–107.

Adner, R. (2017). Ecosystem as structure: An actionable construct for strategy. *Journal of Management, 43*(1), 39–58.

Adner, R., & Kapoor, R. (2010). Value creation in innovation ecosystems: How the structure of technological interdependence affects firm performance in new technological generations. *Strategic Management Journal, 31*(3), 306–333.

Adner, R., & Levinthal, D. (2001). Demand heterogeneity and technology evolution: Implications for product and process innovation. *Management Science, 47*(5), 611–628.

Adner, R., Oxley, J. E., & Silverman, B. S. (Eds.). (2013). *Collaboration and competition in business ecosystems*. Emerald.

Adner, R., & Snow, D. (2010). Old technology responses to new technology threats: Demand heterogeneity and graceful technology retreats. *Industrial and Corporate Change, 19*(5), 1655–1675.

Adner, R., & Zemsky, P. (2006). A demand-based perspective on sustainable competitive advantage. *Strategic Management Journal, 27*(3), 215–239.

Aghion, P., Jones, B. F., & Jones, C. I. (2017). Artificial intelligence and economic growth. NBER Working Paper No. 23928.

Agrawal, A. K., Gans, J. S., & Goldfarb, A. (2018). Prediction, judgment and complexity: A theory of decision making and artificial intelligence. NBER Working Paper No. 24243.

Ahuja, G., Lampert, C. M., & Tandon, V. (2009). Moving beyond Schumpeter: Management research on the determinants of technological innovations. *The Academy of Management Annals, 2*(1), 1–98.

J. Y.-L. Forrest, Y. Liu, *Value in Business*, Contributions to Management Science,
https://doi.org/10.1007/978-3-030-82898-1

Alchian, A. A., & Woodward, S. (1987). Reflections on the theory of the firm. *Journal of Institutional and Theoretical Economics, 143*(1), 110–136.

Aldrich, H. E., & Zimmer, C. (1986). *Entrepreneurship through social networks*. University of Illinois at Urbana-Champaign: Academy for Entrepreneurial Leadership Historical Research Reference in Entrepreneurship. Retrieved from http://ssrn.com/abstract=1497761.

Almeida, F., Adão, D., & Martins, C. (2019). Decision support system for assigning members to agile teams. *International Journal of Information Technologies and Systems Approach, 12*(2), 43–60.

Alshanty, A. M., & Emeagwali, O. L. (2019). Market-sensing capability, knowledge creation and innovation. *Journal of Innovation & Knowledge, 4*(3), 171–178.

Altinay, L., Madanoglu, M., Vita, G. D., Arasli, H., & Ekinci, Y. (2016). The interface between organizational learning capability, entrepreneurial orientation, and SME growth. *Journal of Small Business Management, 54*(3), 871–891.

Amit, R., & Schoemaker, P. (1993). Strategic assets and organizational rent. *Strategic Management Journal, 14*(1), 33–46.

Amit, R., & Zott, C. (2001). Value creation in e-business. *Strategic Management Journal, 22*(6/7), 493–520.

Anderson, J., Kumar, N., & Narus, J. A. (2007). *Value merchants: Demonstrating and documenting superior value in business markets*. Harvard Business Press.

Anderson, J. C., & Narus, J. A. (1998). Business marketing: Understand what customers value. *Harvard Business Review, 76*(6), 53–65.

Anderson, J., Narus, J., & Van Rossum, W. (2006). Customer value propositions in business markets. *Harvard Business Review, 84*(3), 91–99.

Anderson, P. C., & Tushman, M. L. (1990). Technological discontinuities and dominant designs. *Administrative Science Quarterly, 35*(4), 604–633.

Andreoni, A., & Chang, H. J. (2019). The political economy of industrial policy. *Structural Change and Economic Dynamics, 48*(March), 136–150.

Ardyan, E. (2016). Market sensing capability and SMEs performance. *DLSU Business & Economics Review, 25*(2), 79–97.

Arens, Z. G., & Hamilton, R. W. (2018). The substitution strategy dilemma: Substitute selection versus substitute effectiveness. *Journal of the Academy of Marketing Science, 46*(1), 130–146.

Argote, L. (1999). *Organizational learning: Creating, retaining and transferring knowledge*. Kluwer.

Argyres, N. S., Bercovitz, J., & Mayer, K. J. (2007). Complementarity and evolution of contractual provisions: An empirical study of IT services contracts. *Organization Science, 18*(1), 3–19.

Argyres, N. S., & Liebeskind, J. P. (2002). Governance inseparability and the evolution of the biotechnology industry. *Journal of Economic Behavior and Organization, 47*, 197–219.

Arifin, Z., & Frmanzah. (2015). The effect of dynamic capability to technology adoption and its determinant factors for improving firm's performance: Toward a conceptual model. *Procedia - Social and Behavioral Sciences, 207*(October), 786–796.

Aristotle. (2016). *Nicomachean ethics* (Translated by W. D. Ross with an Introduction by R. W. Browne). Digireads.com.

Armstrong, M. (2006). Competition in two-sided markets. *RAND Journal of Economics, 37*(3), 668–691.

Arnold, T., Fang, E., & Palmatier, R. (2011). The effects of customer acquisition and retention orientations on a firm's radical and incremental innovation performance. *Journal of the Academy of Marketing Science, 39*(2), 234–251.

Arocena, R., & Sutz, J. (2018). Re-thinking innovation as a lever for development: Taking into account inequality. In B. Laspra & J. A. L. Cerezo (Eds.), *Spanish philosophy of technology* (pp. 125–138). Springer.

Aronfreed, J. (1968). *Conduct and conscience: The socialization of internalized control over behavior*. Academic Press.

Assink, M. (2006). Inhibitors of disruptive innovation capability: A conceptual model. *European Journal of Innovation Management, 9*(2), 215–233.

Attour, A., & Lazaric, N. (2018). From knowledge to business ecosystems: Emergence of an entrepreneurial activity during knowledge replication. *Small Business Economics*. https://doi.org/10.1007/s11187-018-0035-3

Atuahene-Gima, K. (2005). Resolving the capability-rigidity paradox in new product innovation. *Journal of Marketing, 69*(4), 61–83.

Auclert, A. (2019). Monetary policy and the redistribution channel. *American Economic Review, 109*(6), 2333–2367.

Augier, M., & Teece, D. J. (2008). Strategy as evolution with design: The foundations of dynamic capabilities and the role of managers in the economic system. *Organization Studies, 29*(8–9), 1187–1208.

Auh, S., & Menguc, B. (2009). Broadening the scope of the resource-based view in marketing: The contingency role of institutional factors. *Industrial Marketing Management, 38*(7), 757–768.

Azevedo, E. M., & Leshno, J. D. (2016). A supply and demand framework for two-sided matching markets. *Journal of Political Economy, 124*(5), 1235–1268.

Bai, J. H. (2011). Are government R&D subsidies efficient in China! Evidence from large and medium enterprises. *China Economic Quarterly, 10*(04), 1375–1400. https://doi.org/10.4236/ajibm.2014.49056

Baizakov, S., Forest, J. Y. L., & Baizakov, N. (2019). Modeling the managements of the economies of developing countries. *Advances in Systems Science and Applications, 19*(2), 101–119. https://doi.org/10.25728/assa.2019.19.2.673

Baker, T., & Nelson, R. E. (2005). Creating something from nothing: Resource construction through entrepreneurial bricolage. *Administrative Science Quarterly, 50*(3), 329–366.

Balakrishnan, A., Kumara, S. R. T., & Sundaresan, S. (1999). Manufacturing in the digital age: Exploiting information technologies for product realization. *Information Systems Frontiers, 1*, 25–50.

Ballantyne, D. (2003). A relationship-mediated theory of internal marketing. *European Journal of Marketing, 37*(9), 1242–1260.

Ballantyne, D., Frow, P., Varey, R. J., & Payne, A. (2011). Value propositions as communication practice. *Industrial Marketing Management, 40*(2), 202–210.

Banerjee, A. V. (1992). A simple model of herd behavior. *The Quarterly Journal of Economics, 107*(3), 797–817.

Baregheh, A., Rowley, J., & Sambrook, S. (2009). Towards a multidisciplinary definition of innovation. *Management Decision, 47*(8), 1323–1339.

Barney, J. (1991). Firm resources and sustained competitive advantage. *Journal of Management, 17*(1), 99–120.

Barney, J. (1997). *Gaining and sustaining competitive advantage*. Addison-Wesley.

Barney, J. (2001). Is the resource-based "view" a useful perspective for strategic management research? Yes! *Academy of Management Review, 26*(1), 41–56.

Barney, J., & Arikan, A. (2001). The resource-based view: Origins and implications. In M. Hitt, R. Freeman, & J. Harrison (Eds.), *Handbook of strategic management* (pp. 124–185). Blackwell.

Barney, J., & Clark, D. (2007). *Resource-based theory: Creating and sustaining competitive advantage*. Oxford University Press.

Barney, J., & Hansen, M. H. (1994). Trustworthiness as a form of competitive advantage. *Strategic Management Journal, 15*(Winter Special Issue), 175–190.

Barney, J., & Hesterly, W. (2012). *Strategic management and competitive advantage: Concepts and cases* (4th ed.). Pearson.

Barney, J., Ketchen, D., & Wright, M. (2011). The future of resource-based theory: Revitalization or decline? *Journal of Management, 37*(5), 1299–1315.

Baron, R. A., & Markman, G. D. (2003). Beyond social capital: The role of entrepreneurs' social competence in their financial success. *Journal of Business Venturing, 18*(1), 41–60.

Basu, K. (1993). Consumers' categorization processes: An examination with two alternative methodological paradigms. *Journal of Consumer Psychology, 2*(2), 97–121.

Baum, J. A. C., Calabrese, T., & Silverman, B. S. (2000). Don't go it alone: Alliance network composition and startups' performance in Canadian biotechnology. *Strategic Management Journal, 21*(3), 267–294.

Baum, J. A. C., & Dutton, J. (1996). The embeddedness of strategy. In P. Shrivastava, A. S. Huff, & J. E. Dutton (Eds.), *Advances in strategic management* (Vol. 13, pp. 3–40). JAI Press.

Baum, J. A. C., & Korn, H. J. (1999). Dynamics of dyadic competitive interaction. *Strategic Management Journal, 20*(3), 251–278.

Bayighomog Likoum, S. W., Shamout, M. D., Harazneh, I., et al. (2018). Market-sensing capability, innovativeness, brand management systems, market dynamism, competitive intensity, and performance. *Journal of the Knowledge Economy*. https://doi.org/10.1007/s13132-018-0561-x

Becheikh, N., Landry, R., & Amara, N. (2006). Lessons from innovation empirical studies in the manufacturing sector: A systematic review of the literature from 1993 to 2003. *Technovation, 26* (5), 644–664.

Belu, M., & Caragin, A. R. (2008). Strategies of entering new markets. *The Romanian Economic Journal, 11*(27), 83–98.

Bernard, A. B., & Jensen, J. B. (1995). Exporters, jobs, and wages in U.S. manufacturing: 1976 – 1987. *Brookings Papers on Economic Activity, Microeconomics, 1995*(1), 67–119.

Bernanke, B. S., & Gertler, M. (1999). Monetary policy and asset price volatility. *Economic Review, Federal Reserve Bank of Kansas City*, fourth quarter, pp. 17–51.

Bernheim, B. D., & Whinston, M. D. (1990). Multimarket contact and collusive behavior. *RAND Journal of Economics, 21*, 1–26.

Bernardo, R., Galina, S. V. R., & de Pádua, S. I. D. (2017). The BPM lifecycle: How to incorporate a view external to the organization through dynamic capability. *Business Process Management Journal, 23*(1), 155–175.

Berry, L. L., Carbone, L. P., & Haeckel, S. H. (2002). Managing the total customer experience. *Sloan Management Review, 43*(3), 85–89.

Betancourt, R., & Gautschi, D. (1990). Demand complementarities, household production, and retail assortments. *Marketing Science, 9*(2), 146–161.

Beugelsdijk, S., & Cornet, M. (2002). 'A far friend is worth more than a good neighbour': Proximity and innovation in a small country. *Journal of Management and Governance, 6*(2), 169–188.

Bharadwaj, N., & Dong, Y. X. (2014). Toward further understanding the market-sensing capability–value creation relationship. *Journal of Product Innovation Management, 31*(4), 799–813.

Bilgicer, T., Jedidi, K., Lehmann, D. R., & Neslin, S. A. (2015). Social contagion and customer adoption of new sales channels. *Journal of Retailing, 91*(2), 254–271.

Bilgihan, A. (2016). Gen Y customer loyalty in online shopping: An integrated model of trust, user experience and branding. *Computers in Human Behavior, 61*(August), 103–113.

Bjerknes, V. (1898). Uber einen hydrodynamischen Fundamentalsatz und seine Anwendung besonders auf die Mechanik der Atmosphare und des Weltmeeres. *Kungl. Svenska vetenskapsakademiens handlingar, 31*, 1–35.

Blankenburg, H. D., Eriksson, K., & Johanson, J. (1999). Creating value through mutual commitment to business network relationships. *Strategic Management Journal, 20*(5), 467–486.

Blauberg, I. V., Sadovsky, V. N., & Yudin, E. G. (1977). *Systems theory, philosophical and methodological problems*. Progress Publishers.

Blocker, C. P. (2011). Modeling customer value perceptions in cross-cultural business markets. *Journal of Business Research, 64*(5), 533–540.

Boeckx, J., & de Sola Perea, M., & Peersman, G. (2016). *The transmission mechanism of credit support policies in the Euro Area. National Bank of Belgium Working Paper, No. 302.*

Borrás, S., & Edquist, C. (2013). The choice of innovation policy instruments. *Technological Forecasting and Social Change, 80*(8), 1513–1522.

Boulding, W., & Christen, M. (2003). Sustainable pioneering advantage? Profit implications of market entry order. *Marketing Science, 22*(3), 371–392.

Boush, D. M., & Loken, B. (1991). A process-tracing study of brand extension evaluation. *Journal of Marketing Research, 28*(1), 16–28.

Bowman, C., & Ambrosini, V. (2001). "Value" in the resource-based view of the firm: A contribution to the debate. *Academy of Management Review, 26*, 501–502.

Boyce, W. E., & DiPrima, R. C. (2001). *Elementary differential equations and boundary value problems* (7th ed.). Wiley.

Božič, V., & Cvelbar, L. K. (2016). Resources and capabilities driving performance in the hotel industry. *Tourism and Hospitality Management, 22*(2), 225–246.

Branden, N. (1969). *The Psychology of Self-Esteem*. Nash Publishing Corp.

Brandenburger, A. M., & Stuart, H. (1996). Value-based business strategy. *Journal of Economics and Management Strategy, 5*, 5–25.

Branscomb, L. M. (2001). *Technological innovation. International Encyclopedia of the Social & Behavioral Sciences* (pp. 15498–15502). Elsevier.

Brem, A., & Viardot, É. (2015). Adoption of innovation: Balancing internal and external stakeholders in the marketing of innovation. In A. Brem & É. Viardot (Eds.), *Adoption of innovation* (pp. 1–10). Springer.

Breslin, S. (2009). State led development in historical perspective: From Friedrich List to a Chinese mode of governance? Presented at Beijing University, *The Beijing Forum*, Beijing, November 2009.

Briggs, J. R. (2018). Immersive wearables: Their political and social effects and what both mean for western liberal democracy. *International Journal of Sociotechnology and Knowledge Development, 10*(3), 54–71.

Brown, S. L., & Eisenhardt, K. M. (1995). Product development: Past research, present findings, and future directions. *Academy of Management Review, 20*(2), 343–378.

Brülhart, M., & Mathys, N. A. (2008). Sectoral agglomeration economies in a panel of European regions. *Regional Science and Urban Economics, 38*(4), 348–362.

Bruneel, J., Spithoven, A., & Clarysse, B. (2017). Interorganizational trust and technology complexity: Evidence for new technology-based firms. *Journal of Small Business Management, 55* (S1), 256–274.

Brusoni, S., & Prencipe, A. (2001). Unpacking the black box of modularity: Technologies, products and organizations. *Industrial and Corporate Change, 10*, 179–205.

Brusoni, J., Prencipe, A., & Pavitt, K. (2001). Knowledge specialization, organizational coupling, and the boundaries of the firm: Why do firms know more than they make? *Administrative Science Quarterly, 46*, 597–621.

Brynjolfsson, E., Hitt, L. M. & Kim, H. H. (2011). Strength in numbers: How does data-driven decision-making affect firm performance? SSRN Id.1819486.

Brynjolfsson, E., & Mcafee, A. (2014). *The second machine age*. Norton.

Bstieler, L., & Hemmert, M. (2015). The effectiveness of relational and contractual governance in new product development collaborations: Evidence from Korea. *Technovation, 45-46*, 29–39.

Buffington, J. (2016). The future of manufacturing: An end to mass production. In J. Buffington (Ed.), *Frictionless markets* (pp. 49–65). Springer.

Burt, R. S. (1992). *Structural holes: The social structure of competition*. Harvard University Press.

Burtch, G. (2011). Herding behavior as a network externality. *Proceedings of the International Conference on Information Systems*, ICIS 2011, Shanghai, China, December 4–7, 2011.

Burton, N., & Galvin, P. (2018). When do product architectures mirror organisational architectures? The combined role of product complexity and the rate of technological change. *Technology Analysis & Strategic Management, 30*(9), 1057–1069.

Buss, D. (*2004*). *Evolutionary psychology: The new science of the mind. Pearson Education, Inc.*.

Butler, R. W., Adam, R., & Szromek, A. R. (2019). Incorporating the value proposition for society with business models of health tourism enterprises. *Sustainability, 11*(23), 6711. https://doi.org/10.3390/su11236711

Byun, J. G., Sung, T. F., & Park, H. W. (2018). Technological innovation strategy: How do technology life cycles change by technological area. *Technology Analysis & Strategic Management, 30*(1), 98–112.
Cacciolatti, L., & Lee, S. H. (2016). Revisiting the relationship between marketing capabilities and firm performance: The moderating role of market orientation, marketing strategy and organisational power. *Journal of Business Research, 69*(12), 5597–5610.
Camerer, C. F. (2019). Artificial intelligence and behavioral economics. In A. K. Agrawal, J. Gans, & A. Goldfarb (Eds.), *The economics of artificial intelligence: An agenda*. University of Chicago Press.
Camps, J., Oltra, V., Aldás-Manzano, J., Buenaventura-Vera, G., & Torres-Carballo, F. (2016). Individual performance in turbulent environments: The role of organizational learning capability and employee flexibility. *Human Resource Management, 55*(3), 363–383.
Cao, J., & Zhou, Y. L. (2018). Research progress on the impact of artificial intelligence on economy. *Economic Dynamics, 1*, 103–115.
Cao, Y., & Gruca, T. S. (2005). Reducing adverse selection through customer relationship management. *Journal of Marketing, 69*(4), 219–229.
Cao, Z., & Lumineau, F. (2015). Revisiting the interplay between contractual and relational governance: A qualitative and meta-analytic investigation. *Journal of Operations Management, 33–34*, 15–42.
Caputo, A., Marzi, G., & Pellegrini, M. M. (2016). The internet of things in manufacturing innovation processes: Development and application of a conceptual framework. *Business Process Management Journal, 22*(2), 383–402.
Card, D., & Krueger, A. B. (1994). Minimum wages and employment: A case study of the fast-food industry in New Jersey and Pennsylvania. *American Economic Review, 84*(4), 772–793.
Carlson, C., Polizzotto, L., & Gaudette, G. R. (2019). The "NABC's" of value propositions. *IEEE Engineering Management Review, 47*(3), 15–20.
Carree, M., Piergiovanni, R., Santarelli, E., et al. (2015). Factors favoring innovation from a regional perspective: A comparison of patents and trademarks. *International Entrepreneurship and Management Journal, 11*(4), 793–810.
Carreyrou, J. (2018). *Bad blood: Secrets and lies in a silicon valley startup*. Alfred A. Knopf.
Cavell, M. (1998). Triangulation, one's own mind and objectivity. *The International Journal of Psycho-Analysis, 79*(pt3), 449–467.
Chadwick, C., Super, J. F., & Kwon, K. (2015). Resource orchestration in practice: CEO emphasis on SHRM, commitment-based HR systems, and firm performance. *Strategic Management Journal, 36(3)*, 360–376.
Chang, H. J. (2003). *Kicking away the ladder: Development strategies in historical perspective*. Anthem Press.
Chang, W., Ellinger, A. E., Kim, K., & Franke, G. R. (2016). Supply chain integration and firm financial performance: A meta-analysis of positional advantage mediation and moderating factors. *European Management Journal, 34*(3), 282–295.
Chemers, M. M. (2001). Cognitive, social, and emotional intelligence of transformational leadership: Efficacy and Effectiveness. In R. E. Riggio, S. E. Murphy, & F. J. Pirozzolo (Eds.), *Multiple intelligences and leadership* (pp. 139–160). Lawrence Erlbaum.
Chemmanur, T. J., Loutskina, E., & Tian, X. (2014). Corporate venture capital, value creation, and innovation. *The Review of Financial Studies, 27*(8), 2434–2473.
Chen, C. C., Peng, M. W., & Saparito, P. A. (2002). Individualism, collectivism, and opportunism: A cultural perspective on transaction cost economics. *Journal of Management, 28*(4), 567–583.
Chen, C. J. (2010). The effects of knowledge attribute, alliance characteristics, and absorptive capacity on knowledge transfer performance. *R and D Management, 34*(3), 311–321.
Chen, D. Q., Preston, D. S., & Swink, M. (2015). How the use of big data analytics affects value creation in supply chain management. *Journal of Management Information Systems, 32*(4), 4–39.

Chen, J., Yin, X. M., & Mei, L. (2018). Holistic innovation: An emerging innovation paradigm. *International Journal of Innovation Studies, 2*(1), 1–13.

Cheng, C. P., & Peng, H. (2018). The mechanism of artificial intelligence affecting employment and China's countermeasures. *China Soft Science Magazine, 2018*(10), 62–70.

Cheung, S. N. S. (1992). On the new institutional economics. In L. Werin & H. Wijkander (Eds.), *Contract economics* (pp. 48–65). Blackwell.

Chiou, J. S., Hsiao, C. C., & Chiu, T. Y. (2018). The credibility and attribution of online reviews: Differences between high and low product knowledge consumers. *Online Information Review, 42*(5), 630–646.

Chowdhury, T. G., Murshed, F., & Khare, A. (2018). Flexible flexibility! Food categorization flexibility and utilitarian preference. *Journal of Consumer Marketing, 35*(1), 1–10.

Christensen, C. M. (1997). *Innovator's dilemma: When new technologies cause great firms to fail.* Harvard Business School Press.

Christensen, C. M., Suárez, F. F., & Utterback, J. M. (1998). Strategies for survival in fast-changing industries. *Management Science., 44*(12-part-2), S207–S220.

Christiansen, J. A. (1999). *Competitive innovation management: Techniques to improve innovation performance.* Palgrave MacMillan.

Christoff, K., Irving, Z. C., Fox, K. C. R., Spreng, R. N., & Andrews-Hanna, J. R. (2016). Mind-wandering as spontaneous thought: A dynamic framework. *Nature Reviews Neuroscience, 17*, 718–731.

Ciccone, A., & Hall, R. E. (1996). Productivity and the density of economic activity. *American Economic Review, 86*, 54–70.

Clark, C. L., Shaver, P. R., & Abrahams, M. F. (1999). Strategic behaviors in romantic relationship initiation. *Personality and Social Psychology Bulletin, 25*(6), 709–722.

Clark, K. B. (1985). The interaction of design hierarchies and market concepts in technological evolution. *Research Policy, 14*, 235–251.

Clark, K. B., & Fujimoto, T. (1991). *Product development performance*. Harvard Business School Press.

Clarke, I., III. (2001). Emerging value propositions for M-commerce. ***Journal of Business Strategies, 18*(2)**, 133–148.

Clarysse, B., Wright, M., Bruneel, J., & Mahajan, A. (2014). Creating value in ecosystems: Crossing the chasm between knowledge and business ecosystems. *Research Policy, 43*(7), 1164–1176.

Clemons, E. K., & Row, M. C. (1992). Information technology and industrial cooperation: The changing economics of coordination and ownership. *Journal of Management Information Systems, 9*, 9–28.

Clements, C., Alwang, J., & Achdiawan, R. (2019). Value chain approaches in a stagnant industry: The case of furniture production in Jepara, Indonesia. *Bulletin of Indonesian Economic Studies, 55*(3), 341–365.

Cohen, J. B., & Basu, K. (1987). Alternative models of categorization: Toward a contingent processing framework. *Journal of Consumer Research, 13*(4), 455–472.

Coleman, D. C. (1983). Proto-industrialization: A concept too many. *The Economic History Review (New Series), 36*(3), 435–448.

Collis, D. J., & Montgomery, C. A. (2008). Competing on resources. *Harvard Business Review, 86* (7/8), 140–150.

Conner, K. R., & Prahalad, C. K. (1996). A resource-based theory of the firm: Knowledge versus opportunism. *Organization Science, 7*(5), 477–501.

Cooke, E. F. (1974). *A detailed analysis of the constitution.* Littlefield Adams & Co..

Covey, S. R. (1989). *The 7 habits of highly effective people: Powerful lessons in personal change.* Free Press.

Covin, J. G., Eggers, F., Kraus, S., Cheng, C. F., & Chang, M. L. (2016). Marketing-related resources and radical innovativeness in family and non-family firms: A configurational approach. *Journal of Business Research, 69*(12), 5620–5627.

Crook, T., Ketchen, D., Combs, J., & Todd, S. (2008). Strategic resources and performance: A meta-analysis. *Strategic Management Journal, 29*(11), 1141–1154.

Crossland, P., & Smith, F. I. (2002). Value creation in fine arts: A system dynamics model of converse demand and information cascades. *Strategic Management Journal, 23*(5), 417–434.

Cui, G., & Lui, H. (2005). Order of entry and performance of multinational corporations in an emerging market: A contingent resource perspective. *Journal of International Marketing, 13*(4), 28–56.

Cúrdia, V., & Woodford, M. (2016). Credit frictions and optimal monetary policy. *Journal of Monetary Economics, 84*(December), 30–65.

Cusumano, M., Mylonadis, Y., & Rosenbloom, R. (1992). Strategic maneuvering and mass market dynamics: The triumph of VHS over Beta. *The Business History Review, 66*(1), 51–94.

Damanpour, F. (1991). Organizational innovation: A meta-analysis of effects of determinants and moderators. *The Academy of Management Journal, 34*(3), 555–590.

Damanpour, F. (1996). Organizational complexity and innovation: Developing and testing multiple contingency models. *Management Science, 42*(5), 693–716.

Dance, J. (2008). What is innovation? 30+ definitions lead to one fresh summary. *Fresh Consulting* (blog), May 22, 2008, http://freshconsulting.com/what-is-innovation, Accessed on May 2, 2019.

Datta, A. (2016a). Evaluating the antecedents of foundational innovations: A longitudinal look at patents from information technology industry. *International Journal of Innovation Management, 20*(1). https://doi.org/10.1142/S1363919616500134

Datta, A. (2016b). Antecedents to radical innovations: A longitudinal look at firms in the information technology industry by aggregation of patents. *International Journal of Innovation Management, 20*(7). https://doi.org/10.1142/S1363919616500687

Davenport, T. H. (1993). *Process innovation: Reengineering work through information technology*. Harvard Business School Press.

Day, G. S. (1994). The capabilities of market-driven organizations. *Journal of Marketing, 58* (October), 37–52.

Day, G. S. (2011). Closing the marketing capabilities gap. *Journal of Marketing, 75*(4), 183–195.

De Bono, E. (2015). *The mechanism of mind: Understanding how your mind works to maximize memory and creative potential*. Random House.

De Oliveira, J. F. (2017). The importance of information technology as a strategic resource for hospital management process and the management skills required for managers. *International Journal of Business and Systems Research, 11*(1/2), 182–197.

Debackere, K., Clarysse, B., & Rappa, M. A. (1996). Dismantling the ivory tower: The influence of networks on innovative output in emerging technologies. *Technological Forecasting and Social Change, 53*, 139–154.

Debruyne, M., & Reibstein, D. J. (2005). Competitor see, competitor do: Incumbent entry in new market niches. *Marketing Science, 24*(1), 55–66.

Deiss, K. J. (2004). Innovation and strategy: Risk and choice in shaping user-centered libraries. *Library Trends, 53*(1), 17–32.

Demsetz, H. (2003). Ownership and the externality problem. In T. L. Anderson & F. S. McChesney (Eds.), *Property rights: Cooperation, conflict, and law*. Princeton University Press.

Dennett, D. C. (2015). *Elbow room: The varieties of free will worth wanting. A Bradford book*. MIT Press.

Derbaix, M., & Gombault, A. (2016). Selling the invisible to create an authentic experience: Imagination at work at Cézanne's studio. *Journal of Marketing Management, 32*(15–16), 1458–1477.

Dickson, P. R. (1992). Toward a theory of competitive rationality. *Journal of Marketing, 56*(1), 69–84.

Dirac, P. (1958). *The principles of quantum mechanics* (4th ed.). Oxford at the Clarendon Press.

Dobrev, S. D., & Gotsopoulos, A. (2010). Legitimacy vacuum, structural imprinting, and the first mover disadvantage. *Academy of Management Journal, 53*, 1153–1174.

Dosi, G. (1982). Technological paradigms and technological trajectories. *Research Policy, 11*, 147–162.
Dosi, G., Freeman, C., Nelson, R., Silverberg, G., & Soete, L. (Eds.). (1988). *Technical change and economic theory*. Pinter.
Dotzel, T., & Shankar, V. (2019). The relative effects of business-to-business (vs. business-to-consumer) service innovations on firm value and firm risk: An empirical analysis. *Journal of Marketing, 83*(5). https://doi.org/10.1177/0022242919847221
Dougherty, D. (1992). A practice-centered model of organizational renewal through product innovation. *Strategic Management Journal, 13*(Spring), 77–92.
Drucker, P. F. (1954). *The practice of management*. Harper & Row.
Du, C. J., Hu, J., & Chen, W. X. (2018a). Development model and countermeasures of China's new generation of artificial intelligence industry. *Economic Review Journal, 2018*(4), 41–47.
Du, J. Y., Meng, X. F., Li, J. Q., Wu, X. G., Song, Z. Y., & Ouyang, M. G. (2018b). Insights into the characteristics of technologies and industrialization for plug-in electric cars in China. *Energy, 164*(1), 910–924.
Duan, Y. Q., Edwards, J. S., & Dwivedi, Y. K. (2019). Artificial intelligence for decision making in the era of Big Data – evolution, challenges and research agenda. *International Journal of Information Management, 48*(October), 63–71.
Dutta, S., Narasimhan, O., & Rajiv, S. (1999). Success in high-technology markets: Is marketing capability critical? *Marketing Science, 18*(4), 547–568.
Dutta, S., & Segev, A. (1999). Business transformation on the Internet. *European Management Journal, 17*, 466–476.
Dyer, J. H. (1997). Effective interfirm collaboration: How firms minimize transaction costs and maximize transaction value. *Strategic Management Journal, 18*(7), 535–556.
Dyer, J. H., & Hatch, N. W. (2006). Relation-specific capabilities and barriers to knowledge transfers: Creating advantage through network relationships. *Strategic Management Journal, 27*(8), 701–719.
Dyer, J. H., & Nobeoka, K. (2000). Creating and managing a high-performance knowledge-sharing network: The Toyota case. *Strategic Management Journal, 21*(3), 345–367.
Dyer, J. H., & Singh, H. (1998). The relational view: Cooperative strategy and sources of inter-organizational competitive advantage. *Academy of Management Review, 23*(4), 660–679.
EC. (2014). *For a European Industrial Renaissance: Communication from the Commission to the European Parliament, the Council, the European Economic and Social Committee and the Committee of the Regions*. European Commission. Eurostat, & European Commission.
EC (2018). http://publications.jrc.ec.europa.eu/repository/bitstream/JRC113826/ai-flagship-report-online.pdf, Accessed on April 16, 2019.
Egan, K. (1992). *Imagination in teaching and learning*. University of Chicago Press.
Eggert, A., Ulaga, W., Frow, P., & Payne, A. (2018). Conceptualizing and communicating value in business markets: From value in exchange to value in use. *Industrial Marketing Management, 69*(February), 80–90.
Ehie, I. C., & Olibe, K. (2010). The effect of R&D Investment on firm value: An examination of US manufacturing and service Industries. *International Journal of Production Economics, 128*(1), 127–135.
Einstein, A. (1997). In R. Schulmann et al. (Eds.), *The collected papers of Albert Einstein: The Berlin years* (Vol. 8). Princeton University Press.
Eisenhardt, K., & Martin, J. (2000). Dynamic capabilities: What are they? *Strategic Management Journal, 21*(10/11), 1105–1121.
Endut, N., Morley, J., & Tien, P. L. (2018). The changing transmission mechanism of US monetary policy. *Empirical Economics, 54*, 959–987.
Ethiraj, S. K. (2007). Allocation of inventive effort in complex product systems. *Strategic Management Journal, 28*(6), 563–584.
European Commission (EC). (2014). *For a European Industrial Renaissance: Communication from the Comission to the European Parliament, the Council, the European Economic and*

Social Committee and the Committee of the Regions. European Commission. Eurostat, & European Commission.

Evans, D. S., & Schmalensee, R. (2007). *Catalyst code: The strategies behind the world's most dynamic companies*. Harvard Business School Press.

Evans, J. R. (2020). *Business analytics: Methods, models, and decisions* (3rd ed.). Pearson.

Evans, P. B., & Wurster, T. S. (1999). *Blown to bits: How the new economics of information transforms strategy*. Harvard Business School Press.

Evanschitzky, H. (2007). Market orientation of service networks: Direct and indirect effects on sustained competitive advantage. *Journal of Strategic Marketing, 15*(4), 349–368.

Fang, E., Palmatier, R., & Grewal, R. (2011). Effects of customer and innovation asset configuration strategies on firm performance. *Journal of Marketing Research, 48*(3), 587–602.

Farjoun, M. (1998). The independent and joint effects of the skill and physical bases of relatedness in diversification. *Strategic Management Journal, 19*(7), 611–630.

Felcher, E. M., Malaviya, P., & McGill, A. L. (2001). The role of taxonomic and goal-derived product categorization in, within, and across category judgments. *Psychology & Marketing, 18* (8), 865–887.

Feng, Y. C. (2019). Effects of environmental regulation and FDI on urban innovation in China: A spatial Durbin econometric analysis. *Journal of Cleaner Production, 235*. https://doi.org/10.1016/j.jclepro.2019.06.184

Ferber, U., & Schlappa, H. (2016). Managing brownfield land in stagnant land markets. In W. J. V. Neil & H. Schlappa (Eds.), *Future directions for the European Shrinking City* (pp. 138–154). Routledge.

Fernando, C. S., Sharfman, M. P., & Uysal, V. B. (2017). Corporate environmental policy and shareholder value: Following the smart money. *Journal of Financial and Quantitative Analysis, 52*(5), 2023–2051.

Filippetti, A., Frenz, M., & Ietto-Gillies, G. (2016). The impact of internationalization on innovation at countries' level: The role of absorptive capacity. *Cambridge Journal of Economics, 41* (2), 413–439.

Fiske, S. T. (2018). *Social cognition: Selected works of susan t. fiske*. Routledge.

Fiske, S. T., & Pavelchak, M. A. (1986). Category-based versus piecemeal-based affective responses: Developments in schema-triggered affect. In R. M. Sorrentino & E. T. Higgins (Eds.), *Handbook of motivation and cognition: Foundations of social behavior* (pp. 167–203). Guilford Press.

Fleischman, R. K., & Parker, L. D. (2017). *What is past is prologue: Costing accounting in the British Industrial Revolution, 1760-1850*. Routledge.

Forbes. (2018). *How can businesses adapt to a rapidly changing world?* https://www.forbes.com/sites/quora/2018/01/05/how-can-businesses-adapt-to-a-rapidly-changing-world/#4edbc26a5930, Accessed on September 17, 2019.

Forrest, J. Y. L. (2013). *A systemic perspective on cognition and mathematics*. CRC Press, an imprint of Taylor and Francis.

Forrest, J. Y. L. (2014). *A systems perspective on financial systems*. CRC Press, an imprint of Taylor and Francis.

Forrest, J. Y. L. (2018). *General systems theory: Foundation, intuition and applications in business decision making*. Springer Nature.

Forrest, J. Y. L., & Anderson, M. (2017). Customers' indecisiveness: A pricing strategy approach. *Pennsylvania Economic Review, 24*(2), 46–64.

Forrest, J. Y.-L., Ashimov, A., Ren, L., Gias, S., & Borovskiy, Y. (n.d.-a). (to appear). Conditions under which adopted consumer value propositions are effective. Submitted for publication.

Forrest, J. Y.-L., Baizakov, S., Haile, Y. G., Gyan, A., & Liu, Y. (2020a). Value potentials in innovation, resources, strategic networks and blocks. *Advances in Systems Science and Applications, 20*(3), 50–72.

Forrest, J. Y. L., Buttermore, J., & Wajda, T. A. (2017). At Nash equilibrium when new market competitions appear? *Kybernetes, The International of Cybernetics, Systems and Management Science, 46*(2), 256–271.

Forrest, J. Y.-L., Dang, Y. G., McCarthy, L., Liu, S. F., & Liu, Y. (2019e). Conditions under which synergistic innovations appear. *Proceedings of the 2019 Annual Conference of Decision Sciences Institute* (pp. 2061–2080).

Forrest, J. Y.-L., Dang, Y. G., McCarthy, L., Liu, S. F., & Liu, Y. (2019h). Simultaneous consumer utilities, multi-sided markets and consumer synergies. *Proceedings of the 2019 Annual Meeting of National Association of Business, Economics and Technology* (pp. 101–121).

Forrest, J. Y.-L., Duan, X. J., Zhao, C. L., & Xu, L. D. (2013a). *Systems science: Methodological approaches*. CRC Press, an imprint of Taylor and Francis.

Forrest, J. Y.-L., Gong, Z. W., Scott, E., & Soto Ruiz, G. E. (n.d.-b). (to appear). Market-sensing capabilities, profitability within stagnant industries and crafting of customer value propositions. Submitted for possible publication.

Forrest, J. Y.-L., Hopkins, Z., & Liu, S. F. (2013b). Currency wars and a possible self-defense (I): How currency wars take place. *Advances in Systems Science and Application, 13*, 198–217.

Forrest, J. Y.-L., Jallow, A. K., Wen, C., Shi, J. H., & Guo, H. (n.d.-c). (to appear). The role government policy and supports play to stimulate economic growth. Submitted for publication.

Forrest, J. Y.-L., Li, Z., Solano, R. A., & Doorn, D. (n.d.-d). (to appear). Technological and disruptive innovations: Their roles in a nation's industrialization and international leadership development. Submitted for publication.

Forrest, J. Y. L., Lin, C. C., Mondal, S., & Tucker, R. (2019a). Environmental forces underneath the innovativeness of manufacturing firms. *Theoretical Economics Letters, 9*, 1353–1382.

Forrest, J. Y.-L., Liu, J., Guo, H., Liu, Y., Phipps, T. M., Yang, Y. J., & Cao, Y. R. (n.d.-e). (to appear). Market competition and economic effects of government policies. Submitted for publication.

Forrest, J. Y.-L., Liu, Y., Golden, J., Sissoko, Y., & Qiaoxing Li, Q. X. (2020b). A reformulated value chain framework useful for different kinds of markets. *The Journal of Business and Economic Studies, 24*(1), 33–53.

Forrest, J. Y.-L., Liu, J., Martorell, G., Xu, L., Liu, Y., & Darvishi, D. (2021a). Role of manufacturing sector in the forthcoming national industrial transformations. Accepted for publication by *International Journal of Sociotechnology and Knowledge Development*, to appear.

Forrest, J. Y.-L., Liu, Y., Swift, E., & Karimi, T. (2019f). Exploitation, exploration, development and decay of resources: A systemic view. *Proceedings of the 2019 annual conference of Pennsylvania Economic Association* (pp. 139–165).

Forrest, J. Y.-L., Liu, Y., Valencia, J., Sissoko, Y., & Li, Q. X. (2019i). A generalized value chain framework simultaneously useful for companies that offer either products, or services, or informational goods. *Proceedings of the 2019 annual conference of Pennsylvania Economic Association* (pp. 166–194).

Forrest, J. Y.-L., Liu, Y., Wajda, T. A., Köse, E., & Arik, O. A. (2019g). How challenges and successes of a firm depend on its input-output wcosystem. *Proceedings of the 2019 Annual Conference of Decision Sciences Institute* (pp. 2381–2402). New Orleans, LA, November 23–25, 2019.

Forrest, J. Y.-L., Liu, Y., Wajda, T. A., Köse, E., & Arik, O. A. (2019j). Dependence of a firm's performance on its ecosystem's upstream/downstream challenges. *Proceedings of the 2019 Annual Meeting of National Association of Business, Economics and Technology* (pp. 83–100).

Forrest, J. Y. L., Mondal, S., Tucker, R., & Lin, C. C. (2018a). Effects of manufacturing firms' strategies on innovation: A holistic view. *Proceedings of the 2018 Annual Conference of NABET* (pp. 74–94). November 1–2, 2018, State College, PA.

Forrest, J. Y.-L., Nicholls, J., Schimmel, K., & Liu, S. F. (2019b). *Managerial decision making: A holistic approach*. Springer.

Forrest, J. Y.-L., Nicholls, J., Schimmel, K., & Liu, S. F. (2020d). *Managerial decision making: A holistic approach*. Springer.

Forrest J. Y. L, & Nightingale, J. (2017). Successfully transition into the era of transient competitive advantages. *Proceedings of 2017 NABET Annual Conference*, October 26–27, 2017 (pp. 57–79). State College, PA.

Forrest, J. Y.-L., Novikov, D. A., Larson, S., Wang, F., & Yang, J. (2021b). Competitive advantages and values created and attained out of well-crafted customer value propositions, *Studies in Business and Economics Journal, 16*(2).

Forrest, J. Y.-L., & Orvis, B. (2016). Principles of management efficiency and organizational inefficiency. *Kybernetes: The International Journal of Cybernetics, Systems and Management Sciences, 45*(8), 1308–1322.

Forrest, J. Y.-L., Shao, L., McKinney, R., Shao, D. (2020c). Firm performance and market expansion from the point of view of resources. *International Journal of Business and Systems Research*, accepted for publication.

Forrest, J. Y.-L., & Tallapally, P. (2018). Customers are less patient, sustainable advantage becomes transient and the key for firms to succeed in fast changing markets. *Journal of Business, Economics and Technology, 21*(1), 1–14.

Forrest, J. Y. L., & Tao, L. X. (2014). *Investment and employment opportunities in China*. CRC Press, an imprint of Taylor and Francis.

Forrest, J. Y.-L., Trebing, M. E., Chatterjee, A., & Wagner, J. (2019c). Domestic firms, exporters and importers: A systemic perspective. *Theoretical Economics Letters, 9*, 649–674.

Forrest, J. Y. L., Tucker, R., Lin, C. C., & Mondal, S. (2018b). Impacts of manufacturing firms' culture, structure and leadership on innovation: A systemic approach. *Proceedings of the 41st Annual Meeting of Northeastern Association of Business, Economics and Technology*, State College, PA.

Forrest, J. Y. L., Tucker, R., Lin, C. C., & Mondal, S. (2019d). The innovativeness of manufacturing firms from the angles of organizational culture, structure and leadership. *Journal of Business, Economics and Technology, 22*(1).

Forrest, J. Y.-L., Wagner, J., Anderson, M., Lipinski, J., Liu, Y., & Tian, X. G. (n.d.-f). (to appear). A profit-based view of economic granularity in the perspective of market competition. Submitted for publication.

Forrest, J. Y. L., Zhao, H. C., & Shao, L. (2018c). Engineering rapid industrial revolutions for impoverished agrarian nations. *Theoretical Economics Letters, 8*, 2594–2640. https://doi.org/10.4236/tel.2018.811166

Forrester Research Report (2000). *Global eCommerce Approaches Hypergrowth*. April 2000.

Forsgren, M. (2016). A note on the revisited Uppsala internationalization process model – the implications of business networks and entrepreneurship. *Journal of International Business Studies, 47*(9), 1135–1144.

Fosfuri, A., Lanzolla, G., & Suarez, F. F. (2013). Entry-timing strategies: The road ahead. *Long Range Planning, 46*, 300–311.

Foss, K., Foss, N. J., Klein, P. G., & Klein, S. K. (2007). The entrepreneurial organization of heterogeneous capital. *Journal of Management Studies, 44*(7), 1165–1186.

Frank, A. G. (1978). Multilateral merchandise trade imbalances and uneven economic development. In *Dependent accumulation and underdevelopment*. Palgrave Macmillan.

Fratzscher, M., Lo Duca, M., & Straub, R. (2016). ECB unconventional monetary policy: Market impact and international spillovers. *IMF Economic Review, 64*, 36–74.

Freel, M. S. (2003). Sectoral patterns of small firm innovation, networking and proximity. *Research Policy, 32*, 751–770.

Frow, P., & Payne, A. (2011). A stakeholder perspective of the value proposition concept. *European Journal of Marketing, 45*(1-2), 223–240.

Fuentes-Fuentes, M. D. M., Bojica, A. M., Ruiz-Arroyo, M., & Welter, F. (2017). Innovativeness and business relationships in women-owned firms: The role of gender stereotypes. *Canadian Journal of Administrative Sciences, 34*(1), 63–76.

Fuentelsaz, L., & Gómez, J. (2006). Multipoint competition, strategic similarity and entry into geographic markets. *Strategic Management Journal, 27*(5), 477–499.

Fujitsu Limited & Riken. (2018). Fujitsu and RIKEN Demonstrate AI's Utility in Material Design. https://www.fujitsu.com/global/about/resources/news/press-releases/2018/0316-01.html, Accessed on April 29, 2019.

Fultz, D., Long, R.R., Owens, G.V., Bohan, W., Kaylor, R., & Weil, J. (1959). *Studies of thermal convection in a rotating cylinder with some implications for large-scale atmospheric motion.* : Meteorol. Monographs (American Meteorological Society) *21*(4).

Gabaix, X. (2011). The granular origins of aggregate fluctuations. *Econometrica, 79*, 733–772.

Galaskiewicz, J., & Zaheer, A. (1999). Networks of competitive advantage. In S. Andrews & D. Knoke (Eds.), *Research in the sociology of organizations* (pp. 237–261). JAI Press.

Gans, J. S., MacDonald, G., & Ryall, M. D. (2008). *The two sides of competition and their implications for strategy*. Working paper, Melbourne Business School. https://works.bepress.com/michael_ryall/15/

Gardner, R. W. (1953). Cognitive styles in categorizing behavior. *Journal of Personality, 22*(2), 214–233.

Gary, M. S. (2005). Implementation strategy and performance outcomes in related diversification. *Strategic Management Journal, 26*(7), 634–664.

General Electric. (2012). *GE Global Innovation Barometer: Global Research Report*, 30, http://files.gecompany.com/gecom/innovationbarometer/GE_Global_Innovation_Barometer_Report_January_2012.pdf.

Gerassi, J. (1989). *Jean-Paul Sartre: Hated conscience of his century*. In *Volume 1: Protestant or protester?* University of Chicago Press.

Gerschenkron, A. (1962). *Economic backwardness in historical perspective*. Belknap Press of Harvard University Press.

Gliva-McConvey, G., Nicholas, C. F., & Clark, L. (Eds.). (2020). *Comprehensive healthcare simulation: Implementing best practices in standardized patient methodology*. Springer.

Golder, P. N., & Tellis, G. J. (1993). Pioneer advantage: Marketing logic or marketing legend? *Journal of Marketing Research, 30*, 158–170.

Goldfarb, A., & Trefler, D. (2018). AI and international trade. NBER Working Paper No. 24254.

Goldfarb, B. (2005). Diffusion of general-purpose technologies: Understanding patterns in the electrification of US Manufacturing 1880-1930. *Industrial and Corporate Change, 14*(5), 745–773.

Goldstein, E. B. (*2010*). *Encyclopedia of perception. SAGE Publications.*

Government of Japan. (2015). https://www8.cao.go.jp/cstp/kihonkeikaku/5basicplan_en.pdf, Accessed on April 16, 2019.

Graetz, G., & Michaels. (2015). Robots at work: The impact on productivity and jobs. *Technical Report*, CEP Discussion Paper No.1335.

Granovetter, M. (1985). Economic action and social structure: A theory of embeddedness. *American Journal of Sociology, 91*, 481–510.

Grant, J. I. (2003). *Foundations of economic value added*. Wiley.

Grant, R. M. (1996). Prospering in dynamically-competitive environments: Organizational capability and knowledge integration. *Organization Science, 7*(July), 375–387.

Grant, R. M. (2008). *Contemporary strategy analysis: Concepts, techniques, applications* (6th ed.). Blackwell Publishing.

Greve, H. R. (2000). Market niche entry decisions: Competition, learning, and strategy in Tokyo banking, 1894–1936. *Academy of Management Journal, 43*(5), 816–836.

Guan, J. C., & Gao, X. (2009). *Exploring the h-index at patent level*. John Wiley & Sons.

Gulati, R. (1998). Alliances and networks. *Strategic Management Journal, 19*(4), 293–317.

Gulati, R. (1999). Network location and learning: The influence of network resources and firm capabilities on alliance formation. *Strategic Management Journal, 20*(5), 397–420.

Gulati, R., & Gargiulo, M. (1999). Where do interorganizational networks come from? *American Journal of Sociology, 104*(5), 1439–1493.

Gulati, R., Nohria, N., & Zaheer, A. (2000). Strategic networks. *Strategic Management Journal, 21* (3), 203–215.

Gurhan-Canli, Z., & Maheswaran, D. (1998). The effects of extensions on brand name dilution and enhancement. *Journal of Marketing Research, 35*(4), 464–473.

Gustafsson, R., Jaaskelainen, M., Maula, M., & Uotila, J. (2016). Emergence of industries: A review and future directions. *International Journal of Management Reviews, 18*(1), 28–50. https://doi.org/10.1111/ijmr.12057

Hagel, J., III, & Armstrong, A. G. (1997). *Net gain: Expanding markets through virtual communities*. Harvard Business School Press.

Hall, A. D., & Fagen, R. E. (1956). Definitions of systems. *General Systems, 1*, 18–28.

Hall, L. A., & Bagchi-Sen, S. (2002). A study of R&D, innovation, and business performance in the Canadian biotechnology industry. *Technovation, 22*, 231–244.

Hall, W., & Pesenti, J. (2017). Growing the artificial intelligence industry in the UK. https://assets.publishing.service.gov.uk/government/uploads/system/uploads/attachment_data/file/652097/Growing_the_artificial_intelligence_industry_in_the_UK.pdf, Accessed on October 15, 2017.

Hallberg, N. L. (2015). Uncertainty, judgment, and the theory of the firm. *Journal of Institutional Economics, 11*(3), 623–650.

Han, X. F., Hui, N., & Song, W. F. (2014). Can informantization improve the technology innovation efficiency of Chinese industrial sectors. *China Industrial Economics, 2014*(12), 70–82.

Haraguchi, N., Martorano, B., & Sanfilippo, M. (2019). What factors drive successful industrialization? Evidence and implications for developing countries. *Structural Change and Economic Dynamics, 49*(June), 266–276.

Harmancioglu, N., Droge, C., & Calantone, R. (2009). Strategic fit to resources versus NPD execution proficiencies: What are their roles in determining success? *Journal of the Academy of Marketing Science, 37*(3), 266–282.

Harrigan, K. R. (1985). *Strategies for joint ventures*. Lexington Books.

Harris, P. (2000). *The work of the imagination*. Wiley-Blackwell.

Hatch, N. W., & Mowery, D. C. (1998). Process innovation and learning by doing in semiconductor manufacturing. *Management Science, 44*(11), 1461–1477.

Hauser, J. R., & Shugan, S. M. (1983). Defensive marketing strategies. *Marketing Science, 2*(4), 319–360.

Haveman, H. A., & Nonnemaker, L. (2000). Competition in multiple geographic markets: The impact on growth and market entry. *Administrative Science Quarterly, 45*(2), 232–267.

Hayes, R. H., Wheelwright, S. C., & Clark, K. B. (1988). *Dynamic manufacturing*. The Free Press.

Heaton, H. (1933). Industrial revolution. In: *Encyclopeadia of the Scoial Sciences*, Vol. VIII (1033); and in: Hartwell, R. M. (ed.) (2017), *The Causes of the Industrial Revolution in England* (pp. 30–52). : Routledge.

Helfat, C. E., Campo-Rembado, M. A., Helfat, C. E., & Campo-Rembado, M. A. (2016). Integrative capabilities, vertical integration, and innovation over successive technology lifecycles. *Organization Science, 27*(2), 233–504.

Helfat, C. E., & Lieberman, M. B. (2002). The birth of capabili ties: Market entry and the importance of pre-history. *Industrial and Corporate Change, 11*(4), 725–760.

Helfat, C. E., & Peteraf, M. A. (2003). The dynamic resource-based view: Capability lifecycles. *Strategic Management Journal, 24*(10), 997–1010.

Hellström, M., Tsvetkova, A., Gustafsson, M., & Wikström, K. (2015). Collaboration mechanisms for business models in distributed energy ecosystems. *Journal of Cleaner Production, 102*(1), 226–236.

Helm, R., & Mark, A. (2007). Implications from cue utilisation theory and signaling theory for firm reputation and the marketing of new products. *International Journal of Product Development, 4* (3/4), 396–411.

Helsley, R. W. (2004). Economics of agglomeration: Cities, industrial location and regional growth. *Journal of Economic Geography, 4*(3), 345–349.

Hemphill, T. A. (2014). The US advanced manufacturing initiative: Will it be implemented as an innovation – or industrial – policy? *Innovation: Organization & Management, 16*(1), 67–70.

Henderson, R. M. (1995). Of life cycles real and imaginary: The unexpectedly long old age of optical lithography. *Research Policy, 24*(4), 631–643.

Henderson, R. M., & Clark, K. B. (1990). Architectural innovation: The reconfiguration of existing product technologies and the failure of established firms. *Administrative Science Quarterly, 35* (1), 9–30.

Hendrikse, G., Hippmann, P., & Windsperger, J. (2015). Trust, transaction costs and contractual incompleteness in franchising. *Small Business Economics, 44*(4), 867–888.

Hermann, M., Pentek, T., & Otto, B. (2016). Design principles for Industry 4.0 scenarios. *The 49th Hawaii International Conference on System Sciences (HICSS)*, Koloa, HI (pp. 3928–3937).

Hernandez, A. A. (2017). Green IT adoption practices in education sector: A developing country perspective. *International Journal of Sociotechnology and Knowledge Development, 9*(3), 1–16.

Hide, R. (1953). Some experiments on thermal convection in a rotating liquid. *Quarterly Journal of the Royal Meteorological Society, 79*, 161.

Higgins, R. C. (1981). Sustainable growth under inflation. *Financial Management, 10*(4), 36–40.

Hill, N. (1928). *The law of success: In sixteen lessons (reprint of 2007)*. BN Publishing. www.bnpublishing.com

History. Copenhagen Business School Press DK. ISBN 978-87-630-0194-6

Hitt, M., Bierman, L., Shimizu, K., & Kochhar, R. (2001). Direct and moderating effects of human capital on strategy and performance in professional service firms: A resource-based perspective. *The Academy of Management Journal, 44*(1), 13–28.

Hitt, M., & Ireland, R. D. (2017). The intersection of entrepreneurship and strategic management research. In D. L. Sexton, H. Landstrom, & C. Jones (Eds.), *The Blackwell handbook of entrepreneurship* (pp. 45–63). Blackwell.

Hoetker, G. (2006). Do modular products lead to modular organizations? *Strategic Management Journal, 27*(6), 501–518.

Holcomb, T. R., Holmes, R. M., Jr., & Connelly, B. L. (2009). Making the most of what you have: Managerial ability as a source of resource value creation. *Strategic Management Journal, 30*(5), 457–485.

Holmstrom, J., Holweg, M., Khajavi, S. H., & Partanen, J. (2016). The direct digital manufacturing (r) evolution: Definition of a research agenda. *Operations Management Research, 9*, 1–10.

Holttinen, H. (2014). Contextualizing value propositions: Examining how consumers experience value propositions in their practices. *Australian Marketing Journal, 22*(2), 103–110.

Hooley, G., Greenley, G., Fahy, J., & Cadogan, J. (2005). Market-focused resources, competitive positioning and firm performance. *Journal of Marketing Management, 17*(5/6), 503–520.

Hopkins, E. (2000). *Industrialisation and Society A Social History, 1830-1951*. Routledge.

Hoskisson, R. E., Gambeta, E., Green, C. D., & Li, T. X. (2018). Is my firm-specific investment protected? overcoming the stakeholder investment dilemma in the resource-based view. *Academy of Management Review, 43*(2). https://doi.org/10.5465/amr.2015.0411

Howell, R., Beers, C. V., & Doorn, N. (2018). Value capture and value creation: The role of information technology in business models for frugal innovations in Africa. *Technological Forecasting and Social Change, 131*, 227–239.

Howell, T. R., Noellert, W. A., MacLaughlin, J. H., & Wolff, A. W. (2019). *The microelectronics race: The impact of government policy on international competition*. Routledge.

Howlett, M. (2018). Matching policy tools and their targets: Beyond nudges and utility maximisation in policy design. *Policy & Politics, 46*(1), 101–124.

Hoyt, R. E., & Liebenberg, A. P. (2011). The value of enterprise risk management. *The Journal of Risk and Insurance, 78*(4), 795–822.

Hsieh, K. Y., & Vermeulen, F. (2014). The structure of competition: How competition between one's rivals influences imitative market entry. *Organization Science, 25*(1), 299–319.

Huang, M. C., Yen, G. F., & Liu, T. C. (2014). Reexamining supply chain integration and the supplier;s performance relationships under uncertainty. *Supply Chain Management: An International Journal, 19*(1), 64–78.

Huang, R. (2015). *China: A macro history*. Routledge.
Hughes, M., Martin, S., Morgan, R., & Robson, M. (2010). Realizing product-market advantage in high-technology international new ventures: The mediating role of ambidextrous innovation. *Journal of International Marketing, 18*(4), 1–21.
Hughes, T. P. (1983). *Networks of power: Electrification in Western Society 1880-1930*. Johns Hopkins University Press.
Hult, G. T. (1998). Managing the international strategic sourcing process as a market-driven organizational learning system. *Decision Sciences, 29*(winter), 193–216.
Hult, G. T., & Ketchen, D. J. (2001). Does market orientation matter? A test of the relationship between positional advantage and performance. *Strategic Management Journal, 22*(9), 899–906.
Hult, G. T., Ketchen, D. J., & Slater, S. F. (2005). Market orientation and performance: An integration of disparate approaches. *Strategic Management Journal, 26*(12), 1173–1181.
Hunt, S. (1997). Resource-advantage theory: An evolutionary theory of competitive firm behavior? *Journal of Economic Issues, 31*(1), 59–77.
Hunt, S. (2007). Synthesising resource-based, evolutionary and neoclassical thought. In N. Foss & P. Robertson (Eds.), *Resources, technology, and strategy* (pp. 53–79). Routledge.
Hunt, S., & Morgan, R. (1995). The comparative advantage theory of competition. *Journal of Marketing, 59*(4), 1–15.
Huntington, S. P. (1996). *The clash of civilizations and the remaking of world order*. Simon & Schuster.
Hurley, R. F., & Hult, G. T. M. (1998). Innovation, market orientation, and organizational learning: An integration and empirical examination. *Journal of Marketing, 62*(July), 42–54.
Huselid, M. A., Jackson, S. E., & Schuler, R. S. (1997). Technical and strategic human resources management effectiveness as determinants of firm performance. *Academy of Management Journal, 40*(1), 171–188.
Iansiti, M. (1998). *Technology integration: Making critical choices in a dynamic world*. Harvard Business School Press.
Iansiti, M., & Levien, R. (2004). *The keystone advantage: What the new dynamics of business ecosystems mean for strategy, innovation, and sustainability*. Harvard Business School Press.
II - Industrialization and Informationalization. (2017). Three-Year Action Plan to Promote the Development of a New Generation of AI Industry 2018-2022. http://www.miit.gov.cn/n1146285/n1146352/n3054355/n3057497/n3057507/c5979554/content.html, Accessed on April 22, 2019.
Intel Corporation. (2004). Intel sees unified platform and ecosystem as key to enabling the digital home. Press release 17 February.
Irwin, D. A., & Klenow, P. J. (1994). Learning-by-doing spillovers in the semiconductor industry. *Journal of Political Economy, 102*(6), 1200–1227.
Jaehrling, K., Johnson, M., Larsen, T. P., Refslund, B., & Grimshaw, D. (2018). Tackling precarious work in public supply chains. *Work, Employment and Society, 32*(3), 546–563.
Jap, S. D., & Anderson, E. (2003). Safeguarding interorganizational performance and continuity under expost opportunism. *Management Science, 49*(12), 1684–1701.
Jap, S. D. (2001). Perspectives on joint competitive advantages in buyer-supplier relationships. *International Journal of Research in Marketing, 18*(1), 19–35.
Jaworski, B. J., & Kohli, A. K. (1993). Market orientation: Antecedents and consequences. *Journal of Marketing, 57*(July), 53–70.
Jayachandran, S., Gimeno, J., & Varadarajan, P. R. (1999). The theory of multimarket competition: A synthesis and implications for marketing strategy. *Journal of Marketing, 63*(3), 49–66.
Jeng, D. J., & Pak, A. (2016). The variable effects of dynamic capability by firm size: The interaction of innovation and marketing capabilities in competitive industries. *International Entrepreneurship and Management Journal, 12*(1), 115–130.

Jiang, Y., & Tornikoski, E. T. (2018). Perceived uncertainty and behavioral logic: Temporality and unanticipated consequences in the new venture creation process. *Journal of Business Venturing*. https://doi.org/10.1016/j.jbusvent.2018.06.002
Jiang, Z. J., & Wu, Z. Q. (1979). *Theory of real-valued functions*. Press of People's Education.
Johnson, E., & Russo, J. (1984). Product familiarity and learning new information. *Journal of Consumer Research, 11*(June), 542–550.
Jomo, K. S. (2019). *Southeast Asia's misunderstood miracle: Industrial policy and economic development in Thailand, Malaysia and Indonesia*. Routledge.
Jullien, B. (2005). Two-sided markets and electronic intermediaries. *CESifo Economic Studies, 51* (2–3), 233–260.
Kabue, L. W., & Kilika, J. M. (2016). Firm resources, core competencies and sustainable competitive advantage: An integrative theoretical framework. *Journal of Management and Strategy, 7* (1), 98–108.
Kagermann, H. (2015). Change through digitization: Value creation in the age of industry 4.0. In H. Albach, H. Meffert, A. Pinkwart, & R. Reichwald (Eds.), *Management of permanent change* (pp. 23–48). Springer Gabler.
Kaleka, A. (2011). When exporting manufacturers compete on the basis of service: Resources and marketing capabilities driving service advantage and performance. *Journal of International Marketing, 19*(1), 40–58.
Kalmuk, G., & Acar, A. Z. (2015). The mediating role of organizational learning capability on the relationship between innovation and firm's performance: A conceptual framework. *Procedia - Social and Behavioral Sciences, 210*(2), 164–169.
Kane, R. (2005). *A contemporary introduction to free will*. Oxford University Press.
Kanter, R. M. (2001). *Evolve! Succeeding in the digital culture of tomorrow*. Harvard Business School Press.
Kapoor, R., & Adner, R. (2012). What firms make vs. what they know: How firms' production and knowledge boundaries affect competitive advantage in the face of technological change. *Organization Science, 23*(5), 1213–1522.
Kapoor, R., & Agarwal, S. (2017). Sustaining superior performance in business ecosystems: Evidence from application software developers in the iOS and Android smartphone ecosystems. *Organization Science, 28*(3), 379–595.
Karlsen, A. (2018). Framing industrialization of the offshore wind value chain – A discourse approach to an event. *Geoforum, 88*, 148–156.
Kaser, D. (2011). Editor's notes: Innovation can be fun. *Computers in Libraries, 31*(5), 4.
Kaufmann, A., & Todtling, F. (2001). Science-industry interaction in the process of innovation: The importance of boundary-crossing between systems. *Research Policy, 30*, 791–804.
Kay, D. C. (1994). *College geometry: A discovery approach*. Harper Collins College Publishers.
Keats, B. W., & Hitt, M. A. (1988). A causal model of linkages among environmental dimensions, macro organizational characteristics, and performance. *Academy of Management Journal, 31* (3), 570–598.
Keizer, J. A., Dijkstra, L., & Halman, J. I. M. (2002). Explaining innovative efforts of SMEs. An exploratory survey among SMEs in the mechanical and electrical engineering sector in the Netherlands. *Technovation, 22*, 1–13.
Kellerman, A. (1985). The evolution of service economies: A geographical perspective. *The Professional Geographer, 37*(2), 133–143.
Kenny, A. (2012). *A new history of western philosophy*. Oxford University Press.
Khantimirov, D. (2017). Market share as a measure of performance: Conceptual issues and financial accountability for marketing activities within a firm. *Journal of Research in Marketing, 7*(3), 587–592.
Kim, J. Y., & Finkelstein, S. (2009). The effects of strategic and market complementarity on acquisition performance: Evidence from the U.S. commercial banking industry, 1989–2001. *Strategic Management Journal, 30*(6), 617–646.

Kim, P., Vaidyanathan, R., Chang, H., & Stoel, L. (2018). Using brand alliances with artists to expand retail brand personality. *Journal of Business Research, 85*(April), 424–433.
Kim, Y. J., & Brown, M. (2019). Impact of domestic energy-efficiency policies on foreign innovation: The case of lighting technologies. *Energy Policy, 128*(May), 539–552.
Kirchner, M., & van Wijnbergen, S. (2016). Fiscal deficits, financial fragility, and the effectiveness of government policies. *Journal of Monetary Economics, 80*(June), 51–68.
Kirzner, I. M. (1973). *Competition and entrepreneurship*. University of Chicago Press.
Kirzner, I. M. (2018). The ethics of pure entrepreneurship: An Austrian economics perspective. *The Review of Austrian Economics*. https://doi.org/10.1007/s11138-017-0412-1
Klein, B. (2000). Fisher-General Motors and the nature of the firm. *Journal of Law and Economics, 43*(1), 105–142.
Klein, K., Semrau, T., Albers, S., & Zajac, E. J. (2020). Multimarket coopetition. *Long Range Planning, 53*(1). https://doi.org/10.1016/j.lrp.2019.02.001
Kline, M. (1972). *Mathematical thought from ancient to modern times*. Oxford University Press.
Kline, S. J., & Rosenberg, N. (1986). An overview of innovation. In R. Landau & N. Rosenberg (Eds.), *The positive sum strategy: Harnessing technology for economic growth* (pp. 275–307). National Academy Press.
Klir, G. (1972). *Trends in general systems theory*. Wiley.
Klir, G. (1985). *Architecture of systems problem solving*. Plenum Press.
Klir, G. (2001). *Facets of systems science*. Springer.
Kogut, B. (1988). Joint ventures: Theoretical and empirical perspectives. *Strategic Management Journal, 9*(4), 319–332.
Komkov, N. I., & Kulakin, G. K. (2018). Technological innovations: Development, application and results. *Studies on Russian Economic Development, 29*, 558–572.
Kopalle, P. K., & Lehmann, D. R. (2006). Setting quality expectations when entering a market: What should the promise be? *Marketing Science, 25*(1), 8–24.
Kor, Y. Y., Mahoney, J. T., & Michael, S. C. (2007). Resources, capabilities and entrepreneurial perceptions. *Journal of Management Studies, 44*(7), 1187–1212.
Kotler, P. (1977). From sales obsession to marketing effectiveness. *Harvard Business Review, 55* (6), 67–75.
Kotler, P., & Keller, K. L. (2016). *A framework for marketing management* (6th ed.). Pearson.
Kotter, J. (2012). Barriers to change: The real reason behind the Kodak downfall. *Forbes*, May 2, 2012, www.forbes.com/sites/johnkotter/2012/05/02/barriers-to-change-the-real-reason-behind-the-kodak-downfall.
Kouzes, J., & Posner, B. (2007). *The leadership challenge*. Jossey Bass.
Kowalkowski, C. (2011). Dynamics of value propositions: Insights from service-dominant logic. *European Journal of Marketing, 45*(1/2), 277–294.
Kozlenkova, I. V., Samaha, S. A., & Palmatier, R. W. (2014). Resource-based theory in marketing. *Journal of the Academy of Marketing Science, 42*, 1–21.
Kuhn, T. (1962). *The structure of scientific revolutions*. University of Chicago Press.
Kumar, K. R., & Sudharshan, D. (1988). Defensive marketing strategies: An equilibrium analysis based on decoupled response function models. *Management Science, 34*(7), 805–815.
Kunen, K. (1980). *Set theory: An introduction to independence proofs*. North Holland.
Kuratowski, K., & Mostowski, A. (1976). *Set theory: With an introduction to descriptive set theory*. North-Holland.
Lages, L. F., Silva, G., & Styles, C. (2009). Relationship capabilities, quality, and innovation as determinants of export performance. *Journal of International Marketing, 17*(4), 47–70.
Lambert-Pandraud, R., Laurent, G., Mullet, E., & Yoon, C. (2017). Impact of age on brand awareness sets: A turning point in consumers' early 60s. *Marketing Letters, 28*(2), 205–218.
Landes, D. S. (1999). *The wealth and poverty of nations: Why some are so rich and some so poor*. W. W. Norton & Company.
Landler, M., & Barbaro, M. (2006). International business: No, not always. *New York Times, 8*(2), A1–A3.

Landry, R., Amara, N., & Lamari, M. (2002). Does social capital determine innovation? To what extent? *Technological Forecasting and Social Change, 69*, 681–701.

Lane, V. R., & Fastoso, F. (2016). The impact of repeated ad exposure on spillover from low fit extensions to a global brand. *International Marketing Review, 33*(2), 298–318.

Langston, D. C. (2001). *Conscience and other virtues: From bonaventure to MacIntyre*. Penn State University Press.

Lanning, M. (1998). *Delivering profitable value*. Perseus Publishing.

Lanning, M. (2003). An introduction to the market-focused philosophy, framework and methodology called delivering profitable value. Retrieved February 24, 2012 from http://www.exubrio.com/whitepapers/DPVIntro-eXubrio.pdf.

Lanning, M., & Michaels, E. (1988). A business is a value delivery system. *McKinsey Staff Paper* No. 41, July.

Laukkanen, T. (2016). Consumer adoption versus rejection decisions in seemingly similar service innovations: The case of the Internet and mobile banking. *Journal of Business Research, 69*(7), 2432–2439.

Le, N., Nguyen, H. M., & Schultz, E. (2020). Financial development and productivity growth in stagnant industries. SSRN: https://ssrn.com/abstract=3275778, Accessed May 4, 2020.

Lecours, M. (2017). Developing a framework to craft a value proposition. *Journal of Financial Planning, 30*(3), 23–25.

Lecun, Y., Bengio, Y., & Hinton, G. (2015). Deep learning. *Nature, 521*(7553), 436.

LeDoux, J. E., & Brown, R. (2017). A higher-order theory of emotional consciousness. *Proceedings of the National Academy of Sciences of the United States of America, 114*(10), E2016–E2025.

Lee, C. Y., Kim, J., Noh, M. S., Woo, H. G., & Gang, K. W. (2017). Patterns of technology life cycles: Stochastic analysis based on patent citations. *Technology Analysis & Strategic Management, 29*(1), 53–67.

Lee, J. (1995). Small firms' innovation in two technological settings. *Research Policy, 24*, 391–401.

Lee, R., & Grewal, R. (2004). Strategic responses to new technologies and their impact on firm performance. *Journal of Marketing, 68*(4), 157–171.

Lei, D., Hitt, M. A., & Bettis, R. (1996). Dynamic core competences through meta-learning and strategic context. *Journal of Management, 22*, 549–596.

Lemoussu, S., Chaudemar, J. C., & Vingerhoeds, R. A. (2018). Systems engineering and project management process modeling in the aeronautics context: Case study of SMEs. *International Journal of Mechanical and Mechatronics Engineering, 12*(2), 88–96.

Leonard, D. A., & Swap, W. C. (1999). *When sparks fly: Igniting creativity in groups*. Harvard Business School Press.

Leonard-Barton, D. (1992). Core capabilities and core rigidities: A paradox in managing new product development. *Strategic Management Journal, 13*(Summer), 111–125.

Lévesque, M., Minniti, M., & Shepherd, D. (2013). How late should Johnny-come-lately come? *Long Range Planning, 46*, 369–386.

Levinthal, D. A., & Myatt, J. (1994). Co-evolution of capabilities and industry: The evolution of mutual fund processing. *Strategic Management Journal, 15*(Winter Special Issue), 45–62.

Li, M., & Nguyen, B. (2017). When will firms share information and collaborate to achieve innovation? *The Bottom Line, 30*(1), 65–86. https://doi.org/10.1108/BL-12-2016-0039

Li, S. X., & Greenwood, R. (2004). The effect of within industry diversification on firm performance: Synergy creation, multi-market contact and market structuration. *Strategic Management Journal, 25*(12), 1131–1153.

Li, T., & Calantone, R. (1998). The impact of market knowledge competence on new product advantage: Conceptualization and empirical examination. *Journal of Marketing, 62*(4), 13–29.

Li, T., & Ma, J. H. (2015). Complexity analysis of dual-channel game model with different managers' business objectives. *Communications in Nonlinear Science and Numerical Simulation, 20*, 199–208.

Libet, B. (1999). Do we have free will? *Journal of Consciousness Studies, 6*(8-9), 47–57.

Lieberman, M. B. (1984). The learning curve and pricing in the chemical processing industries. *RAND Journal of Economics, 15*(2), 213–228.
Lieberman, M. B. (1989). The learning curve, technological barriers to entry, and competitive survival in the chemical processing industries. *Strategic Management Journal, 10*(5), 431–447.
Lieberman, M. B., & Montgomery, D. B. (1988). First-mover advantages. *Strategic Management Journal, 9*, 41–58.
Lieberman, M. B., & Montgomery, D. B. (2013). Conundra and progress: Research on entry order and performance. *Long Range Planning, 44*, 312–324.
Lin, Y. (1987). A model of general systems. *Mathematical Modelling: An International Journal, 9* (2), 95–104.
Lin, Y. (1988a). Can the world be studied in the viewpoint of systems?? *Mathematical and Computer Modelling, 11*, 738–742.
Lin, Y. (1988b). On periodic linked time systems. *International Journal of Systems Science, 19*, 1299–1310.
Lin, Y. (1988c). An application of systems analysis in sociology. *Cybernetics and Systems: An International Journal, 19*, 267–278.
Lin, Y. (1989). A multi-relation approach of general systems and tests of applications. *Synthese - An International Journal for Epistemology, Methodology and Philosophy of Science, 79*, 473–488.
Lin, Y. (1999). *General systems theory: A mathematical approach*. Plenum and Kluwer Academic Publishers.
Lin, Y. (2007). Systemic yoyo model and applications in Newton's, Kepler's laws, etc. *Kybernetes: The International Journal of Cybernetics, Systems and Management Science, 36*(3-4), 484–516.
Lin, Y. (2008). Systematic Studies: The Infinity Problem in Modern Mathematics. *Kybernetes: The International Journal of Cybernetics, Systems and Management Sciences, 37*(3-4), 385–542. (guest editor).
Lin, Y. (2009). *Systemic Yoyos: Some impacts of the second dimension*. Auerbach Publication, an imprint of Taylor and Francis.
Lin, Y., Duan, X. J., Zhao, C. L., & Xu, L. D. (2012). *Systems science: Methodological approaches*. CRC Press, An Imprint of Taylor and Francis.
Lin, Y., & Fan, T. H. (1997). The fundamental structure of general systems and its relation to knowability of the physical world. *Kybernetes: The International Journal of Systems and Cybernetics, 26*(3), 275–285.
Lin, Y., & Forrest, B. (2011). *Systemic structure behind human organizations: From civilizations to individuals*. Springer.
Lin, Y., & Ma, Y. H. (1987). Some properties of linked time systems. *International Journal of General Systems, 13*(2), 125–134.
Lin, Y., Ma, Y., & Port, R. (1990). Several epistemological problems related to the concept of systems. *Mathematical and Computer Modelling, 14*, 52–57.
Lin, Y., & OuYang, S. C. (2010). *Irregularities and prediction of major disasters*. CRC Press, an imprint of Taylor and Francis.
Lindblom, A., Olkkonen, R., Mitronen, L., & Kajalo, S. (2008). Market-sensing capability and business performance of retail entrepreneurs. *Contemporary Management Research, 4*(3), 219–236.
Lindstrom, M. (2005). *Brand sense: Build powerful brands through touch, taste, smell, sight, and sound*. Free Press.
Lins, K. V. (2003). Equity ownership and firm value in emerging markets. *The Journal of Financial and Quantitative Analysis, 38*(1), 159–184.
Lippman, S. A., & Rumelt, R. P. (1982). Uncertain imitability: An analysis of interfirm differences in efficiency under competition. *Bell Journal of Economics, 13*(2), 418–438.
Lipton, M. (1977). *Why poor people stay poor: Urban bias in developing countries*. Temple Smith.
Liu, S. F., & Lin, Y. (2010). *Grey systems: Theory and applications*. Springer.

Liu, Y., Li, K. J., Chen, H. P., & Balachander, S. (2018). The effects of products' aesthetic design on demand and marketing-mix effectiveness: The role of segment prototypicality and brand consistency. *Journal of Marketing, 81*(1), 83–102.

Liu, Y., & Liang, L. T. (2015). Evaluating and developing resource-based operations strategy for competitive advantage: An exploratory study of Finnish high-tech manufacturing industries. *International Journal of Production Research, 53*(4), 1019–1037.

Lockett, A., O'Shea, R. P., & Wright, M. (2008). The development of the RBV: Reflections from Birger Wernerfelt. *Organization Studies, 29*(8–9), 1125–1141.

Lockett, A., Thompson, S., & Morgenstern, U. (2009). The development of the resource-based view of the firm: A critical appraisal. *International Journal of Management Reviews, 11*(1), 9–28.

Lockwood, T., & Papke, E. (2017). *Innovation by design: How any organization can leverage design thinking to produce change, drive new ideas, and deliver meaningful solutions*. Career Press.

Loken, B., & Ward, J. (1987). Measures of the attribute structure underlying product typicality. In M. Wallendorf & P. Anderson (Eds.), *NA - Advances in Consumer Research Volume 14* (pp. 22–26). Association for Consumer Research.

Lonial, S. C., & Carter, R. E. (2015). The impact of organizational orientations on medium and small firm performance: A resource-based perspective. *Journal of Small Business Management, 53*(1), 94–113.

Lucas, R. E. (1977). Understanding business cycles. *Carnegie-Rochester Conference Series on Public Policy, 5*, 7–29.

Luo, X., Griffith, D., Liu, S., & Shi, Y. (2004). The effects of customer relationships and social capital on firm performance: A Chinese business illustration. *Journal of International Marketing, 12*(4), 25–45.

Lv, T., & Han, N. (2015). Intelligent manufacturing: The global trend and Chinese strategy. *Frontiers, 2015*(11), 6–17.

Lyson, T. A., & Falk, W. W. (Eds.). (1993). *Forgotten places: Uneven development in rural america*. Rural America Series. University Press of Kansas.

Madhok, A., Li, S., & Priem, R. L. (2010). The resource-based view revisited: Comparative firm advantage, willingness-based isolating mechanisms and competitive heterogeneity. *European Management Review, 7*(2), 91–100.

Makadok, R. (2001). Toward a synthesis of the resource-based and dynamic-capability views of rent creation. *Strategic Management Journal, 22*(5), 387–401.

Makhija, A. K., & Spiro, M. (2000). Ownership structure as a determinant of firm value: Evidence from newly privatized czech firms. *Financial Review, 35*(3), 1–31.

Makridakis, S. (2017). The forthcoming artificial intelligence (AI) revolution: Its impact on society and firms. *Futures, 90*(June), 46–60.

Malecki, E. J. (1997). *Technology and economic development*. Addison-Wesley Longman.

Mamun, A. A., Mohiuddin, M., Fazal, S. A., & Ahmad, G. B. (2018). Effect of entrepreneurial and market orientation on consumer engagement and performance of manufacturing SMEs. *Management Research Review, 41*(1), 133–147.

Mandler, G. (1982). The integration and elaboration of memory structures. In F. Klix, J. Hoffman, & E. van der Meer (Eds.), *Cognitive research in psychology* (pp. 33–40). North-Holland.

Manteghi, A., Manteghi, M., Aslani, A., & Mohaghar, A. (2018). Catch-up process in aircraft industry: A model based on experiences of six latecomer countries. *International Journal of Sociotechnology and Knowledge Development, 10*(1), 37–56.

Mantovani, A. (2013). The strategic effect of bundling: A new perspective. *Review of Industrial Organization, 42*, 25–43.

Marakhimov, A., Joo, J., & Jaegeol Yim, J. (2018). Customer participation in business ecosystems: An integrated approach of system dynamics and fuzzy sets. *International Journal of Business and Systems Research, 12*(3), 290–316.

Marc, S., Yann, B., Charly, B, et al. (2018). Mission Villani: For a meaningful artificial intelligence towards a French and European strategy. https://www.aiforhumanity.fr/pdfs/MissionVillani_Report_ENG-VF.pdf, Accessed on March 8, 2018.

Marchall, T. (2016). *Prisoners of geography: Ten maps that explain everything about the world*. Scribner.

Marketing Science Institute. (2010). *MSI research priorities 2010–12*. MSI.

Marketing Science Institute. (2014). *MSI research priorities 2014–16*. MSI.

Markides, C. C., & Geroski, P. A. (2005). *Fast second: How smart companies bypass radical innovation to enter and dominate new markets*. Jossey Bass/Wiley.

Markman, G. D., & Gartner, W. B. (2002). Is extraordinary growth profitable? A study of Inc. 500 high-growth companies. *Entrepreneurship: Theory and Practice, 27*(1), 65–75.

Markman, G. D., & Waldron, T. L. (2014). Small entrants and large incumbents: A framework of micro entry. *Academy of Management Perspectives, 28*(2), 179–197.

Markoff, J. (2016). *Machines of loving grace: The quest for common ground between humans and robots*. HarperCollins.

Martin, O. (2001). Growth and Agglomeration. *International Economic Review, 42*(4), 947–968.

Marzi, G., Dabic, M., Daim, T., & Garces, E. (2017). Product and process innovation in manufacturing firms: A 30-year bibliometric analysis. *Scientometrics*. https://doi.org/10.1007/s11192-017-2500-1

Mashey, J. R. (1998). Big data ... and the next wave of infrastress. *Slides from invited talk*. https://static.usenix.org/event/usenix99/invited_talks/mashey.pdf. Accessed on February 12, 2020.

Maslow, A. H. (1943). A theory of human motivation. *Psychological Review, 50(4)*, 370–396.

Maslow, A. H. (1954). *Motivation and personality*. Harper.

Mayer, J. (2018). Digitalization and industrialization: Friends or foes? United Nations Conference on Trade and Development: UNCTAD Research Paper No. 25 UNCTAD/SER.RP/2018/7, https://unctad.org/en/PublicationsLibrary/ser-rp-2018d7_en.pdf, Accessed on December 22, 2019.

Mayer, K. J., & Argyres, N. S. (2004). Learning to contract: Evidence from the personal computer industry. *Organization Science, 15*(4), 394–410.

Mazzei, M. J., Ketchen, D. J., Jr., & Shook, C. L. (2017). Understanding strategic entrepreneurship: A "theoretical toolbox" approach. *International Entrepreneurship and Management Journal, 13*(2), 631–663.

McFarland, C. P., Primosch, M., Maxson, C. M., & Stewart, B. T. (2017). Enhancing memory and imagination improves problem solving among individuals with depression. *Memory & Cognition, 45*(6), 932–939.

McGrath, R. G. (2013). *The end of competitive advantage: How to keep your strategy moving as fast as your business*. Harvard Business Review Press.

McLeod, A. (2019). Employment and technology in manufacturing. *International Journal of Sociotechnology and Knowledge Development, 11*(3), 14–29.

McMullen, J. S. (2015). Entrepreneurial judgment as empathic accuracy: A sequential decision-making approach to entrepreneurial action. *Journal of Institutional Economics, 11*(3), 651–681.

Melitz, M. J. (2003). The impact of trade on intra-industry reallocations and aggregate industry productivity. *Econometrica, 71*(6), 1695–1725.

Melitz, M. J., & Redding, S. J. (2014). Heterogeneous firms and trade. *Handbook of International Economics, 4*, 1–54.

Mendels, F. F. (1972). Proto-industrialization: The first phase of the industrialization process. *The Journal of Economic History, 32*, 241–261. https://doi.org/10.1017/S0022050700075495

Merati, M., Salehi, V., & Rafiei, A. (2019). A unique mathematical programming algorithm for performance optimization of organizational indicators in manufacturing sector. *Journal of Industrial and Production Engineering, 36*(7), 461–474.

Mervis, C. B., & Rosch, E. (1981). Categorization of natural objects. *Annual Review of Psychology, 32*, 89–115.

Mesarovic, M. D., & Takahara, Y. (1975). *General systems theory: Mathematical foundations*. Academic Press.
Messinger, P. R., & Narasimhan, C. (1997). A model of retail formats based on consumers' economizing on shopping time. *Marketing Science, 16*(1), 1–23.
Meyer, D. R. (2003). *The roots of American industrialization*. The Johns Hopkins University Press.
Meyers-Levy, J. (1986). *Gender differences in information processing: A selectivity interpretation*. Northwestern University.
Meyers-Levy, J., & Tybout, A. M. (1989). Schema congruity as a basis for product evaluation. *Journal of Consumer Research, 16*(1), 39–54.
Miao, C. L., Fang, D. B., Sun, L. Y., Luo, Q. L., & Yu, Q. (2018). Driving effect of technology innovation on energy utilization efficiency in strategic emerging industries. *Journal of Cleaner Production, 170*(1), 1177–1184.
MIIT. (2015). Circular of Ministry of Industry and Information Technology of the People's Republic of China on Publishing the List of Pilot Demonstration Projects of Intelligent Manufacturing in 2015. http://www.miit.gov.cn/n1146285/n1146352/n3054355/n3057585/n3057589/c3590738/content.html, Accessed on July 21, 2015.
MIIT. (2016a). Circular of MIIT and MOF on the Issuance of Development Plan of Intelligent Manufacturing (2016-2020). http://www.miit.gov.cn/n1146285/n1146352/n3054355/n3057585/n3057590/c5406038/content.html, Accessed on December 8, 2016.
MIIT. (2016b). Circular of Ministry of Industry and Information Technology of the People's Republic of China on Publishing the List of Pilot Demonstration Projects of Intelligent Manufacturing in 2016. http://www.miit.gov.cn/n1146285/n1146352/n3054355/n3057585/n3057590/c5032698/content.html, Accessed on July 4, 2016.
MIIT. (2017). Circular of Ministry of Industry and Information Technology of the People's Republic of China on Publishing the List of Pilot Demonstration Projects of Intelligent Manufacturing in 2017. http://www.miit.gov.cn/n1146285/n1146352/n3054355/n3057585/n3057590/c5861997/content.html, Accessed on October 16, 2017.
Molineux, P. (2002). *Exploiting CRM: Connecting with customers*. Hodder & Stoughton.
Mondragon, A. E. C., & Mondragon, C. E. C. (2018). Managing complex, modular products: How technological uncertainty affects the role of systems integrators in the automotive supply chain. *International Journal of Production Research*. https://doi.org/10.1080/00207543.2018.1424362
Moorman, C., & Slotegraaf, R. (1999). The contingency value of complementary capabilities in product development. *Journal of Marketing Research, 36*(2), 239–257.
Moran, G. (2018). Chaos theory and psychoanalysis: The fluidic nature of the mind. *Zeitschrift für Psychosomatische Medizin und Psychoanalyse, 40*(4), 384–403.
Moran, P., & Ghoshal, S. (1999). Markets, firms, and the process of economic development. *Academy of Management Review, 24*, 390–412.
Moreau, C. P., & Dahl, D. W. (2005). Designing the solution: The impact of constraints on consumers' creativity. *Journal of Consumer Research, 32*(1), 13–23.
Morgan, I., & Rao, J. (2003). Making routine customer experiences fun. *Sloan Management Review, 45*(1), 93–95.
Morgan, N., Slotegraaf, R., & Vorhies, D. (2009a). Linking marketing capabilities with profit growth. *International Journal of Research in Marketing, 26*(4), 284–293.
Morgan, N. A., Vorhies, D. W., & Mason, C. (2009b). Market orientation, marketing capabilities, and firm performance. *Strategic Management Journal, 30*(8), 909–920.
Moss, G. (2009). *Gender, design and marketing: How gender drives our perception of design and marketing*. Routledge.
Moyo, C., & Jeke, L. (2019). Manufacturing sector and economic Growth: A panel study of selected African countries. *Journal of Business and Economics Review, 4*(3), 114–128.
Narasimhan, C. (1988). Competitive promotional strategies. *Journal of Business, 61*(4), 427–429.
Nason, R. S., & Wiklund, J. (2018). An assessment of resource-based theorizing on firm growth and suggestions for the future. *Journal of Management, 44*(1), 32–60.

Naudé, W., & Nagler, P. (2018). Technological innovation, entrepreneurship and productivity in Germany, 1871-2015. SWPS 2018-02. Available at SSRN: https://doi.org/10.2139/ssrn.3107759, Accessed December 20, 2019.

Naudé, W., & Szirmai, A. (2012). The importance of manufacturing in economic development: Past, present and future perspectives. United Nations University-MERIT UNU-MERIT Working Papers, Accessed on December 10, 2019.

Nedungadi, P. (1990). Recall and consumer consideration sets: Influencing choice without altering brand evaluations. *Journal of Consumer Research, 17*(3), 263–276.

Niraj, R., Gupta, M., & Narasimhan, C. (2001). Customer profitability in a supply chain. *Journal of Marketing, 65*(3), 1–16.

Noble, C. H., Sinha, R. K., & Kumar, A. (2002). Market orientation and alternative strategic orientations: A longitudinal assessment of performance implications. *Journal of Marketing, 66* (October), 25–39.

Nohria, N., & Garcia-Pont, C. (1991). Global strategic linkages and industry structure. *Strategic Management Journal, 12*(Summer Special Issue), 105–124.

Norman, R. (2000). *Cultivating imagination in adult education. Proceedings of the 41st annual adult education research* (pp. 319–323).

North, D. C. (1992). *Transaction costs, institutions, and economic performance*. ICS Press.

O'Brien, C. (2008). Sustainable happiness: How happiness studies can contribute to a more sustainable future. *Canadian Psychology, 49*(4), 289–295.

O'Reilly, C., III, & Tushman, M. L. (2004). The ambidextrous organization. *Harvard Business Review, 82*(April), 74–81.

O'Sullivan, A., & Sheffrin, S. M. (2003). *Economics: Principles in action*. Pearson Prentice Hall.

Öberg, C. (2018). The role of business networks for innovation. *Journal of Innovation & Knowledge*. https://doi.org/10.1016/j.jik.2017.10.001

OECD. (1997). *The measurement of scientific and technological activities, proposed guidelines for collecting and interpreting technological innovation data*. Organization for Economic Cooperation and Development.

Olivera, M. (2018). The possibilities of industrialization and structural change for the periphery in the context of globalization. In V. Fernandez & G. Brondina (Eds.), *Development in Latin America* (pp. 157–177). Palgrave Macmillan.

Oliveira, T., Thomas, M., Baptista, G., & Campos, F. (2016). Mobile payment: Understanding the determinants of customer adoption and intention to recommend the technology. *Computers in Human Behavior, 61*, 404–414.

Ombaka, B. E., Awino, Z. B., Machuki, V. N., & Wainaina, G. (2015). Exploring resources and performance relationships in commercial enterprises: An empirical perspective. *Journal of Management and Strategy, 6*(4), 12–28.

Osterwalder, A., Pigneur, Y., Bernarda, G., & Smith, A. (2014). *Value proposition design: How to create products and services customers want*. Wiley.

OSTP. (2012). A national strategic plan for advanced manufacturing. https://www.manufacturing-policy.eng.cam.ac.uk/documents-folder/policies/usa-a-national-strategic-plan-for-advanced-manufacturing-eop-nstc/view, Accessed on February 15, 2012.

OWH. (2016a). *Preparing for the Future of Artificial Intelligence*. https://obamawhitehouse.archives.gov/sites/default/files/whitehouse_files/microsites/ostp/NSTC/preparing_for_the_future_of_ai.pdf, Accessed on April 20, 2019.

OWH. (2016b). *The National Artificial Intelligence Research and Development Strategic Plan*. https://www.nitrd.gov/PUBS/national_ai_rd_strategic_plan.pdf, Accessed on April 16, 2016.

Palich, L. E., Cardinal, L. B., & Miller, C. C. (2000). Curvilinearity in the diversification-performance linkage: An examination of over three decades of research. *Strategic Management Journal, 21*(2), 155–174.

Papadakis, V., & Bourantas, D. (1998). The chief executive officer as corporate champion of technological innovation: An empirical investigation. *Technology Analysis & Strategic Management, 10*(1), 89–98.

Parasuraman, A. (1997). Reflections on gaining competitive advantage through customer value. *Journal of the Academy of Marketing Science, 25*, 154. https://doi.org/10.1007/BF02894351

Parthasarthy, R., & Hammond, J. (2002). Product innovation input and outcome: Moderating effects of the innovation process. *Journal of Engineering and Technology Management, 19*, 75–91.

Patterson, W., & Ambrosini, V. (2015). Configuring absorptive capacity as a key process for research intensive firms. *Technovation, 36-37*, 77–89.

Pavelchak, M. A. (1989). Piecemeal and category-based evaluation: An idiographic analysis. *Journal of Personality and Social Psychology, 56*(3), 354–363.

Payne, A., & Frow, P. (2005). A strategic framework for customer relationship management. *Journal of Marketing, 69*(4), 167–176.

Payne, A., Frow, P., & Eggert, A. (2017). The customer value proposition: Evolution, development, and application in marketing. *Journal of the Academy of Marketing Science, 45*, 467–489.

PCAST. (2012). Report to the President on capturing domestic competitive advantage in advanced manufacturing. https://permanent.access.gpo.gov/gpo34116/pcast_amp_steering_committee_report_final_july_17_2012.pdf, Accessed on July, 2012.

Penrose, E. (1959). *The theory of the growth of the firm*. Blackwell.

Peteraf, M. A. (1993). The cornerstones of competitive advantage: A resource-based view. *Strategic Management Journal, 14*(3), 179–191.

Peteraf, M., & Barney, J. (2003). Unraveling the resource-based tangle. *Managerial and Decision Economics, 24*(4), 309–323.

Petrovslaya, I., Zaverskiy, S., & Kiseleva, E. (2017). Attribute to entrepreneurship in Russia. *Advances in Systems Science and Applications, 17*(2), 29–43.

Pfaff, D. W. (2007). *The neuroscience of fair play: Why we (usually) follow the golden rule*. Dana Press, The Dana Foundation.

Pike, A., & Tomaney, J. (2009). The state and uneven development: The governance of economic development in England in the post-devolution UK. *Cambridge Journal of Regions, Economy and Society, 2*(1), 13–34.

Piskorski, M. & Nohria, N. (1999). Allocation to open and closed portfolios. *Harvard Business School working paper.*

Podolny, J. (1993). A status based model of market competition. *American Journal of Sociology, 98*, 829–872.

Porse, H., & Rudolph, B. (2017). The seaweed hydrocolloid industry: 2016 updates, requirements, and outlook. *Journal of Applied Phycology, 29*, 2187–2200.

Porter, M. (1979). How competitive forces shape strategy. *Harvard Business Review, 57*(2), 137–145.

Porter, M. E. (1985). *Competitive advantage: Creating and sustaining superior performance*. Free Press.

Powell, T. C. (1992). Strategic planning as competitive advantage. *Strategic Management Journal, 13*(7), 551–558.

Prajogo, D. I., & Oke, A. (2016). Human capital, service innovation advantage, and business performance: The moderating roles of dynamic and competitive environments. *International Journal of Operations & Production Management, 36*(9), 974–994.

Priem, R. L. (2007). A consumer perspective on value creation. *Academy of Management Review, 32*, 219–235.

Priem, R. L., & Butler, J. E. (2001a). Is the resource-based "view" a useful perspective for strategic management research? *The Academy of Management Review, 26*(1), 22–40.

Priem, R. L., & Butler, J. E. (2001b). Tautology in the resource-based view and the implications of externally determined resource value: Further comments. *The Academy of Management Review, 26*(1), 57–66.

Priem, R. L., Wenzel, M., & Koch, J. (2018). Demand-side strategy and business models: Putting value creation for consumers center stage. *Loong-Range Planning, 51*(1), 22–31.

Punj, G., & Stewart, D. W. (1983). Cluster analysis in marketing research: Review and suggestions for application. *Journal of Marketing Research, 20*(2), 134–148.

Pyper, K., Doherty, A. M., Gounaris, S., & Wilson, A. (2019). Investigating international strategic brand management and export performance outcomes in the B2B context. *International Marketing Review, 37*(1), 98–129.

Qu, Y. (1985). *The songs of the South: An anthology of ancient chinese poems by Qu Yuan and Other Poets* (translated by David Haykes). Penguin Classics.

Ramaswami, S., Srivastava, R., & Bhargava, M. (2009). Market-based capabilities and financial performance of firms: Insights into marketing's contribution to firm value. *Journal of the Academy of Marketing Science, 37*(2), 97–116.

Randall, W. S., Defee, C. C., & Brady, S. P. (2010). Value Propositions of the U.S. Trucking Industry. *Transportation Journal, 49(3)*, 5–23.

Rathod, G. H., Puranik, V. S., & Hiremath, V. (2019). Developing agility system in supply chains. *International Journal of Business and Systems Research, 13(2)*, 247–258.

Rayport, J. F., & Sviokla, J. J. (1995). Exploiting the virtual value chain. *Harvard Business Review, 73*(6), 75–85.

Rebei, N. (2017). Evaluating changes in the transmission mechanism of government spending shocks. *International Monetary Fund,* WP/17/49.

Reeves, R. (1961). *Reality in advertising*. Alfred A. Knopf.

Reinartz, W., & Kumar, V. (2000). The CRM process: Its measurement and impact on performance. *Journal of Marketing Research, 64*(4), 17–35.

Reinartz, W., Thomas, J., & Kumar, V. (2005). Balancing customer acquisition and retention resources to maximize customer profitability. *Journal of Marketing, 69*(1), 63–79.

Reinert, E. (1995). Competitiveness and its predecessors: A 500-year cross-national perspective. *Structural Change and Economic Dynamics, 6*(1), 23–42.

Reinert, E. (2019). *How rich countries got rich ... and why poor countries stay poor.* Public Affairs.

Reiss, M. (2011). Complementor relationship management – missing link in supply chain management. In R. M. Samson (Ed.), *Supply-chain management: Theories, activities/functions and problems* (pp. 139–156). Nova Science Pub.

Richard, O. C., & Johnson, N. B. (2001). Strategic human resource management effectiveness and firm performance. *International Journal of Human Resource Management, 12*(2), 299–310.

Richey, G., Tokman, M., & Dalela, V. (2010). Examining collaborative supply chain service technologies: A study of intensity, relationships, and resources. *Journal of the Academy of Marketing Science, 38*(1), 71–89.

Ridley, M. (2016). *The evolution of everything: How new ideas emerge*. Harper Perennial.

Rintamäki, T., & Kirves, K. (2017). From perceptions to propositions: Profiling customer value across retail contexts. *Journal of Retailing and Consumer Services, 37*(July), 159–167.

Rintamaki, T., Kuusela, H., & Mitronen, L. (2007). Identifying competitive customer value propositions in retailing. *Managing Service Quality, 17*(6), 621–634.

Ritalam, P., Heiman, B., & Hurmelinna-Laukkanen, P. (2016). The need for speed—unfamiliar problems, capability rigidity, and ad hoc processes in organizations. *Industrial and Corporate Change, 25*(5), 757–777.

Ritter, T., & Gemunden, H. G. (2003). Network competence: Its impact on innovation success and its antecedents. *Journal of Business Research, 56*, 745–755.

Rivkin, J. W. (2000). Imitation of complex strategies. *Management Science, 46*(6), 824–844.

Robins, J. A. (1992). Organizational considerations in the evaluation of capital assets: Toward a resource-based view of strategic investment by firms. *Organization Science, 3*(4), 442–562.

Robins, J. A., & Wiersema, M. F. (2003). The measurement of corporate portfolio strategy: Analysis of the content validity of related diversification indexes. *Strategic Management Journal, 24*(1), 39–59.

Roca, J. B., Vaishnav, P., Morgan, M. G., Mendonça, J., & Erica Fuchs, E. (2017). When risks cannot be seen: Regulating uncertainty in emerging technologies. *Research Policy, 46*(7), 1215–1233.

Rochet, J. C., & Tirole, J. (2003). Platform competition in two-sided markets. *Journal of the European Economic Association, 1*(4), 990–1029.

Rochet, J. C., & Tirole, J. (2006). Two-sided markets: A progress report. *RAND Journal of Economics, 37*(3), 645–667.

Romijn, H., & Albaladejo, M. (2002). Determinants of innovation capability in small electronics and software firms in southeast England. *Research Policy, 31*, 1053–1067.

Roodman, D. (2009). How to do Xtabond2: An introduction to difference and system GMM. *Stata Journal, 9*(1), 86–136.

Roos, G. (2015). Servitization as innovation in manufacturing: A review of the literature. In *The handbook of service innovation* (pp. 403–435). London. https://link.springer.com/chapter/10.1007/978-1-4471-6590-3_19

Rosch, E. (1978). Principles of categorization. In E. Rosch & B. B. Lioyd (Eds.), *Cognition and categorization* (pp. 27–48). Lawrence Erlbaum.

Rosch, E., Mervis, C. B., Gray, W. D., Johnson, D. M., & Boyes-Braem, P. (1976). Basic objects in natural categories. *Cognitive Psychology, 8*(3), 382–439.

Rosenberg, N. (1972). Factors affecting the diffusion of technology. *Explorations in Economic History, 10*(1), 3–33.

Rosenthal, S. R. (1992). *Effective product design and development*. Irwin.

Rostow, W. W. (1960). *The stages of economic growth: A non-communist manifesto*. Cambridge University Press.

Rowley, J. (2011). Should your library have an innovation strategy? *Library Management, 32*(4/5), 251–265.

Rudinsky, J., & Hvannberg, E. T. (2017). Transferability of voice communication in games to virtual teams training for crisis management. *International Journal of Sociotechnology and Knowledge Development, 9*(1), 1–25.

Ruiz-Ortega, M., & García-Villaverde, P. (2008). Capabilities and competitive tactics influences on performance: Implications of the moment of entry. *Journal of Business Research, 61*(4), 332–345.

Rumelt, R. P. (1974). *Strategy, structure, and economic performance*. Harvard University Press.

Ryals, L. (2005). Making customer relationship management work: The measurement and profitable management of customer relationships. *Journal of Marketing, 69*(4), 252–261.

Sahal, D. (1981). *Patterns of technological innovation*. Addison-Wesley.

Sahni, N. S. (2016). Advertising spillovers: Evidence from online field experiments and implications for returns on advertising. *Journal of Marketing Research, 53*(4), 459–478.

Sakhartov, A. V., & Folta, T. B. (2014). Getting beyond relatedness as a driver of corporate value. *Strategic Management Journal, 36*(13), 1939–1959.

Salamzadeh, A., & Kawamorita, K. H. (2015). Startup companies: Life cycle and challenges. *The 4th International Conference on Employment, Education and Entrepreneurship* (EEE), Belgrade, Serbia, 2015. Available at SSRN: https://ssrn.com/abstract=2628861 or https://doi.org/10.2139/ssrn.2628861

Santalo, J., & Becerra, M. (2008). Competition from specialized firms and the diversification-performance linkage. *Journal of Finance, 63*(2), 851–883.

SAP Corporation. (2006). SAP ecosystem unites to bring innovation to chemicals industry. Press release, 8 May.

Sawhney, M., Verona, G., & Prandelli, E. (2005). Collaborating to create: The Internet as a platform for customer engagement in product innovation. *Journal of Interactive Marketing, 19*(4), 4–17.

Scherer, F. M., & Ross, D. (1990). *Industrial market structure and economic performance*. Houghton Mifflin.

Scholz, R. (2018). *Industrializing Africa: Understanding the uniqueness of the manufacturing sector*. Nomos Verlag.

Schrift, R. Y., Parker, J. R., Zauberman, G., & Srna, S. (2018). Multistage decision processes: The impact of attribute order on how consumers mentally represent their choice. *Journal of Consumer Research, 44*(6), 1307–1324.

Schumpeter, J. A. (1934). *The theory of economic development: An inquiry into profits, capital, credit, interest, and the business cycle*. Harvard University Press.

Schumpeter, J. A. (1942). *Capitalism, socialism, and democracy*. Harper.

Schwartz, L. (2008). *Mathematics for the physical sciences*. Dover.

Selstad, T. (1990). The rise of the quaternary sector. The regional dimension of wisdom or knowledge-based services in Norway, 1970–1985. *Norsk Geografisk Tidsskrift - Norwegian Journal of Geography, 44*(1), 21–37.

Shah, S. K., & Tripsas, M. (2007). The accidental entrepreneur: The emergent and collective process of user entrepreneurship. *Strategic Entrepreneurship Journal, 1*(1–2), 123–140.

Shannon, C. E. (1949). *The mathematical theory of communication*. Illinois University Press.

Shapiro, C., & Varian, H. R. (1999). *Information rules: A strategic guide to the network economy*. Harvard Business School Press.

Sheehan, N. T., & Bruni-Bossio, V. (2015). Strategic value curve analysis: Diagnosing and improving customer value propositions. *Business Horizons, 58*(3), 317–324.

Sheikhzadeh, M., & Elahi, E. (2013). Product bundling: Impacts of product heterogeneity and risk considerations. *International Journal of Production Economics, 144*, 209–222.

Shi, X. A., Li, L. S., Cheng, Z. H., & Liu, J. (2018). Impact analysis of "internet+" on the value-chain improvement of China's manufacturing. *Studies in Science of Science, 36*(08), 1384–1394.

Shin, J. E., Suh, E. M., Eom, K., & Kim, H. S. (2018). What does "happiness" prompt in your mind? Culture, word choice, and experienced happiness. *Journal of Happiness Studies, 19*(3), 649–662.

Shulga, L. V., & Busser, J. A. (2019). Customer acceptance of four types of hospitality value propositions. *International Journal of Hospitality Management, October*. https://doi.org/10.1016/j.ijhm.2019.102388

Siebert, R. (2015). Entering new markets in the presence of competition: Price discrimination versus cannibalization. *Journal of Economics and Management Strategies, 24*(2), 369–389.

Singh, L. (2019). Competitiveness, skill formation and industrialization: The South Asian experience. In S. Chaturvedi & S. Saha (Eds.), *Manufacturing and Jobs in South Asia* (pp. 213–227). Springer.

Sirmon, D. G., Gove, S., & Hitt, M. A. (2008). Resource management in dyadic competitive rivalry: The effects of resource bundling and deployment. *The Academy of Management Journal, 51*(5), 919–935.

Sivathanu, B., & Pillai, R. (2019). Leveraging technology for talent management: Foresight for organizational performance. *International Journal of Sociotechnology and Knowledge Development, 11*(2), 16–30.

Skilton, P. F., & Bernardes, E. (2015). Competition network structure and product market entry. *Strategic Management Journal, 36*(11), 1688–1696.

Slater, S. F., & Narver, J. C. (2000). Intelligence generation and superior customer value. *Journal of the Academy of Marketing Science, 28*(1), 120–127.

Slotegraaf, R., & Dickson, P. (2004). The paradox of a marketing planning capability. *Journal of the Academy of Marketing Science, 32*(4), 371–385.

Slotegraaf, R., Moorman, C., & Inman, J. (2003). The role of firm resources in returns to market deployment. *Journal of Marketing Research, 40*(3), 295–309.

Smith, A. (1776). *The Wealth of Nations, books I-III, (1986 printing)*. Penguin Books.

Smith, A. M. (2013). The value co-destruction process: A customer resource perspective. *European Journal of Marketing, 47*(11/12), 1889–1909.

Smith, S., & Wheeler, J. (2002). *Managing the customer experience: Turning customers into advocates*. Financial Times Prentice Hall.

Smith, W. K., & Tushman, M. L. (2005). Managing strategic contradictions: A top management model for man-aging innovation streams. *Organization Science, 16*(5), 522–536.
Snyder, K. A., Sulle, E., Massay, D. A., Petro, A., Qamara, P., & Brockington, D. (2020). "Modern" farming and the transformation of livelihoods in rural Tanzania. *Agriculture and Human Values*, ***37***, 33–46.
Sobel, R. (1999). *When giants stumble: Classic business blunders and how to avoid them*. Prentice Hall.
Soros, G. (2003). *The alchemy of finance*. Wiley.
Souitaris, V. (2002). Technological trajectories as moderators of firm-level determinants of innovation. *Research Policy, 31*, 877–898.
Spence, M. (1981). The learning curve and competition. *Bell Journal of Economics, 12*(1), 49–70.
Srinivasan, R., Lilien, G., & Rangaswamy, A. (2002). Technological opportunism and radical technology adoption: An application to e-business. *Journal of Marketing, 66*(3), 47–60.
Srivastava, R. K., Shervani, T. A., & Fahey, L. (1999). Marketing, business processes, and shareholder value: An organizationally embedded view of marketing activities and the discipline of marketing. *Journal of Marketing, 63*(4), 168–179.
Stabell, C. B., & Fjeldstad, O. D. (1998). Configuring value for competitive advantage: On chains, shops, and networks. *Strategic Management Journal, 19*(5), 413–437.
Starch, D. (1914). *Advertising: Its Principles, Practice, and Technique*. Scott, Foresman.
State Council of the PRC. (2015). Notice of State Council of the PRC on printing and distributing "Made in China 2025". http://www.gov.cn/zhengce/content/2015-05/19/content_9784.htm, Accessed on May 8, 2015.
State Council of the PRC. (2016). Guidance of State Council of the PRC on deepening the integration of manufacturing industry and internet. http://www.gov.cn/zhengce/content/2016-05/20/content_5075099.htm, Accessed on May 20, 2016.
Stayman, D. M., Alden, D. L., & Smith, K. H. (1992). Some effects of schematic processing on consumer expectations and disconfirmation judgments. *Journal of Consumer Research, 19*(2), 240–255. https://doi.org/10.1086/209299
Steffens, P., Davidsson, P., & Fitzsimmons, J. (2009). Performance configurations over time: Implications for growth- and profit-oriented strategies. *Entrepreneurship Theory and Practice, 33*(1), 125–148.
Stein, B., & Arnold, D. (Eds.). (2010). *A history of India* (2nd ed.). Wiley-Blackwell.
Stephan, J., Murmann, J. P., Boeker, W., & Goodstein, J. (2003). Bringing managers into theories of multimarket competition: CEOs and the determinants of market entry. *Organization Science, 14*(4), 403–421.
Steward, J. (1987). *Calculus*. Brooks/Cole.
Stidham, M., McCaffrey, S., Toman, E., & Shindler, B. (2014). Policy tools to encourage community-level defensible space in the United States: A tale of six communities. *Journal of Rural Studies, 35*(July), 59–69.
Stigler, G. (1961). The economics of information. *Journal of Political Economy, 65*(June), 213–225.
Stock, G. N., Greis, N. P., & Fischer, W. A. (2002). Firm size and dynamic technological innovation. *Technovation, 22*, 537–549.
Stryja, C., & Satzger, G. (2018). Try before you buy: How to design information systems to enhance consumer willingness to test sustainable innovations. *Journal of Technology Management & Innovation, 13*(1). https://doi.org/10.4067/S0718-27242018000100019
Studwell, J. (2013). *How Asia works: Success and failure in the world's most dynamic region*. Grove Press.
Sugiyarti, G., & Ardyan, E. (2017). Market sensing capability and product innovation advantages in emerging markets: The case of market entry quality and marketing performance of Batick Industry in Indonesia. *DLSU Business & Economics Review, 27*(1), 175–189.
Sujan, M. (1985). Consumer knowledge effects on evaluation strategies mediating consumer judgments. *Journal of Consumer Research, 12*(June), 31–46.

Sujan, M., & Bettman, J. R. (1989). The effects of brand positioning strategies on consumers' brand and category perceptions: Some insights from schema research. *Journal of Marketing Research, 26*(4), 454–467.

Sujan, M., & Dekleva, C. (1987). Product categorization and inference making: Some implications for comparative advertising. *Journal of Consumer Research, 14*(3), 372–378.

Sujan, M., & Tybout, A. M. (1988). Applications and extensions of categorization research in consumer behavior. In M. J. Houston & U. T. Provo (Eds.), *NA - Advances in consumer research* (Vol. 15, pp. 50–54). Association for Consumer Research.

Sun, M., & Tse, E. (2009). The resource-based view of competitive advantage in two-sided markets. *Journal of Management Studies, 46*(1), 45–64.

Sun, Z., Xiao, L. P., & Liu, L. H. (2017). The changes of industrial ownership structure and innovation: Is the state-owned enterprise dominant in favor of innovation? *Nankai Economic Studies, 2017*(06), 3–19.

Sung, S. Y., & Choi, J. N. (2014). Do organizations spend wisely on employees? Effects of training and development investments on learning and innovation in organizations. *Journal of Organizational Behavior, 35*(3), 393–412.

Supriyadi, S. (2017). Analysis effect of resources and dynamic capabilities to sustainable competitive advantages and its implications to the firms performance. *International Business Research, 10*(10), 209–222.

Sutcliffe, K. M., & Zaheer, A. (1998). Uncertainty in the transaction environment: An empirical test. *Strategic Management Journal, 19*(1), 1–23.

Swift, E. (2013). Emotional Intelligence of Entrepreneurs. EdD Dissertation of Duquesne University, http://etd.library.duq.edu/cdm-etd/document.php?CISOROOT=/etd&CISOPTR=162542&REC=1.

Szirmai, A., & Verspagen, B. (2015). Manufacturing and economic growth in developing countries, 1950-2005. *Structural Change and Economic Dynamics, 34*, 46–59.

Szymanski, D. M., Troy, L. C., & Bharadwaj, S. G. (1995). Order of entry and business performance: An empirical synthesis and reexamination. *Journal of Marketing, 59*(4), 17–33.

Takeishi, A. (2002). Knowledge partitioning in the interfirm division of labor: The case of automotive product development. *Organization Science, 13*(3), 321–338.

Tanriverdi, H., & Lee, C. H. (2008). Within-industry diversification and firm performance in the presence of network externalities: Evidence for the software industry. *The Academy of Management Journal, 51*(2), 381–397.

Tanriverdi, H., & Venkatraman, N. (2005). Knowledge relatedness and the performance of multibusiness firms. *Strategic Management Journal, 26*(2), 97–119.

Teece, D. (1987). Profiting from technological innovations: Implications for integration, collaboration, licensing, and public policy. In D. J. Teece (Ed.), *The competitive challenge: Strategies for industrial innovation and renewal* (pp. 185–219). Ballinger.

Teece, D., Pisano, G., & Shuen, A. (1997). Dynamic capabilities and strategic management. *Strategic Management Journal, 18*(7), 509–533.

Terho, H., Haas, H., Eggert, A., & Ulaga, W. (2012). 'It's almost like taking the sales out of selling'—Towards a conceptualization of value-based selling in business markets. *Industrial Marketing Management, 41*(1), 174–185.

Thomas, A. B. (1988). Does leadership make a difference to organizational performance? *Administrative Science Quarterly, 33*(3), 388–400.

Timsit, J. P., Castiaux, A., Truong, Y., Athaide, G. A., & Klink, R. R. (2015). The effect of market-pull vs resource-push orientation on performance when entering new markets. *Journal of Business Research, 68*(9), 2005–2014.

Tinbergen, N. (1951). *The study of instinct*. Oxford University Press.

Tiwari, S. K., Mishra, R. K., Ha, S. K., & Huczko, A. (2018). Evolution of graphene oxide and graphene: From imagination to industrialization. *ChemNanoMat: Chemistry of Nanomaterials for Energy, Biology and More, 4*(7), 598–620.

Tomer, J. (2003). Personal capital and emotional intelligence: An increasingly important intangible source of economic growth. *Eastern Economic Journal, 29*, 453.

Tournois, L. (2016). When markets stagnate: Finding new territory through reverse innovation. *Journal of Business Strategy, 37*(6), 18–27.

Trada, S., & Goyal, V. (2017). The dual effects of perceived unfairness on opportunism in channel relationships. *Industrial Marketing Management, 64*(July), 135–146.

Trajtenberg, M. (2018). AI as the next GPT: A political-economy perspective. NBER working paper no. 24245.

Tripsas, M. (2008). Customer preference discontinuities: A trigger for radical technological change. *Managerial and Decision Economics, 29*(2–3), 79–97.

Tsai, W. (2001). Knowledge transfer in intraorganizational networks: Effects of network position and absorptive capacity on business unit innovation and performance. *The Academy of Management Journal, 44*(5), 996–1004.

Tushman, M., & Anderson, P. (1986). Technological discontinuities and organizational environments. *Administrative Science Quarterly, 31*(3), 439–465.

Uchida, Y., Norasakkunkit, V., & Kitayama, S. (2004). Cultural constructions of happiness: Theory and empirical evidence. *Journal of Happiness Studies, 5*(3), 223–239.

Ugur, H., Constantinescu, P. M., & Stevens, M. J. (2015). Self-awareness and personal growth: Theory and application of Bloom's taxonomy. *Eurasian Journal of Educational Research, 60*, 89–110.

Ulaga, W., & Eggert, A. (2006). Value-based differentiation in business relationships: Gaining and sustaining key supplier status. *Journal of Marketing, 70*(1), 119–136.

Upson, J. W., Sanchez, M. S., & SMITH, W. J. (2017). Competitive dynamics of market entry. *Management and Economics Review, 2*(1), 118–132.

Utterback, J. M., & Abernathy, W. J. (1975). A dynamic model of process and product innovation. *Omega, 3*(6), 639–656.

Uzuegbunam, I. S., Geringer, J. M., & Oberst, C. (2018). Creative destruction within and across countries: Do culture and connectedness matter. *Academy of Management Proceedings, 2018* (1). https://doi.org/10.5465/AMBPP.2018.97

VanderWerf, P. A., & Mahon, J. F. (1997). Meta-analysis of the impact of research methods on findings of first-mover advantage. *Management Science, 43*, 1510–1519.

Vaughan, J. (2013). *Technological innovation: Perceptions and definitions*. ALA TechSource, a publishing unit of the American Library Association.

Vendrell-Herrero, F., Bustinza, O. F., Parry, G., & Georgantzis, N. (2017). Servitization, digitization and supply chain interdependency. *Industrial Marketing Management, 60*(January), 69–81.

Verhoef, P. C. (2003). Understanding the effect of customer relationship management efforts on customer retention and customer share development. *Journal of Marketing, 67*(4), 30–45.

Veugelers, R., & Cassiman, B. (1999). Make and buy in innovation strategies: Evidence from Belgian manufacturing firms. *Research Policy, 28*(1), 63–80.

Villasalero, M. (2017). A resource-based analysis of realized knowledge relatedness in diversified firms. *Journal of Business Research, 71*(Feb.), 114–124.

Visnjic, I., Wiengarten, F., & Neely, A. (2016). Only the brave: Product innovation, service business model innovation, and their impact on performance. *Journal of Product Innovation Management, 33*(1), 36–52.

von Bertalanffy, L. (1924). Einfuhrung in Spengler's werk. *Literaturblatt Kolnische Zeitung*, May.

von Bertalanffy, L. (1968). *General Systems Theory*. George Braziller.

von Bertalanffy, L. (1972). The history and status of general systems theory. In G. Klir (Ed.), *Trends in general systems theory* (pp. 21–41). Wiley.

Vorhies, D., & Morgan, N. (2005). Benchmarking marketing capabilities for sustainable competitive advantage. *Journal of Marketing, 69*(1), 80–94.

Vu, K. M. (2018). Embracing globalization to promote industrialization: Insights from the development of Singapore's petrochemicals industry. *China Economic Review, 48*, 170–185.

Wagner, J. (2012a). International trade and firm performance: A survey of empirical studies since 2006. *Review of World Economics, 148*(2), 235–267.

Wagner, J. (2012b). Exports, imports and profitability: First evidence for manufacturing enterprises. *Open Economies Review, 23*, 747–765.

Wagner, J. (2016). A survey of empirical studies using transaction level data on exports and imports. *Review of World Economics, 152*(1), 215–225.

Wagner, P., Olsen, J., & Thach, L. (2007). *Wine marketing & sales: Success strategies for a saturated market* (2nd ed.). Board and Bench Publishing.

Wajda, T., Forrest, J. Y.-L., Hu, M. Y., Li, Z., & Liu, Y. (n.d.). (to appear). The inclusiveness dimension of human cognition and categorization paradigm. Submitted for publication.

Walsh, L. C., Boehm, J. K., & Lyubomirsky, S. (2018). Does happiness promote career success? Revisiting the evidence. *Journal of Career Assessment, 26*(2), 199–219.

Wang, L. J., & Wang, Q. L. (2019). Study on the relationship of input and output of re-innovation after digesting the introduced technology in high-tech industry: An empirical study based on different industry data. *China Soft Science, 01*, 184–192.

Wang, R., Yan, B., & Deng, W. G. (2010). The impact of FDI on the independent innovation capability of Chinese indigenous industries: From the perspective of industrial linkages. *China Industrial Economics, 2010*(11), 16–25.

Wang, Z. H., & Yang, Z. (2017). Thoughts on artificial intelligence technology research and future intelligent information service system. *Telecommunications Science, 33*(05), 1–11.

Webster, F. E. (2002). *Market-driven management: How to define, develop and deliver customer value* (2nd ed.). Wiley.

Weitzman, M. L. (1998). Recombinant growth. *Quarterly Journal of Economics, 113*(2), 331–360.

Welch, J. (2019). Innovation questions & answers. https://innovationcompany.co.uk/innovation-questions, Accessed on December 3, 2019.

Wen, Y. (2016). *The making of an economic superpower: Unlocking china's secret of rapid industrialization*. World Scientific.

Wernerfelt, B. (1984). A resource-based view of the firm. *Strategic Management Journal, 5*(2), 171–180.

Westney, D. E. (1993). Cross-Pacific internationalization of R&D by U.S. and Japanese firms. *R&D Management, 23*(2), 171–181.

WH. (2018). https://www.fedscoop.com/white-house-artifical-intelligence-committee-kratsios/, Accessed on April 16, 2019.

White, M. (2017). Keep Calm and simulate on: Faculty experiences and insights into implementing best practices in simulation. *Teaching and Learning in Nursing, 12*(1), 43–49.

Wieczorek, A. J., & Hekkert, M. P. (2012). Systemic instruments for systemic innovation problems. *Science and Public Policy, 39*(1), 74–87.

Williams, P., & Naumann, E. (2011). Customer satisfaction and business performance: A firm-level analysis. *Journal of Services Marketing, 25*(1), 20–32.

Williamson, O. E. (1979). Transaction cost economics: The governance of contractual relations. *The Journal of Law and Economics, 22*(2), 233–261.

Williamson, O. E. (1983). Organizational innovation: The transaction cost approach. In J. Ronen (Ed.), *Entrepreneurship* (pp. 101–133). Lexington Books.

Winter, S. G. (2000). The satisficing principle in capability learning. *Strategic Management Journal, 21*(10-11), 981–996.

Woodridge, J. M. (2002). *Econometric analysis of cross section and panel data*. The MIT Press.

Wouters, M. (2010). Customer value propositions in the context of technology commercialization. *International Journal of Innovation Management, 14*(6), 2099–1127.

Wouters, M., Anderson, J. C., & Kirchberger, M. (2018). New technology startups seeking pilot customers: Crafting a pair of value propositions. *California Management Review, 60*(4), 101–124.

Wouters, M., & Kirchberger, M. A. (2015). Customer value propositions as interorganizational management accounting to support customer collaboration. *Industrial Marketing Management, 46*(April), 54–67.

Wright, M., Piva, E., Mosey, S., & Lockett, A. (2009). Academic entrepreneurship and business schools. *The Journal of Technology Transfer, 34*, 560–587.

Wu, D., Rosen, D. W., Wang, L., & Schaefer, D. (2015). Cloud-based design and manufacturing: A new paradigm in digital manufacturing and design innovation. *Computer-Aided Design, 59*, 1–14.

Wu, X. M. (1990). *The pansystems view of the world*. People's University of China Press.

Wu, Y., & Lin, Y. (2002). *Beyond nonstructural quantitative analysis: Blown-ups, spinning currents and modern science*. World Scientific.

Xi, J. P. (2017). http://www.gov.cn/zhuanti/2017-10/27/content_5234876.htm, Accessed on April 16, 2019.

Yadav, P. L., Han, S. H., & Kim, H. (2017). Sustaining competitive advantage through corporate environmental performance. *Business Strategy and the Environment, 26*(3), 345–357.

Yan, R. L., Myers, C., & Wang, J. (2014). Bundling products to success: The influence of complementarity and advertising. *Journal of Retailing and Consumer Services, 21*, 48–53.

Yang, C. Y. (2018). *Extension innovation method*. CRC Press.

Yang, C. Y., & Cai, W. (2013). *Extenics: Theory, method and application*. Science Press.

Yazdanparast, A., & Spears, N. (2018). The new me or the me I'm proud of?: Impact of objective self-awareness and standards on acceptance of cosmetic procedures. *European Journal of Marketing, 52*(1-2), 279–301.

Ye, G. L., Priem, R. L., & Alshwer, A. A. (2012). Achieving demand-side synergy from strategic diversification: How combining mundane assets can leverage consumer utilities. *Organization Science, 23*(1), 207–224.

Yeo, C. S., & Saboori-Deilami, V. (2017). Strategic challenges of outsourcing innovation in global market. *Asia Pacific Journal of Innovation and Entrepreneurship, 11*(1), 5–16.

Yi, J., Hong, J., & Hsu, W. C. (2017). The role of state ownership and institutions in the innovation performance of emerging market enterprises: Evidence from China. *Technovation, 2017*, S016649721730247X.

Ying, Y. R., & Forrest, J. Y. L. (2015). *Capital account liberation: Methods and applications*. CRC Press, an imprint of Taylor and Francis.

Yrjölä, M. (2015). Uncovering executive prioritization: Evaluating customer value propositions with the pairwise comparison method. *Journal of Service Science and Management, 8*(1). https://doi.org/10.4236/jssm.2015.81001

Yu, T., & Cannella, A. A., Jr. (2012). A comprehensive review of multimarket competition research. *Journal of Management, 39*, 76–109.

Zachary, M. A., Gianiodis, P. T., Payne, G. T., & Markman, G. D. (2015). Entry timing: Enduring lessons and future directions. *Journal of Management, 41*(5), 1388–1415.

Zaheer, A., & Zaheer, S. (1997). Catching the wave: Alertness, responsiveness, and market influence in global electronic networks. *Management Science, 43*(11), 1493–1509.

Zaheer, A., & Zaheer, S. (1999). The structure of global competition: A network approach. Paper presented at the *SMJ* Special Issue Conference, Northwestern University.

Zahra, S. A., & Nambisan, S. (2012). Entrepreneurship and strategic thinking in business ecosystems. *Business Horizons, 55*(3), 219–229.

Zeidman, P., & Maguire, E. A. (2016). Anterior hippocampus: The anatomy of perception, imagination and episodic memory. *Nature Reviews Neuroscience, 17*, 173–182.

Zhang, H. M., Zheng, Y., Zhou, D. Q., & Long, X. L. (2018). Selection of key technology policies for Chinese offshore wind power: A perspective on patent maps. *Marine Policy, 93*, 47–53.

Zhang, H. S. (2006). Analysis of technological innovation capability of China's manufacturing. *China Soft Science Magazine, 2006*(4), 15–23.

Zhang, T. C., Gu, H. M., & Jahromi, M. F. (2019). What makes the sharing economy successful? An empirical examination of competitive customer value propositions. *Computers in Human Behavior, 95*(June), 275–283.

Zhang, Y. C., & Li, X. T. (2015). Measurement of technological innovation efficiency and its influencing factors in China's high-tech transformation of traditional industries: An empirical analysis based on the transcendental logarithmic stochastic frontier model. *Technology Economics, 34*(03), 18–26+111.

Zhang, Z. H., & Peng, C. Q. (2009). Market analysis of the determinants of regional technological innovativeness: Based on the panel data of China's 30 provinces and districts. *China Industrial Economics, 2009*(11), 35–44.

Zhao, H. C., Forrest, J. Y.-L., & Jirasakuldech, B. (2018). A game analysis of trade dumping and antidumping. *Theoretical Economics Letters, 8*, 2860–2881.

Zheng, C. D., Yu, X. F., & Jin, Q. M. (2017). How user relationships affect user perceived value propositions of enterprises on social commerce platforms. *Information Systems Frontiers, 19*, 1261–1271.

Zheng, W., Yang, B., & McLean, G. N. (2010). Linking organizational culture, structure, strategy, and organizational effectiveness: Mediating role of knowledge management. *Journal of Business Research, 63*, 763–771.

Zhou, J. (2012). Digitization and intelligentization of manufacturing. *China Mechanical Engineering, 23*(20), 2398–2400.

Zollo, L., Marzi, G., Boccardi, A., & Ciappei, C. (2016). Gli effetti della Stampa 3D sulla competitivita aziendale. Il caso delle imprese orafe del distretto di Arezzo. *Piccola Impresa/Small Business, 2*(2), 80–100.

Zona, F., Gomez-Mejia, L. R., & Withers, M. C. (2016). Board interlocks and firm performance: Toward a combined agency–resource dependence perspective. *Journal of Management, 44*(2), 589–618.

Zou, S., Fang, E., & Zhao, S. (2003). The effect of export marketing capabilities on export performance: An investigation of Chinese exporters. *Journal of International Marketing, 11*(4), 32–55.

Index

J. Y.-L. Forrest, Y. Liu, *Value in Business*, Contributions to Management Science,
https://doi.org/10.1007/978-3-030-82898-1

The manufacturer's authorised representative in the EU is Springer Nature Customer Service Centre GmbH, Europaplatz 3, 69115 Heidelberg, Germany. If you have any concerns regarding our products, please contact ProductSafety@springernature.com

Printed and bound by CPI Group (UK) Ltd, Croydon, CR0 4YY

30/09/2026

02225695-0018